Contents

Introduction   v

Unit 1   Solid mechanics   01

Richard Brooks

1.1   Basic design analysis
1.2   Stress, strain and elasticity
1.3   Beam bending
1.4   Multi-axial stress and strain
1.5   Torsion

Unit 2   Materials and processing   59

Andrew Kennedy and Philip Shipway

2.1   Introduction
2.2   The structure and properties of materials
2.3   Properties of materials
2.4   Selection of materials in engineering design
2.5   Materials processing
2.6   Failure of materials

Unit 3   Fluid dynamics   135

Stephen Pickering

3.1   Introductory concepts
3.2   Fluids at rest – hydrostatics
3.3   Fluids in motion
3.4   Fluids in motion – linear momentum

Unit 4   Thermodynamics   213

Paul Shayler

4.1   Introduction
4.2   The first law of thermodynamics, conservation of energy, work and heat transfer
4.3   The second law of thermodynamics, heat engines, the Clausius inequality, entropy and irreversibility
4.4   The properties of perfect gas, water and steam
4.5   Types of process and their analysis for work and heat transfer
4.6   Modes of heat transfer and steady-state heat transfer rates
4.7   Cycles, power plant and engines

Unit 5   Electrical and electronic systems   283

Alan Howe

5.1   Introduction
5.2   Direct current circuits
5.3   Electromagnetic systems
5.4   Capacitance
5.5   Alternating current circuits
5.6   Three-phase circuits
5.7   Semiconductor rectifiers
5.8   Amplifiers
5.9   Digital electronics
5.10 Transformers
5.11 AC induction motors

Unit 6   Machine dynamics   405

Stewart McWilliam

6.1   Introduction
6.2   Basic mechanics
6.3   Kinematics of a particle in a plane
6.4   Kinematics of rigid bodies in a plane
6.5   Kinematics of linkage mechanisms in a plane
6.6   Mass properties of rigid bodies
6.7   Kinematics of a rigid body in a plane
6.8   Balancing of rotating masses
6.9   Geared systems
6.10 Work and energy
6.11 Impulse, impact and momentum

Questions   491
Index   505

An Introduction to
Mechanical
Engineering

Part 1

**Michael Clifford, Richard Brooks, Alan Howe,
Andrew Kennedy, Stewart McWilliam,
Stephen Pickering, Paul Shayler & Philip Shipway**

HODDER
EDUCATION
An Hachette UK Company

Orders: please contact Bookpoint Ltd, 130 Milton Park, Abingdon, Oxon OX14 4SB.
Telephone: (44) 01235 827720. Fax: (44) 01235 400454. Lines are open 9.00 – 5.00, Monday to Saturday, with a 24-hour message answering service. You can also order through our website at www.hoddereducation.co.uk

If you have any comments to make about this, or any of our other titles, please send them to educationenquiries@hodder.co.uk

British Library Cataloguing in Publication SD
A catalogue record for this title is available from the British Library

ISBN: 978 0 340 93995 6

First Edition Published 2009
Impression number   10 9 8 7 6 5 4 3 2 1
Year   2013 2012 2011 2010 2009

Hachette UK's policy is to use papers that are natural, renewable and recyclable products and made from wood grown in sustainable forests. The logging and manufacturing processes are expected to conform to the environmental regulations of the country of origin.

Cover photo from Raw Paw Graphics/Digital vision
Typeset by Tech-Set Ltd., Gateshead, Tyne & Wear
Printed in Italy for Hodder Education, an Hachette UK Company, 338 Euston Road, London NW1 3BH

The publishers's would like to thank the following for use of photographs in this volume:
Figure 2.17a © J Orr/Alamy; Figure 2.17b © Kernal Eksen Photography, photographersdirect.com;
Figure 2.46 © Alexis Rosenfield/Science Photo Library; Figure 2.48 © David Hoffman Photo Library/Alamy;
Figure 2.53 © aviation-images.com

All illustrations in this volume by Barking Dog Art.

Every effort has been made to trace and acknowledge ownership of copyright. The publishers will be glad to make suitable arrrangements with any copyright holders whom it has not been possible to contact.

Introduction

Engineering is not merely knowing and being knowledgeable, like a walking encyclopaedia; engineering is not merely analysis; engineering is not merely the possession of the capacity to get elegant solutions to non-existent engineering problems; engineering is practicing the art of the organized forcing of technological change.

Dean Gordon Brown

This book is written for undergraduate engineers and those who teach them. It contains concise chapters on solid mechanics, materials, fluid mechanics, thermodynamics, electronics and dynamics, which provide grounding in the fundamentals of mechanical engineering science. An introduction to mathematics is covered in the companion publication, *An Introduction to Mathematics for Engineers* by Stephen Lee, also published by Hodder Education.

The material in this book is supported by an accompanying website: www.hodderplus.co.uk/mechanicalengineering.

The authors have over 120 years' experience of teaching undergraduate engineers between them, mostly, but not exclusively, at the University of Nottingham. The material contained within this textbook has been derived from lecture notes, research findings and personal experience from within the lecture theatre and tutorial sessions.

We gratefully acknowledge the support, encouragement and occasional gentle prod from Stephen Halder and Gemma Parsons at Hodder Education, without whom this book would still be a figment of our collected imaginations.

Dedicated to past, present and future engineering students at the University of Nottingham.

Unit 1

Solid Mechanics

Richard Brooks

UNIT OVERVIEW

- Basic design analysis
- Stress, strain and elasticity
- Beam bending
- Multiaxial stress and strain
- Torsion

1.1 Basic design analysis

Forces, moments and couples

> A **force** arises from the action (or reaction) of one body on another.

Although a force cannot be directly observed, its effect can be. A typical example is a force arising from the surface contact between two bodies, e.g. one pushing against the other. Two forces actually occur in this situation as shown in Figure 1.1. One is the 'action' of the man on the wall and the other is the 'reaction' of the wall on the man.

> Newton's third law tells us that the action and reaction forces in this situation (and generally) are equal and opposite.

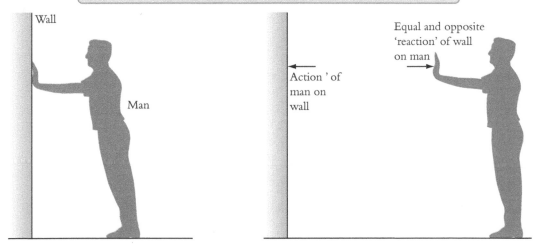

Figure 1.1 **Newton's third law**

Such contact forces occur where bodies interact with each other; however, they can also occur internally within a single body. In this case, it is the microscopic particles, e.g. molecules, atoms, etc. which contact each other and interact with forces between themselves. For this chapter, we will generally be dealing with macroscopic bodies where the interaction forces occur at external surface contacts.

Another type of force occurring is that which arises from the remote influence of one body on another, such as the force of gravity. The Earth's gravity acting on a person gives rise to a force acting at his or her centre of mass. This type of force is termed the person's **weight** and acts vertically downwards or towards the centre of the Earth. Magnetic attraction is another example of a remote (or non-contact) force arising from the influence of a magnetic field on a body.

> The SI unit of force is the newton (N).

A force of 1 N is that force which, when applied to a mass of 1 kg, will result in an acceleration of the mass of 1 m s^{-2}. Thus, in general, a force applied to a body tends to change the state of rest or motion of the body, and the relationship between the resulting motion (acceleration, a) and the applied force, F, is given by Newton's second law, i.e. $F = ma$ where m is the mass of the body. However, in this chapter, we will generally be concerned with bodies in equilibrium, where there is no motion, i.e. static situations. For this to be the case, all forces acting on the body must balance each other out so that there is no resultant force (see the next section on 'equilibrium').

A force has both a **magnitude** and a **direction** and is therefore a vector quantity which can be represented by an arrow as shown in Figure 1.2. The magnitude of the force is represented by a label, e.g. 5 N as shown, or, alternatively, when solving problems graphically, by the length of the arrow. The direction of the force is clearly represented by the orientation of the arrow in space such as the angle θ to the x-direction.

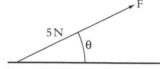

Figure 1.2 Force as a vector

Thus, when considering problems in two dimensions, two scalar quantities are required to describe a force, i.e. its magnitude and direction – in the above case 5 N and $\theta°$ respectively. To aid the analysis of systems with several forces, the forces are often resolved into their *components* in two perpendicular directions, as shown in Figure 1.3 for the force F. The x- and y-directions are commonly chosen, although resolving in other (perpendicular) directions relevant to the boundaries of a body may be more convenient for a specific problem. From Figure 1.3 the magnitudes of the two components in the x- and y-directions are given by:

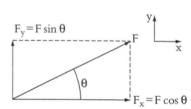

Figure 1.3 Resolving the force vector into components

$$F_x = F \cos \theta$$
$$F_y = F \sin \theta \qquad (1.1)$$

With this representation there are still two scalar quantities describing the force, in this case, F_x and F_y.

> The **moment** of a force about a point is equal to the product of the magnitude of the force and the perpendicular distance from the point to the line of action of the force.

This is illustrated in Figure 1.4, where the moment, M, of force F, about point O, is given by:

$$M = F.d \qquad (1.2)$$

An example of a device which creates a moment is a spanner, also shown in Figure 1.4. The hand applies the force, F, at one end and imparts a moment, $M = F.d$, on the nut at the other end, O.

A **couple** is a special case of a moment of a force and arises from a pair of equal and opposite parallel forces acting on a body but not through the same point, as shown in Figure 1.5. If the two forces, F, act at a distance d apart, then the magnitude of couple C, about any point, is given by:

$$C = F.d \qquad (1.3)$$

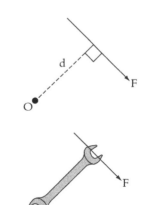

Figure 1.4 Moment of a force applied by a spanner

Solid mechanics

As the two forces, F, in Figure 1.5, are equal and opposite, their sum is zero and the body on which they act is *not* translated. However, they do create a couple which tends to *rotate* the body. Therefore, a consequence of a couple acting on a body is to impart pure rotation. For this reason, the term 'pure moment' is often used instead of 'couple'.

An example of a device which creates a couple is a wheel nut wrench, also shown in Figure 1.5. Here, the hands apply forces, F, in and out of the page at both ends of one arm of the wrench, imparting a turning couple on a locked nut at O.

When a couple or moment is applied at a point on a body its effect is 'felt' at all other points within the body. This can be illustrated with the cantilever beam shown in Figure 1.6 where a couple of 5 kN m is applied at end A. If we assume that the couple is created by the application of two equal and opposite 5 kN forces, 1 m apart, acting through a rigid bar attached to the beam at A, we can determine the influence that these forces also have at points B and C, at 5 m and 10 m from A respectively.

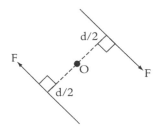

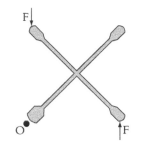

Taking moments about B:

$M_B = 5\,\text{kN}.(5\,\text{m} + 0.5\,\text{m}) - 5\,\text{kN}.(5\,\text{m} - 0.5\,\text{m})$

$\qquad = 27.5 - 22.5 = 5\,\text{kN m}$

Taking moments about C:

$M_C = 5\,\text{kN}.(10\,\text{m} + 0.5\,\text{m}) - 5\,\text{kN}.(10\,\text{m} - 0.5\,\text{m})$

$\qquad = 52.5 - 47.5 = 5\,\text{kN m}$

In both cases the effect, i.e. a 5 kN m turning moment, is felt at B and C. In other words, the turning moment felt on the bar is independent of the distance from A.

Moment/couple of 5 kNm felt at both B and C and all points along the beam

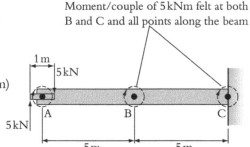

Figure 1.5 Couple and wheel nut wrench (forces act in and out of page)

Figure 1.6 Influence of a moment or couple acting at a point

Conditions of equilibrium

For a body to be in equilibrium, it must not translate or rotate. Considering movement in one plane only (i.e. a two-dimensional system), this means the body must not move in the x- or y-directions or rotate about its position. Three conditions are required of the applied forces for this to be the case.

> These three conditions of equilibrium are:
>
> (i) the sum of all the acting forces in the x-direction must be zero, i.e. $\Sigma F_x = 0$.
>
> (ii) The sum of all the acting forces in the y-direction must be zero, i.e. $\Sigma F_y = 0$.
>
> (iii) The sum of all the moments about any point must be zero.

Resultants of forces

When a number of forces act at a point on a body, their resultant force can be determined either algebraically or graphically.

Algebraic method

The algebraic method for determining the resultant of a number of forces has the following steps:

(i) Resolve all forces into their x- and y-components.

(ii) Sum the x-components (ΣF_x) and the y-components (ΣF_y).

(iii) Determine the magnitude and direction of the resultant force from ΣF_x and ΣF_y.

The following example illustrates the method.

Figure 1.7 shows three forces F_A, F_B and F_C acting at a point A. Determine the magnitude and direction of the resultant force at A.

The components of the forces are,

$$F_{Ax} = 0 \qquad\qquad F_{Ay} = 4\,kN$$
$$F_{Bx} = -8\,kN \qquad\qquad F_{By} = 0$$
$$F_{Cx} = -6.\cos60° = -3\,kN \qquad F_{Cy} = -6.\sin60° = -5.196\,kN$$

Summing these components in the x- and y-directions,

$$\Sigma F_x = 0 - 8 - 3 = -11\,kN$$
$$\Sigma F_y = 4 + 0 + -5.196 = -1.196\,kN$$

(note the −ve values indicating that the resultant forces act in the −ve x and −ve y directions)

The magnitude, F_R, of the resultant of ΣF_x and ΣF_y is,

$$F_R = \sqrt{(\Sigma F_x)^2 + (\Sigma F_y)^2}$$
$$= \sqrt{(-11)^2 + (-1.196)^2}$$
$$= 11.064\,kN$$

The angle, θ (with respect to the x-axis), of the resultant force is,

$$\theta = \tan^{-1}\left(\frac{\Sigma F_y}{\Sigma F_x}\right)$$
$$= \tan^{-1}\left(\frac{-1.196}{-11}\right)$$
$$= 6.2° \text{ to the negative } x\text{-direction as shown in Figure 1.7.}$$

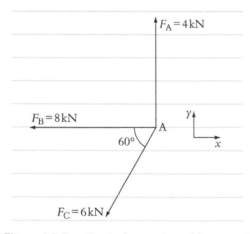
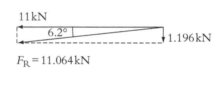

Figure 1.7 **Resultant of a number of forces acting at a point**

Graphical Method

The procedure for the graphical method of determining the resultant of a number of forces is shown in Figure 1.8 for the problem given above.

Firstly, draw to scale each of the three vector forces, F_A, F_B and F_C, following on from each other, as shown in the figure. The resultant force, F_R, is the single vector force that joins the start point A to the finishing point B, i.e. that closes the polygon of forces. Its magnitude and direction (θ) may be measured off from the scale vector diagram.

(NB: it does not matter in which order the three vectors are drawn in the diagram.)

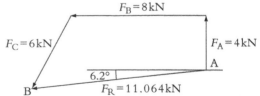

Figure 1.8 **Resultant of forces acting at a point – graphical method**

The graphical method is useful to give a quick approximate solution, whereas the algebraic method normally takes longer but will yield an exact result.

Frictional forces

Consider a solid body, i.e. a block, weight W, resting on the ground but in equilibrium under the action of an applied force, F_A, as shown in Figure 1.9. In general, where the body contacts the ground there will be a reaction force (from the ground) acting on the body. This reaction force has two components as follows:

(i) a tangential force, F, termed the *friction force*;

(ii) a normal force, N.

As the body is in equilibrium, these two components of the reaction force counterbalance the applied force, F_A, and the weight of the body, W, to prevent any movement. (NB: the body's weight is given by its mass $\times$ the acceleration of gravity, i.e. Mg and acts at the centre of mass.)

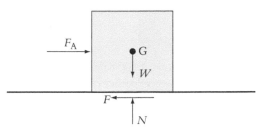

Figure 1.9 **Frictional force (F) and normal force (N) at point of contact between a block and the ground**

The frictional force, F, exists because of the rough nature of the contact surface between the body and the ground. In some cases, where the contact is smooth or lubricated, the frictional force will be negligible and there will be a normal reaction force only. This is a special case only found under certain circumstances, e.g. contact surfaces in a lubricated bearing.

If the applied force, F_A, is slowly increased, the frictional component, F, will also increase to maintain equilibrium. At some point the applied force will become sufficiently large to overcome the frictional force and cause movement of the body. Up to this point of 'slip' between the surfaces, a relationship exists between the frictional force and the normal reaction force as follows:

$$F \leq \mu N \qquad (1.4)$$

Note the 'less than or equal to' sign indicates that a limiting condition can occur. This limiting condition is the point of slip, at which point $F = \mu N$. Thus, the maximum value of F, i.e. the limiting frictional force, is proportional to N. The constant of proportionality, μ, is termed the **coefficient of static friction** and its value depends on the roughness of the two contacting surfaces and hence the contacting materials. Typical values are in the range $0.1 - 1.0$, where a lower value indicates a smoother surface and reduced friction. Values outside this range can occur for some material contact surfaces e.g. lubricated surfaces can have values lower than 0.1 while stick–slip surfaces, such as rubber on a hard surface, can have values in the range 1–10.

A number of important observations can now be stated about the frictional force, F:

(i) F cannot exceed μN;

(ii) the direction of F always *opposes* the direction in which subsequent motion would take place if slip occurred;

(iii) the magnitude of F is independent of the size of the contact area between the contacting surfaces;

(iv) if slip does occur, the magnitude of F is independent of the velocity of sliding between the two contact surfaces.

Although in this chapter we will be concerned primarily with static friction up to the point of slip, if slip does occur, the coefficient of dynamic friction (also called the kinetic frictional coefficient, μ_k) is usually marginally lower than the coefficient of static friction. In the sliding (i.e. slipping) condition the limiting form of equation (1.4) still applies, i.e. $F = \mu_k N$.

Free body diagrams

To analyse the forces in more complex systems, such as assemblies of components or structures containing many different elements, it is normal to break down the problem into separate free bodies.

Figure 1.10 shows two bodies, body A positioned on top of body B which itself is located on the ground. To analyse this problem for forces, we separate the two bodies and draw on each all the external forces acting as shown in the figure. The aim is to solve for the unknown reaction forces between the two bodies and between body B and the ground.

Thus, for body A, the external forces are its weight, W_A, acting at its centre of mass ($W_A = M_A.g$) and the vertical reaction force, R_B, from body B. There is no horizontal friction force at the contact between the bodies because there are no horizontal forces acting.

For body B, there is also its weight, $W_B = M_B.g$, again acting at its centre of mass, the action force, R_A, acting downwards from A and the reaction force, R_G, acting upwards from the ground.

Newton's third law tells us that $R_A = R_B$, i.e. 'for every action there is an equal and opposite reaction'.

We can now look at the *equilibrium* of each body in turn:

For body A, $\qquad \Sigma F_y = 0 \qquad \therefore R_B = W_A$

and for body B $\qquad \Sigma F_y = 0 \qquad \therefore R_G = R_A + W_B = R_B + W_B = W_A + W_B$

It is no surprise that the reaction force at the ground is equal to the sum of the weights of the two bodies. This is necessary to maintain the system in equilibrium.

Although this is a simple problem, it clearly illustrates the value of separating the two bodies, allowing us to solve for the unknown reaction force between the bodies. The diagrams of each separate body are referred to as **freebody diagrams** (FBDs).

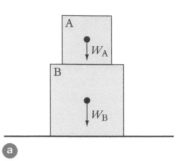

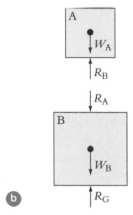

Figure 1.10 **Free body diagrams**

> Key points about free body diagrams:
>
> (i) A free body diagram, as the name implies, is a diagram of a free body which shows *all* the external forces acting on the body.
>
> (ii) Where several bodies (or subcomponents) interact as part of a more complex system, each body should be drawn separately, and interacting bodies should be replaced at their contact points with suitable reaction forces and/or moments.

General design principles

A number of general principles related to force analysis can be applied in design to simplify problems. In this section we will consider several of these principles.

Principle of transmissibility

A force can be moved along its line of action without affecting the static equilibrium of the body on which it acts. This principle of transmissibility is illustrated in Figure 1.11 where the equilibrium of the body is the same whether it is subjected to a pushing force or a pulling force acting along the same line of action. It should be pointed out that, although static equilibrium is the same in each case, the internal forces within the body will be different.

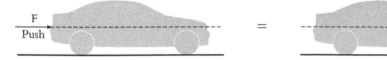

Figure 1.11 **Principle of transmissibility**

Statically equivalent systems

A load system can be replaced by another one, provided the static behaviour of the body on which they act is the same. Such load systems are termed **statically equivalent**.

Figure 1.12 shows a number of loads (five in total) each of 5 kN acting on a beam structure in such a way as to be evenly distributed along the length of the beam. If we are not interested in the internal forces developed within the beam but only the equilibrium of the beam as a whole, then this loading system can be replaced by a simpler point load of 25 kN applied at the centre of the beam. The two load systems are statically equivalent and the equilibrium conditions for the beam will apply, whichever of the loading systems is assumed.

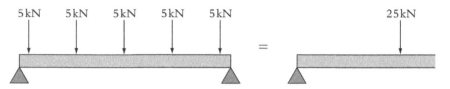

Figure 1.12 Statically equivalent

Two-force principle

> The **two-force principle** states that, for a two-force body (i.e. a body with forces applied at two points only) to be in equilibrium, both forces must act along the same line of action.

This is illustrated in Figure 1.13 where Body A is subjected to two forces, F_1 and F_2, *not* acting along the same line of action. Taking moments about point A, the application point for F_1, it is clear that there is a resultant moment and the body cannot be in equilibrium. For it to be in equilibrium, the distance d must be zero. This is the case for Body B, where F_1 and F_2 act along the same line and cannot therefore generate a moment. In addition, F_1 must equal F_2.

Three-force principle

> The **three-force principle** states that for a three-force body (i.e. a body with forces applied at three points only) to be in equilibrium, the lines of action of these forces must pass through a common point.

This is illustrated in Figure 1.14 where Body A is subjected to three forces, F_1, F_2 and F_3 not acting along the same line of action. Taking moments about point O, where the lines of action of F_1 and F_2 meet, it is clear that there is a resultant moment arising from F_3 and the body cannot be in equilibrium. For it to be in equilibrium, the distance d must be zero. This is the case for Body B, where the lines of action of F_1, F_2 and F_3 meet at O and there cannot be a resulting moment. In addition, the vector sum of F_1, F_2 and F_3 must be zero.

Pin-jointed structures

A pin-jointed structure, as shown in Figure 1.15, comprises an assembly of several members, which are joined together by frictionless pin joints. Such a joint cannot transmit moments due to the free rotation of the pin. This simplification is found in practice to be valid for many structures and enables the analysis of forces within the structure to be significantly simplified. The objective is usually to determine the forces occurring at each of the pin joints in the structure, and this is achieved by considering equilibrium of individual members and the structure as a whole. A solution can be obtained algebraically or graphically.

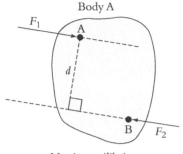

Body A

Not in equilibrium

ⓐ

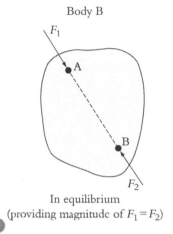

Body B

In equilibrium
(providing magnitude of $F_1 = F_2$)

ⓑ

Figure 1.13 Two-force principle

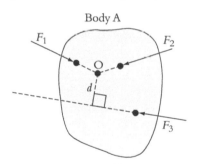

Body A

Not in equilibrium

ⓐ

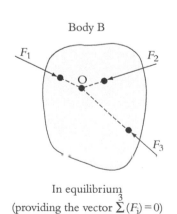

Body B

In equilibrium
(providing the vector $\sum_1^3 (F_i) = 0$)

ⓑ

Figure 1.14 Three-force principle

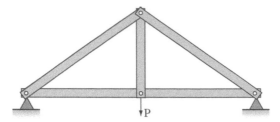

Figure 1.15 A pin-jointed structure

Algebraic solution to a pin-jointed structure problem

Consider a wall bracket comprising a simple two-bar pin-jointed structure, ABC, as shown in Figure 1.16. Both members are of equal length and inclined to the horizontal at 45°. The joints at A, B and C are all pin joints, and the lower member, BC, is subjected to a vertical downward load, P, acting half way along its length. The weights of the members may be ignored. The aim is to determine the forces at A, B and C in terms of the applied load P.

The first stage is to draw the freebody diagrams for the two members, also shown in Figure 1.16.

Member AB is a two-force member as forces act only at the two ends of the member. For equilibrium of a two-force member, the directions of the forces, R_A and R_B, must be along the same line, i.e. along the axis of the member. As AB is clearly in tension, the directions are as indicated in the figure. Note that for some problems it may not be possible to establish on simple inspection whether a member is in tension or compression. This is not a problem, because if the forces are drawn incorrectly, say in compression rather than tension, the analysis will, in that case, result in a negative magnitudes for the forces.

Moving to member BC, this is a three-force member with forces acting at both ends and the third force, P, acting at the centre of the member. The three-force principle could be used for this member to establish the directions of the forces; however, we will not do so, as we are solving the problem algebraically.

R_B acting on BC at joint B must be equal and opposite to R_B acting on AB at joint B (Newton's third law). As we do not know the direction of the force at C, we will give it two unknown components, H_C and V_C, as shown in Figure 1.16. Now looking at the equilibrium of BC:

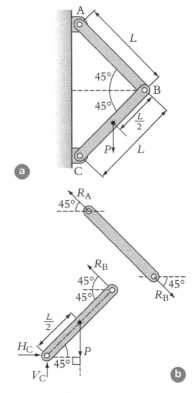

Figure 1.16 Pin-jointed structure – algebraic solution

horizontal forces 　　　$H_C - R_B \cos45° = 0$

$\therefore\ H_C = 0.707\ R_B$ 　　　　　　(1.5)

vertical forces 　　　$V_C - P + R_B \sin45° = 0$

$\therefore\ V_C = P - 0.707\ R_B$ 　　　　(1.6)

moments about C 　　　$P.\left(\dfrac{L}{2}\right).\cos45° - R_B.L = 0$

$\therefore\ R_B = 0.354P$

and $R_A = R_B = 0.354P$

Both R_A and R_B act at 45° to the horizontal, i.e. along the line of AB.

Substituting for R_B into (1.5) and (1.6) gives,

$$H_C = 0.25P \text{ and } V_C = 0.75P$$

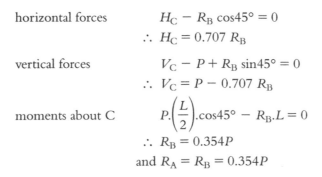

Solid mechanics

The magnitude of the resultant force R_C is:

$$R_C = \sqrt{(H_C)^2 + (V_C)^2} = \sqrt{(0.25P)^2 + (0.75P)^2} = 0.791P$$

and the angle of R_C with respect to the horizontal is:

$$\tan \theta_C = \frac{V_C}{H_C} = \frac{0.75P}{0.25P} = 3$$

$$\theta_C = 71.6°$$

Thus, the magnitudes and directions of all three forces at the joints A, B and C have been determined.

Geometrical solution to a pin-jointed structure problem

Figure 1.17 shows a schematic model of a crane boom supported by a pin-joint at one end, A, and a cable attached to the other end, C, inclined at an angle of 30° to the horizontal. The member is assumed to be weightless but carries a vertical downwards load, P, part way along its length at position B. The aim is to determine the tension, T, in the cable and the magnitude and direction of the reaction force, R_A.

The first stage is to draw the freebody diagram (to a suitable scale) of the member ABC, also shown in Figure 1.17. The freebody diagram should include all forces acting on the member. The directions of P and the cable tension, T, are known and can be drawn in immediately. However, the direction of R_A is not known. To find this, the three-force principle can be used because the member has three forces acting on it and, for equilibrium, all three forces must meet at a point. Thus, the line of action of P should be extended to intersect the line of action of T at point O. Then the line of action of R_A must also pass through the point O to satisfy the three force principle. AO therefore defines the direction of R_A, shown in the figure as the angle α to the horizontal. From a scale drawing, α can be measured as 44°, or alternatively this value can be calculated by trigonometry in triangles ABO and BCO.

The above stage only yields the direction of the unknown force, R_A. Its magnitude and the magnitude of the tension T are still required. To find these, a 'force polygon' is drawn. In this case it is actually a force triangle as there are only three forces, but the general term is 'force polygon'. The triangle comprises the vectors of the three forces, P, T and R_A, drawn following on from each other to form a closed triangle as shown in Figure 1.17. The triangle must close as the member, ABC, is in equilibrium under the action of the three forces and their sum must be zero. To draw the triangle, firstly draw the vector representing P vertically downwards. Its end points define two points of the triangle. Next, at one end of the vector P, draw a line representing the direction of T. At the other end of P, draw a line representing the direction of R_A (NB: it does not matter which end each of the force directions are drawn from, as long as it is a different end for each force). These two lines intersect at the third point of the triangle and the lengths of the other two sides (not P) give the magnitudes of T and R_A respectively. Note also that the direction of R_A is upwards as it must close the triangle.

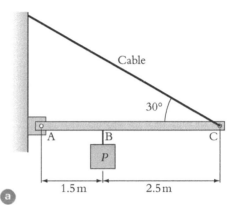

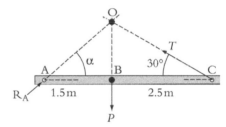

Free body diagram

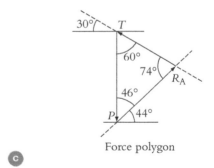

Force polygon

Figure 1.17 Pin-jointed structure – graphical solution

Measuring the force triangle gives the magnitudes of T and R_A, in terms of P, as,

$$T = 0.75P \text{ and } R_A = 0.9P$$

These magnitudes can alternatively be found by using trigonometry as one side and three angles of the triangle are known, allowing the two unknown sides to be determined using the sine rule (see 'Trigonometry' overleaf).

Trigonometry

Trigonometry is often needed to solve pin-jointed structure problems. Both the 'sine rule' and the 'ccosine rule' may be needed to solve for unknown sides and angles in vector polygons (or triangles). These rules are given in Figure 1.18.

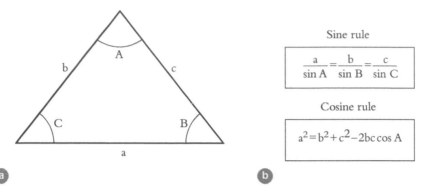

Sine rule

$$\frac{a}{\sin A} = \frac{b}{\sin B} = \frac{c}{\sin C}$$

Cosine rule

$$a^2 = b^2 + c^2 - 2bc \cos A$$

(a) (b)

Figure 1.18 Useful trigonometric relationships

Free body diagrams

The inclined edge of block B rests on top of the inclined edge of block A, as shown in Figure 1.19. A horizontal force, P, is applied to block A. Block A can slide horizontally along the ground and block B can slide vertically against the wall. All surfaces are frictional contacts with a coefficient of friction μ. The masses of the blocks are M_A and M_B respectively.

Draw the free body diagrams for both blocks when they are just on the point of slipping under the action of the horizontal force P.

The solution is also given in Figure 1.19. The two blocks are separated and drawn as free bodies. All contact surfaces contain a normal reaction force and a parallel friction force. As the system is on the point of slip, the frictional force = $\mu \times$ normal reaction force at each contact surface. Note that where the two blocks contact each other, there are equal and opposite reaction and friction forces (Newton's third law). In this problem, there are two pairs of equal and opposite forces N and μN where the two bodies contact. Using the principle of statically equivalent systems, all forces are drawn as point forces for simplification. In reality, they will be distributed over the contact surfaces. The weight of each body is a force acting at its centre of mass.

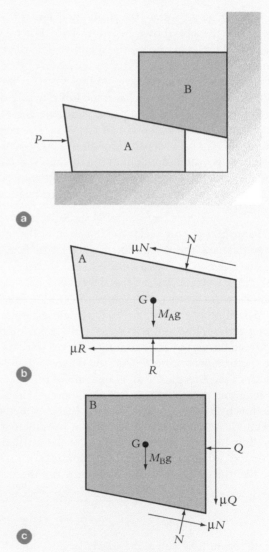

Figure 1.19 Free body diagrams worked example

Equilibrium and frictional forces

Figure 1.20 shows a ladder of weight 500 N resting on two surfaces at points A and B. Dimensions are as indicated. The coefficient of friction at all contacting surfaces is 0.4.

Determine the maximum height to which a man of weight 1200 N can walk up the ladder before slip occurs.

General solution procedure (see Figure 1.20):

(i) Draw the freebody diagram of the ladder.

(ii) At the point of slip, assume the man has climbed to a height, h, above the base point, B.

(iii) Draw frictional forces acting opposite to the direction of slip.

(iv) Assume both points A and B slip at the same time.

(v) At slip, assume friction force $= \mu \times$ normal force.

(vi) Use equilibrium of forces and moments to solve for unknown forces and height h.

Figure 1.20 shows the free body diagram of the ladder with all forces acting, when on the point of slip.

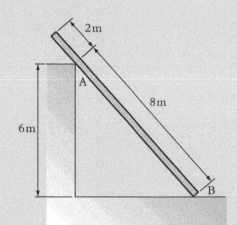

Solving for the angle θ

$$\sin \theta = \frac{6}{8} = 0.75$$
$$\therefore \theta = 48.6°$$
$$\text{and } \cos \theta = 0.661$$

Equilibrium of vertical forces
$$-1200 - 500 + R \cos \theta + \mu R \sin \theta + N = 0$$
$$-1700 + R.(0.661) + 0.4R.(0.75) + N = 0$$
$$\therefore N = 1700 - 0.961R \tag{1.7}$$

Equilibrium of horizontal forces
$$-\mu N + R \sin \theta - \mu R \cos \theta = 0$$
$$-0.4N + R.(0.75) - 0.4R.(0.661) = 0$$
$$\therefore N = 1.214R \tag{1.8}$$

From equations (1.7) and (1.8), we obtain
$$N = 948.9 \text{ N and } R = 781.6 \text{ N}$$

Taking moments about B
$$-500 \cos \theta . 5 - 1200 \cos \theta . (a) + R.(8) = 0$$
$$\therefore a = 5.8 \text{ m (i.e. distance up the ladder)}$$

Therefore, height of man, h is
$$h = a \sin \theta = 5.8 (0.75) = 4.35 \text{ m}$$

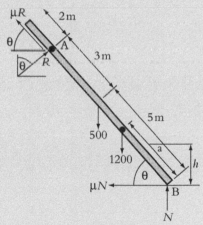

Figure 1.20 Equilibrium and friction worked example

Pin-jointed structure – algebraic solution

The pin-jointed structure, shown in Figure 1.21, supports a block, E, of weight 14.4 N.

Neglecting the weights of the members of the structure, use an algebraic solution to determine the vertical and horizontal components, and the magnitudes and directions of the resultant forces at the pin joints A and C on member AFC.

Firstly, draw the free body diagrams of each member in the structure. These are shown in Figure 1.21. Here tension in the cable, T = weight of the block E = 14.4 N.

Equilibrium of the pulley:

vertical forces $\quad V_D - T = 0 \qquad \therefore V_D = T = 14.4$ N

horizontal forces $\quad H_D - T = 0 \qquad \therefore H_D = T = 14.4$ N

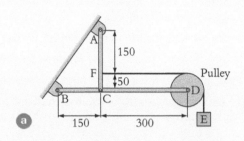

Equilibrium of member BCD:

vertical forces $\quad V_B - V_C - V_D = 0 \quad \therefore V_B = V_C + 14.4$

$$(1.9)$$

horizontal forces $\quad H_B + H_C - H_D = 0 \quad \therefore H_B = -H_C + 14.4$

$$(1.10)$$

Moments about D $\quad V_B(450) - V_C(300) = 0 \quad \therefore V_B = 0.667\, V_C$

$$(1.11)$$

From (1.9) and (1.11), $\quad V_C = -43.2$ N

(the negative sign indicates that V_C acts in the opposite direction to that drawn)

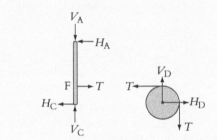

Equilibrium of member FBD:

vertical forces $\quad V_C - V_A = 0 \qquad \therefore V_A = -43.2$ N

horizontal forces $\quad T - H_C + H_A = 0 \quad \therefore H_C + H_A = 14.4$

$$(1.12)$$

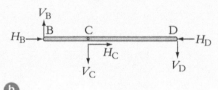

Moments about F $\quad H_C(50) - H_A(150) = 0 \quad \therefore H_C = 3H_A$

$$(1.13)$$

From (1.12) and (1.13), $H_A = 3.6$ N and $H_C = 10.8$ N

The magnitude of the resultant forces at A and C are:

$$R_A = \sqrt{(H_A)^2 + (V_A)^2} = \sqrt{(3.6)^2 + (43.2)^2} = 43.35 \text{ N}$$
$$R_C = \sqrt{(H_C)^2 + (V_C)^2} = \sqrt{(10.8)^2 + (43.2)^2} = 44.52 \text{ N}$$

The angles of these resultants with respect to the horizontal are:

$$\tan \theta_A = \frac{V_A}{H_A} = \frac{43.2}{3.6} = 12$$

$$\theta_A = 85.24°$$

$$\tan \theta_C = \frac{V_C}{H_C} = \frac{43.2}{10.8} = 4$$

$$\theta_C = 75.96°$$

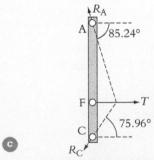

Figure 1.21 Pin-jointed structure – algebraic solution worked example

These directions are also illustrated in Figure 1.21. Note that the member AFC is a three-force system and the resultants, R_A and R_C, and the tension, T, must all meet at a point, as indicated.

Pin-jointed structure – graphical solution

The pin-jointed structure shown in Figure 1.22 supports a weight of 60 kN.

Neglecting the weights of the members of the structure, use a graphical solution to determine the reaction force at point C and the tension in the cable.

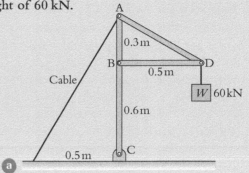

Since the directions of the weight, W, and the cable tension, T, are known, a graphical method provides a quick solution. Referring to Figure 1.22, the steps in the solution procedure are:

(i) Draw the free body diagram of the whole structure (to a suitable scale) showing all forces acting on it.

(ii) Using the three-force principle, the forces T, W and R_C must meet at a point. Thus extend the lines of W and T to meet at point O. Joining points O and C, gives the direction of force R_C.

(iii) Either measure or use trigonometry to determine the angles α and β, giving $\alpha = 29.05°$ and $\beta = 15.52°$. Note that, so far, these angles give the directions of T and R_C but not their magnitudes.

(iv) Draw the force polygon (see Figure 1.22) as follows:

– First, draw a vertical vector representing W (the known force).

– From the top and bottom end of W, draw lines representing the directions of T and R_C (NB: it does not matter which end each of the force directions are drawn from, as long as it is a different end for each force).

– Extend the lines of T and R_C until they intersect at a point. This defines the force polygon (or triangle in this case because there are only three forces). The intersection point gives the lengths of the vectors and hence the magnitudes of T and R_C.

– Note that R_C must act upwards to close the polygon, i.e. ensure equilibrium.

(v) From the force polygon, measure the magnitudes of T and R_C as,

$$T = 68.6 \text{ kN} \quad \text{and} \quad R_C = 124.5 \text{ kN}$$

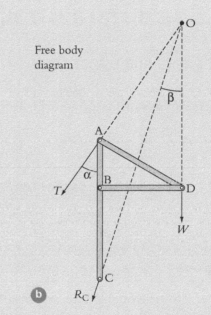

Free body diagram

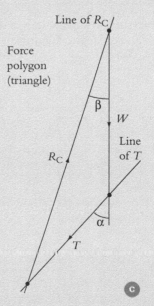

Force polygon (triangle)

Figure 1.22 **Pin-jointed structure – graphical solution worked example**

1.2 Stress, strain and elasticity

Direct stress and direct strain

Figure 1.23 shows a force, F, acting perpendicularly to a cross-sectional area, A. The average **direct stress** acting on the area, denoted by the symbol σ (sigma), is given by

$$\sigma = \frac{F}{A}$$

The units of direct stress are the units of force divided by area, i.e. $N\,m^{-2}$, also known as pascal (Pa).

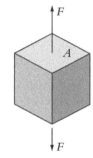

Figure 1.23 Direct stress caused by the action of a force on an area

Typical magnitudes of stress applied to engineering bodies are in the region of mega ($\times 10^6$) pascals (MPa) or even as high as giga ($\times 10^9$) pascals (GPa). If a stress causes the body on which it acts to elongate or stretch then it is a **positive** stress and is referred to as **tensile**. If the stress causes contraction of the body, then it is **negative** and referred to as **compressive**.

When a direct stress is applied to a body, it will cause the body to change its dimensions. **Direct strain**, denoted by the symbol ε (epsilon), is a measure of this change and is defined as the ratio of the change in length to the original length (see Figure 1.24). Thus

$$\varepsilon = \frac{\delta L}{L}$$

Because strain is a ratio of lengths, it has no units.

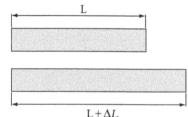

Figure 1.24 Strain as a measure of change in length, δL, divided by the original length L

As for direct stress, a **positive** strain refers to an **elongation** of the body, while a **negative** strain is a **contraction** of the body.

Elasticity, Hooke's law and Young's modulus

When a body is subjected to a force or forces, it deforms. If then, upon removal of the applied forces, the body returns to its original shape, it is considered to be **elastic**.

Thus, elastic materials are fully recoverable upon unloading.

Solid mechanics

Observations of the deformation and recovery of **elastic** materials under load, by the eminent 17th-century scientist/engineer Robert Hooke (1635–1703), showed that, in **uniaxial** loading (see Figure 1.25), there is a relationship between the direct stress, σ, and the direct strain, ε, in the test specimen. He found that, within the elastic limit (i.e. the elastic deformation region), stress is proportional to strain, or

$$\sigma \propto \varepsilon$$

This can be shown if one plots stress against strain, as in Figure 1.25, where a straight-line relationship is observed.

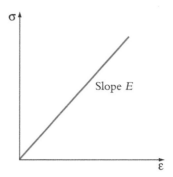

Figure 1.25 **Stress v. strain in the elastic region**

> The ratio of $\dfrac{\sigma}{\varepsilon}$, i.e. the gradient of this line, is constant and called Young's modulus (after the scientist Thomas Young, 1773–1829).

Young's modulus is also referred to as the **elastic modulus** of the material and is given the symbol E. Thus, in mathematical terms:

$$\sigma = E.\varepsilon \qquad (1.14)$$

> This equation is known as **Hooke's law** and, because it describes a straight line, materials obeying it are termed **linear** elastic (NB: there are some exceptions to this linear behaviour but most engineering solids, at least at small deformations, do obey the law).

Hooke's law can also be written in terms of the applied force, F, and the change in length, δL, as follows

$$\frac{F}{A} = \frac{E\,\delta L}{L}$$

or rearranging

$$\delta L = \frac{FL}{AE} \qquad (1.15)$$

which is a common form of expressing Hooke's law.

From equation (1.14), it can be seen that the units of Young's modulus, E, are those of stress divided by strain. As strain has no units, the units of E are in fact the same as those of stress i.e. Pa. Values of E for some typical engineering materials are given in Table 1.1.

As can be seen, there is quite a range of values for E, from high values for stiff materials, such as steel, to low values for flexible materials, such as rubber. The higher the value of E, the more stiff or rigid the material and the steeper the gradient in the stress–strain graph (Figure 1.25).

Material	Young's modulus (E) GPa
Steel	210
Aluminium	69
Concrete	14
Nylon	3
Rubber	0.01

Table 1.1 Some typical values of Young's modulus for engineering materials

Poisson's ratio

So far, we have considered *uniaxial* loading and deformation only. If, however, a bar is loaded along one axis with a resulting stress as shown in Figure 1.26, and we consider deformation in two dimensions, then there will actually be two strains. As expected, one occurs in the direction of the applied load/stress along the axis (termed the **longitudinal strain**) and the other occurs in the transverse direction (termed the **lateral strain**). In general, the lateral strain is of the opposite sign to the longitudinal strain. Thus, a stretching of the bar longitudinally will result in a contraction laterally, whereas a compression longitudinally will give rise to an expansion in the lateral direction. It is interesting to note that strain in the lateral direction is created *without* stress actually being applied in that direction.

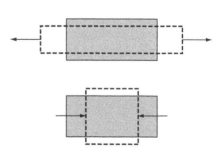

Figure 1.26 **Longitudinal and lateral strains**

The ratio of lateral strain to longitudinal strain depends on the material and, like Young's modulus, is an important material property. It is called **Poisson's ratio**, denoted by the symbol ν (nu), and named after the French mathematician Simeon Denis Poisson (1781–1840).

Thus:

$$\nu = \left| \frac{lateral\ strain}{longitudinal\ strain} \right|$$

As the two strains are of opposite sign, it is their absolute ratio which is calculated. This always results in a positive value for ν. Also, as a ratio of strains is calculated, ν has no units.

Values of Poisson's ratio for some typical engineering materials are given in Table 1.2.

Concrete deforms very little laterally when a longitudinal load is applied; hence, it has a low value of ν. Metals, on the other hand, have values in an intermediate range. Rubber is a special case, in that it has a value of ν very close to but just less than 0.5. A value of $\nu = 0.5$ exactly indicates a material which deforms at constant volume i.e. one which is incompressible. In this case, when a longitudinal strain is applied to the material, the deformation in the two transverse directions (for three-dimensional deformation) is such as to maintain constant volume. The theoretical maximum value for Poisson's ratio is 0.5.

Material	Poisson's ratio (ν)
Concrete	0.1
Most metals	0.25–0.35
Steel	0.29
Nylon	0.4
Rubber	~0.5

Table 1.2 Some typical values of Poisson's ratio for engineering materials

Stress–strain curve

It is important for design that material properties, such as Young's modulus and Poisson's ratio, are measured accurately so that they can be used in calculations to avoid excessive deformation and/or component failure. One of the simplest tests to conduct, which can yield a significant amount of information about a material, is the **uniaxial tensile test**.

Consider such a test on a round steel bar as shown in Figure 1.27. Load is applied axially to the ends of the bar and monitored continuously during the test by a load cell in the test machine crosshead. This load can then be used to calculate the stress on the bar during the test by dividing by the original cross-sectional area of the bar. Although the cross-sectional area of the bar does change during the test, e.g. due to Poisson's effects, it is more convenient to use the original cross-sectional area for the calculation. The resulting stress is termed the **nominal stress**. During the test, one end of the bar is fixed while the other end moves with the machine crosshead. The deformation of the bar can be measured in a number of ways, e.g. using a clip-on extensometer to monitor stretching of the bar, or by bonding electrical resistance strain gauges to the bar to give a direct measure of strain or just simply monitoring the movement of the machine cross head. Whichever method is used, a continuous measure of **strain** in the bar can be obtained during the test.

Figure 1.28 shows the variation of **nominal stress versus strain** during a tensile test on a mild steel bar. Firstly, the material deforms in a linear elastic fashion and the slope of the straight line gives **Young's modulus** for the bar material. There is then some loss of linearity as the **yield point** is approached.

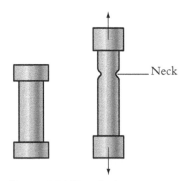

Figure 1.27 **Tensile test on a round steel bar**

Neck

The yield point is the point beyond which permanent deformations occur.

If the load were removed in the region beyond the yield point, unlike in the elastic region, the material would not fully recover, i.e. the unloading curve would not be a reverse of the loading curve back down to the origin (see Figure 1.28). The stress at which the yield point occurs is termed the **yield stress** (or **yield strength**) of the material, an important property in design.

Solid mechanics

The yield point itself is not always clearly defined. Beyond yield there may be some levelling out of the curve, i.e. a plateau region, where plastic deformation occurs resulting in further permanent deformation until the curve starts to rise again in a region of **strain hardening**. Eventually, the curve reaches a peak which defines the **ultimate tensile stress** (or **strength**) **(UTS)** of the material, i.e. the maximum stress that the material can withstand before failure occurs. Beyond the UTS is a region of **necking** of the bar where the cross-section thins down significantly until finally the curve falls off to a point at which **fracture** and complete **failure** occurs. The strain at which failure occurs, i.e. the **strain to failure**, of the material is another useful quantity as it indicates the level of ductility of the material or its ability to deform under load. The apparent reduction in stress before failure occurs is actually a consequence of the plotting of nominal stress based on the original cross-sectional area. If the actual cross-sectional area could be measured during necking and used to calculate stress, the curve would continue to rise as shown by the dashed line in Figure 1.28. This is the **true stress versus strain** curve; however, it is rarely plotted due to the difficulty in continuously monitoring changes in the cross-sectional area.

The simple uniaxial tensile test can therefore provide useful materials data for design, including Young's modulus, yield strength, ultimate tensile strength and strain to failure. The example discussed is a typical curve for mild steel. Other materials will exhibit different curves. Cast iron, for instance, which is a brittle material, shows none of the post yield plastic response of steel or the high level of ductility (large strain to failure). It is essentially linear elastic almost to failure, with no yield point, and a catastrophic brittle fracture at a relatively low level of strain. Other materials, such as rubber, for instance, show very high degrees of non-linear recoverable elastic deformation to very large strains.

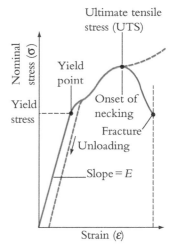

Figure 1.28 Stress–strain data from a tensile test on a mild steel bar

Shear stress, shear strain and shear modulus

Whereas direct stress occurs when a force is applied perpendicular to an area …

> **Shear stress** results when a force is applied parallel to an area as shown in **Figure 1.29**.

The applied shear force tends to cause one part of the body to *slide* relative to another. The average **shear stress** acting on the area, denoted by the symbol τ (tau), is given by

$$\tau = \frac{F}{A}$$

The units of shear stress are therefore the same as for direct stress, i.e. $N\,m^{-2}$ or Pa.

> **Shear strain**, denoted by the symbol γ (gamma), is a measure of the angle of distortion resulting from the applied shear stress as shown in **Figure 1.29**.

Because shear strain is an angle, it is measured in radians or degrees. Usually this angle is small (for deformation within the elastic range) and therefore $\gamma \approx \tan \gamma$ and, referring to Figure 1.29, shear strain is given by

$$\gamma \approx \tan \gamma = \frac{a}{b}$$

As with direct strain, shear strain is also a ratio and has no units.

Again, as with direct stress and strain, within the elastic limit, shear stress and strain are related through the shear form of Hooke's law, which is written as

$$\boxed{\tau = G\gamma} \qquad\qquad (1.16)$$

> Thus, shear stress is proportional to shear strain and the constant of proportionality, **G**, is termed the **shear modulus**.

The units of G are the same as Young's modulus, E, i.e. stress divided by strain or Pa.

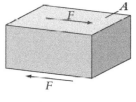

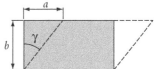

Figure 1.29 Shear stress and strain caused by the action of a force parallel to an area

A typical situation in engineering where shear occurs is the twisting of a shaft, as shown in Figure 1.30. When a torque is applied through the shaft, it twists about its axis. An element of material on the surface, as shown, is subjected to a shear stress, τ, and distorts through an angle, i.e. shear strain, γ. (Torsion theory is dealt with in detail in section 1.5 of this chapter and shows how to calculate the shear stress from the applied torque and the shear strain from the angle of twist.) If the torque is increased steadily, within the elastic limit, a plot of τ versus γ is a straight line, as shown in Figure 1.30, and expected from equation (1.16). The slope of this line is G, the shear modulus.

A typical value for the shear modulus of steel is 80 GPa and for aluminium is 26 GPa.

By considering the distortion of an element of material under pure shear, it can be shown that there is a relationship between the shear modulus and Young's modulus (this is only the case for an isotropic material, i.e. a material whose properties are the same in all directions). The relationship is as follows

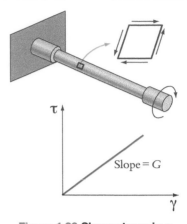

Figure 1.30 Shear stress in a twisted shaft

$$G = \frac{E}{2(1 + v)} \tag{1.17}$$

Thus the three elastic constants, Young's modulus (E), Poisson's ratio (v) and shear modulus (G) are related by equation (1.17). Only two of these constants are therefore independent. Assuming a typical value of $v = 0.3$ for a metal, the shear modulus is given by

$$G = \frac{E}{2(1 + 0.3)} = \frac{E}{2.6} \approx \frac{E}{3} \tag{1.18}$$

One-third × Young's modulus is therefore a rough estimate of the shear modulus of a metal. As tensile data is more often available, mainly because shear testing can be quite difficult, an estimate of the shear modulus of the material can be obtained by using equation (1.17).

Solving stress–strain problems

Stress–strain problems fall into one of two categories depending on the type of system. These are problems involving **statically determinate** systems and **statically indeterminate** systems respectively.

Statically determinate systems

Figure 1.31(a) illustrates axial loading of a statically determinate system where two dissimilar bars of different materials (A) and (B) are welded together in series, and the structure is loaded at its ends with a known load, F. In such systems, all forces and stresses can be solved for by using the equations of statics alone, i.e. the equilibrium equations. The stress–strain relationship will then yield the strains from which the extensions of each part of the bar and the total extension can be determined.

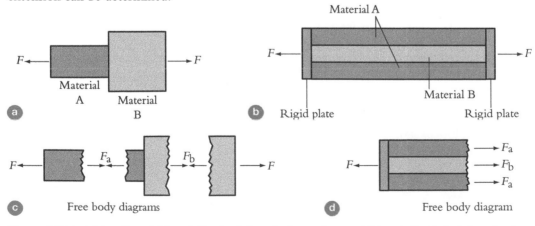

Figure 1.31 Axial loading of (a) a statically determinate system and (b) a statically indeterminate system

Solid mechanics

Thus, from the free body diagrams in Figure 1.31(c), which show the bar cut in several places:

Equilibrium gives: $F = F_a = F_b$

and the stresses follow: $\sigma_a = \dfrac{F_a}{A_a}$ and $\sigma_b = \dfrac{F_b}{A_b}$

Thus, the forces and stresses in the bar are found from equilibrium alone.

The stress–strain law tells us that: strain, $\varepsilon = \dfrac{\sigma}{E}$ or extension, $\delta L = \dfrac{FL}{AE}$

Thus, the extension of part (a) $\delta L_a = \dfrac{F_a L_a}{A_a E_a}$

and the extension of part (b) $\delta L_b = \dfrac{F_b L_b}{A_b E_b}$

and the total extension of the bar $\delta L_{tot} = \delta L_a + \delta L_b = F\left(\dfrac{L_a}{A_a E_a} + \dfrac{L_b}{A_b E_b}\right)$

Statically indeterminate systems

Figure 1.31(d) illustrates axial loading of a statically indeterminate system where three strips are bonded together in parallel and the structure is loaded at its ends, through rigid plates connecting the strips, with a known load, F. The outer two strips are of material A and the inner strip is of a different material B. In such systems, the forces and stresses *cannot* be solved for by using the equations of statics alone. Additional conditions are required. In this case, knowledge of the way in which the strips deform or elongate are used. It is known, for instance, that the extensions are the same for all strips. This is called a **compatibility** condition, meaning that the extensions of the strips must be compatible. The process is better illustrated by analysis as follows:

From the free body diagram in Figure 1.31(b), which shows the left-hand part of the bar after cutting:

Equilibrium gives

$$F = 2F_a + F_b \tag{1.19}$$

Thus, as there are two unknowns, F_a and F_b, the forces and stresses in the bar cannot be found from equilibrium alone.

Compatibility tells us that

$$\delta L_a = \delta L_b \tag{1.20}$$

The *stress–strain law* tells us that

$$\text{strain, } \varepsilon = \frac{\sigma}{E} \quad \text{or extension, } \delta L = \frac{FL}{AE} \tag{1.21}$$

Thus, from (1.20) and (1.21)

$$\frac{F_a L_a}{A_a E_a} = \frac{F_b L_b}{A_b E_b} \tag{1.22}$$

(NB: in this problem, A_a is the area of one strip only.)

Now, substituting for F_b from (1.19) into (1.22)

$$F_a L_a \backslash A_a E_a = (F - 2F_a) L_b \backslash A_b E_b$$

Rearranging and knowing that $L_a = L_b$

$$F_a\left[\frac{1}{A_a E_a} + \frac{2}{A_b E_b}\right] = \frac{F}{A_b E_b} \tag{1.23}$$

Thus, given the applied force, F, we can find F_a from (1.23), and F_b from (1.22).

The stresses, as before, are given by

$$\sigma_a = \frac{F_a}{A_a} \quad \text{and} \quad \sigma_b = \frac{F_b}{A_b}$$

and the total extension of the bar

$$\delta L_{tot} = \frac{F_a L_a}{A_a E_a} = \frac{F_b L_b}{A_b E_b}$$

To summarize, this problem is statically indeterminate, in that the additional compatibility condition of equal extension of the strips is necessary to solve for the unknown forces and hence stresses.

Principle of Superposition

In reality, engineering systems generally have more than a single load acting upon them and this adds to the complexity of problem solving. To overcome this, the **Principle of Superposition** is often used, which can be stated as follows:

$$\begin{bmatrix} \text{The total effect of combined} \\ \text{loads applied to a body} \end{bmatrix} = \sum \begin{bmatrix} \text{The effects of the individual} \\ \text{loads applied separately} \end{bmatrix}$$

Thus, a problem can be solved by analysing for each individual loading case and superposing (i.e. summing) the results to determine the total effect of all the loads.

However, there are two conditions which must be satisfied for the principle of superposition to apply. These are:

(i) Deformation is linearly dependent on the loads (linear elastic).

(ii) The deformations produced are small.

Figure 1.32 illustrates an example of the use of superposition.

The example shows a case of loading applied in two directions simultaneously. Stresses and strains can be determined in the two directions separately and added to obtain the total stress and strain arising from the combined loading

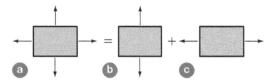

Figure 1.32 The principle of superposition

Thermal stress and strain

Stresses and strains generally arise from the application of a mechanical load to a body or structure. However, they can also arise when no mechanical load is applied. A typical situation is when a temperature change occurs.

A temperature change, ΔT, causes a thermal strain, ε_T, that is proportional to the magnitude of the temperature change. Thus

$$\varepsilon_T = \alpha.\Delta T \tag{1.24}$$

The constant of proportionality, α, is termed the **coefficient of thermal expansion** and has units degrees K^{-1}. α varies from material to material depending on the molecular structure and is generally small. Values for some typical engineering materials are given in Table 1.3.

For a bar of length, L, subjected to a temperature rise ΔT, it can be seen from equation (1.24) that the thermal extension, $\delta L_{thermal}$, is given by:

$$\delta L_{thermal} = L \alpha \Delta T \tag{1.25}$$

Material	Thermal expansion coefficient (α)
Concrete	$10 \times 10^{-6} K^{-1}$
Steel	$11 \times 10^{-6} K^{-1}$
Aluminium	$23 \times 10^{-6} K^{-1}$
Nylon	$144 \times 10^{-6} K^{-1}$
Rubber	$162 \times 10^{-6} K^{-1}$

Table 1.3 Coefficient of thermal expansion for some typical engineering materials

Using the principle of superposition, thermal extensions can simply be added to elastic (i.e. mechanical) extensions to give the total extension as follows

$$\delta L_{total} = \delta L_{elastic} + \delta L_{thermal}$$

Thus,

$$\delta L_{total} = \frac{FL}{AE} + L \alpha \Delta T$$

Resistive heat expansion of a bar

The bar shown in Figure 1.33 is subjected to a temperature rise of ΔT and restricted from expanding by rigid ends.

Since the bar cannot extend

$$\delta L_{\text{total}} = \delta L_{\text{elastic}} + \delta L_{\text{thermal}} = 0$$

i.e.

$$\delta L_{\text{total}} = \frac{FL}{AE} + L\alpha\Delta T = 0 \qquad (1.26)$$

Cancelling L in equation (1.26) and rearranging to find the force, F, in the bar, we find

$$F = -AE\alpha\Delta T \qquad (1.27)$$

and the stress in the bar is

$$\sigma = \frac{F}{A} = -E\alpha\Delta T \qquad (1.28)$$

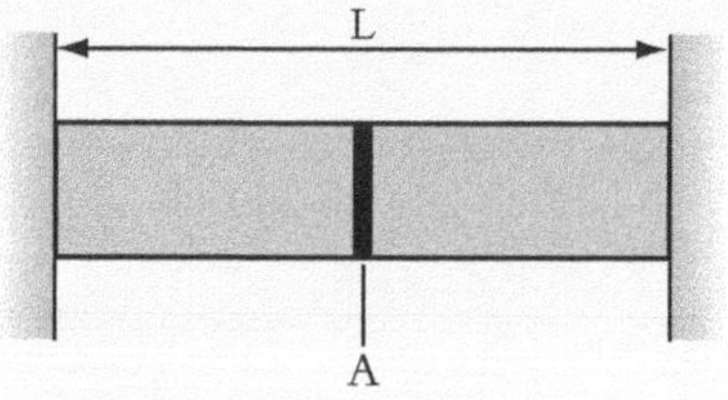

Figure 1.33 Resistive heat expansion of a bar

Note that the force and hence the stress are negative, indicating that the bar is under compression. As the bar heats up, it wants to expand but is prevented from doing so by the rigid end constraints. The outcome is that a compressive force/stress sets up in the bar with magnitudes given by equations (1.27) and (1.28).

Heating of a compound assembly

Figure 1.34 shows a compound assembly comprising a steel and an aluminium bar of the same dimensions held between rigid end blocks that are free to slide (without friction) between guide rails. The temperature of the whole assembly is then raised by ΔT.

After heating will the bars be in tension or compression?

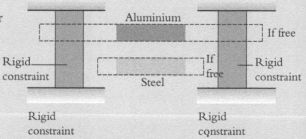

Figure 1.34 Heating of a compound assembly

(Intuitive)

Because of the end constraints, the compatibility condition tells us that the extension of the two bars must be the same

$$\delta L_{\text{steel}} = \delta L_{\text{al}}$$

Since $\alpha_{\text{al}} > \alpha_{\text{steel}}$, the aluminium bar is wanting to extend more than the steel bar. However, it is constrained from doing so by the rigid blocks attached to the steel bar. Therefore, the *aluminium bar will be in compression*. Also, the steel bar is wanting to extend less than the aluminium bar but is forced to extend further than it would do if free by the rigid blocks attached to the aluminium bar. Therefore, the *steel bar will*

Figure 1.35 The effect of the constraints by each bar on the other

be in tension. Figure 1.35 illustrates the relationship between the free extension of each bar and the actu al constrained extension in this example. Thus, although both bars expand as a result of the temperature change, one is in compression (aluminium) and one is in tension (steel).

(Analytical)

The *compatibility* condition is: $\delta L_{steel} = \delta L_{al}$

$$\therefore \qquad \frac{F_s L}{A_s E_s} + L\alpha_s \Delta T = \frac{F_a L}{A_a E_a} + L\alpha_a \Delta T \qquad (1.29)$$

The *equilibrium* condition is obtained from the freebody diagram shown in Figure 1.36 as follows:

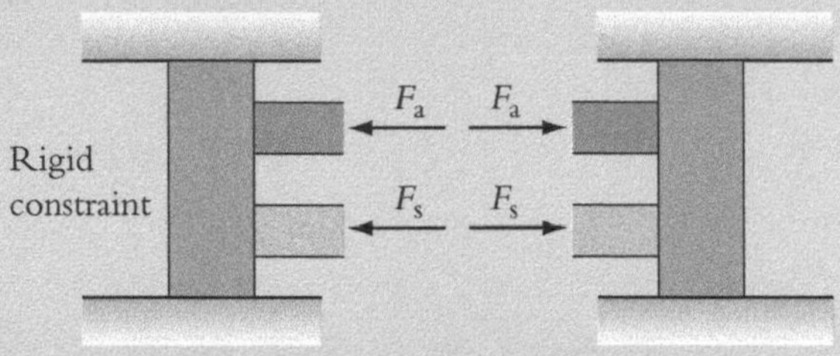

Equilibrium condition

$$F_s = -F_a \qquad (1.30)$$

Substituting for F_s from equation (1.30) into equation (1.29) and rearranging gives

Figure 1.36 Free body diagram of the compound assembly

$$L\Delta T(\alpha_s - \alpha_a) = F_a L \left[\frac{1}{A_a E_a} + \frac{1}{A_s E_s} \right]$$

$$\therefore \qquad \sigma_a = \frac{F_a}{A_a} = \frac{\Delta T(\alpha_s - \alpha_a)}{\left[\frac{1}{E_a} + \frac{A_a}{A_s E_s} \right]} \qquad (1.31)$$

As $\alpha_s < \alpha_a$, $\sigma_a < 0$

i.e. the aluminium bar is in compression.

Substituting for F_s from equation (1.30) into equation (1.29) and rearranging gives

$$\frac{F_s}{A_a} = \frac{\Delta T(\alpha_a - \alpha_s)}{\left[\frac{1}{E_a} + \frac{A_a}{A_s E_s} \right]}$$

$$\therefore \qquad \sigma_s = \frac{F_s}{A_s} = \frac{\Delta T A_a}{A_s} \frac{(\alpha_a - \alpha_s)}{\left[\frac{1}{E_a} + \frac{A_a}{A_s E_s} \right]}$$

As $\alpha_s < \alpha_a$, $\sigma_s > 0$

i.e. the *steel bar is in tension*.

Direct stress and strain

In a uniaxial tensile test experiment, a steel rod, initially 12 mm diameter and 50 mm gauge length, changes to 11.997 mm diameter and 50.044 mm gauge length under a load of 20 kN.

Determine the material properties of Young's modulus and Poisson's ratio.

Change in length

$$\Delta L = 50.044 - 50.0 = 0.044 \, \text{mm}$$

Strain

$$\varepsilon = \frac{\Delta L}{L} = \frac{0.044}{50.0} = 8.8 \times 10^{-4}$$

Stress

$$\sigma = \frac{Force}{Area} = \frac{20 \times 10^3}{\left(\frac{\pi}{4}\right).(0.012)^2} = 176.84 \times 10^6\,\text{Nm}^{-2} = 176.84\,\text{MPa}$$

Young's modulus

$$E = \frac{\sigma}{\varepsilon} = \frac{176.84 \times 10^6}{8.8 \times 10^{-4}} = 200.95 \times 10^9\,\text{Nm}^{-2} = 200.95\,\text{GPa}$$

Lateral strain

$$\varepsilon_T = \frac{\Delta d}{d} = \frac{12.0 - 11.997}{12.0} = 2.5 \times 10^{-4}$$

Poisson's ratio

$$\nu = \frac{Lateral\ strain}{Longitudinal\ strain} = \frac{2.5 \times 10^{-4}}{8.8 \times 10^{-4}} = 0.284$$

Shear stress and strain

The rubber-bearing pad, shown in Figure 1.37, consists of two rigid steel plates bonded to an elastomer. The pad has dimensions $a = 120$ mm, $b = 150$ mm and thickness, $t = 40$ mm and is subjected to a shear force of 5.4 kN.

Given that the top plate displaces 6 mm with respect to the bottom plate, determine the shear modulus, G, of the elastomer (it may be assumed that the rigid steel plates do not deform)

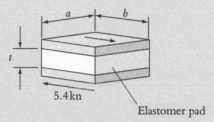

Figure 1.37 **Shear force acting on a rubber bearing pad**

Shear stress

$$\tau = \frac{F}{A} = \frac{5.4 \times 10^3}{120.150} = 0.3\,\text{Nmm}^{-2} = 0.3\,\text{MPa}$$

Shear strain

$$\tan \gamma = \frac{6}{40} = 0.15$$

$$\therefore \quad \gamma = 8.53^\circ = 0.149\,\text{rad}$$

Shear modulus

$$G = \frac{\tau}{\gamma} = \frac{0.3}{0.149} = 2.01\,\text{MPa}$$

Thermal stress and compatibility

The rigid casting, marked by A in Figure 1.38, is held by two steel bolts, B and C, of diameter 20 mm and length 400 mm, against one end of an aluminium alloy tube D, which has internal and external diameters of 50 mm and 60 mm respectively and length 300 mm. The other end of the tube is in contact with a rigid

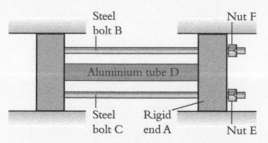

Figure 1.38 **Aluminium tube and bolt assembly**

surface to which the steel bolts are also attached. The pitch of each bolt thread is 1 mm. After the casting has been brought just into contact with the tube, each of the two nuts E and F are tightened by a further quarter of a turn. If the temperature of the whole assembly is then raised by 30°C, calculate the stresses in the bolts and the tube.

The material of the casting may be assumed to have a negligibly small coefficient of thermal expansion. The following material properties may be used:
$E_{steel} = 207$ GPa, $\alpha_{steel} = 11 \times 10^{-6}$ °C^{-1}, $E_{aluminium} = 68.9$ GPa, $\alpha_{aluminium} = 23 \times 10^{-6}$ °C^{-1}.

The solution for the stresses can be arrived at using the principle of superposition by solving for stresses before any temperature rise, then for stresses for a temperature rise only and finally adding the two cases together.

(a) Stresses before the temperature rise (quarter turn of the nuts)

Quarter turn $= 0.25 \times 1$ mm pitch $= 0.25$ mm

Equilibrium $\qquad\qquad\qquad F_a + 2F_s = 0$ $\qquad\qquad\qquad$ (1.32)
(as shown from the FBD in Figure 1.39(a))

Compatibility $\qquad\qquad\qquad \delta_s - \delta_a = 0.25 \times 10^{-3}$ $\qquad\qquad$ (1.33)
(as the net deformation to the left is 0.25 mm as shown in Figure 1.39(b))

Stress–strain $\qquad\qquad\qquad \delta = \dfrac{FL}{AE}$ $\qquad\qquad\qquad$ (1.34)

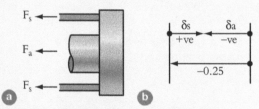

Figure 1.39 (a) **Equilibrium of forces and (b) compatibility conditions in the assembly**

Substituting (1.34) into (1.33)

$$\frac{F_s L_s}{A_s E_s} - \frac{F_a L_a}{A_a E_a} = 0.25 \times 10^{-3}$$

Now, substitute $F_a = -2F_s$ from (1.32)

$$F_s \left[\frac{L_s}{A_s E_s} + \frac{2L_a}{A_a E_a} \right] = 0.25 \times 10^{-3}$$

Substituting in the given values

$$L_s = 0.4 \text{ m}, L_a = 0.3 \text{ m}$$

$$A_s = \left(\frac{\pi}{4} \right)(0.02)^2 = 3.14 \times 10^{-4} \text{ m}^2$$

$$A_a = \left(\frac{\pi}{4} \right)(0.06^2 - 0.05^2) = 8.64 \times 10^{-4} \text{ m}^2$$

We obtain

$$F_s = 15.40 \text{ kN} \quad \text{and} \quad F_a = -30.81 \text{ kN}$$

Note that the negative sign indicates that the forces act in opposing directions, the aluminium tube being in compression and the steel bolts in tension.

The stresses are given by

$$\sigma_s = \frac{F_s}{A_s} = +49.02 \text{ MPa (tensile)}$$

$$\sigma_a = \frac{F_a}{A_a} = -35.66 \text{ MPa (compressive)}$$

(b) Stresses caused only by the temperature rise ($\Delta T = 30°C$)

Equilibrium

$$F_a^\star + 2F_s^\star = 0 \tag{1.35}$$

Compatibility

$$\delta_s^\star = \delta_a^\star \tag{1.36}$$
(since both are rigidly attached)

Stress–strain

$$\delta^\star = \frac{F^\star L}{AE} + L\alpha\Delta T \tag{1.37}$$

Substituting (1.37) into (1.36)

$$\frac{F_s^\star L_s}{A_s E_s} + L_s \alpha_s \Delta T = \frac{F_a^\star L_a}{A_a E_a} + L_a \alpha_a \Delta T$$

Now, substituting $F_a^\star = -2F_s^\star$ from (1.35) and rearranging:

$$F_s^\star \left[\frac{L_s}{A_s E_s} + \frac{2L_a}{A_a E_a} \right] = (L_a \alpha_a - L_s \alpha_s)\Delta T$$

Now substituting in the given values, we obtain

$$F_s^\star = 4.62 \text{ kN}$$

and

$$F_a^\star = -9.24 \text{ kN}$$

and the stresses due to the temperature rise only are

$$\sigma_s^\star = \frac{F_s^\star}{A_s} = +14.71 \text{ MPa (tensile)}$$

$$\sigma_a^\star = \frac{F_a^\star}{A_a} = -10.69 \text{ MPa (compressive)}$$

(c) Stresses caused by both tightening of the nuts and the temperature rise

Using superposition, the stresses in (a) and (b) can simply be added to give

$$\sigma_{s \text{ tot}} = \sigma_s + \sigma_s^\star = 49.02 + 14.71 = 63.73 \text{ MPa (tensile)}$$
$$\sigma_{a \text{ tot}} = \sigma_a + \sigma_a^\star = -35.66 - 10.69 = -46.35 \text{ MPa (tensile)}$$

Alternative solution

It is possible to arrive at the final stresses in (c) above in a single step without using superposition. So, if the problem requires the final stresses only, the following solution can be used:

Equilibrium

$$F_a + 2F_s = 0 \tag{1.38}$$
(as in solution (a))

Compatibility

$$\delta L_s - \delta L_a = 0.25 \times 10^{-3} \tag{1.39}$$
(as in solution (a))

Stress–strain

$$\delta L = \frac{FL}{AE} + L\alpha\Delta T \tag{1.40}$$

Substituting (1.40) into (1.39)

$$\left[\frac{F_s L_s}{A_s E_s} + L_s \alpha_s \Delta T\right] - \left[\frac{F_a L_a}{A_a E_s} + L_a \alpha_a \Delta T\right] = 0.25 \times 10^{-3}$$

Using (1.38) and rearranging gives

$$F_s\left[\frac{L_s}{A_s E_s} + 2\frac{L_a}{A_a E_a}\right] = (L_a \alpha_a - L_s \alpha_s)\Delta T + 0.25 \times 10^{-3}$$

which gives the final forces

$$F_s = 63.73\,\text{kN}$$

and

$$F_a = -44.9\,\text{kN}$$

and the final stresses

$$\sigma_{s\,\text{tot}} = 63.73\,\text{MPa (tensile) as before}$$
$$\sigma_{a\,\text{tot}} = -46.35\,\text{MPa (tensile) as before}$$

Learning summary

By the end of this section you will have learnt:

✔ the definitions of *direct* stress and strain and shear stress and strain;

✔ the principles of linear elasticity and Hooke's law which relates stress and strain through the material property Young's modulus in direct stress conditions, and shear modulus in shear conditions, both a measure of the stiffness of a material;

✔ how lateral deformations are quantified by the material property, Poisson's ratio;

✔ that there are generally two types of stress–strain problem, namely 'statically determinate' and 'statically indeterminate' problems. Statically determinate problems can be solved using the equations of statics alone (i.e. equilibrium and the stress–strain relationship), while statically indeterminate problems require further information such as the compatibility of strains or deformations;

✔ how temperature changes can give rise to stresses and strains in bodies or structures even in the absence of any applied loading. In such cases, an important material property, determining the magnitude of these 'thermal' stresses and strains, is the 'coefficient of thermal expansion' of the material;

✔ how to use the principle of superposition to solve problems involving both mechanical and thermal loading.

1.3 Beam bending

Beams are slender members, often horizontal (but not always), loaded by forces in one plane, perpendicular to the beam axis. In this section, we will deal with prismatic beams, i.e. those with a constant cross-section, operating in the linear elastic region where deflections and strains are small.

Types of beam, supports and loading

We will consider two main types of beam as shown in Figures 1.40(a) and **(b)**. The first is a **simply supported** beam which is supported at both ends by knife-edge simple supports. (NB: one of the knife-edge supports is shown mounted on rollers to indicate that there is no longitudinal constraint acting on the beam.) The second is a **cantilever** beam which is supported at one end only by a built-in support.

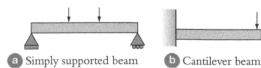

a Simply supported beam **b** Cantilever beam

Figure 1.40 Types of beam: (a) simply supported and (b) cantilever

Solid mechanics

The two types of support can be modelled as shown in Figures 1.41(a) and (b).

> The knife edge **simple support**, Figure 1.41(a), is modelled with a vertical reaction force which constrains vertical movement but allows free rotation of the beam at the support.

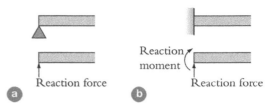

Figure 1.41 Types of beam support: (a) simple support and (b) built-in support

> The **built-in support**, Figure 1.41(b), is modelled with a vertical reaction force and a reaction moment which constrain vertical movement and rotation respectively at the support. This type of support is also referred to as **encastré**.

A beam may be subjected to several different types of loading, either separately or combined. The common types of loading are illustrated in Figure 1.42.

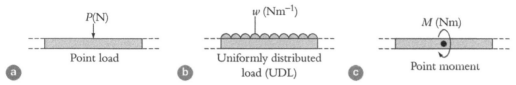

Figure 1.42 Types of beam loading: (a) point load, (b) uniformly distributed load (UDL) and (c) point moment

> A **point load** has units of newtons (N) and acts vertically at a point along the beam span. A **uniformly distributed load**, shortened to **UDL**, has units of newtons per metre length of beam (Nm^{-1}), and acts over the full span or part of the span. Finally, a **point moment** has units of Nm and again acts at a point along the beam span.

Various combinations of support and type of loading can exist on any specific beam and a general method of analysis is needed to deal with all possibilities.

Shear force and bending moment diagrams

When a beam bends under load, it creates internal forces and moments which are present at every point along the beam span. Consider a simply supported beam, with a point load, W, acting at mid span, as shown in Figure 1.43. The simple supports can be represented as vertical reaction forces and, because of symmetry, each must have a magnitude equal to $\dfrac{W}{2}$.

By sectioning the beam part way in from the right-hand end and drawing the free body diagrams of each half of the beam, we can deduce that, in order to satisfy equilibrium, an internal force and moment must exist at the cross-section.

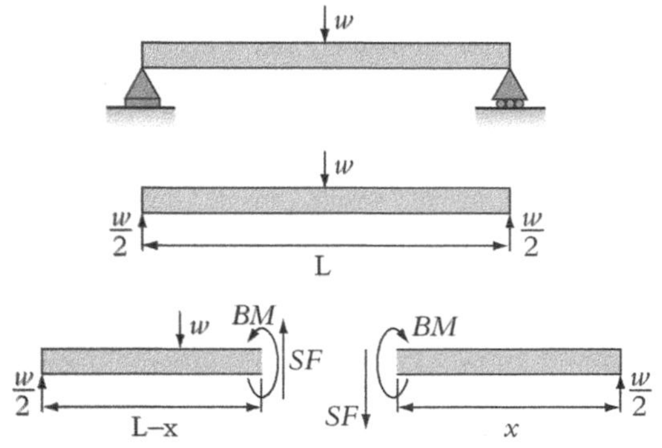

Figure 1.43 Shear force and bending moment in a simply-supported beam with central point load

Looking more closely at the right-hand part of the beam:

A **shear force (SF)** must exist on the internal face to balance the vertical reaction force, $\dfrac{W}{2}$. It acts as a 'shear' force trying to shear the plane perpendicular to the beam axis.

A **bending moment (BM)** must also exist to satisfy equilibrium and prevent the beam from rotating under the action of $\dfrac{W}{2}$ and SF. It acts to 'bend' the cross-section of the beam.

Looking at the left-hand part of the beam, a similar shear force and bending moment must exist on the internal face to maintain this part in equilibrium. This shear force and bending moment are equal and opposite to those acting on the right-hand face.

Similar shear forces and bending moments exist internally at all sections along the beam span although their magnitudes will vary at different sections. It is important to be able to determine the variation (or distribution) of these internal shear forces and bending moments along the beam if we are to be able to analyse for bending stresses within the beam.

Sign convention

Before looking at the methodology that we use for determining shear force and bending moment distributions, it is important to define a sign convention for each. The sign convention we use is illustrated in Figure 1.44.

Sign convention

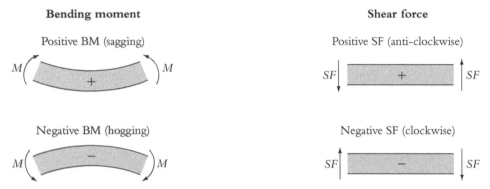

Figure 1.44 **Sign conventions for shear force and bending moment**

A *positive shear* force results in *anticlockwise rotation* of the face on which it acts.
A *negative shear* force results in *clockwise rotation* of the face on which it acts.

A *positive bending moment* results in *sagging* of the beam.
A *negative bending moment* results in *hogging* of the beam.

Sign conventions do differ from textbook to textbook, so it is important to be clear about the specific sign convention used when reading alternative texts. The above sign convention is used throughout this textbook.

Point load on a simply supported beam

Referring to Figure 1.45, we look at the example of a simply supported beam, ABC, with a point load, W, at mid span, position B.

Firstly, the reaction forces at A and C are determined by considering equilibrium of the whole beam and symmetry, giving

$$R_A = R_C = \frac{W}{2}$$

Next, cut the beam at a cross-section X–X between A and B, at a distance x from A. The cross-section carries an unknown shear force, S, and unknown bending moment, M. These have both been drawn as positive unknowns using our sign convention. This part of the beam can now be considered as a free body for which equilibrium conditions must hold.

Equilibrium of vertical forces

$$S = -\frac{W}{2} \tag{1.41}$$

Equilibrium of moments about X–X

$$M = +\frac{W}{2}x \tag{1.42}$$

Solid mechanics

Equation (1.41) gives the distribution of S at any point along the beam between A and B and equation (1.42) that of M.

At point B, at mid span, where $x = \dfrac{L}{2}$, from equations (1.41) and (1.42)

$$S\left(x = \frac{L}{2}\right) = -\frac{W}{2} \quad \text{and} \quad M\left(x = \frac{L}{2}\right) = \frac{WL}{4}$$

Now, cut the beam at a cross-section between B and C, at a distance x from A. The cross-section again carries an unknown shear force, S, and unknown bending moment, M. Equilibrium conditions for this part of the beam give

$$S = W - \frac{W}{2} = +\frac{W}{2} \tag{1.43}$$

and

$$M = \frac{W}{2}x - W\left(x - \frac{L}{2}\right) = \frac{W}{2}(L - x) \tag{1.44}$$

Using equations (1.41) to (1.44) we can now plot the shear force and bending moment distribution diagrams as shown in Figure 1.45.

It can be seen that the shear force is constant between A and B, and again between B and C, but shows a step change in sign from $-\dfrac{W}{2}$ to $+\dfrac{W}{2}$ at the mid span, position B.

The bending moment is not constant but varies linearly with x between A and B, and again between B and C. It is zero at the simple supports but rises to a maximum value of $+\dfrac{WL}{4}$ at the midspan, position B.

Determining the magnitude and position of the maximum bending moment is important and, as we will see in a later section this is where the maximum bending stresses occur in the beam.

Uniformly distributed load (UDL) on a simply supported beam

Figure 1.46 shows a simply supported beam carrying a UDL of w per unit length.

The unknown reaction forces at A and B can be found by considering equilibrium of the whole beam

Vertical forces

$$R_A + R_B = wL$$

Moments about A

$$R_B L - wL \cdot \left(\frac{L}{2}\right) = 0$$

∴

$$R_B = \frac{wL}{2}$$

$\left(\text{alternatively, symmetry could have been used to show that } R_A = R_B = \dfrac{wL}{2}\right).$

Next, cut the beam at a cross-section X–X distance x from the right-hand end. The cross-section carries an unknown shear force, S, and unknown bending moment, M, both drawn as positive unknowns. Considering the cut part of the beam as a free body (NB: the point load case was analysed by looking at the left-hand part of the beam as a free body. This case looks at the right-hand part. Either can be used in any particular analysis.) Equilibrium conditions show:

Equilibrium of vertical forces

$$S = \frac{wL}{2} - wx \tag{1.45}$$

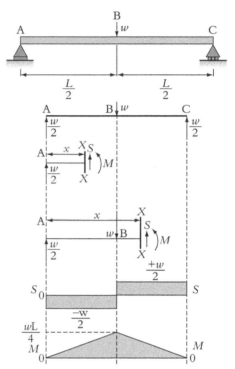

Figure 1.45 Determining the shear force and bending moment diagrams for a simply supported beam subject to a central point load

Equilibrium of moments about section X–X

$$M = \left(\frac{wL}{2}\right) \cdot x - (wx) \cdot \left(\frac{x}{2}\right)$$

∴
$$M = \left(\frac{wx}{2}\right) \cdot (L - x)$$
(1.46)

Equations (1.45) and (1.46) can now be used to plot the shear force and bending moment distribution diagrams as shown in Figure 1.46.

From equation (1.45), it can be seen that S varies linearly with x (compare this with the fact that S was constant for a point load). When $x = 0$, $S = +\dfrac{wL}{2}$ and when $x = L$, $S = -\dfrac{wL}{2}$. At mid span, when $x = \dfrac{L}{2}$, S is zero.

From equation (1.46), it can be seen that M varies parabolically with x (i.e. with x^2) (not linearly as it did for a point load). It is zero at the simple supports, i.e. when $x = 0$ and L. To find the position of the maximum bending moment, the gradient of M must be zero. From equation (1.46):

$$\frac{dM}{dx} = \frac{wL}{2} - wx$$

Thus,
$$\frac{dM}{dx} = 0 \text{ when } x = \frac{L}{2}$$

Hence, the position at which the maximum bending moment occurs is at the mid-span and its magnitude, given by substituting $x = \dfrac{L}{2}$ into equation (1.46), is $\dfrac{wL^2}{8}$. Thus,

$$M_{\text{max}} = \frac{wL^2}{8}$$

Note that the shear force is zero at the point of maximum bending moment. A relationship does in fact exist between shear force and bending moment as can be seen in the following subsection.

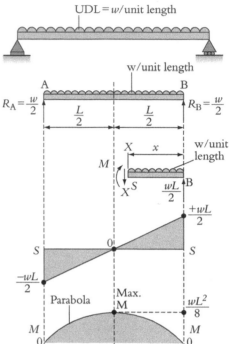

Figure 1.46 Determining the shear force and bending moment diagrams for a simply-supported beam subjected to a uniformly distributed load (UDL)

Relationship between shear force and bending moment

For the beam subjected to a UDL, w per unit length, consider a small element, length dx, of the beam span, as shown in Figure 1.47. In general, along such a length dx, it can be assumed that the shear force changes from S to $S + dS$ and the bending moment changes from M to $M + dM$.

Equilibrium of moments about the right-hand end of the element, i.e. point A, gives

$$-Sdx - (wdx) \cdot \left(\frac{dx}{2}\right) + M - (M + dM) = 0$$

Rearranging gives

$$dM = -Sdx - \frac{w}{2}dx^2$$

Since dx is small, dx^2 is negligible and the above expression can be simplified to:

$$dM = -Sdx$$

$$\text{or } S = -\frac{dM}{dx}$$

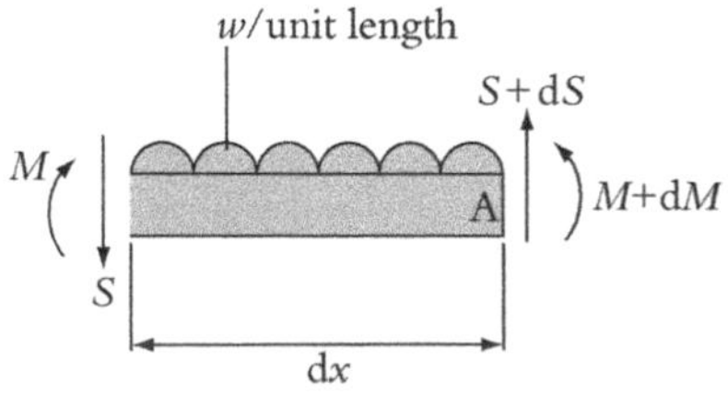

Figure 1.47 Element of beam length subjected to a changing shear force and bending moment

The *shear force* is therefore equivalent to the *negative* of the *slope* of the *bending moment distribution*.

This relationship can be used to check shear force and bending moment distribution diagrams for consistency.

Solid mechanics

Concentrated moment on a simply supported beam

Figure 1.48 shows a simply supported beam carrying a concentrated moment, M, part way along the span.

The unknown reaction forces at A and C can be found by considering the equilibrium of the whole beam:

Vertical forces

$$R_A + R_C = 0$$

Moments about A

$$-R_C L - M_B = 0$$

$$\therefore \qquad R_C = \frac{-M_B}{L}$$

As R_C is negative, it must act vertically downwards as shown in Figure 1.48.

Next, cut the beam at a cross-section X–X distance x from the right-hand end, C. The cross-section carries an unknown shear force, S, and unknown bending moment, M, both drawn as positive unknowns. Considering the cut part of the beam as a freebody equilibrium conditions show:

Equilibrium of vertical forces

$$S = -\frac{M_B}{L} \qquad (1.47)$$

Equilibrium of moments

$$M = \frac{M_B x}{L} \qquad (1.48)$$

Now, cut the beam at another cross-section between A and B, again at a distance x from the right-hand end C. The cross-section again carries an unknown shear force, S, and unknown bending moment, M. Equilibrium conditions for this part of the beam give

$$S = \frac{-M_B}{L} \qquad (1.49)$$

and

$$M = \frac{M_B x}{L} + M_B \qquad (1.50)$$

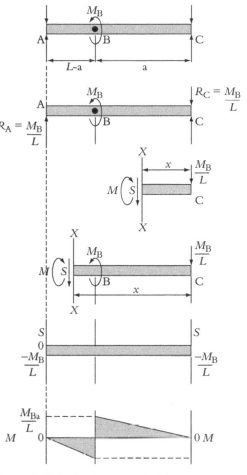

Figure 1.48 A simply supported beam subjected to a concentrated moment part way along its span

Using equations (1.47) to (1.50) we can now plot the shear force and bending moment distribution diagrams as shown in Figure 1.48.

It can be seen that S has a constant negative value along the length of the beam. By contrast, the bending moment varies linearly with distance x with a positive slope (as x increases to the left) of magnitude $\dfrac{M_B}{L}$ which is the negative of the shear force as expected. At position B, where the concentrated moment acts, there is a step change in the bending moment equivalent to the magnitude of M_B.

General observations on shear force and bending moment distribution diagrams

The following general observations can help in achieving and checking for correct diagrams:

(a) Always use a consistent sign convention – see the previous section

(b) At simple supports, the bending moment is zero. At built-in ends it is not zero.

(c) There is usually a step function in the shear force diagram under a point load.

(d) There is usually a step function in the bending moment diagram under a concentrated moment.

(e) Between point loads the bending moment usually varies in a linear way.

(f) Under a UDL, the bending moment diagram is parabolic (i.e. quadratic), even if the UDL is applied only over part of the span.

(g) The position where the maximum bending moment occurs is usually where the shear force is zero. This is a consequence of the fact that the shear force is the negative of the slope of the bending moment curve.

Beam-bending theory

The aim of this section is to develop a simple theory of bending which will enable stresses, arising in the beam from the applied loading, to be calculated. Whereas the previous section has shown that, in general, both a shear force and a bending moment exist at any particular section along the beam span, for relatively slender beams (where span/depth ratio is > 16), with which we will be dealing, the shear forces only have limited effect. Shear stresses arising from these forces are low in such systems and we only need to consider bending stresses arising from the bending moments. (NB: for a thicker beam, i.e. one of reduced span/depth ratio, this is not the case and shear forces and shear stresses should be taken into account.) This section will develop a general equation for bending, called the 'beam-bending equation', which will enable bending stresses to be calculated for a specific beam geometry and applied loading.

Assumptions

Referring to Figure 1.49, and to simplify the analysis, we make a number of reasonable assumptions:

(a) The beam is initially straight.

(b) After bending, the curvature of the beam is small, i.e. the radius of curvature, R, is large.

(c) Plane transverse sections remain plane after bending. Thus, in Figure 1.49, lines A_1A_2 and B_1B_2 remain straight.

(d) The beam material remains linear elastic during bending.

(e) $E_{tension} = E_{compression}$

(f) The bending moment acting at all sections along the part of the beam span under consideration is *constant* – this is termed 'pure bending'

(g) Axial direct stress (termed 'bending stress') is significantly larger than all other stresses, e.g. shear stresses.

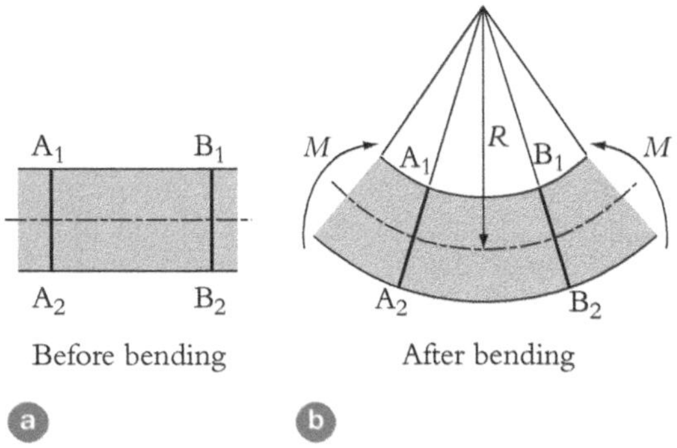

Before bending   After bending

Figure 1.49 Pure bending of a section of beam span

The neutral axis (NA)

Figures 1.50(a) and (b) show part of a beam under (a) positive bending (sagging) and (b) negative bending (hogging). It is assumed that the beam is under pure bending (constant bending moment) in each case. For the sagging beam, the upper layers of the beam are clearly in compression and subject to compressive bending stress in the axial direction, while the lower layers are in tension and subject to tensile bending stress. The opposite is true of the hogging beam, i.e. upper layers in tension, lower layers in compression. There must also be a layer or position through the thickness where the material in the beam is neither in tension nor compression. This layer is termed the **neutral surface** or, more commonly, when looking at a cross-section of the beam, as shown in Figure 1.50, the **neutral axis** (**NA** for short). It is the position in the cross-section where the tensile stress changes to compressive stress, i.e. where the bending stress is zero. In pure bending such as this, the NA coincides with the centroid of the beam cross-section as we will prove later. The bending stress does in fact change gradually through the thickness. For the sagging beam it changes linearly from an extreme compressive stress at the upper surface through zero at the NA, to an extreme tensile stress at the lower surface.

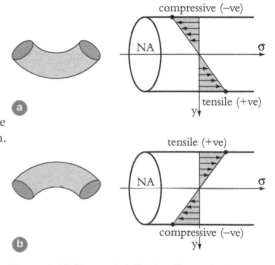

Figure 1.50 Stress distribution through the thickness of a beam under bending conditions: (a) positive (sagging) bending and (b) negative (hogging) bending

Solid mechanics

Geometry of deflection

Figure 1.51 shows in more detail what happens to two vertical sections in the beam, before and after bending. The sections are defined by points A_0A_1 and B_0B_1 before bending. A_0 and B_0 are on the neutral axis and A_1 and B_1 are in the lower half of the beam. The bending is caused by a positive bending moment (causing sagging) which is constant along the part of the beam of interest. This causes the line $A'_1B'_1$, which is at distance y vertically down from the NA, to stretch to $A'_1B'_1$. At the same time, A_0B_0 remains the same length during this bending as it is positioned on the NA. The NA bends to a radius of curvature, R, and the line $A'_0B'_0$ subtends an angle $\Delta\theta$.

$\therefore$ $\qquad A_1B_1 = A_0B_0 = A'_0B'_0 = R\Delta\theta$

since R is very large (see assumptions in Figure 1.51)

and $\qquad A'_1B'_1 = (R + y)\Delta\theta$

Thus, the strain at position y

$$\varepsilon = \frac{A'_1B'_1 - A_1B_1}{A_1B_1} = \frac{(R + y)\Delta\theta - R\Delta\theta}{R\Delta\theta} = \frac{y}{R} \qquad (1.51)$$

However, the strain

$$\varepsilon = \frac{\sigma}{E} \text{ from Hooke's law} \qquad (1.52)$$

Hence

$$\frac{\sigma}{E} = \frac{y}{R}$$

or rearranged,

$$\frac{\sigma}{y} = \frac{E}{R} \qquad (1.53)$$

Equation (1.53) shows that, for constant E and R, the bending stress, σ, is proportional to the distance, y, from the neutral axis and varies linearly with y. Also, $\sigma = 0$ at the NA where $y = 0$, and the maximum stress occurs where y is a maximum, at the surfaces of the beam. When y is a maximum positive, at the lower surface, σ is a maximum positive or tensile. When y is a maximum negative, at the upper surface, σ is a maximum negative or compressive. Equation (1.53) therefore confirms the discussion in the neutral axis section above, on the variation of the bending stress through the thickness of the beam.

Bending stress

Figure 1.52 shows a beam subjected to a positive constant bending moment i.e. pure bending. Consider a small strip of area, dA, in the cross-section. For equilibrium, net axial force on the whole cross-section must be zero.

$$\int_A \sigma dA = 0$$

However, from the previous section

$$\sigma = \frac{E}{R}y$$

$\therefore \qquad \int_A \frac{Ey}{R} dA = \frac{E}{R}\int_A y dA = 0$

$\therefore \qquad \int_A y dA = 0$

This proves that the NA coincides with the centroid of the cross-section.

Figure 1.51 **Geometry of deflection during bending**

Before bending

After bending

Figure 1.52 **Determining stresses through the thickness of a beam subjected to positive bending**

Now, taking moments about the NA, the elemental moment, dM, arising from the stress on the elemental area, dA, is

$$dM = (\sigma dA).y = \frac{Ey}{R}.dA.y$$

The total moment arising from stresses on all elemental areas in the cross-section is

$$M = \int_A dM = \frac{E}{R}\int_A y^2 dA \qquad (1.54)$$

> We now define the geometric term, $I = \int_A y^2 dA$, called the **second moment of area** about the NA. This quantity is a measure of the distribution of the area within the beam cross-section and is an important structural term in bending.

Then incorporating I in equation (1.54), we have

$$M = \frac{E}{R}I \quad or \quad \frac{M}{I} = \frac{E}{R} \qquad (1.55)$$

Combining equation (1.55) with equation (1.53), we have

$$\frac{M}{I} = \frac{\sigma}{y} = \frac{E}{R} \qquad (1.56)$$

This is the **beam-bending equation**. The key variables in this equation and their units are:

M = bending moment at the position in the beam (Nm)

I = second moment of area about NA (m^4)

σ = bending stress at distance y from NA (Nm^{-2})

y = distance from the NA (m) $-$ NB: y is +ve downwards

E = Young's modulus (Nm^{-2})

R = radius of curvature (m)

The two terms on the left-hand side of the beam-bending equation, i.e. $\frac{M}{I} = \frac{\sigma}{y}$, allow the calculation of the bending stress, σ, at distance y from the NA, knowing the applied bending moment, M, and the second moment of area of the section, I.

The radius of curvature of bending, R, can also be calculated from $\frac{M}{I} = \frac{E}{R}$.

In order to calculate bending stresses from equation (1.56), it is necessary to evaluate the second moment of area for the particular beam cross-section of interest. The next section explains how to calculate I for several shapes of cross-section.

Second moments of area

Rectangular cross-section

Consider the rectangular cross-section, width, B, and depth, D, as shown in Figure 1.53. XX is the NA which, from symmetry, passes through the centroid, G, of the cross-section. Considering the elemental strip of area, dA, distance y from the NA and dy in thickness, as shown. Then, the second moment of area about the NA (XX) is:

$$I_{XX} = \int_A y^2 dA = \int_{-D/2}^{+D/2} y^2 (Bdy)$$

$$= B\left[\frac{y^3}{3}\right]_{-D/2}^{+D/2} = \frac{B}{3}\left[\frac{D^3}{8} - \left(-\frac{D^3}{8}\right)\right]$$

$$\therefore \quad I_{XX} = \frac{BD^3}{12} \qquad (1.57)$$

Solid mechanics

Thus, the second moment of area for a rectangular section depends only on the dimensions of the section, B and D. It has units m^4 as expected. I_{XX} is relevant to bending about the XX axis. If the section was being bent about a vertical axis, YY say, then the value of I_{YY} would be needed. The expression for I_{YY} is similar to equation (1.57), but with B and D transposed, as follows:

$$I_{YY} = \frac{DB^3}{12} \qquad (1.58)$$

Figure 1.53 **Determining the 2nd moment of area of a solid rectangular cross-section**

Two theorems

To aid in the calculation of I for more complex sections, two theorems are introduced:

(a) The parallel axis theorem:

Consider the arbitrary cross-section shown in Figure 1.54, which has the axis XX passing through its centroid and a parallel axis $X'X'$, distance h from XX. A general element of area, dA, is at a distance, y, from the XX axis. The second moment of area, $I_{X'X'}$, about $X'X'$ is given by:

$$I_{X'X'} = \int_A (y + h)^2 dA = \int_A y^2 dA + \int_A h^2 dA + \int_A 2yh dA$$

$$= \int_A y^2 dA + h^2 \int_A dA + 2h \int_A y dA \qquad (1.59)$$

As XX passes through the centroid of the section, $\int_A y dA$ is zero, and equation (1.59) simplifies to:

$$I_{X'X'} = I_{XX} + Ah^2 \qquad (1.60)$$

Equation (1.60) is a statement of the **parallel axis theorem** for second moments of area. The minimum value of I is always the value about an axis through the centroid of the section, i.e. I_{XX}. Values of I about axes parallel to XX are always larger by the second term, Ah^2, which is always positive.

(b) Perpendicular axis theorem:

Consider the arbitrary cross-section shown in Figure 1.55. Two perpendicular axes, XX and YY, are shown passing through the centroid. A third perpendicular axis, ZZ, also passing through the centroid, is drawn coming out of the page. Thus the origin of all three axes is at the centroid. A general element of area, dA, is also shown, positioned at distance, r, from the centroid (origin).

Figure 1.54 **The parallel axis theorem**

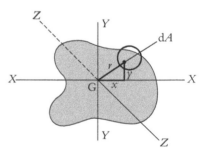

Figure 1.55 **The perpendicular axis theorem**

We now define a new quantity, J, the **polar second moment of area**, as follows:

$$J = \int_A r^2 dA$$

J is effectively, the second moment of area about the third axis, ZZ.

However

$$r^2 = x^2 + y^2$$

$$J = \int_A x^2 dA + \int_A y^2 dA$$

$$\text{i.e.} \quad J = I_{XX} + I_{YY} \qquad (1.61)$$

Equation (1.61) is a statement of the **perpendicular axis theorem** for second moments of area. In words, it states that, for a cross-section, the sum of the second moments of area about two perpendicular axes is equal to the second moment of area about the third perpendicular axis.

Circular cross-section

The perpendicular axis theorem enables us to obtain an expression for the second moment of area of a circular cross-section. Figure 1.56 shows such a section, of diameter, D, with the perpendicular axes, XX and YY, drawn through the centroid, i.e. centre of the circle. An elemental ring of area, dA, is shown, at a radius, r, and with a thickness, dr. The polar second moment of area is given by

$$J = \int_A r^2 dA$$

As dr is very small, $dA = 2\pi r\,dr$, and

$$J = \int_0^{D/2} r^2 (2\pi r dr) = 2\pi \int_0^{D/2} r^3 dr$$

$$= 2\pi \left[\frac{r^4}{4} \right]_0^{D/2} = \frac{\pi}{2} \left[\frac{D^4}{16} \right]$$

$$\therefore \quad J = \frac{\pi D^4}{32} \qquad\qquad\qquad (1.62)$$

This is the polar second moment of area of the circular section and is an important parameter when considering torsion or twisting of the section (see later section in this chapter on torsion of shafts). For bending, the two perpendicular second moments of area, I_{XX} and I_{YY}, are more relevant. Knowing J, they can now be obtained using the perpendicular axis theorem as follows

$$J = I_{XX} + I_{YY}$$

But, because of symmetry of the circular section, $I_{XX} = I_{YY}$, and therefore

$$J = 2I_{XX}$$

or

$$I_{XX} = I_{YY} = \frac{J}{2} = \frac{\pi D^4}{64} \qquad\qquad (1.63)$$

Thus, the second moment of area for a circular section depends only on the diameter, D. It has units m^4 as expected. Also, because of the symmetry of the section, it does not matter if you are bending about the XX or YY axes, the second moment of area is the same. This is not the case for a rectangular section where I_{YY} is different from I_{XX}.

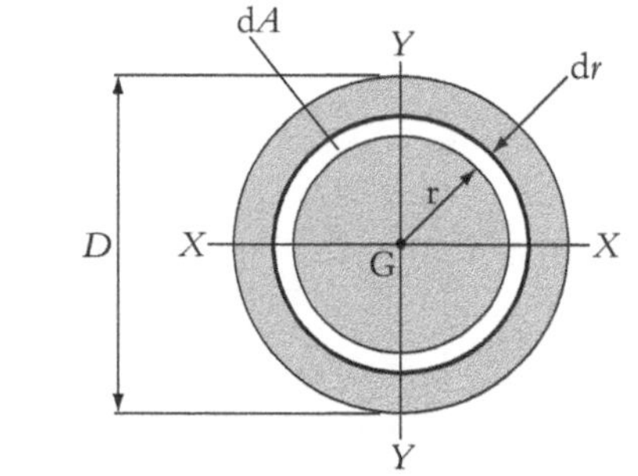

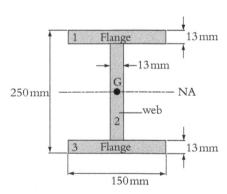

Figure 1.56 Determining the second moment of area of a solid circular cross-section

I-section (symmetrical)

Consider the symmetrical I-section shown in Figure 1.57. I-beams are common structural members, precisely because their sections have relatively high values for second moment of area. This is a consequence of a large portion of the cross-sectional area being in the flange regions which are positioned a significant distance from the neutral axis resulting in a high I-value. From the beam equation (1.56), a high I-value reduces the stress level for a particular applied moment. Furthermore, a high I-value also results in a more rigid structure, i.e. reduced deformations under load.

As the section shown in Figure 1.57 is symmetric, the NA passes through the centroid of the section. To find the second moment of area about the NA, we divide the section into convenient subsections, in this case, the flanges and the web, as shown. Because each of these subsections is a rectangle, we can use equation (1.57) to determine the I-value for each subsection about axes through each subsection's centroid. These values are then transposed onto the NA of the complete section using the parallel axis theorem. The procedure is best performed using a table to avoid error and permit ease of checking when the calculation is complete.

Figure 1.57 A symmetrical I-section

Solid mechanics

Table 1.4 I-section second moment of area calculation Subsection I for subsection

Subsection	I for subsection mm^4	A mm^2	h mm	Ah^2 mm^4	$I_{NA} = I + Ah^2$ mm^4
1. Flange	$\frac{(150 \times 13^3)}{12} = 2.746 \times 10^4$	$(150 \times 13) = 1950$	$\left(\frac{13}{2} + 112\right) = 118.5$	$= 2.738 \times 10^7$	$= 2.74 \times 10^7$
2. Web	$\frac{(13 \times 224^3)}{12} = 1.217 \times 10^7$	$(224 \times 13) = 2912$	$= 0$	$= 0$	$= 1.217 \times 10^7$
3. Flange	$\frac{(150 \times 13^3)}{12} = 2.746 \times 10^4$	$(150 \times 13) = 1950$	$\left(\frac{13}{2} + 112\right) = 118.5$	$= 2.738 \times 10^7$	$= 2.74 \times 10^7$

The total I-value about the NA is given by summing the final column:

$$\text{Total } I_{NA} = 6.70 \times 10^7 \text{ mm}^4$$
$$= 6.70 \times 10^{-5} \text{ m}^4$$

Alternative method:

Because of symmetry of this I-section, an alternative, simpler approach may be used. Figure 1.58 shows that I_{NA} may be found by determining the value for a rectangular section and subtracting values for the two shaded rectangular sections. Thus

$$I_{NA} = I_{\text{rect. section}} - I_{\text{shaded area}}$$

$$= \frac{150 \times 250^3}{12} - 2 \times \left[\frac{68 \times 5.224^3}{12}\right]$$

$$= 6.70 \times 10^7 \text{ mm}^4 \quad \textit{as before}$$

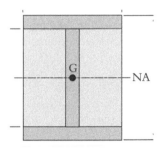

Figure 1.58 Simplified approach for determining the 2nd moment of area of a symmetric I-section

T-section (nonsymmetrical)

Unlike the I-section, the T-section, shown in Figure 1.59, is not symmetrical about a horizontal axis. The position of the NA must therefore first be determined. The NA passes through the centroid of the section, which is positioned at some unknown distance from the datum at the bottom surface. The section is divided into two subsections, in this case, the top flange and the web, as shown. Now, the *first* moment of the total cross-sectional area about the datum should equate to the sum of the first moments of each subsection area about the datum. Thus

$$(\text{Total area})\bar{y} = \Sigma(A_i.y_i) \text{ for each subsection}$$

$$(200 \times 50 + 125 \times 50)\bar{y} = (200 \times 50)\left(125 + \frac{50}{2}\right) + (125 \times 50)\left(\frac{125}{2}\right)$$

$$\bar{y} = 116.35 \text{ mm from the bottom surface datum}$$

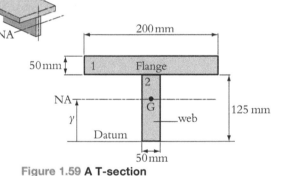

Figure 1.59 A T-section

Knowing the position of G and hence the NA, we can now use the tabulated method to determine the second moment of area for the T-section:

Table 1.5 T-section second moment of area calculation

Subsection	I for subsection mm^4	A mm^2	h mm	Ah^2 mm^4	$I_{NA} = I + Ah^2$ mm^4
1. Flange	$\frac{(200 \times 50^3)}{12} = 2.08 \times 10^6$	$(200 \times 50) = 10\,000$	$(125 + 25 - 116.35) = 33.65$	$= 1.132 \times 10^7$	$= 1.34 \times 10^7$
2. Web	$\frac{(50 \times 125^3)}{12} = 8.14 \times 10^6$	$(125 \times 50) = 6250$	$\frac{125}{2} - 116.35 = -53.85$	$= 1.805 \times 10^7$	$= 2.64 \times 10^7$

The total I-value about the NA is given by summing the final column:

$$\text{Total } I_{NA} = 3.96 \times 10^7 \text{ mm}^4$$
$$= 3.96 \times 10^{-5} \text{ m}^4$$

Solving beam problems

Based on the above sections, the general procedure for solving beam problems, i.e. calculating bending stresses, can now be summarized:

(a) For a given load acting on the beam, draw the bending moment diagram.

(b) Determine the positions along the span of maximum +ve or maximum −ve bending moment (BM).

(c) From the shape and dimensions of the beam cross-section at the position of max BM, determine the centroid of area and hence the position of the neutral axis (NA) which passes through the centroid.

(d) Calculate the second moment of area of the beam cross-section about the neutral axis, i.e. I_{NA}.

(e) At the positions of maximum bending moment, use part of the beam bending equation (1.56) to determine the maximum stresses at the top and bottom of the cross-section, i.e. furthest from the NA.

(f) From the loading on or deflected shape of the beam determine which of the above stresses are +ve (tensile) or −ve (compressive).

Point-loaded beam – Hollow rectangular section

For the simply supported beam, shown in Figure 1.60, determine the position and magnitude of the maximum tensile bending stress.

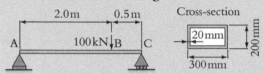

Figure 1.60 **A simply supported beam carrying a point load at a position part way along its span**

Reaction forces

Consider Figure 1.61(a), which shows the beam's simple supports replaced by point reaction forces, R_A and R_C. The magnitudes of these reaction forces are found by considering the equilibrium of the beam as a whole as follows

Vertical forces

$$R_A + R_C = 100$$

Clockwise moments about A

$$100 \times 2 = R_C \times 2.5$$

∴
$$R_C = 80 \, \text{kN}$$
$$R_A = 20 \, \text{kN}$$

Bending moment diagram

Cutting the beam at distance x from the left-hand end, we have an unknown internal shear force (S) and bending moment (M) at the cut section, as shown in Figure 1.61(b). We will not consider the shear force any further in this example. Now, considering equilibrium of moments about section X–X for the left-hand part of the span as a free body, we find

$$M = 20x$$

This expression applies between A and B and enables the bending moment diagram to be drawn in that region, as shown in Figure 1.61(c). The bending moment rises from zero at the simple support, A, to a maximum at the load point, B. Here, $x = 2$ and the maximum bending moment is $M_{max} = 20 \times 2 = 40 \, \text{kNm}$.

Next, if we cut the beam at a distance x from the RH end, and considered equilibrium of moments on the RH part of the span as a free body, we would find:

$$M = 80x$$

This expression enables the second part of the bending moment diagram to be drawn for region B to C, also shown in Figure 1.61(c). In this case the bending moment again rises from zero at the simple support, C, to a maximum at the load point, B. Here, $x = 0.5$ from the RH end and the maximum bending moment is $M_{max} = 80 \times 0.5 = 40$ kNm, as before. Thus, the full bending moment diagram can now be drawn as shown in Figure 1.61(c). The critical section, where the maximum stresses occur, is at the position of the maximum bending moment, i.e. below the 100 kN load, position B.

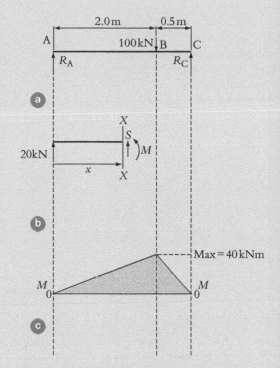

Figure 1.61 **Determining the shear force and bending moment distribution for the simply supported beam carrying a point load**

Second moment of area

To calculate the bending stresses, we need the second moment of area, I, of the cross-section about its neutral axis. As it is a hollow rectangular section, as shown in Figure 1.60, we can use the expression for a solid rectangular section, derived earlier on page 34. We treat the hollow section as two solid sections, and subtract the I-value for one from the other. Thus:

$$I = \frac{b_o d_o^3}{12} - \frac{b_i d_i^3}{12}$$

where o = outer dimensions of the section

i = inner dimensions of the section

$$\therefore \qquad I = \frac{300 \times 200^3}{12} - \frac{260 \times 160^3}{12}$$

$$= 111.25 \times 10^6 \, \text{mm}^4 = 111.25 \times 10^{-6} \, \text{m}^4$$

Maximum tensile stress

Looking at the way the beam bends in a sagging manner, the maximum tensile stress occurs on the lower surface of the cross-section, i.e. at a position furthest away from the neutral axis (NA). It will also occur at position B along the span, where the maximum bending moment exists. Due to symmetry of the cross-section, the NA is at the centre of the section, 100 mm above the lower surface. Thus, the maximum tensile stress, as given by the beam-bending equation (1.56), is

$$\sigma_{T \, max} = \frac{M_{max} y_{max}}{I} = \frac{40 \times 10^3 \, 100 \times 10^{-3}}{111.25 \times 10^{-6}} = 35.96 \, \text{MPa}$$

Uniformly distributed load – I-section

Figure 1.62 shows the cross-section of an I-section beam which has an overall depth of 250 mm, a top flange 120 mm wide and 10 mm thick, a bottom flange 80 mm wide and 10 mm thick, and a web of thickness 6 mm. The beam is simply supported over a span of 5 metres and carries a uniformly distributed load of $5\,\text{kNm}^{-1}$ over its full span. Assume that the beam is made of steel with E = 200 GPa.

Calculate the maximum compressive and tensile stresses in the beam and the radius of curvature at the point of maximum bending moment.

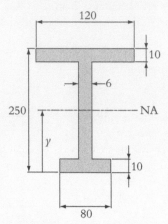

Figure 1.62 An unsymmetric I-section

Maximum bending moment

For a simply supported beam under a UDL of $w\,\text{N/m}$, the maximum bending moment occurs at the centre of the span and is given by (see the section on 'Shear force and bending moment diagrams'):

$$M_{max} = \frac{wL^2}{8} = \frac{5 \times 5^2}{8} = 15.625\ \text{kNm}$$

Position of the neutral axis

The position of the neutral axis from the bottom surface, y, is determined by taking area moments about the bottom surface as follows:

$$A_{tot}.\bar{y} = \Sigma A_i y_i$$

$$[(120 \times 10) + (6 \times 230) + (80 \times 10)]\bar{y} = (120 \times 10) \times 245 + (6 \times 230) \times 125 + (80 \times 10) \times 5$$

giving

$$\bar{y} = 139.2\ \text{mm from the bottom surface}$$

Second moment of area of the cross-section

Table 1.4 I-section second moment of area calculation

Subsection	I for subsection mm^4	A mm^2	h mm	Ah^2 mm^4	$I_{NA} = I + Ah^2$ mm^4
1. Flange	$\dfrac{(120 \times 10^3)}{12}$ $= 1.0 \times 10^4$	(120×10) $= 1200$	$= 105.8$	$= 1.343 \times 10^7$	$= 1.344 \times 10^7$
2. Web	$\dfrac{(6 \times 230^3)}{12}$ $= 6.08 \times 10^6$	(6×230) $= 1380$	$= 14.2$	$= 0.278 \times 10^6$	$= 6.36 \times 10^6$
3. Flange	$\dfrac{(80 \times 10^3)}{12}$ $= 6.67 \times 10^3$	(80×10) $= 800$	$= 134.2$	$= 1.44 \times 10^7$	$= 1.441 \times 10^7$

The total I-value about the NA is given by summing the final column:

$$\text{Total } I_{NA} = 3.421 \times 10^7\ \text{mm}^4$$
$$= 3.421 \times 10^{-5}\ \text{m}^4$$

Bending stresses

As the beam is sagging, the top surface is in compression and the bottom surface in tension.

The maximum tensile stress on the bottom surface, i.e. at $y = 139.2$ mm, is given by:

$$\sigma_{bottom} = \frac{M_{max}y_{max}}{I} = \frac{(15.625 \times 10^3) \times 0.1392}{3.421 \times 10^{-5}} = 63.58 \text{ MPa tensile}$$

The maximum compressive stress on the top surface, i.e. at $y = -110.8$ mm, is given by:

$$\sigma_{top} = \frac{M_{max}y_{max}}{I} = \frac{(15.625 \times 10^3) \times 0.1108}{3.421 \times 10^{-5}} = 50.61 \text{ MPa compressive}$$

Radius of curvature

To obtain the radius of curvature, the other part of the beam bending equation (1.56) is used:

$$\frac{M}{I} = \frac{E}{R}$$

Thus,
$$R = \frac{EI}{M} = \frac{(200 \times 10^9) \times (3.421 \times 10^{-5})}{15.625 \times 10^3} = 437.9 \text{ mm}$$

Learning summary

By the end of this section you will have learnt:

- ✔ various types of beam, their supports and loading conditions;
- ✔ how to calculate and draw shear force and bending moment distribution diagrams for beams under point loading, UDL and concentrated moments;
- ✔ beam-bending theory and the derivation of the beam bending equation relating the bending moment to bending stress and radius of curvature;
- ✔ how to calculate the second moments of area of simple and more complex sections such as I-sections and T-sections;
- ✔ the general procedure for calculating stresses within beams.

1.4 Multi-axial stress and strain

So far in this chapter we have considered **uniaxial** stress and strain. Even in beam bending, introduced in the previous section, although the stress and strain vary through the thickness of the beam, from maximum values at the surface to zero at the neutral axis, they essentially act along a single axis, i.e. the beam axis. In more general engineering situations, stresses and strains can act simultaneously in more than one direction at a point in the structure. Firstly, we will consider stresses and strains acting in a single plane, i.e. in **two dimensions**, and develop a graphical technique, called **Mohr's circle**, for analysing in detail the state of plane stress. This will enable critical stresses within the plane, which might cause material failure, to be determined. Many engineering situations can be analysed as plane stress problems. Next, the section will look at the more general state of **three-dimensional** stress and strain and introduce methods of analysis based on **generalized (i.e. three-dimensional) Hooke's law**.

The general state of plane stress (two-dimensional stress)

Figure 1.63 illustrates the **general state of plane stress** acting on a small element of material at a point in a structure. This plane element is assumed to have unit thickness. The plane in which the element exists is the x–y plane but it could equally be one of either the x–z or y–z planes. In addition to direct stresses acting in the x and y directions, i.e. σ_x and σ_y respectively, in general, shear stresses also act on the element as shown in Figure 1.63. Some definitions and nomenclature are called for:

> An x-plane is defined as a plane whose *normal* acts in the x-direction. Thus, the right-hand edge and the left-hand edge of the element are both x-planes.

> A y-plane is defined as a plane whose normal acts in the y-direction. Thus, the top edge and the bottom edge of the element are both y-planes.

(Note that an x-plane is parallel to the y-axis while a y-plane is parallel to the x-axis.)

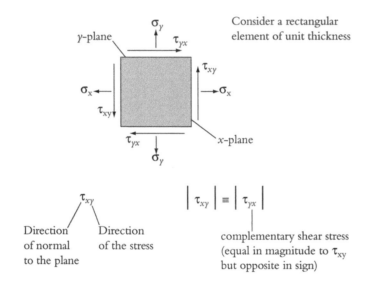

NB: +ve shear causes anticlockwise rotation

Figure 1.63 **The general state of plane stress**

For the direct stresses, σ_x and σ_y, the subscripts x and y refer to the direction in which (or axis along which) the stresses act. Also, σ_x acts on an x-plane and σ_y acts on a y-plane. A direct stress with a positive magnitude is a tensile stress, while a direct stress with a negative magnitude is a compressive stress. In Figure 1.63, σ_x and σ_y are both drawn as positive tensile stresses.

The nomenclature for shear stresses is a little more complicated. As with the direct stresses, the shear stresses also act on the x and y planes. Consider a shear stress, τ_{xy}, acting upwards on the right-hand edge of the element. For the element to be in equilibrium in the *vertical sense*, an equal and opposite shear stress, τ_{xy}, must act downwards on the left-hand edge of the element. Again, for the element to be in equilibrium with respect to *rotation*, shear stresses, τ_{yx}, must also act on the top and bottom edges of the element as shown in Figure 1.63. These latter shear stresses are termed **complementary shear stresses,** i.e. τ_{yx} are complementary to τ_{xy}.

The shear stresses are given two subscripts. The first subscript denotes the *plane* on which the stress acts while the second subscript denotes the *direction* in which the stress acts. Thus, τ_{xy} acts on the x-plane in the y-direction.

Solid mechanics

The sign convention for shear stresses is governed by the sense of rotation that the shear stress is trying to impart on the element. Thus:

> A *positive shear stress* is trying to rotate the element *anticlockwise*. τ_{xy} is therefore positive on both the right-hand and left-hand edges.

> A *negative shear stress* is trying to rotate the element clockwise. τ_{yx} is therefore negative on both the top and bottom edges.

In general, the shear stresses and complementary shear stresses are equal in magnitude but opposite in sign.

Stress transformation in two dimensions

The stresses acting within a body are specified with respect to given coordinate axes, i.e. the stresses are known to act on certain defined planes. For example, σ_x is a *normal* stress acting *parallel* to the x-axis on the x-plane *normal* to the x-axis; τ_{xy} is a *shear* stress acting parallel to the y-axis on the x-plane normal to the x-axis.

The question arises – 'What are the stresses on other planes inclined to those given?'

Consider the small element PQRS shown in Figure 1.64, with direct stresses σ_x and σ_y, and shear stresses τ_{xy} and τ_{yx} acting on faces normal to the x and y directions respectively. (All stresses are assumed to act in this x–y plane, i.e. a 'plane stress' problem. As before, we assume that the thickness of the element normal to the plane of the paper is unity.)

The stresses on the plane RT, inclined at an angle α to the plane carrying σ_x and τ_{xy} are required; they will in general include both a direct (i.e. normal) stress σ_α and a shear stress τ_α as shown. To determine the values of σ_α and τ_α, we consider the equilibrium of forces on the prism RTQ and the following expressions can be derived

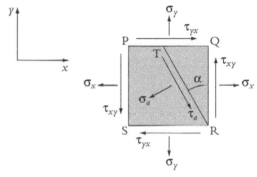

Figure 1.64 Stresses on a plane inclined at α to the x-plane

$$\sigma_\alpha = \frac{\sigma_x + \sigma_y}{2} + \frac{\sigma_x - \sigma_y}{2}\cos 2\alpha + \tau \sin 2\alpha$$

$$\tau_\alpha = \frac{\sigma_x - \sigma_y}{2}\sin 2\alpha + \tau \sin 2\alpha$$

where

$$\tau = \tau_{xy} = -\tau_{yx}.$$

These transformation equations can be used to determine the values of σ_α, and τ_α, at any specified angle α, directly from known values of σ_x, σ_y, τ_{xy}.

Mohr's circle for plane stress

> Mohr's circle is a graphical construction representing the above transformation equations.

A set of axes is set up in terms of direct stress and shear stress as shown in Figure 1.65. Points B and E, which represent the stresses on the element faces QR and PQ respectively (see Figure 1.64) are plotted with coordinates (σ_x, τ_{xy}) and (σ_y, τ_{yx}). *Positive shear stresses* acting on the face of an element tend to rotate the element *anticlockwise* and are plotted *downwards*, i.e. τ_{xy} on QR is positive and τ_{yx} on PQ is negative. The centre of the Mohr's circle, C, is the mid-point of EB $\left(\text{coordinates: } \left(\frac{(\sigma_x + \sigma_y)}{2}\right)\right)$ and the circle is drawn, with a radius CB (or CE), to pass through both B and E.

Consider a point N on the circle circumference, obtained by drawing the radius, CN, at an angle 2α anticlockwise from CB. From trigonometry it can be shown that the coordinates of N, (OM, MN) are given by $(\sigma_\alpha, \tau_\alpha)$. Thus, N represents the state of stress (direct and shear) acting on a plane oriented at an angle α anticlockwise from QR, i.e. N represents the state of stress on RT.

As α can take on any value, points on the circumference of the circle represent the states of stress acting on planes defined by their angular displacement from CB. Points E and B were obtained from the known stresses. This diagram is known as the **'Mohr's circle for plane stress'**.

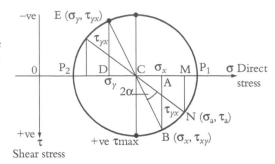

Figure 1.65 Mohr's Circle construction

Important observations on the Mohr's circle for plane stress

1. For a rotation of α in coordinate space the radius of Mohr's circle turns through 2α, i.e. the angle between radii to points representing stresses on faces with included angle α is 2α.

2. The rotational sense in coordinate space is the same as that around the Mohr's circle. This is a consequence of the sign adopted for direct and shear stresses and the Mohr's circle axes.

3. There are two points on the circle for which the shear stress components are zero – points P_1 and P_2.
   P_1 represents the maximum direct stress on any plane through the element, and is termed the 'maximum principal stress'.
   P_2 represents the minimum direct stress on any plane through the element, and is termed the 'minimum principal stress'.
   (Maximum is here taken to mean 'algebraically greater', or the right-hand end of the diameter, i.e. $10 > 0$ and $0 > -20$).

4. Points P_1 and P_2 lie $180°$ apart on the Mohr's circle. Hence, the principal stresses act on planes which are $90°$ apart through the element. These two planes are termed the 'principal planes'.

5. The shear stress is a maximum at two points on the circle (+ve and −ve) – the value of the maximum shear stress is equal to the radius of the circle.
   The planes which carry maximum shear stress are $180°$ apart on the circle, and hence $90°$ apart through the element. They are inclined at $45°$ to the principal planes ($90°$ on the circle).
   The direct stress on the planes carrying maximum shear stress is $\left(\dfrac{\sigma_x + \sigma_y}{2}\right)$.

6. Sign conventions can differ. It is important when reading alternative texts to establish the sign convention if confusion is to be avoided.

Mohr's circle example

An element of material within a loaded body is known to carry the stresses shown in Figure 1.66. Determine:

(i) the principal stresses;

(ii) the positions of the principal planes (illustrate with a sketch);

(iii) the maximum shear stress.

Graphical solution

The known stresses on the element are:
$$\sigma_x = 10 \text{ MPa}$$
$$\sigma_y = -70 \text{ MPa}$$
$$\tau_{xy} = 30 \text{ MPa}$$
$$\tau_{yx} = -30 \text{ MPa}$$

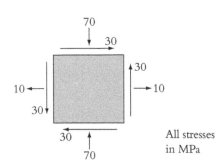

Figure 1.66 An element of material subjected to direct and shear stresses

Solid mechanics

Figure 1.67 shows the Mohr's circle for this stress system. To draw the circle, firstly draw the point B which represents stresses on the x-plane (coordinates: 10, 30). Next draw point E which represents stresses on the y-plane (coordinates: $-70, -30$). Join the two points with the line BE which intersects the x-axis at the centre of the circle, C. The circle can now be drawn and the following quantities measured

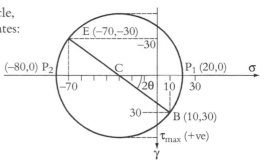

$$\text{Principal stretch: } \sigma_1 = 20 \text{ MPa}$$
$$\sigma_2 = -80 \text{ MPa}$$
$$\tau_{max} = 50 \text{ MPa}$$
$$2\theta \approx 37°$$

Figure 1.67 Mohr's circle for the example

On the element, the angle of the principal plane (P1) from the x-plane is $\theta \approx 18.5°$ anticlockwise as shown in Figure 1.68.

Analytical solution

The centre of the circle is given by $\quad C = \dfrac{(\sigma_x + \sigma_y)}{2} = -30$

The radius of the circle is given by $\quad R = \sqrt{\left(\dfrac{\sigma_x - \sigma_y}{2}\right)^2 + \tau_{xy}{}^2} = 50$

The principal stresses are:
$$\sigma_1 = C + R = 20 \text{ MPa}$$
$$\sigma_2 = C - R = -80 \text{ MPa}$$
$$\tau_{max} = R = 50 \text{ MPa}$$

The angle of the principal planes: $\quad \tan 2\theta = \dfrac{\tau_{xy}}{\left(\dfrac{\sigma_x - \sigma_y}{2}\right)} = 0.75$

$$2\theta = 36.87°$$
$$\theta = 18.4°$$

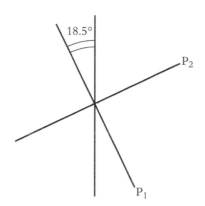

Figure 1.68 Orientation of the principal planes

Plane strain and two-dimensional Hooke's law

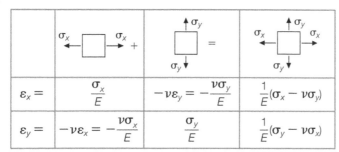

			=	
$\varepsilon_x =$	$\dfrac{\sigma_x}{E}$	$-v\varepsilon_y = -\dfrac{v\sigma_y}{E}$		$\dfrac{1}{E}(\sigma_x - v\sigma_y)$
$\varepsilon_y =$	$-v\varepsilon_x = -\dfrac{v\sigma_x}{E}$	$\dfrac{\sigma_y}{E}$		$\dfrac{1}{E}(\sigma_y - v\sigma_x)$

Table 1.6 Two-dimensional direct stress and strain

Disregarding shear stresses, **Table 1.6** uses the principle of superposition and Hooke's Law to show how plane strains may be calculated from the direct stresses acting in the plane. The strain in the x-direction, ε_x, comprises a direct strain due to σ_x and a Poisson's strain due to σ_y. Similarly, the strain in the y-direction, ε_y, comprises a direct strain due to σ_y and a Poisson's strain due to σ_x. The result is an expression of Hooke's Law in two-dimensions as follows

$$\varepsilon_x = \frac{1}{E}(\sigma_x - v\sigma_y)$$
$$\varepsilon_y = \frac{1}{E}(\sigma_y - v\sigma_x)$$

Three-dimensional stresses and strains

The approach in the previous section may be extended to three dimensions as shown in **Table 1.7**.

Table 1.7 Three-dimensional stresses and strains

$\varepsilon_x =$	$\dfrac{\sigma_x}{E}$	$-\dfrac{\nu\sigma_y}{E}$	$-\dfrac{\nu\sigma_z}{E}$	$\dfrac{1}{E}(\sigma_x - \nu(\sigma_y + \sigma_z))$
$\varepsilon_y =$	$-\dfrac{\nu\sigma_x}{E}$	$\dfrac{\sigma_y}{E}$	$-\dfrac{\nu\sigma_z}{E}$	$\dfrac{1}{E}(\sigma_y - \nu(\sigma_x + \sigma_z))$
$\varepsilon_z =$	$-\dfrac{\nu\sigma_x}{E}$	$-\dfrac{\nu\sigma_y}{E}$	$\dfrac{\sigma_z}{E}$	$\dfrac{1}{E}(\sigma_z - \nu(\sigma_x + \sigma_y))$

In this case, the strain in a particular direction (x, y or z) comprises a direct strain, due to the stress in that direction, and Poisson's strains due to stresses in the other two directions. The result is an expression of Hooke's law in three dimensions, called **Generalized Hooke's Law** as follows:

$$\varepsilon_x = \frac{1}{E}(\sigma_x - \nu(\sigma_y + \sigma_z))$$

$$\varepsilon_y = \frac{1}{E}(\sigma_y - \nu(\sigma_x + \sigma_z))$$

$$\varepsilon_z = \frac{1}{E}(\sigma_z - \nu(\sigma_x + \sigma_y))$$

These expressions can be used to determine the direct strains in any direction from direct stresses acting simultaneously in up to three directions.

Example 1 – Membrane stresses in thin cylinders and spheres

Figure 1.69 shows a thin-walled cylinder, mean radius (to centre of wall), R, and wall thickness, t, where $\dfrac{t}{R} < 0.1$. It may be assumed that the cylinder has closed ends and is under an internal pressure, P. This model is a good representation of engineering applications such as steam boilers, chemical reactors and heat exchanges, etc. The objective is to determine the stresses in the cylinder wall.

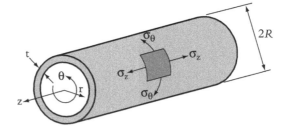

Figure 1.69 Membrane stresses in the wall of a thin cylinder under internal pressure

For convenience, it is more appropriate to work in cylindrical coordinates, r, z, θ, as shown, instead of cartesian coordinates x, y, z. Because the wall is thin, we may consider a small plane element in the wall, as shown. For an internally pressurised cylinder (with closed ends), three stresses act on this element. These are:

(i) **radial stress** (σ_r) which acts normal to the in-plane element and varies from zero at the outer free surface to a compressive stress equivalent to the internal pressure at the inner surface. For a thin-walled cylinder, it can be shown that this level of stress is small compared to the other in-plane stresses in the wall. Thus, we can assume that $\sigma_r \approx 0$.

Solid mechanics

(ii) **axial stress** (σ_z) which acts in the axial direction in the plane of the wall. Sectioning the cylinder vertically as shown in Figure 1.70 and considering equilibrium of forces, the force arising from the axial stress within the wall must equate to the force arising from the pressure acting on the end cap. Thus

$$P \times \text{(end cap area)} = \sigma_z \times \text{(wall area)}$$

$$P \times \pi R^2 = \sigma_z \times 2\pi R t$$

giving

$$\sigma_z = \frac{PR}{2t} \qquad (1.57)$$

(iii) **Hoop (i.e. circumferential) stress** (σ_θ) which acts around the circumference of the cylinder in the 'hoop' direction. Sectioning the cylinder horizontally as shown in Figure 1.71 and again considering equilibrium of forces, the force arising from the hoop stress in the wall must equate to the vertical force arising from the pressure acting on the projected area of the section. Thus

$$P \times \text{(projected area)} = \sigma_\theta \times \text{(wall area)}$$

$$P \times 2RL = \sigma_\theta \times 2tL$$

giving

$$\sigma_\theta = \frac{PR}{t} \qquad (1.58)$$

Equations (1.57) and (1.58) are the fundamental expressions for the axial and hoop stresses in a thin walled, closed-end cylinder under internal pressure. Note that the hoop stress is the maximum stress and has a magnitude $2\times$ the axial stress. These two stresses act within the plane of the wall and consequently are called membrane stresses.

Membrane strains in the wall of the thin cylinder can be determined using generalized Hooke's law as follows

hoop strain $\quad \varepsilon_\theta = \dfrac{\Delta(\text{circumference})}{\text{circumference}} = \dfrac{\Delta(2\pi R)}{2\pi R} = \dfrac{\Delta R}{R}$

$$= \frac{1}{E}(\sigma_\theta - v\sigma_z) = \frac{1}{E}\left(\frac{PR}{t} - v\frac{PR}{2t}\right) = \frac{PR}{Et}\left(1 - \frac{v}{2}\right)$$

axial strain $\quad \varepsilon_\theta = \dfrac{\Delta L}{L}$

$$= \frac{1}{E}(\sigma_z - v\sigma_\theta) = \frac{1}{E}\left(\frac{PR}{2t} - v\frac{PR}{t}\right) = \frac{PR}{Et}\left(\frac{1}{2} - v\right)$$

Thin spheres under internal pressure, P, can be analysed in a similar way to thin cylinders. Consider the thin sphere shown in Figure 1.72, mean radius, R, and wall thickness, t. Considering equilibrium of forces on a section of the sphere, we have

$$P \times \text{(projected area)} = \sigma_\varphi \times \text{(wall area)}$$

$$P \times \pi R^2 = \sigma_\varphi \times 2\pi R t$$

giving

$$\sigma_\phi = \frac{PR}{2t} \qquad (1.59)$$

Due to symmetry, the membrane stress, σ_φ, in a thin sphere under pressure is the same in all directions in the plane of the wall.

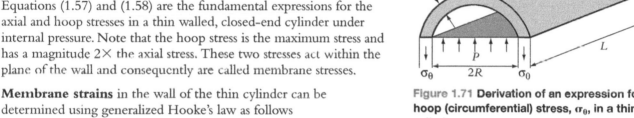

Figure 1.70 Derivation of an expression for the axial stress, σ_z, in a thin cylinder

Figure 1.71 Derivation of an expression for the hoop (circumferential) stress, σ_θ, in a thin cylinder

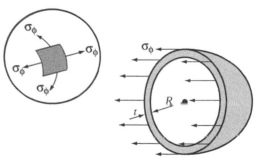

Figure 1.72 Membrane stress in a thin sphere under internal pressure

Example 2 – Hydrostatic stress and volumetric strain

When a body is subject to a stress which is equal in all directions, the stress is called a **hydrostatic stress**. For the element shown in Figure 1.73, $\sigma_x = \sigma_y = \sigma_z = \sigma_H$, the hydrostatic stress. A typical example of hydrostatic stress is when a body is underwater, in which case the hydrostatic stress is compressive (−ve) and equal to the underwater pressure.

When a body is subjected to a hydrostatic stress, in general, it will undergo a change in volume. The corresponding **volumetric strain**, ε_V, is given by $\frac{\Delta V}{V}$, i.e. the change in volume divided by the original volume.

For an elastic body, a plot of hydrostatic stress, σ_H, against volumetric strain, ε_V, is a straight line as shown in Figure 1.74. The slope of the line is a measure of the **bulk modulus, K** of the material. Thus,

$$K = \frac{\sigma_H}{\varepsilon_V}$$

The bulk modulus, K, is an elastic material property similar to Young's modulus, E, shear modulus, G, and Poisson's ratio, ν. In section 1.2, we saw that the material properties, E, G and ν are related. It will be shown below that K is also related to these properties.

Consider a block of elastic material, shown in Figure 1.75, which deforms under stresses acting in three directions. Consider the volume change of this block. Thus,

initial volume $\qquad\qquad\qquad V = xyz$

final volume $\qquad\qquad\qquad V' = x'y'z'$ $\qquad$ (1.60)

However,

$$x' = x + \Delta x = x\left(1 + \frac{\Delta x}{x}\right) = x(1 + \varepsilon_x) \qquad (1.61)$$

Similarly

$$y' = y(1 + \varepsilon_y) \qquad (1.62)$$

and

$$z' = z(1 + \varepsilon_z) \qquad (1.63)$$

Substituting (1.61), (1.62) and (1.63) into (1.60), we obtain

$$V' = xyz(1 + \varepsilon_x)(1 + \varepsilon_y)(1 + \varepsilon_z)$$

The volumetric strain is $\qquad \varepsilon_V = \dfrac{\Delta V}{V} = \dfrac{V' - V}{V} = \dfrac{xyz(1 + \varepsilon_x)(1 + \varepsilon_y)(1 + \varepsilon_z) - xyz}{xyz}$

$$\varepsilon_V = (1 + \varepsilon_x)(1 + \varepsilon_y)(1 + \varepsilon_z) - 1$$

$$= (1 + \varepsilon_x + \varepsilon_y + \varepsilon_x\varepsilon_y)(1 + \varepsilon_z) - 1$$

$$\varepsilon_V \approx \varepsilon_x + \varepsilon_y + \varepsilon_z \text{ (neglecting multiples of strain)}$$

Now, using generalized Hooke's law, and substituting for $\varepsilon_x = \dfrac{1}{E}(\sigma_x - \nu(\sigma_y + \sigma_z))$, etc. we obtain

$$\varepsilon_V = \frac{1}{E}(1 - 2\nu)(\sigma_x + \sigma_y + \sigma_z)$$

Assuming, hydrostatic stress, i.e. $\sigma_x = \sigma_y = \sigma_z = \sigma_H$, then,

$$\varepsilon_V = \frac{1}{E}(1 - 2\nu)(3\sigma_H) = \frac{\sigma_H}{K} \qquad (1.64)$$

Thus,

$$E = 3K(1 - 2\nu)$$

or

$$K = \frac{E}{3(1 - 2\nu)}$$

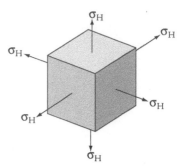

Figure 1.73 Hydrostatic stress

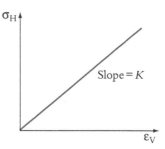

Figure 1.74 Linear relationship between hydrostatic stress, σ_H, and volumetric strain, ε_V

Solid mechanics

This proves that the bulk modulus, K, is related to the other two elastic constants, E and v. As the shear modulus, G, is also related to E and v $\left(G = \dfrac{E}{2(1 + v)} \right)$, there are in fact only two independent elastic constants for an isotropic material. Also, note from equation (1.64), that when $v = 0.5$, $\varepsilon_v = 0$, i.e. the material is incompressible. Rubber is a material which has these properties.

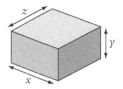

Figure 1.75 **A block of material, volume *xyz***

Mohr's circle

An element of material in a loaded structure is subjected to the stresses shown in Figure 1.76.

(a) Draw the Mohr's circle for the element.

(b) Determine the values of the maximum and minimum principal stresses and the maximum shear stress.

(c) Sketch the orientation of the planes on which the principal stresses act.

The stresses on the element are: $\sigma_x = 75$ MPa
$$\sigma_y = -25 \text{ MPa}$$
$$\tau_{xy} = 45 \text{ MPa}$$

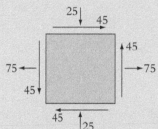

Figure 1.76 **An element of material subjected to direct and shear stresses**

Mohr's circle

centre of the circle $= \dfrac{(\sigma_x + \sigma_y)}{2} = 25$ MPa

radius of the circle $= \sqrt{\left(\dfrac{75 + 25}{2} \right)^2 + 45^2} = 67.3$ MPa

The Mohr's circle is shown in Figure 1.77.

Principal stresses

The maximum principal stress,
$\sigma_1 = $ centre + radius = 92.3 MPa

The minimum principal stress,
$\sigma_2 = $ centre - radius = −42.3 MPa
(i.e. compressive)

The maximum shear stress,
$\tau_{max} = \pm$ radius $= \pm 67.3$ MPa

Angle of the principal planes

The angle, 2θ, of the maximum principal plane with respect to the x-plane, on the Mohr's circle, is given by

$$\tan 2\theta = \frac{\tau_{xy}}{\dfrac{(\sigma_x - \sigma_y)}{2}} = \frac{45}{\dfrac{(75 - (-25))}{2}} = 0.9$$

$\therefore \quad 2\theta = 42°$

On the element, the angle of the maximum principal plane is $\theta = 21°$ anticlockwise from the x-plane. The minimum principal plane is oriented 90° from this angle.

Figure 1.78 shows the principal planes drawn at these angles.

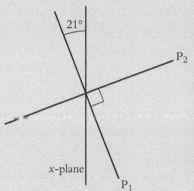

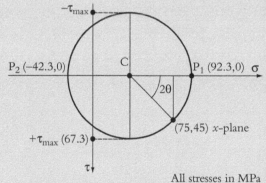

Figure 1.77 **Mohr's Circle for the example**

Figure 1.78 **Orientation of the principal planes**

Stresses in a pressurised cylinder

A closed thin-walled cylinder of inner diameter 0.5 m and length 3 m is to contain gas at a pressure of 1.5 MPa.

If the maximum allowable direct stress in the cylinder wall is not to exceed 125 MPa, calculate the minimum required wall thickness.

What will be the resulting change in diameter and length of the cylinder? Ignore end effects and assume $E = 209$ GPa and $\nu = 0.3$.

Minimum wall thickness

For a thin-walled cylinder, the hoop stress, σ_θ, and the axial stress, σ_z, are given by

$$\sigma_\theta = \frac{PR}{t} \quad \text{and} \quad \sigma_z = \frac{PR}{2t}$$

Since the hoop stress is larger than the axial stress, the maximum value of σ_θ must not exceed 125 MPa. Thus

$$\sigma_{\theta\text{max}} = 125 \times 10^6 = \frac{(1.5 \times 10^6)(0.25)}{t}$$

$$\therefore \quad t = 3 \text{ mm}$$

Change in diameter

The axial stress is given by

$$\sigma_z = \frac{PR}{2t} = \frac{(1.5 \times 10^6)(0.25)}{2(0.003)} = 62.5 \text{ MPa}$$

From Hooke's law, the hoop (circumferential) strain is given by

$$\varepsilon_\theta = \frac{\Delta D}{D} = \frac{1}{E}[\sigma_\theta - \nu(\sigma_z + \sigma_r)]$$

$$\frac{\Delta D}{0.5} = \frac{1}{209 \times 10^9}[125 \times 10^6 - 0.3(62.5 \times 10^6 + 0)]$$

$$\Delta D = 0.254 \text{ mm}$$

Change in length

Again, from Hooke's law, the axial strain is given by

$$\varepsilon_z = \frac{\Delta L}{L} = \frac{1}{E}[\sigma_z - \nu(\sigma_\theta + \sigma_r)]$$

$$\frac{\Delta L}{0.3} = \frac{1}{209 \times 10^9}[62.5 \times 10^6 - 0.3(125 \times 10^6 - 0)]$$

$$\Delta L = 0.359 \text{ mm}$$

Learning summary

By the end of this section you will have learnt:

✔ about stresses and strains acting in a single plane (two-dimensional) and how we define the general state of plane stress;

✔ the general equations for the angular transformation of stresses in two dimensions and how these equations may be represented by a graphical construction called 'Mohr's circle for plane stress';

✔ the use of Mohr's circle to analyse stresses at a point in a material and to determine the planes of maximum direct stress, i.e. the principal planes, on which the principal stresses act, and the planes of maximum shear stress;

✔ the derivation of 'generalised Hooke's law' relating direct stresses to direct strains in three dimensions and its application in solving three-dimensional problems, including stresses in thin cylinders and spheres and hydrostatic stress/volumetric strain problems.

1.5 Torsion

We have seen in section 1.1 of this chapter, that the **moment** of a force about a point is equal to the product of the magnitude of the force and the perpendicular distance from the point to the line of action of the force.

> When a moment is applied about the axis of a shaft or bar, as shown in **Figure 1.79**, it is termed a **torque** and the shaft is said to be under a state of **torsion**.

The units of torque are the same as the units of a moment, i.e. force × distance — Nm. A torque produces internal **shear stresses** in the shaft and causes the shaft to deform by **twisting** about its axis. A common example of a shaft under torsion is a rotating circular shaft transmitting mechanical power. This might be the drive shaft of a car or a propeller shaft in a ship.

Figure 1.79 A shaft under torsion

For design reasons, it is important to be able to determine both the magnitude of the internal shear stresses and the twisting deformation of a shaft under the action of an applied torque. This section will develop a general equation for torsion, called the 'torsion equation', which will enable torsional shear stresses and twisting deformations to be calculated for a specific shaft geometry, material and applied torque.

Assumptions

To simplify the analysis of torsion, we make a number of reasonable assumptions:

(a) The shaft is straight and has a uniform cross-section along its length.

(b) Plane cross-sections remain plane after twisting.

(c) The applied torque is constant over the length of the shaft.

(d) The material is elastic and obeys Hooke's law.

(e) During twisting, radial lines remain radial.

For shafts deforming in the elastic range, experimental evidence indicates that these assumptions are justified.

Geometry of twisting

Figure 1.80 shows a circular shaft clamped at one end, under torsion. The twist generated in the shaft can be studied by considering lines AB and OA:

line AB, along the length of the shaft, twists to A′B, developing an angle of shear, γ.

line OA, a radial line, twists to OA′, developing a twist angle, θ.

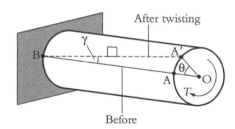

Figure 1.80 Geometry of twisting of a shaft

Since γ and θ are usually very small (at least in the elastic range), the arc AA′ can be assumed to be a straight line. Then

$$\tan \theta = \frac{AA'}{OA} \quad \text{and} \quad \tan \gamma = \frac{AA'}{AB}$$

However, because the angles are small

$$\tan \theta \approx \theta \quad \text{and} \quad \tan \gamma \approx \gamma \text{ (the shear strain)}$$
$$AA' = \theta OA \quad \text{and} \quad AA' = \gamma AB$$

As OA = radius of the shaft, r and AB = length of the shaft, L

$$\theta r = \gamma L \tag{1.65}$$

Now, by definition, the shear modulus, G, is given by,

$$G = \frac{\tau}{\gamma}$$

(1.66)

Substituting (1.66) into (1.65) and rearranging, we obtain

$$\frac{\tau}{r} = \frac{G\theta}{L}$$

(1.67)

Equation (1.67) is the first part of the torsion equation which relates the magnitude of the internal shear stress, τ, at the radial position, r, to the degree of twist, θ. This equation effectively relates the geometry of twisting to the internal shear stresses.

Shear stresses in torsion

Consider a small element of area, dA, at radial position, r, in the circular cross-section of the shaft under torsion. This element of area carries an unknown shear stress, τ, due to the twisting of the shaft, as shown in Figure 1.81.

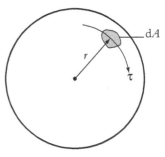

The shear force, F_s, acting on the element is given by the product of the shear stress and the area of the element:

$$F_s = \tau dA$$

Now, the torque, T, carried by the shaft must be equal to the sum of all the torques arising from shear forces on all the small elements of area in the cross-section:

$$T = \Sigma_A F_s r = \Sigma_A (\tau \, dA)r$$

Figure 1.81 Shear stress acting on an element of area

In the limit, this becomes an integral as follows:

$$T = \int_A (\tau dA) \, r$$

However, from before

$$\frac{\tau}{r} = \frac{G\theta}{L} \quad \text{or} \quad \tau = \frac{G\theta}{L}r$$

$$\therefore \qquad T = \int_A \left(\frac{G\theta}{L}r\right)dAr$$

$$T = \frac{G\theta}{L}\int_A r^2 dA$$

(1.68)

> We define the geometric quantity, **J = polar second moment of area** given by
>
> $$J = \int_A r^2 dA$$
>
> J is an important term in torsion and describes the distribution of the elements of area about the axis of the shaft. Its units are m⁴.

We can now rewrite (1.68) as

$$T = \frac{G\theta}{L}J$$

or

$$\frac{T}{J} = \frac{G\theta}{L}$$

(1.69)

Equation (1.69) is the second half of the torsion equation which relates the applied torque, T, to the degree of twist, θ.

We can now combine equations (1.67) and (1.69) into a single equation as follows:

$$\frac{T}{J} = \frac{\tau}{r} = \frac{G\theta}{L}$$

(1.70)

Solid mechanics

This is the **torsion equation**. The key variables in this equation and their units are:

T = applied torque about the shaft axis (Nm)

J = polar second moment of area (m⁴)

τ = shear stress at radius r (Nm⁻²)

r = radial position from the axis, i.e. centre, of the shaft (m)

G = shear modulus (Nm⁻²)

θ = angle of twist (radians)

L = length of shaft (m)

The two terms on the left-hand side of the torsion equation, i.e. $\dfrac{T}{J} = \dfrac{\tau}{r}$, allow the calculation of the shear stress, τ, at a radial distance r from the axis, knowing the applied torque, T, and the polar second moment of area of the section, J.

The twist angle, θ, for a specific length of shaft, L, can also be calculated from $\dfrac{T}{J} = \dfrac{G\theta}{L}$.

> Note that for a given T and J, the shear stress, τ, is proportional to r and is a maximum at r = outer radius, i.e. on the surface of the shaft, and a minimum, $\tau = 0$, at the centre of the shaft. The shear stress varies linearly with radial position.

In order to calculate torsional shear stresses from equation (1.70), it is necessary to evaluate the polar second moment of area for the particular shaft cross-section of interest. The next section explains how to calculate J for several shapes of cross-section.

Polar second moment of area, J

Example 1 – circular section

Consider the circular cross-section, radius R, shown in Figure 1.82, and the annular element of area, dA, at radial position, r. For a thin annulus, its area can be closely approximated by:

$$dA \approx 2\pi r dr$$

The polar second moment of area, J, is given by:

$$J = \int_A r^2 dA = \int_{r=0}^{r=R} r^2 2\pi r dr$$

∴

$$J = 2\pi \int_0^R r^3 dr = 2\pi \left[\frac{r^4}{4} \right]_0^R = \frac{\pi}{2} R^4$$

$$\therefore \quad J = \frac{\pi}{2} R^4 = \frac{\pi}{32} D^4 \tag{1.71}$$

for a circular section (radius R, diameter D).

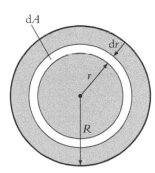

Figure 1.82 Annular element of area in a solid circular section

Example 2 – Hollow circular section

For a hollow circular cross-section, as shown in Figure 1.83, the same method is used as for the circular section, but the integration limits are changed to R_1 and R_2 to represent the inner and outer radii respectively. The expression for J is then given by:

∴

$$J = \frac{\pi}{2} R_2^4 - \frac{\pi}{2} R_1^4 = \frac{\pi}{32} [D_2^4 - D_1^4] \tag{1.72}$$

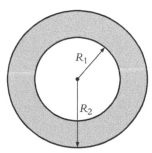

Figure 1.83 A hollow circular cross-section

Example 3 – thin-walled circular section

When the hollow circular cross-section has a thin wall, thickness t, as shown in Figure 1.84, the inner radius is then $R_1 = R$, and the outer radius $R_2 = R + t$, and equation (1.73) can be simplified as follows:

$$J = \frac{\pi}{2}(R_2{}^4 - R_1{}^4)$$

$$= \frac{\pi}{2}(R_2{}^2 - R_1{}^2)(R_2{}^2 + R_1{}^2)$$

$$= \frac{\pi}{2}(R_2 - R_1)(R_2 + R_1)((R_1 + t)^2 + R_1{}^2)$$

$$= \frac{\pi}{2}(t)(R_1 + t + R_1)((R_1 + t)^2 + R_1{}^2) \tag{1.73}$$

$As\ t \ll R_1$

$$J \approx \frac{\pi}{2}(t)\,(2R_1)\,(2R_1{}^2)$$

$$J \approx 2\pi R^3 t \approx \frac{\pi}{4}D^3 t \tag{1.74}$$

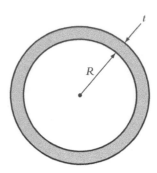

Figure 1.84 A thin-walled hollow circular cross-section

Note that for $t \approx 0.2R$, the error in using equation (1.74) instead of equation (1.73) is $\sim 1.2\%$.

Solving problems in torsion applications

Power transmission

Consider a force, F, acting on a particle moving around the circumference of a circle, radius R, as shown in Figure 1.85. In time δt, the force moves through an arc, length δs, subtending an angle $\delta\theta$.

The work done, W, by the force is,

$$W = F\,\delta s = FR\,\delta\theta$$

However,

Torque $T = FR$

$\therefore$

$$W = T\,\delta\theta$$

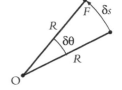

Figure 1.85 Work done by a force moving along a circular arc

The power, P, is the work done in unit time

$$P = W\,\delta t^{-1} = T\,\delta\theta\,\delta t^{-1}$$

$$\boxed{P = T\omega} \tag{1.75}$$

where P is power (J s^{-1} or watts), T is torque (Nm), and ω is angular velocity (rad s^{-1}).

This is the power equation for transmission of power along a shaft. In words, it states that

$$\text{power transmitted} = \text{torque} \times \text{angular velocity}$$

Torque in a stepped shaft

The stepped shaft, shown in Figure 1.86, comprises two sections, A and B, of different radii and is subjected to an applied torque, T. To solve problems of this type of shaft the following conditions apply:

equilibrium: applied torque $T = T_A = T_B$

i.e. the torque is the same in each section of the shaft.

compatibility: total twist $\theta_{total} = \theta_A + \theta_B$

i.e. the total twist is the sum of the twists in the two sections of the shaft.

Figure 1.86 A stepped shaft subject to a torque

Solid mechanics

Torque in a composite shaft

Consider the composite shaft, shown in Figure 1.87, comprising two concentric parts, outer A and inner B. Provided there is no slipping at the interface between the parts, the following conditions can be used to solve problems of this type:

equilibrium: applied torque $T = T_A + T_B$

i.e. the torque is the sum of the torques in the two parts of the shaft.

compatibility: total twist $\theta_{total} = \theta_A = \theta_B$

i.e. the twist is the same in each part of the shaft.

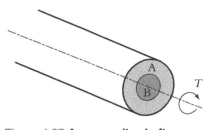

Figure 1.87 A composite shaft subjected to a torque

Note carefully the difference between the composite shaft and the previous stepped shaft. For the composite shaft, the twist is the same in both parts while the torque is the sum of the two individual torques. For the stepped shaft, the torque is the same in both sections while the twist is the sum of the two individual twists.

Coupling of shafts

Power is transmitted from one shaft to another by a coupling. A typical flanged coupling is shown in Figure 1.88 and comprises two flanges on the shaft ends joined by n bolts on a pitch radius, R. Shear forces act on the bolts as the torque, T, is transmitted through this coupling. If the shear force on one bolt is F, then the torque transmitted is given by:

$$T = nFR \qquad (1.76)$$

The average shear stress in a bolt, τ_{av}, is:

$$\tau_{av} = \frac{F}{A_{bolt}} \qquad (1.77)$$

where A_{bolt} is the cross-sectional area of a bolt.

From (1.76) and (1.77)

$$\tau_{av} = \frac{T}{nA_{bolt}R} \qquad (1.78)$$

Equation (1.78) can be used to determine the number, size and pitch radius of the bolts required to transmit a specified torque while maintaining the shear stress below a specified level.

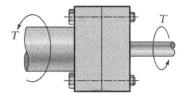

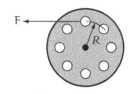

Figure 1.88 Bolted coupling between shafts

Torsion in a shaft

The stepped steel shaft, comprising sections A and B, shown in Figure 1.89 is subjected to a torque $T = 2\,kNm$.

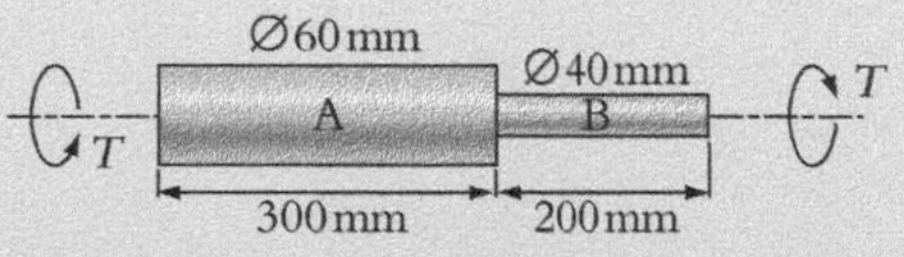

Figure 1.89 A stepped steel shaft under torsion

Taking a value of $G_{steel} = 70\,GPa$, determine the maximum shear stress in section A and section B of the shaft and calculate the total angle of twist (in degrees).

Polar second moments of area

$$J_A = \frac{\pi D_A{}^4}{32} = \frac{\pi (60 \times 10^{-3})^4}{32} = 1.272 \times 10^{-6}\,m^4$$

$$J_B = \frac{\pi D_B{}^4}{32} = \frac{\pi (40 \times 10^{-3})^4}{32} = 2.213 \times 10^{-7}\,m^4$$

Maximum shear stress

The maximum shear stress occurs on the outer surface of each section where the radius is a maximum. Using part of the torsion equation (1.70), the stresses are given by:

$$\tau_{A\,max} = \frac{TR_A}{J_A} = \frac{2 \times 10^3 \times 30 \times 10^{-3}}{1.272 \times 10^{-6}} = 47.2 \, \text{MPa}$$

$$\tau_{B\,max} = \frac{TR_B}{J_B} = \frac{2 \times 10^3 \times 20 \times 10^{-3}}{2.513 \times 10^{-7}} = 159 \, \text{MPa}$$

Therefore the smaller shaft B has the greater shear stress.

Twist angle

Using the other part of the torsion equation (1.70), the twist angle in each section is:

$$\theta_A = \frac{TL_A}{GJ_A} = \frac{2 \times 10^3 \times 300 \times 10^{-3}}{70 \times 10^9 \times 1.272 \times 10^{-6}} = 6.739 \times 10^{-3} \, \text{rad} = 0.386°$$

$$\theta_B = \frac{TL_B}{GJ_B} = \frac{2 \times 10^3 \times 200 \times 10^{-3}}{70 \times 10^9 \times 2.513 \times 10^{-7}} = 2.27 \times 10^{-2} \, \text{rad} = 1.303°$$

The total twist
$$\theta_{tot} = \theta_A + \theta_B = 1.69°$$

Power transmission in a shaft

A hollow circular shaft of outer diameter 100 mm and wall thickness 5 mm is required to transmit power at a frequency of 2 Hz.

If the shear stress is not to exceed 100 MPa, determine the maximum power that can be transmitted by this shaft.

If the hollow shaft is replaced by a solid shaft operating under the same conditions and of the same material and length, what will be the percentage increase in weight of material used?

Hollow shaft

Firstly, the angular velocity must be converted from hertz to radians per second:

$$1 \, Hz = \frac{1 \, \text{cycle}}{\text{sec}} \times \frac{2\pi \, \text{radians}}{1 \, \text{cycle}} = 2\pi \, \text{rad s}^{-1}$$

The polar second moment of area of the hollow cylinder is given by:

$$J = \frac{\pi}{32}(D_o^4 - D_i^4) = \frac{\pi}{32}(0.1^4 - 0.09^4) = 3.376 \times 10^{-6} \, \text{m}^4$$

Using the torsion equation, and noting that the maximum shear stress, τ, occurs at the maximum radius, i.e. at the outer surface:

$$\frac{T}{J} = \frac{\tau_{max}}{r_{max}} \Rightarrow T = \frac{\tau_{max} J}{r_{max}} = \frac{(100 \times 10^6)(3.376 \times 10^{-6})}{\left(\frac{0.1}{2}\right)} = 6.752 \, \text{kNm}$$

Hence, the maximum power that can be transmitted by this shaft is given by:

$$\text{Power} = T\omega = (6.752 \times 10^3)(2 \times 2\pi) = 84.85 \, \text{kW}$$

Solid shaft

Since the power transmitted is the same with the solid shaft and ω is the same, then T must also be the same.

The polar second moment of area is given by

$$J = \frac{\pi D^4}{32}$$

Using the torsion equation

$$\frac{T}{J} = \frac{\tau_{max}}{r_{max}} \Rightarrow \frac{6.752 \times 10^3}{\frac{\pi d^4}{32}} = \frac{100 \times 10^6}{\frac{d}{2}} \Rightarrow d = 0.0701 \text{ m} = 70.1 \text{ mm}$$

Note that the diameter of the solid shaft is less than the outer diameter of the hollow shaft, but the weight of the solid shaft is greater than that of the hollow shaft.

Let ρ be the density, A the cross-sectional area and L the length of the shaft. Then, the weight $W = \rho A L$.

$$\% \text{ weight increase} = \frac{W_{solid} - W_{hollow}}{W_{hollow}} \times 100 = \frac{\rho A_{solid} L - \rho A_{hollow} L}{\rho A_{solid} L} \times 100$$

$$= \frac{A_{solid} - A_{hollow}}{A_{hollow}} \times 100 = \frac{d^2 - (d_o^2 - d_i^2)}{(d_o^2 - d_i^2)} \times 100$$

$$= \frac{(0.0701)^2 - (0.1^2 - 0.09^2)}{(0.1^2 - 0.09^2)} \times 100$$

$$= 158.6\%$$

Learning summary

By the end of this section you will have learnt:

✔ the definition of torque and how it gives rise to twist and shear stresses within a shaft;

✔ the derivation of the 'torsion equation' and how it is used to calculate twist angle and shear stresses in a shaft subjected to a known torque;

✔ how to calculate the polar second moment of area for solid and hollow circular cross-sections;

✔ how to solve torsion problems including power transmission, stepped and composite shafts.

Unit 2

Materials and processing

Andrew Kennedy and Philip Shipway

UNIT OVERVIEW

- Introduction
- The structure and properties of materials
- Properties of materials
- Selection of materials in engineering design
- Materials processing
- Failure of materials

2.1 Introduction

All objects are made of something which we generally call 'materials'. Materials are made up of atoms (sometimes – but rarely – one atom type, but more commonly a mixture of atom types) and the nature of these atoms and the way that they are bonded together to make a macroscopic material dictates the properties of the material.

The properties of the macroscopic materials are the features of interest to the engineer. How we measure and define the properties of materials is the first task to be considered. However, the engineer needs also to understand that the properties of the material are a complex function of attributes of the material ranging from the atomic level upwards. By understanding some of the basic attributes of materials at these scales, we not only understand the attributes of a material, but also understand how to control and modify the attributes of a material.

To make materials useful to engineers, they need to be shaped into components. There are restrictions as to what processes can be used to shape materials, based upon the attributes of those materials and the nature of the component being made. We do not use the same manufacturing processes for high-melting-point ceramics as we do for low-melting-point polymers; similarly, the same processes would not be used to make the metallic interconnects on a flash drive and a metallic cylinder block for a large diesel engine. In addition, the processing of materials often changes the structure of the material at the atomic scale and thus changes the attributes of the material. As such, there is a need to understand the processing of materials and the effects that such processing may have on the attributes of the material in the final component.

There are certain Underpinning Principles that need to be understood as you move through the chapter. Rather than interrupt the flow of the main chapter, these have been inserted as stand-alone sections throughout the chapter, placed in blue boxes to distinguish them from the main flow. These may need to be referred to a number of times and are integral principles upon which your knowledge and understanding will be based.

2.2 The structure and properties of materials

Before we can select a material, or design with it, we need to understand the basic requirements (or properties it must have) for it to fulfil its function (for example, it might need to have a high melting point and absorb lots of energy on impact). With this understanding and a knowledge of how these properties vary for different types (or classes) of material, we can make a broad choice of material that would be suitable (a metal would be most suitable in this instance). With data for the properties of different materials and the equations that govern the behaviour under the appropriate conditions, we can select a specific material and define the geometry required.

This section gives a broad introduction to materials and their properties. First, the structure of different classes of material (metals, ceramics and polymers) is described and this is then related to their characteristic properties. A number of important materials properties are then defined and their relevance to engineering is placed in context. Methods for measuring these properties are given, along with the origin of these properties (understanding this can help us to create new and improved materials). Finally, worked examples for designing with materials and how to select the best engineering material for a particular application are presented.

Classification of materials

Broadly speaking, we can place the thousands of materials available into several categories. These categories, or classes, contain materials with similar types of bond (see Underpinning Principles 1) which hold together the basic building blocks (atoms or molecules) of the material. Since the nature of the bonding defines the physical and mechanical properties, materials in the same class share similar properties and are suitable for similar applications. It should be noted that while some of the characteristic properties of materials in a particular class might be broadly the same (i.e. they might all be brittle or they might all be good electrical conductors), there can also be a wide variation in other basic properties (for example, mercury and tungsten are both metals, but have very different melting points). Materials are commonly classified into the following four groups: (i) metals, (ii) ceramics and glasses, (iii) polymers and elastomers and (iv) composite materials.

Underpinning Principles 1: Atomic structure and bonding

The structure of the atom can be described in a simplified way using the Bohr model. This model, shown in Figure UP 1.1 depicts the atom as a small, positively charged nucleus containing protons (which are positively charged) and neutrons (with no charge) surrounded by electrons (negatively charged) that travel in circular orbits around the nucleus. The atomic number, N, is the number of protons or electrons that the atom possesses (they must be the same for the atom to have no overall net charge; if charged, it is called an ion). The atomic mass, Z, is the total number of protons and neutrons in the nucleus (the number of protons and neutrons is similar but not always the same – isotopes of an element, some of which are radioactive, have the same number of protons and electrons but different numbers of neutrons).

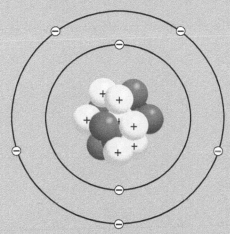

Figure UP 1.1 The Bohr model of the (nitrogen) atom showing electrons orbiting the nucleus

The example in Figure UP 1.1 is for nitrogen, with atomic number 7. The model shows a nucleus containing seven protons and, since it has atomic mass 14, seven neutrons. Seven electrons can be seen orbiting the nucleus. The number of electrons in the outermost (or valence) shell increases from one to a maximum of eight when traversing from left to right across the groups in the Periodic Table of Elements. For nitrogen, in group V, two electrons occupy the inner orbit (or shell) and five occupy the valence shell.

The number of electrons in the valence shell is important in determining how an element reacts chemically with other elements: the fewer valence electrons an atom has, the less stable it becomes and the more likely it is to react. The most stable configuration is when the valence shell is full of electrons (for elements with an atomic number greater than three, a full shell would contain eight electrons – below this it is full with two electrons).

It is worth noting that this early model is a simplification. For example, Figure UP 1.1 is not to scale; for a nucleus of that size, the electrons are much too close. Real atoms are mostly empty space. The figure also represents the electron orbits as circles. In reality, scientists cannot tell exactly where an electron is at a given moment or where it is going. They can calculate the probability that an electron will be found in a given volume of space, but that is not the same as knowing where that electron is. Despite the limitations of this model, it is sufficient to help our understanding of how atoms bond and pack together to form structures.

Bonding and packing in metals (metallic bonding)

In metallic bonding, the valence electrons form a delocalized 'sea' or 'cloud' around the close-packed, positively charged metal cations (a schematic of this is shown in Figure UP 1.2). The non-directional nature of this bonding enables metal atoms to pack closely together, resulting in high-density structures. An example of one type of close-packed structure (called face-centred cubic or FCC) is also shown in Figure UP 1.2, where the metal atoms are represented by spheres.

Bonding and packing in ceramics (ionic and covalent bonding)

In ceramic materials the bonding is ionic or covalent. With these bonding types the valence electrons are either donated or shared respectively. An example of ionic bonding is shown in Figure UP 1.3 for the chemical compound sodium chloride, NaCl. In this process the sodium (Na) atom gives up its single valence electron to become a positively charged sodium ion (cation), Na^+, and the chlorine atom accepts this electron to become a negatively charged chlorine ion (anion), Cl^-. In doing so, the sodium atom empties its valence shell and the chlorine atom receives the one electron it needed to fill its valence shell. The resulting stable positively and negatively charged ions are strongly attracted to each other and hence bonded together.

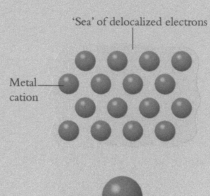

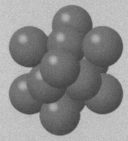

Figure UP 1.2 Schematic illustrations of metallic bonding and the packing of metal atoms

Figure UP 1.3 also shows a schematic of how the sodium and chlorine atoms pack to form the crystal structure NaCl. Because the ions have different charges, neighbouring ions must have opposite (attractive) charges. The figure clearly shows the alternating small, Na, and large, Cl, ions in the crystal structure.

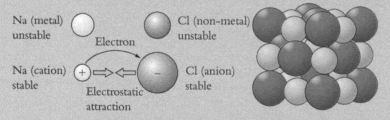

Figure UP 1.3 Schematic illustrations of ionic bonding and the packing in NaCl

Figure UP 1.4 also shows schematics for covalent bonding. Here the valence electrons are shared rather than donated. The example shown is for methane gas but the principle is the same for ceramics such as diamond and silicon carbide (SiC). In the case of methane, carbon has four valence electrons and hydrogen only one. In order to produce a stable structure, carbon needs to share electrons with four hydrogen atoms to fill its outer shell (forming the molecule or compound CH_4). The shared electrons from the carbon atoms also fill each of the hydrogen atom's valence shells (it only needs two). The sharing of electrons from four atoms to a single central atom dictates discrete angles between the neighbours (in this case the atoms adopt a tetrahedral structure with bonds at angles of approximately 110°).

Figure UP 1.4 also shows the three-dimensional packing structure for the covalently bonded Si and C atoms in silicon carbide (SiC), where an open structure is observed due to the directional nature of the bonds. In the case of silicon carbide, both the silicon and carbon atoms have four electrons in their valence shell. Each silicon atom, therefore, shares one of its valence electrons with that from one of four neighbouring carbon atoms. For the carbon atoms to fill their valence shell, they too must share electrons with four neighbouring silicon atoms.

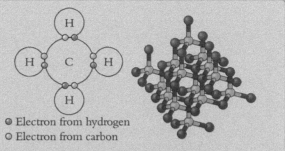

○ Electron from hydrogen
○ Electron from carbon

Figure UP 1.4 Schematic illustrations of covalent bonding in methane and the packing in SiC

Bonding and packing in polymers (covalent and secondary bonding)

There are two types of bonding in polymers, the covalent bonds between the carbon atoms that comprise the polymer backbone and their neighbours, and the weaker secondary bonding between neighbouring polymer chains. Figure UP 1.5 shows examples of the directional covalent bonding in the polymer chains, in this case for poly(ethylene) (also called polyethylene), in both two and three dimensions.

Figure UP 1.5 Schematic illustrations, in two and three dimensions, of the bonding in polyethylene

Weaker bonding can also occur between neighbouring polymer chains due to different charges that are associated with different chemical species that form the side groups (in the case of polyethylene, the side group is simply hydrogen). Different combinations can lead to different types of bonding with different strengths (for example, ionic, hydrogen, dipole–dipole, van der Waals bonding). Figure UP 1.6 shows examples of dipole–dipole and hydrogen bonding in HCl (hydrochloric acid) and H_2O (water) respectively that are typical of the interactions observed in more complicated polymer systems. Bonding occurs due to the development of small charges ($\delta+$ or $\delta-$) associated with individual atoms in the molecule and the resulting attraction between oppositely charged atoms in neighbouring molecules.

**Figure UP 1.6
Schematic depictions of (a) dipole–dipole bonding and (b) hydrogen bonding**

Polymer chains often arrange themselves in tangles with no regularity, particularly if the bonding between the chains is strong. Polymers with weak van der Waals bonding between the chains, polyethylene for example, are capable of forming ordered structures (crystallizing) if cooled slowly from the liquid state. Figure UP 1.7 shows a schematic illustration of a polymer that is partially crystalline, that is to say, it has areas that form a regular, repeating structure (crystalline) and areas with tangled chains that are amorphous (without shape).

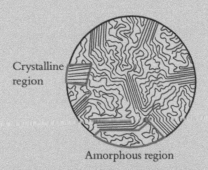

Crystalline region

Amorphous region

Figure UP 1.7 Schematic depiction of crystalline and amorphous regions in a partially crystalline polymer

Metals

Metals have metallic bonding in which the outer (valence) electrons form a delocalized 'sea or cloud' around the close-packed metal cations (see Underpinning Principles 1). The non-directional nature of this bonding allows metal atoms to slide past each other, resulting in their characteristic property of ductility, and also enables metal atoms to pack closely together, resulting in high-density structures. The delocalized sharing of free electrons enables them to move easily under an applied electrical potential difference, giving metals their characteristic high electrical conductivity. The metallic bond is generally strong and stiff, meaning that metals normally exhibit high stiffness and a high melting point.

Metals are usually used in the form of an alloy, a mixture of two or more elements in which the major component is a metal. Most pure metals are either too soft, brittle or chemically reactive to use on their own, and so alloying can be used to increase strength and hardness and improve corrosion resistance. Examples of common engineering metals are:

Steels: Carbon steels are alloys of iron and carbon. Steels with additions of other elements (for example, chromium, molybdenum and nickel) are called alloy steels and have higher strengths. Stainless steels contain very high additions of chromium which gives them their excellent corrosion resistance. Owing to the combination of an impressive set of properties and low cost, steels are among the most commonly used metals and have a wide variety of applications, such as in transport and construction. They are available as plate, sheet, tube, bar, wire, castings and forgings.

Aluminium alloys: Aluminium is a light metal (i.e. one with relatively low density) with a high strength-to-weight ratio and good corrosion resistance. As a result it is often used for containers, packaging, construction, transportation and sports equipment. Aluminium is available in a wide variety of cast and wrought (mechanically worked) shapes. There are two main types of alloy: those used for casting (mainly aluminium–silicon alloys) and those suitable for mechanical working (mainly aluminium–copper and aluminium–magnesium alloys).

Copper alloys: Copper is ductile (it can be deformed to a significant degree before fracture) and is an excellent conductor of heat and electricity. As a result, copper is used for electrical and electronic components and for plumbing. Copper alloys can achieve a wide range of properties by the addition of different alloying elements. Brass is an alloy of copper and zinc and has a much higher strength than pure copper. Bronze is an alloy of copper and tin and is commonly used for bearings.

Nickel superalloys: Superalloys are metals with excellent heat resistance and good high-temperature mechanical properties. These alloys were mostly developed for use as turbine blades in jet engines and can operate at over 1000°C under high stresses. Current jet engine superalloys contain mostly nickel with large additions of chromium and cobalt. They have trade names such as Inconel™ and Hastelloy™.

Ceramics and glasses

Ceramic materials are held together by ionic or covalent bonds (see Underpinning Principles 1). With these bonding types, the valence electrons are either shared between neighbouring atoms (covalent) or donated from one atom to its neighbour (ionic). As such, the electrons are not free to move throughout the structure in the way that they are for materials with metallic bonding, and this means that ceramics are poor conductors of electricity and heat. The bonds are very strong and stiff, resulting in ceramics that generally exhibit very high stiffness and very high melting points. The nature of their bonding means that it is very difficult for atoms to slide past each other in the way that they do in metals under stress. Their inability to do this means that when ceramics are put under mechanical stress, rather than deforming, they are susceptible to failure in a catastrophic way (fracture).

Materials and processing

Engineering ceramics are usually used in compression to avoid problems with fracture in tension. Ceramics resist oxidation and corrosion and are frequently used for their ability to withstand high temperature and for their high hardness. They are the least widely used group of engineering materials. Examples of engineering ceramics are:

Alumina: Alumina (aluminium oxide) is hard and brittle with poor electrical and thermal conductivity. It is used for thermal insulation, as an oxidation barrier, in spark plugs, in cutting tools and as an abrasive.

Silicon nitride: Silicon nitride has a high thermal conductivity and toughness (compared to other ceramics). It is used in cutting tools, grit-blasting nozzles, turbocharger rotors, turbine blades and shroud rings (for small jet engines).

Carbon fibre: Carbon fibre has a high strength-to-weight ratio and a low coefficient of thermal expansion. It is most commonly used to reinforce composite materials and is also used in filters, electrodes and antistatic devices.

Glasses are amorphous materials (see Underpinning Principles 2) without a regular crystal structure, usually produced when a viscous molten material cools very rapidly. Its atoms are arranged in a somewhat random fashion, as they are in a liquid. Metals can be formed into a glassy state, but when we think of glasses, we are normally referring to inorganic materials. The non-crystalline (see Underpinning Principles 2) nature of inorganic glasses along with their low electrical conductivity, allows light to be transmitted through them. The bonding in glass is covalent, resulting in a brittle material which is very susceptible to failure in the presence of defects. Glasses contain mainly silicon dioxide (also called silica), with other oxides added to impart specialist properties, for example to improve durability, colour or lustre, light absorption or transmission. The main uses for glass are based on its optical transparency and it being non-reactive. It is also used in the form of fibres for optical cables, insulation and reinforcement for polymers. Examples of common glasses are:

Soda–lime glasses: Sodium oxide (soda) and calcium oxide (lime) are added to silica to produce low-melting-point glasses that are easily formed and very widely used for windows, bottles and light bulbs.

Borosilicate glasses: Boron oxide is added to make heat-resistant and low-expansion borosilicate glasses (Pyrex™) used for cookware and laboratory equipment.

E-glass: E-glass contains aluminium oxide, calcium oxide and magnesium oxide additions to the silica base. It has excellent fibre-forming capabilities and is used almost exclusively as the reinforcing phase in glass-fibre-reinforced polymer composites (fibreglass).

Polymers

Polymer is a term used to describe a chain of thousands of monomers (see Underpinning Principles 1) that are linked together by the carbon atoms of the polymer backbone. For example, the formation of poly(ethylene) (also called polyethylene) involves thousands of ethylene molecules (these molecules are the monomers) bonded together to form a long chain. The structure of polymers can be visualized as tangled chains which form low-density structures with no regularity (see Underpinning Principles 1). The attractive forces between polymer chains play a large part in determining a polymer's structure and properties. Some polymers, such as polyethylene, have weak forces between the chains. This low bond strength between the chains gives these polymers a low strength and melting temperature but they are capable of forming ordered structures (crystallizing) if cooled slowly from the liquid state (after which they become opaque – since light is scattered by the regular polymer structure). These polymers are deemed thermoplastic and can be repeatedly heated and cooled to form a viscous melt and a solid structure. Polymers with stronger, chemical crosslinking between the chains have higher tensile strengths, are incapable of forming ordered structures (so remain amorphous and transparent, like glasses) and do not soften with heating (they burn instead and are termed thermosetting).

Rubber is an elastic hydrocarbon polymer (elastomer) that occurs as a milky colloidal suspension (known as latex) in the sap of several varieties of plant. Rubber can also be produced synthetically through the polymerization of a variety of monomers. In its relaxed state, rubber consists of long, coiled chains that are interlinked at a few points. When rubber is stretched, the coiled chains unravel, when relaxed the chains re-coil. Vulcanization of rubber creates more bonds between chains, making each free section of chain shorter, making the rubber stiffer and less extendable.

The use of polymers is widespread and exploits their reasonable strength, low density (light weight), low cost and ease of manufacture. Some examples are:

Nylon: Nylon has good mechanical properties and abrasion resistance; it is self-lubricating and resistant to most chemicals. It is commonly used for gears and bearings.

Kevlar: Kevlar has a very high tensile strength and stiffness and is commonly used in the form of fibres to reinforce tyres, bullet-proof jackets and sports equipment.

Perspex: Perspex has moderate strength but good optical properties (it can be transparent or opaque), it weathers well and is resistant to chemicals. It is frequently used for lenses, windshields and windows.

Polyethylene: Polyethylene has moderate strength but is easily formed. Low-density polyethylene is used for houseware, bottles and car bumpers; the high-density version is used for canoes and machine parts.

Epoxies: Epoxy resins, such as Araldite™, have excellent mechanical properties and good adhesive properties. They are resistant to heat and chemical attack and are used as matrices for polymer composites.

Phenolics: Phenolics, such as Bakelite™, are brittle but have a high resistance to heat, electricity and chemicals. They are used as electrical insulators and as connection blocks.

Rubbers: Natural and synthetic rubbers (for example neoprene or silicone) are flexible and can be made hard wearing with the addition of fillers such as sand or graphite. They are used for car tyres, tubing, 'O' rings and gaskets and for insulating electric cables.

Composite materials

Composite materials (or composites for short) are engineered materials made from two or more constituent materials (normally with significantly different physical or chemical properties) which remain separate and distinct within the finished structure. There are normally two types of constituent material within a composite, namely the matrix and the reinforcement. The matrix material surrounds and supports the reinforcements by maintaining their relative positions. The reinforcements enhance the matrix properties and the resulting mechanical and physical properties are thus intermediate between those of the matrix and those of the reinforcement, determined by the fraction of reinforcement added.

Natural composites include bone and wood. The most widely used engineering composites are polymer matrices reinforced with ceramic (glass or carbon) fibres but metal–ceramic and ceramic–ceramic composites, containing reinforcement in the form of either particles or fibres, are also common. Composites containing fibres (including wood) have the complication of exhibiting different properties in different directions with respect to the direction of the fibres (they are stronger and stiffer in the direction of the fibres). We term this 'anisotropy'.

Composites offer the possibility of tailoring the properties of a material, in order to meet very demanding service conditions, through the type and addition level of reinforcement. Composites are often designed to take advantage of the combination of the desirable properties of the constituent materials while circumventing their drawbacks. For example, carbon fibres display very high stiffness and strength, but can only be manufactured in the form of thin fibres, with limited use as engineering components. However, carbon-fibre composites display the high strength and stiffness associated with the carbon fibres and are bonded together with a polymeric matrix so that complex three-dimensional shapes can be manufactured. Such materials are used to meet the very demanding service conditions encountered in military aviation and motor sport.

Learning summary

By the end of this section you will have learnt:

✔ The different classes of materials;
✔ The different bonding and structure observed in different material classes;
✔ Some basic relationships between structure and properties for different classes of material;
✔ The composition, properties and uses of some common engineering materials.

Underpinning Principles 2
Structure of matter

Liquids, crystals and glasses

In the liquid state, most materials (including metals) have no long-range order; the atoms have no regular pattern within the structure.

Solid metals normally have a crystalline structure. The crystalline structure implies that atoms are packed within the solid in a regular array. The main crystalline structures observed in metals are known as (i) face-centred cubic, (ii) body-centred cubic and (iii) hexagonal close packed.

On solidification, the reorganization of atoms from the randomness of the liquid state to the ordered state of a crystal requires time. The amount of time required depends upon the mobility of the atoms and the complexity of the crystalline structure in the solid. If there is not enough time for this reorganization on cooling, the liquid can become solid without forming crystals; the solid essentially has the randomly ordered structure that was seen in the liquid. Such a solid with no crystalline structure is known as an amorphous solid or a glass. Glasses easily form on cooling of mixtures of inorganic oxides based upon silica sand (SiO_2) and do not require high cooling rates (this is the basis of the formation of window glass). However, to get metals to form amorphous solids normally requires very high cooling rates from the melt. Such amorphous metals are used as the cores of highly efficient transformers where low power losses are required.

Solidification and microstructure

When a metal begins to solidify from the melt, regions of crystalline solid are formed in the liquid. This crystal will grow by the addition of atoms to it. In normal conditions, a large number of these crystals will grow as the liquid solidifies and when two crystals impinge upon each other, there is a boundary. In the fully solidified material, each of these crystals is known as a grain, and the region where grains come into contact with other grains is known as the grain boundary (ses Figure UP 2.1). A grain boundary is a region of misorientation between the grains either side of it (the crystals on either side of it may be of different types in certain systems, but do not necessarily need to be so). The grain boundary is an important feature of common metals as it has a strong influence on strength and high–temperature properties. A metal that is made up of a large number of grains is known as a polycrystalline metal.

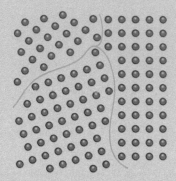

Figure UP 2.1 **Schematic diagram of a polycrystalline metal. Atoms are indicated by the circles, with the grain boundaries marked by the red lines**

While most metals and ceramics that you will encounter will be polycrystalline, materials can be processed deliberately to form a bulk component that is made up of one crystal (i.e. there will be no grain boundaries). One application of this is single crystal turbine blades which have a high resistance to creep.

The process of crystallization involves the formation of an ordered (crystalline) solid from a more randomly structured liquid. Although most materials (liquids and solids) tend to show a decrease in specific volume as the temperature is lowered, there is a dramatic change as a liquid solidifies to a crystalline form (see Figure UP 2.2). As well as a change in volume, the formation of the new crystalline structure releases energy from the solidifying material in the form of heat. Thus, if a pure metal is liquefied and then heat is extracted, its temperature will go down until the melting point is reached, whereupon, although heat is still being extracted, its temperature will remain constant until all the liquid has solidified; only then will the temperature fall again. An illustration of the temperature history of a pure metal on solidification from the liquid is shown in Figure UP 2.3.

The solidification of alloys is more complex and will be dealt with in Underpinning Principles 4, 'Phase diagrams'.

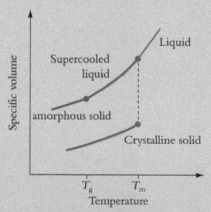

Figure UP 2.2 **Changes in specific volume on solidification from the melt to a crystalline or amorphous solid**

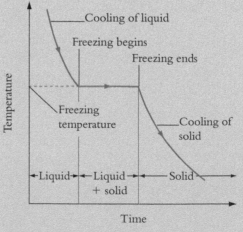

Figure UP 2.3 **Temperature history during the solidification of a pure metal**

2.3 Properties of materials

Elastic modulus (stiffness)

Definition

The elastic modulus, modulus of elasticity or stiffness, of an object is defined as its resistance to elastic deformation. The deformation is elastic if when the applied load (which causes extension or deflection of the sample) is removed, the material returns to its original shape and dimensions. This behaviour occurs at low stresses (and therefore strains). Above a certain stress and strain, termed the elastic limit, the deformation is no longer elastic and permanent shape change occurs.

The elastic modulus is defined mathematically as the ratio, for elastic strains below the elastic limit (typically below 0.1 per cent), of the rate of change of stress with strain. A material that requires a high stress to produce a given strain will, therefore, have a high stiffness. In the SI system, the units of elastic modulus are newtons per square metre ($N\,m^{-2}$) or pascals (Pa). Given the large values of elastic modulus that are typical of many common materials, figures are usually quoted in GPa ($10^9\,Pa$).

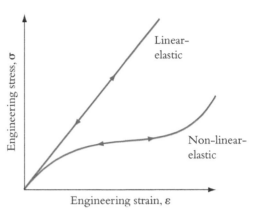

Figure 2.1 Schematic relationship between stress and strain for a linear-elastic material that obeys Hooke's Law (most materials behave in this way) and for a non-linear elastic material (rubber)

For most materials, the elastic modulus is a constant over a range of strains (up to the elastic limit). These materials are described as being linear-elastic and obey Hooke's law. Some materials, rubber for example, are non-linear elastic materials. Figure 2.1 shows both types of elastic behaviour. The area under the stress–strain curve, for any stress up to the elastic limit, gives the elastic energy stored per unit volume of material.

How it is measured

The Young's modulus can be experimentally determined from tensile tests on a sample of the material. The configuration of this test is shown in Figure 2.2. A tensile test produces data for the load developed during the extension of a sample at a constant speed. The sample is prepared so that it has thick sections to fit into the grips of the machine and a thinner parallel section of defined (gauge) length in which the deformation is concentrated (since the stress in this thinner section is larger). Load versus displacement data can be converted to produce a plot of engineering stress, σ, (the force causing the deformation, F, divided by the original cross-sectional area of the sample, A_0) versus the engineering strain, ε, (the ratio of the extension caused by a given stress, ΔL, to the original length of the object, L_0). The Young's modulus, E, can then be calculated from the slope of the stress–strain curve (from the origin to the point where the curve deviates from linearity – this is shown in detail in Figure 2.3) or directly using equation (2.1):

$$E = \frac{\text{stress}}{\text{strain}} = \frac{\sigma}{\varepsilon} = \frac{\left(\dfrac{F}{A_0}\right)}{\left(\dfrac{\Delta L}{L_0}\right)} = \frac{FL_0}{A_0\Delta L} \qquad \textbf{(2.1)}$$

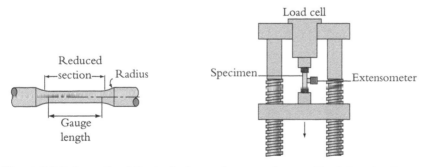

Figure 2.2 Schematic of the typical sample geometry used to determine the properties of a material in tension (left) and the tensile testing apparatus (right)

The work done (or elastic energy stored) per unit volume, W, as the stress is raised from zero to any stress (up to the elastic limit) is given by the area under the stress–strain curve up to this stress. This is also shown in Figure 2.3. Since the Young's modulus is the ratio of stress to strain, the energy stored can also be determined by considering either the applied stress or applied strain. These relationships are shown as:

$$W = \frac{\sigma\varepsilon}{2} = \frac{\sigma^2}{2E} = \frac{E\varepsilon^2}{2} \qquad (2.2)$$

It should be noted that because of the small elastic strains involved, measurement of strain from the displacement of the crosshead of the tensile testing machine is often highly inaccurate owing to displacement in the machine itself. To measure the strain more accurately, an extensometer (an instrument to measure extension) is attached directly to the gauge length of the sample, and data are recorded from this.

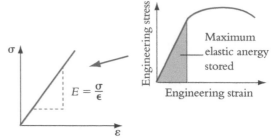

Figure 2.3 Schematic engineering stress–strain curve, derived from a load–displacement curve from tensile testing (right), showing the maximum elastic energy stored (area under the linear part of the curve) and how the Young's Modulus is determined from the slope of the curve (left)

Origin of properties

The bonds between atoms (see Underpinning Principles 1) can be thought of as springs; the 'stiffness' of the bond can be expressed in terms of the force required to stretch the bond a unit distance (the 'spring constant'). The Young's modulus (or stiffness) of a material can be approximated to the bond 'spring constant' divided by the length of the bond separating neighbouring atoms. Highest stiffness is observed for materials with high 'spring constant' and small separations between atoms (for small or closely packed atoms).

The covalent bond has a high stiffness (the spring constant is between 20–$200\,\text{N m}^{-1}$) and as a result, ceramics with this type of bonding, diamond for example, are very stiff. Metals are bound together by metallic bonding; the spring constant is 15–$100\,\text{N m}^{-1}$, making them stiff, but less stiff than ceramics. Polymers deform by untangling and alignment of the polymer chains in the tensile direction, rather than by stretching of the covalently bonded chains that comprise the backbone of the polymer. Since the stiffness of the bonding between (rather than along) the polymer chains is weak (the spring constant is 0.5–$5\,\text{N m}^{-1}$), polymers have low stiffness.

As little can be done to vary the stiffness of the bonds between atoms, the elastic modulus of a given material varies very little with alloying, heat treatment or mechanical working. The most common way of increasing the stiffness of a material is to reinforce it with a stiffer material, in the form of particles or fibres, to form a composite.

When a material is heated, the spacing between the atoms increases (we will see by how much, later in this section), the spring constant for the bond between the atoms decreases and, in the case of polymers, the bond between polymer chains weakens. As a result, the stiffness decreases with increasing temperature. In fact, the stiffness of most materials decreases linearly with heating up to the melting point.

Relevance to engineering applications

The elastic modulus enables the extension or deflection of a material under load to be calculated. It is applicable to situations where the applied load produces reversible elastic rather than permanent plastic deformation. In most engineering situations this is the case, as permanent plastic deformation would lead to unacceptable distortion of the component. For example, the elastic modulus would be used to predict the amount a wire will elastically extend under tension and the deflection of structures such as beams under loading. The elastic modulus can also be used to determine the energy stored in a structure that has been elastically deformed (and hence can be recovered to do work), such as a spring.

Summary: elastic modulus

✔ Elastic modulus, modulus of elasticity or stiffness, is defined as an object's resistance to elastic deformation;

✔ Units are the pascal (Pa); most materials have elastic moduli $\sim 10^9$ Pa (GPa);

✔ The Young's modulus can be determined from the slope of the straight line portion of stress–strain curves produced by tensile testing;

✔ The origin of the elastic modulus of a material lies predominantly in the stiffness of the bonding between the atoms that comprise the material. Stiff bonding gives a high modulus. As little can be done to change the stiffness of the bonds, the elastic modulus varies very little with alloying, heat treatment and mechanical working;

✔ The elastic modulus enables the extension, compression, deflection and energy stored in structures under load to be determined (as long as the load produces reversible elastic deformation).

Material	Young's modulus (GPa)
Rubber (small strain)	0.01–0.1
Low-density polyethylene	0.2
High-density polyethylene	0.7
Polycarbonate	2.6
Nylon	3
Common woods (along grain)	9–16
Fibreglass (glass fibre–epoxy, GFRP)	35–45
Aluminium and alloys	70
Soda glass	70
Titanium	110
Carbon-fibre-reinforced plastic (CFRP)	70–200
Steels	210
Silicon carbide (SiC)	450
Diamond (C)	1000

Table 2.1 Data for the Young's modulus (stiffness) of materials. The range of values reported for some materials reflects the different degrees of crosslinking (in rubbers), the variable densities for different woods and the different volume fractions of fibres in a composite

Stiffness

The load–displacement curve (Figure 2.4) has been produced from a sample 10 mm in diameter, with a gauge length of 40 mm. Using this plot, determine the stiffness of the material and the maximum elastic energy stored per unit volume.

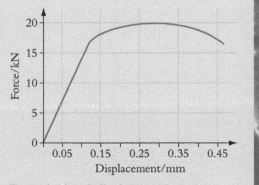

Figure 2.4 Load–displacement curve

Using Figure 2.4, determine the force and displacement values corresponding to the end of the linear part of the curve (this is shown in Figure 2.5 – any point on the linear part of the curve would do, but you need to find the end of the linear part of the curve for the next part!). The force and displacement are 16.5 kN and 0.12 mm respectively, corresponding to engineering stress and strain values of 210 MPa and 3×10^{-3} respectively. The Young's modulus is the stress divided by the corresponding strain and is 70 GPa (it is an aluminium alloy). The maximum stored energy per unit volume is the product of stress and strain values corresponding to the end of the linear section (the limit of the elastic behaviour) divided by 2, and is 315,000 J m^{-3} or 0.315 MJ m^{-3}.

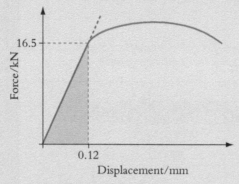

Figure 2.5 Load–displacement curve showing the construction for determining the elastic limit from which the stiffness and the maximum stored energy can be calculated

A 28 mm diameter steel cable is used to raise and lower a passenger lift through a height of 50 m. The cab has a mass of 200 kg and the maximum mass of passengers that may be transported is 1000 kg. If the deformation in the cable is purely elastic (and we ignore the mass of the cable), what is the maximum extension in the cable? (The stiffness of steel is 210 GPa.)

The load on the cable is $(200 + 1000 \, \text{kg}) \times 9.81$ and is 11,772 N. The maximum stress is the load divided by the cross-sectional area of the cable ($6.158 \times 10^{-4} \, \text{m}^2$) and is 19.1 MPa. Since the stiffness is 210 GPa (210×10^9 Pa) and the stress is 19.1 MPa (19.1×10^6 Pa), the strain is 9.1×10^{-5}. The maximum extension is the strain multiplied by the length when the cable is fully extended (to 50 m) and is 4.55×10^{-3} m or 4.55 mm.

A slingshot has two linear-elastic cords, each 4 mm in diameter and 100 mm long. The cords are extended to twice their original length before being released. How much energy is stored in the slingshot before release? (The modulus of the material is 0.01 GPa.)

The change in length of the cord is 100 mm; the original length was also 100 mm. The strain in each cord is, therefore, 1.

The energy stored per unit volume, W, in each cord is obtained from the equation

$$W = \frac{E\varepsilon^2}{2}$$

and is 5×10^6 J m^{-3}. The total energy stored is the energy per unit volume, multiplied by the total volume of the two cords. Each cord has a volume of 1.26×10^{-6} m^3, giving a total stored energy of 12.6 J.

Yield and tensile strength

Definition

As a material is continuously strained, the stress upon it increases. Prior to the yield stress (also called the yield strength or elastic limit) the material will deform elastically and will return to its original shape when the applied stress is removed. Once the yield stress is exceeded, the deformation will be non-reversible (or plastic), resulting in permanent shape change.

The tensile strength, or ultimate tensile stress, is a measure of the maximum tensile stress a material can withstand before failure. Both the yield and the tensile strength are measured in units of force per unit area. In the SI system, the units are newtons per square metre ($N\,m^{-2}$) or pascals (Pa). Typical values for many common materials are quoted in MPa (10^6 Pa).

How it is measured

As in the case for measuring stiffness, the yield and tensile strengths can be determined from engineering stress–strain curves obtained from uniaxial tensile tests. In simple terms, the yield strength corresponds to the stress at which the slope of the stress–strain curve starts to deviate from linearity. The tensile strength is the maximum stress on this curve. The measurement of these values from a typical stress–strain curve is shown in Figure 2.6.

In real materials it is often difficult to determine accurately the point at which the stress–strain curve becomes non-linear. Because of this, the yield point is often defined as the stress at some arbitrary value of plastic strain (typically 0.2 per cent or 0.002). The yield strength at 0.2 per cent offset (or more usually termed the 0.2 per cent proof stress) is determined by finding the intersection of the stress–strain curve with a line parallel to the initial slope of the curve and which intersects the x-axis at 0.002 (0.2 per cent). This construction is also shown in Figure 2.6.

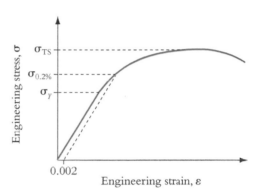

Figure 2.6 Schematic stress–strain plot with the yield stress (σ_y), 0.2 per cent proof stress ($\sigma_{0.2}$) and tensile stress (σ_{TS}) marked

Origin of properties

It should be possible to determine the strength of a material from the strength of the bonds between the atoms and the distance between atoms (see Underpinning Principles 1). However, such a calculation yields values that are much higher than those measured experimentally. This is because real materials contain defects.

In metals, the non-directional nature of the metallic bond allows metal atoms to slide past each other. Helped by packing defects or dislocations (see Underpinning Principles 3) in the atomic structure, this occurs at much lower stresses than those required to break the bonds. In covalent ceramics, the directional bonding means that it is very difficult for atoms to slide past each other (since they must keep the same spatial relationship with their neighbours) and this limits their capacity to undergo plastic flow. As a result, there is no concept of a yield point in either tension or compression in ceramics, i.e. no stress above which plastic deformation can occur, simply a failure stress. The strength of the bonds in ceramics is high and they exhibit high failure strengths so long as they do not contain cracks or defects. Crack defects result in failure at much lower stresses than the tensile stress and since most ceramics contain small cracks, ceramics are best used in compression for engineering applications (so the cracks do not open). Data for the fracture stress for ceramics are always quoted in compression for this reason. Polymers deform by the alignment and untangling of the long chain structure; the more tangled the structure, the more resistant it is to deformation. Polymers, with weak forces between the chains, result in low-strength materials; polymers with stronger, chemical crosslinking have high tensile strengths. Unlike the stiffness of these materials, the yield and tensile strength can be greatly affected by chemical composition, heat treatment and processing.

With increasing temperature, dislocation motion becomes easier, material flow can occur by creep, assisted by diffusion and the bonds between polymer chains weaken. As a result, the yield strength of materials decreases with increasing temperature and, unlike the stiffness, the yield strength decreases rapidly with temperature, in a non-linear fashion.

Relevance to engineering applications

Knowledge of the yield strength of a material is vital when designing a component, since it generally represents an upper limit to the stress that can be applied. Exceeding this stress will, of course, lead to permanent distortion of the component. Examples of applications include calculating the maximum loads on wires and cables in tension, spinning discs and pressurized vessels. Very few designs allow the plastic deformation of components to occur and for this reason designing for stresses up to the tensile stress is not common. It is, however, useful to appreciate the interval of stress between yielding and failure for a material so that appropriate safety factors can be considered to avoid catastrophic failure as a result of overloading the component.

Learning summary

Summary: yield and tensile strength

✔ The yield stress is the stress above which non-reversible (or plastic) deformation occurs. The tensile strength, or ultimate tensile stress, is the maximum tensile stress a material can withstand before failure;

✔ Units are the pascal (Pa), and most materials have yield and tensile strengths $\sim 10^6$ Pa (MPa);

✔ The yield and tensile strength can be determined from stress–strain curves, obtained from tensile tests, by taking the stress at which the stress–strain curve deviates from linearity and the maximum value of the stress on the curve respectively;

✔ The origin of plastic deformation in metals is the shearing of planes of atoms past each other, rather than the breaking of bonds between neighbouring atoms. Ceramics do not show plastic deformation. In tension the fracture of ceramics occurs due to the presence of cracks at stresses below the tensile stress. In polymers, plastic deformation occurs as the chains slide past each other. As the strength of the bonding between chains increases, so does the yield strength;

✔ The yield strength defines an upper limit to the tensile or compressive stress that can be applied to a component without irreversible plastic deformation.

Material	Yield strength (MPa)	Tensile strength (MPa)
Low-density polyethylene	6–20	20
Rubber	–	30
High-density polyethylene	20–30	37
Common woods (along grain)	–	35–55
Polycarbonate	55	60
Nylon	40–90	100
Fibreglass (glass fibre–epoxy, GFRP)	–	100–300
Mild steel	220	430
Aluminium alloys	100–600	300–700
Carbon-fibre-reinforced plastic (CFRP)	–	640–670
Titanium alloys	180–1320	300–1400
Low-alloy steels	500–1900	680–2400
Silica glass	–	(7200)
Silicon carbide (SiC)	–	(10,000)
Diamond (C)	–	(50,000)

Table 2.2 Data for the yield and failure strengths of materials (values in brackets are in compression). The wide range of values for a given material type reflects the ability to change the properties through changes in the chemical composition, mechanical working and heat treatment

Yield and tensile strength

Using the stress–strain curve in Figure 2.7, determine the yield, 0.2 per cent proof stress and tensile strength of the material.

Using the plot, construct lines for the appropriate stresses at the end of the linear section (the elastic limit or yield stress – 215 MPa), the stress at an offset strain of 0.002 (the 0.2 per cent proof stress – 240 MPa) and the maximum stress (the tensile stress – 250 MPa). These constructions are shown in Figure 2.8.

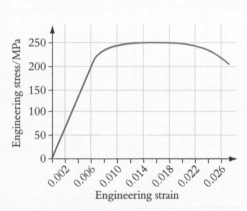

Figure 2.7 **Schematic stress–strain curve**

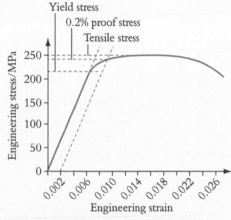

Figure 2.8 **Schematic stress–strain curve with construction lines to calculate the appropriate stresses**

Four mild steel tie bars are to be used to clamp caps onto the ends of a cylinder with an internal diameter of 50 mm, which must contain a gas pressure of 100 bar (10 MPa). What diameter should these rods be to ensure that they can provide the required clamping force without yielding? (Assume the yield strength of the mild steel used to be 220 MPa.)

First calculate the clamping force needed to clamp the ends onto the cylinder. The force is pressure $\times$ area and is 19,635 N. This is shared among the four tie rods each carrying 4909 N. Knowing the load and the yield stress of the steel, we can calculate the cross-sectional area (force/stress) and hence the diameter for the rods. The cross-sectional area is 2.23×10^{-5} m^2 and the diameter is 5.3×10^{-3} m or 5.3 mm. Since the rods will be overtightened to ensure sufficient clamping force is provided, this represents a minimum diameter (an appropriate safety factor should also be included).

A mild steel bolt with a 4 mm diameter shank and a (approximately) 7 mm diameter head and nut are used to clamp two nylon plates together. A clamping force of 1000 N is required to ensure that the assembly functions in service. If the bolt is overtightened to give a force of 2000 N, will the bolt plastically deform and will the nylon plate be damaged? (Assume the yield strength of mild steel and nylon to be 220 and 45 MPa respectively.)

Calculate the maximum stress in the bolt to see if it yields. The diameter is 4 mm, the applied load is 2000 N and the stress is therefore, 159 MPa, well below the yield stress, and so the bolt will deform elastically. This same force is transmitted through the head of the bolt and nut to the nylon plate (they are 7 mm in diameter). The contact stress on the nylon is 52 MPa, so the plates will be plastically deformed. In reality, washers larger than 7 mm would be used to increase the contact area and decrease the contact stress.

Underpinning Principles 3: Dislocations and metal deformation

If a stress higher than the yield stress is applied to a metal, then permanent shape change takes place – plastic deformation. While we do not see this, at an atomic scale, planes of atoms are moving with respect to each other.

As we have seen, most materials are crystalline in nature. Most have grain boundaries where the atomic packing is less than perfect (see Underpinning Principles 2). In addition, most metals contain dislocations within the individual crystals (grains) themselves. These dislocations can be visualized as missing rows of atoms. These missing rows of atoms are crucial in terms of permanent deformation of metals.

Figure UP 3.1 shows a crystal with a missing row of atoms (a dislocation). If a shear stress is applied as indicated, not all the bonds have to break at once, but instead, rows of bonds can break one at a time to allow easy passage of the dislocation through the material. Macroscopically, if we observe plastic deformation, then on an atomic scale, dislocations have moved. Conversely, if we can change materials at the atomic scale so that dislocations are prevented from moving, then macroscopically, we observe an increase in yield stress. There is a direct link here between atomic scale processes and the macroscopic material's properties.

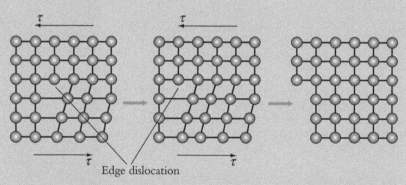

Edge dislocation

Figure UP 3.1 A dislocation in a metal being caused to move by the application of stress

Strengthening mechanisms

To make a material more resistant to plastic deformation (i.e. increase the yield stress), we need either to remove all the dislocations (impractical) or make it harder for dislocations to move. There are four main ways to impede the motion of dislocations.

Work strengthening (dislocation strengthening)

When a metal is deformed plastically, not only do dislocations move, but they also multiply in number, leading to an increased dislocation density. As the dislocation density increases, the dislocations get tangled up on each other as they move (and can indeed lock each other in position). As the dislocation mobility decreases, so the yield stress increases.

In a carefully prepared metal, the dislocation density may be as low as $10^7 \, \text{m m}^{-3}$ (length of dislocation per metre cubed of material), but in a heavily deformed sample it may rise to as much as $10^{16} \, \text{m m}^{-3}$.

As the dislocation density rises, the yield strength increases, but the ductility (elongation to failure) decreases.

Solid solution strengthening

Local strain can be induced in a crystal structure by alloying. The alloying elements will dissolve into the main crystal (a solid solution) but will disrupt the perfect structure, since the atoms of the main crystal are of a different size. This makes it more difficult for dislocations to move through this structure. As such, alloying results in an increase in yield strength.

Precipitation strengthening

Small particles of a second phase within a crystal may be able to hinder dislocation motion. To do this effectively, the precipitates need to be small and very close together, and this can be achieved by very careful heat treatment of well-designed alloys.

Grain boundary strengthening

Grain boundaries are regions of crystal imperfection in a structure, and as such, the motion of dislocations is impeded by the presence of grain boundaries. More grain boundary area is present in a structure as the grain size decreases, and an increase in the yield strength is observed.

The increase in yield strength with decreasing grain size is given by the Hall–Petch equation:

$$\sigma_{\text{yield}} = \sigma_0 + k_y d^{-\frac{1}{2}}$$

where d is the grain size, and σ_0 and k_y are constant for a given alloy.

Ductility and toughness

Definition

Ductility is a measure of the plastic strain at failure. The larger the strain at failure, the more ductile the material. The opposite of ductile is brittle. Ductility, a strain, is unitless and is usually described in terms of a percentage elongation or reduction in cross-sectional area at failure.

The toughness is the resistance to fracture of a material when stressed. It is defined as the amount of energy that a unit volume of material can absorb before fracture. Toughness is measured in units of joules per cubic metre ($J\,m^{-3}$).

How it is measured

The ductility can be determined from a stress–strain curve produced by tensile testing. A rough indication of the strain to failure or ductility can be obtained from the strain corresponding to the breaking stress (marked by the cross at the end point on the curve shown in Figure 2.9), but this strain represents both the plastic and elastic deformation of the sample. The elastic part of the deformation is recovered, as the sample contracts after fracture, but this contraction is not recorded via the displacement of the crosshead on the tensile testing machine. From the stress–strain curve, the elastic contraction can be removed by drawing a line parallel to the linear part of the curve, backwards from the breaking point. Where this intersects the x-axis is the actual plastic deformation (this construction is also shown in Figure 2.9). The ductility can also be measured by reassembling the fractured tensile specimen and measuring the extension of the gauge length. The engineering strain to failure is then given by:

$$\varepsilon_f = \frac{L_f - L_0}{L_f} \tag{2.3}$$

where L_f and L_0 are the final and original gauge lengths of the sample respectively.

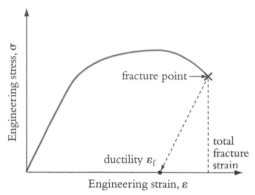

Figure 2.9 Schematic stress–strain plot with the fracture point, total fracture strain and ductility (plastic strain to fracture) shown

The toughness of a material can be determined from the stress–strain curve by finding the area underneath the curve (and subtracting the elastic component). Another measure of the fracture energy or toughness can be obtained from a Charpy impact test. In this simple test, a swinging pendulum with a hammer attached, impacts plain or notched samples of the material located at the bottom point of the swing path. The height to which the pendulum swings through the fractured sample is used to calculate the potential energy that is lost during the fracture process and hence the fracture energy (in this case in $J\,m^{-2}$).

Origin of properties

The origin of ductility comes from a material's ability to undergo plastic deformation. Plastic deformation readily occurs in metals owing to the relative ease with which planes of atoms can slide past one another and hence they have high ductility. Materials with high yield strengths, in which atomic sliding is more difficult, tend to have low ductilities. In general, as the strength of a given material is increased, either as a result of alloying or processing, the ductility will decrease. Ceramics have limited capacity for plastic deformation and are brittle (the opposite of ductile). This behaviour is typified by the fact that the undeformed surfaces of fractured ceramics (such as vases or cups) can be glued back together to recreate the original shape. Polymers are able to show high ductilities if their chains are able to slide past each other. Polymers with high levels of crosslinking or strong bonding between the chains show lower ductilities.

The toughness of a material is in part determined by the degree to which a material can undergo plastic deformation before it fails, but also the stress at which deformation occurs. Tough materials, such as metals, have a combination of high strength and high ductility.

Figure 2.10 shows typical stress–strain curves for a polymer, a metal and a ceramic (in compression). It is clear that although the ceramic has a higher tensile strength and the polymer a much larger ductility than the metal, from the size of the areas under the respective curves it is clear that the metal (which has a good balance of strength and ductility) has the highest toughness.

Figure 2.11 shows the dramatic effect that the testing temperature has on the energy required to fracture different materials. The change in fracture energy is small for metals such as aluminium, copper, nickel and stainless steel (which have a particular atomic structure: face-centred cubic or FCC) and the failure process is ductile (fracture requires a lot of energy). The change in fracture energy with temperature is also small for high-strength (non-metallic) materials; it remains low and can be considered brittle (fracture requires very little energy). Of most interest (and concern to designers, as it introduces uncertainty) is the behaviour of materials that do not have a face-centred cubic structure (which is most steels) and polymers, where a change from a ductile to a brittle mode of fracture occurs as the temperature decreases.

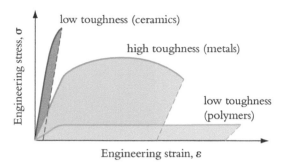

Figure 2.10 Schematic stress–strain curves for a ceramic (in compression) a metal and a polymer. The areas under the curves show the metal to have the highest toughness

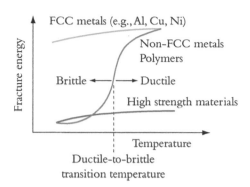

Figure 2.11 Schematic showing the effect of temperature on the energy required to fracture different types of materials

In non-FCC metals, the decrease in toughness is a result of dislocation motion becoming much more difficult as the temperature is decreased. In polymers the ductile–brittle transition (more commonly called the rubber–glass transition) occurs because, above the transition temperature, the weak bonds between the polymer chains 'melt' (the strong crosslinks, however, remain intact). Above the glass transition temperature, polymers behave like viscous liquids or rubbers (with high ductility but low strength and stiffness); below it they behave like true solids (with low ductility but high strength and stiffness). The temperature for the transition varies with material: for example, for perspex it is at 100°C, so at room temperature it is glassy; for low-density polyethylene it is −20°C and for natural rubber it is −40°C, so at room temperature they are rubbery. The ductile–brittle transition temperature for some steels can be as high as 0°C.

Relevance to engineering applications

The ductility of a material is not often considered in the design of a structure or component; in fact, efforts are usually made to avoid plastic deformation. An appreciation of the toughness of a material is, however, relevant in a number of engineering applications. The toughness is important in determining the energy that is absorbed during fracture. This may be appropriate for designing materials for bike locks to resist impact from a hammer or to select materials that will intentionally absorb impact such as roadside safety barriers.

Learning summary

✔ Ductility is a measure of the plastic strain at failure and is unitless. The toughness is the amount of energy a material absorbs during fracture and has units of $J\,m^{-3}$;

✔ The ductility can be determined from a stress–strain curve produced by tensile testing and is roughly the strain corresponding to the breaking stress. The toughness of a material can be determined from a stress–strain curve by finding the area underneath the curve;

✔ The origin of the ductility of a material comes from its ability to undergo plastic deformation. The toughness of a material is in part determined by the degree to which the material can undergo plastic deformation before it fails but also the stress required to produce plastic deformation;

✔ The ductility of a material is not often considered in the design of a structure or component. The toughness is important in determining the energy that is absorbed during fracture either in an attempt to resist fracture or to absorb as much energy as possible during impact.

Material	Ductility (%)
Rubber	800
Low-density polyethylene	100–650
Nylon	15–80
Copper	55
Mild steel	18–25
Aluminium alloys	5–40
Titanium alloys	6–30
Low-alloy steels	2–30
Glass (all types)	4.3
Carbon-fibre-reinforced plastic (CFRP)	0.3
Ceramics	nil

Table 2.3 Data for the ductility of materials. The wide range of values for a given material type reflects the ability to change the ductility through chemical composition, mechanical working and heat treatment

Ductility and toughness

Using the stress–strain curve in Figure 2.12, determine both the total and the plastic strain at fracture. Estimate the toughness of the material in $J\,m^{-3}$.

Determine the total and plastic strains from the stress–strain plot (using the construction shown in Figure 2.13). The total strain to failure is 0.026 (2.6 per cent) and the plastic strain is 0.020. Estimate an average stress above the yield point such that the net area of the region between the average stress and the curve is close to zero. The area of the parallelogram is simply the base multiplied by the height. The average stress is roughly 240 MPa, the plastic strain is 0.020 and the toughness is $4.8 \times 10^6\,J\,m^{-3}$.

If a metal slab is reduced in thickness from 20 mm to 18 mm by a metalworking process, what is the elongation imposed? If the elongation to failure is 25 per cent, what is the minimum thickness that can be achieved?

The reduction in thickness is 2 mm and hence the engineering strain is 0.1 (or 10 per cent). If the maximum elongation is 25 per cent (or 0.25) then the maximum reduction is 5 mm and the final thickness would be 15 mm.

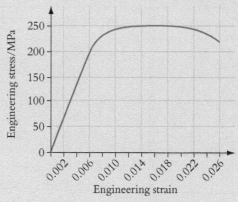

Figure 2.12 Schematic stress–strain curve

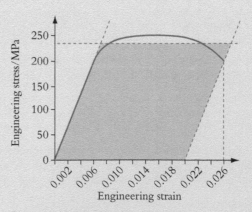

Figure 2.13 Schematic stress–strain curve with construction lines drawn

Hardness

Definition

The hardness of a material is an expression of its resistance to indentation and, more specifically, its resistance to permanent plastic deformation. The harder a material is, the greater its resistance to indentation and plastic deformation. The hardness is determined by the load over the projected area of the indentation. Vickers hardness has non-SI units of force divided by area (in $kgf\,mm^{-2}$).

How it is measured

Hardness is normally measured using an indentation method. In this technique, an indenter of specified geometry (usually made from diamond) is forced into the material under a controlled load. By measuring the size of the resulting indent in the material, the hardness can be measured. There are a number of different ways of conducting this test, employing different shapes of indenter, which require different conversion factors between indent size and hardness. The main methods are the Vickers, Brinell and Rockwell hardness tests.

The Vickers hardness test uses a diamond indenter, in the shape of a square-based pyramid with an angle of 136° between opposite faces. The geometry of the indenter is shown in Figure 2.14.

The Vickers pyramid number (HV) or hardness is then determined by the ratio $\dfrac{F}{A}$, where F is the force applied to the diamond and A is the projected surface area of the resulting indentation, as seen from above. The projected surface area, A, and the hardness can be determined using equations (2.4) and (2.5), where l is the average length of the diagonal left by the indenter:

$$A = \frac{\frac{1}{2}l^2}{\sin(136°/2)} \tag{2.4}$$

$$HV = \frac{F}{A} = \frac{1.854F}{l^2} \tag{2.5}$$

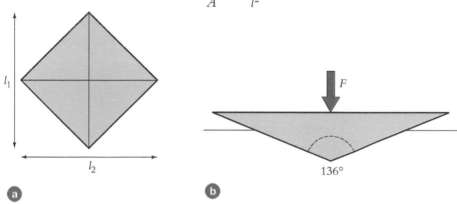

a   b

Figure 2.14 Schematic illustration of the geometry of the Vickers indent (left) and indenter (right)

Vickers hardness numbers are reported as $xxx\,HV_{yy}$, e.g. $120\,HV_{10}$, where 120 is the hardness number, HV gives the hardness scale (Vickers), and 10 indicates the load used in kgf. This represents a hardness of $120\,kgf\,mm^{-2}$. To convert a Vickers hardness number to SI units (Pa) the force applied must be converted from kgf to newtons and the area from mm^2 to m^2. A practical method to convert HV to MPa is to multiply by 9.81. The hardness, HV (after conversion to MPa) can be approximated as three times the yield strength of the material (in MPa). For example, 60 HV corresponds to a yield stress (σ_y) of approximately 200 MPa.

For all methods, the depth of the indent is reasonably shallow and, as a result, the surface properties of the material are measured. For most materials these properties are representative of the bulk. Since little damage to the material occurs, the testing method is largely non-destructive and is ideal for quality control. By using a small indent and an appropriately reduced load, the properties of coatings and hardened surface layers can be measured (if the indent depth is less than 10 per cent of the coating thickness).

Origin of properties

The hardness of a material is dictated by its resistance to plastic deformation. Hence those materials that have high yield strengths will have high hardness. Since the test applies a largely compressive force, the hardness of ceramics can be measured by this method. Ceramics have high hardness; metals, which have lower compressive yield stresses than ceramics, show intermediate levels of hardness; and polymers, which are weak, have low hardness. Since hardness is related to the resistance to plastic deformation and the yield strength, as the temperature increases, the hardness decreases.

Relevance to engineering applications

While the selection of a material in a design may not be driven by the hardness, the hardness of a material is a good indicator of other properties that are highly relevant. A high hardness indicates a high yield strength and a high resistance to indentation and deformation (and forming). A hardness indent on a component may be used as a means of quality control to ensure that the desired strength has been achieved, for example after heat treatment.

Learning summary

By the end of this section you will have learnt:

✔ The hardness of a material is an expression of its resistance to indentation and plastic deformation;

✔ Hardness is normally measured using an indentation method. For a given load, the smaller the indent, the higher the hardness;

✔ The hardness of a material is dictated by its resistance to plastic deformation. Those materials that have high yield strengths will have high hardness. The hardness of metals, ceramics, polymers and coatings can be measured using this method;

✔ Hardness is a good indicator of a material's ability to resist yield and deformation and can be used as a means of quality control.

Material	Hardness (kgf mm^{-2})	Yield stress (MPa)
Diamond	8400	54,100
Alumina	2600	11,300
Tungsten carbide	2100	7000
Mild steel (normalised)	210	700
Annealed copper	47	150
Annealed aluminium	22	60
Lead	6	16

Table 2.4 Data for hardness of materials (and comparison with actual yield stress data – compressive failure stress for ceramics)

Hardness

A Vickers hardness test is performed on a material. A load of 30 kgf is applied and it produces a square indent of diagonal length 0.654 mm. Estimate the hardness from the indent size. What is the approximate yield stress of the material?

Using the expression below, substituting the force (in kgf) and diagonal length (in mm), a hardness of 130 HV$_{30}$ is obtained. The hardness is 1275.3 MPa and the yield stress is approximately 425 MPa:

$$HV = \frac{1.854F}{l^2}$$

Density

Definition

Density (ρ) is a measure of the mass of a material per unit volume. The SI unit of density is kilograms per cubic metre ($kg\,m^{-3}$). Most commonly, the bulk density of a material is considered (for a lump of material). For powder materials and foams an apparent density is often defined which takes into account the air spaces between the powder particles or within the material.

How it is measured

The simplest way to determine the density of an object of regular shape is to measure its mass and divide this by its volume (obtained by accurate measurement of the sample dimensions). For irregular shapes, their volume can be measured by either the volume of water (Archimedes' method) or gas that they displace.

Origin of properties

The density of a material originates from the weight, size and the packing of the atoms or molecules that comprise the material. The atomic size does not vary that much from the largest to the smallest atom but the weight does and increases (along with the density) as the atomic number increases (see Underpinning Principles 1). Compounds and alloys with closely packed structures, such as metals, tend to have high densities; those with more open structures, as a result of directional, covalent bonding, as observed in ceramics, have intermediate densities; and polymers, which are composed of light atoms and have open, tangled chain structures, have low densities. Heating a material will cause it to expand (the atoms will become less closely packed) and the density will decrease.

As is the case with the stiffness of a material, little can be done to change the density of a material by processing or heat treatment. The mixing of two materials (for example, a glass fibre with a polymer matrix) to make a composite does, however, change the density of the new material. The density of the composite can be described by a simple law of mixtures, according to equation (2.6), where ρ is the density of the fibre (f) or matrix (m) and V_f is the volume fraction of fibres:

$$\rho_{composite} = V_f\rho_f + (1 - V_f)\rho_m \qquad (2.6)$$

Relevance to engineering applications

The density of a material is often very important in design as it will affect the mass of the component. If a material with the same mechanical properties but with a lower density can be used, the component will be lighter. This may be of high importance if the component is moving (either in an engine or machine, or as part of an aircraft), since the energy required to move it will be less, thereby giving scope for increasing performance or saving fuel.

Learning summary

By the end of this section you will have learnt:

✔ Density (ρ) is a measure of the mass of a material per unit volume. The SI unit of density is kilograms per cubic metre ($kg\,m^{-3}$);

✔ The simplest way to determine the density of an object of regular shape is to measure its mass and divide this by its volume;

✔ The density of a material originates from the weight and packing of the atoms or molecules that comprise the material. Materials with closely packed structures will have high densities; those with open structures, low densities;

✔ The density of the material used in a component will affect its mass. The use of low-density materials is important to save weight and energy in structures that move.

Material	Density (kg m^{-3})
Platinum	21,450
Gold	19,300
Tungsten	19,250
Copper	8960
Iron	7870
Steel	7850
Titanium	4500
Diamond	3500
Aluminium	2700
Graphite	2200
Fibreglass (glass fibre–epoxy, GFRP)	2000
Magnesium	1740
Carbon-fibre-reinforced plastic (CFRP)	1500
PVC	1300
Nylon	1100
Water	1000
High-density polyethylene	970
Rubber	850
Wood	600

Table 2.5 Data for the density of materials (at 20°C).

Density

A metal disc, 50 mm in diameter and 30 mm thick, has a mass of 265 g. What is the density of this material?

The density is the mass per unit volume. The volume is 5.89×10^{-5} m^3 and the mass 0.256 kg. The density is 4499 kg m^{-3} (it is titanium).

What is the density (in kg m^{-3}) of a composite containing 60 per cent by volume of glass fibres (density $= 2.7$ g cc^{-1}) in a polymer matrix (density $= 1.0$ g cc^{-1})?

$$\rho_{composite} = V_f \rho_f + (1 - V_f)\rho_m$$

Using the equation above, where $V_f = 0.6$ (60 per cent) and 1 g cm$^{-3} = 1000$ kg m^{-3}, the density of the composite is 2020 kg m^{-3}.

An aluminium casing for the electronics to control a robot weighs 810 g. What are the masses for boxes of the same size made from polyethylene and steel? (The densities for aluminium, polyethylene and steel are 2700, 970 and 7850 kg m^{-3} respectively).

From the mass divided by the density, the aluminium box has a volume of 3×10^{-4} m^3. The polyethylene and steel boxes will have the same volumes. Their masses will be 291 g and 2355 g respectively.

Thermal expansion

Definition

Thermal expansion is the tendency of matter to increase in volume when heated. For liquids and solids the amount of expansion will vary depending on the material's coefficient of thermal expansion (CTE). The linear strain produced due to a change in temperature, ΔT, can be given as

$$\varepsilon_{thermal} = \alpha \Delta T \qquad \text{(2.7)}$$

where α is the coefficient of linear thermal expansion which has units of K^{-1}. For most (isotropic) materials, the volumetric thermal expansion coefficient can be approximated to three times the linear coefficient (3α) and would replace α in equation (2.7) when calculating the volumetric strain.

How it is measured

The linear thermal expansion coefficient for a material can be measured by heating a material in a controlled manner (at a constant and slow heating rate) and measuring the length dilation of the sample. This dilation can be measured using contacting displacement transducers (usually employing an intermediate silica 'push rod' that does not expand so that the transducer remains cold) or non-contact techniques using induced currents or lasers. From a plot of extension (as strain) versus temperature, the linear coefficient of expansion can be determined from the gradient of the slope. The coefficient of expansion (slope) does vary with temperature (and so is usually quoted at a given temperature) but can be considered to be constant over a short temperature interval.

Origin of properties

When a material is heated, the energy that is stored in the bonds increases, the atoms vibrate with increasing amplitude and the bond length between atoms increases. As a result, solids expand in response to heating (reducing in density) and contract on cooling. This response to temperature change is expressed as its coefficient of thermal expansion. The open structure of polymers and the ease with which the weakly bonded polymer chains can unravel means that polymers have very high coefficients of expansion. The higher bond stiffness in metals and ceramics means they have much lower expansion coefficients. Since thermal expansion depends mainly on the bond stiffness, very little can be done to change the thermal expansion behaviour for a material.

Relevance to engineering applications

Heat-induced expansion has to be taken into account in many engineering and manufacturing applications. Expansion-induced strains can cause the distortion of components and structures. Without the use of expansion joints, structures such as pipelines may buckle. In most cases thermal strains should be minimized so that structures remain dimensionally stable. If different materials are joined, differential thermal strains can result in large stresses and distortion or failure of the component.

Learning summary

By the end of this section you will have learnt:

✔ Thermal expansion is the tendency of matter to increase in volume when heated. The coefficient of thermal expansion, multiplied by the change in temperature, gives the thermal strain. The coefficient of linear thermal expansion has units of K^{-1};

✔ The linear thermal expansion coefficient for a material can be measured by heating a material in a controlled manner and measuring the length dilation of the sample as a function of temperature;

✔ When a material is heated, the energy in the bonds between atoms increases, as does the bond length. As a result, solids expand in response to heating and contract on cooling;

✔ Thermal-expansion-induced strains can cause the distortion of components and structures. In most cases thermal strains should be minimized.

Material	α (10^{-6} K^{-1})
Mercury	60
Aluminium	23
Brass	19
Stainless steel	17.3
Copper	17
Nickel	13
Carbon steel	11
Glass	8.5
Tungsten	4.5
Glass, Pyrex	3.3
Silicon	3
Diamond	1
Quartz, fused	0.59

Table 2.6 Data for the coefficient of linear thermal expansion of materials (at 20°C)

Thermal expansion

A 100 mm long copper bar is heated by 150°C. Calculate the strain and expansion in the heated bar. If the material is not allowed to expand, what is the (compressive) stress in the bar? (The coefficient of thermal expansion (CTE) for copper is 17×10^{-6} K^{-1} and its stiffness at 150°C is 110 GPa).

Using the expression below, assuming that the CTE is constant over the temperature interval of interest, the thermal strain is 2.55×10^{-3} and the expansion is 0.255 mm. When constrained, assuming Hooke's law applies, the stress is the stiffness multiplied by the strain and is 281 MPa.

$$\varepsilon_{\text{thermal}} = \alpha \Delta T$$

Copper and steel sheets of the same size are riveted together. Calculate the misfit strain between the two if they are both heated by 150°C ($\alpha_{\text{copper}} = 17 \times 10^{-6}$ K^{-1}, $\alpha_{\text{steel}} = 11 \times 10^{-6}$ K^{-1}).

Calculate the strains in the two materials: for Cu it is 2.55×10^{-3} for Fe it is 1.65×10^{-3}. They are joined, therefore they cannot move independently. The steel extends less, so is stretched by the copper (the steel is in tension) and the Cu is pulled back by Fe (the copper is in compression). The misfit strain is the difference and is 9.0×10^{-4}. The result is that the unstable structure curls up (a large displacement compared to the small misfit strain!). Materials are used in this way to make bimetallic strips that work as switches (in kettles, for example).

Cost

Definition

The cost of a material is defined as the price to the consumer to purchase a unit quantity, commonly in terms of the mass. Described in this way it has units (in the UK) of £ kg^{-1}.

How it is measured

Prices for materials can be obtained directly from suppliers or from the commodities markets such as the London Metal Exchange (for metals and plastics). The daily prices for many materials are also listed in the *Financial Times*.

Origin of properties

The cost of a material, as delivered to the consumer, has many components to it. These include extraction of the raw material (metal ore, oil, ceramic oxides) which can vary considerably with their abundance, the quality (purity) of the raw material and any socio-economic aspects relating to the country from which it is being extracted (export laws or restrictions, civil wars). There are energy and processing costs associated with the conversion of the raw material and then processing of the material into the appropriate form or shape. In addition to all these issues is the cost of energy relating to transportation between processing stages and delivery to the user. Finally, prices fluctuate due to supply and demand and the abundance of scrap or recycled material. Figure 2.15 shows how the price of primary aluminium (after it has been extracted from its ore) has fluctuated over a six-year period.

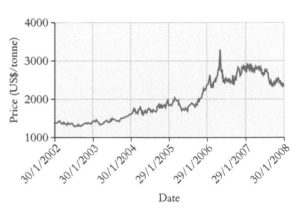

Figure 2.15 Fluctuation in price (in US$/tonne) for primary aluminium over a six-year interval. Data from the London Metal Exchange Limited and produced with permission

Relevance to engineering applications

While the cost of a material is not a property governed by its physical behaviour, it is an important aspect to consider in the selection of a material. Often material selection is constrained by the target cost of the product and the profit that must be made. The best material for a particular application may be too expensive or in too limited a supply to be used.

Learning summary

By the end of this section you will have learnt:

✔ The cost of a material is defined as the price to the consumer to purchase a unit quantity, usually in terms of the mass. Described in this way it has units (in the UK) of £ kg^{-1};

✔ Values for materials prices can be determined directly from suppliers or from the commodities markets such as the London Metal Exchange;

✔ The cost of a material depends on the extraction, processing and transportation costs; it is sensitive to the cost of energy relating to processing and transportation and the abundance of the raw material and of scrap or recycled material;

✔ Cost is an important aspect to consider as the choice of material is often constrained by the cost of the product and the best material for a particular application may be too expensive or in too limited a supply to be used.

Material	Cost (£ kg^{-1})
Diamond (industrial)	1,000,000
Platinum	25,000
Gold	10,000
Carbon-fibre-reinforced plastic (CFRP)	100
Nickel alloys	100
Titanium	50
Silicon carbide (fine ceramic)	35
Alumina (fine ceramic)	15
Nylon	7.5
Fibreglass (glass fibre–epoxy, GFRP)	5
Stainless steel	3
Glass	2
Copper (ingot or tube)	2
Aluminium alloy (ingot/tubes)	1.5/2
Natural rubber	1.5
Hard woods	1.25
Polyethylene	1
Low-alloy steels	0.65
Mild steel (sheet and bars)	0.5
Cast iron	0.45
Soft woods	0.35
Fuel oil	0.25

Table 2.7 Data for the cost of materials (this is only a rough indication as prices fluctuate on a daily basis)

Cost

An aluminium casing for the electronics to control a robot weighs 2.0 kg. What is the cost of the raw material for this part if aluminium prices are £1.5 kg^{-1}? If the casting process used to make this part has a scrap rate of 25 per cent, a combined labour and overhead rate of £40 per hour and a productivity rate of 20 parts per hour, what is the minimum price for which the part could be sold?

The cost of the Al ingot is £1.5 kg^{-1} and hence for the part is £3.00. A scrap rate of 25 per cent would require the use of 2.5 kg of material, costing £3.75 and the labour and overhead rate per part is £2.00 (20 parts can be made in an hour, costing the company £40). The part should not be sold for less than £5.75. This price still does not include delivery to the customer or any profit.

Learning summary

By the end of this section you will have learnt:

✔ The definitions for a number of important material properties and the testing methods which enable us to measure these properties;
✔ The origins of the properties for different material types;
✔ Typical values for these properties for different materials;
✔ The relevance of these different properties to engineering applications and how to use property data to solve design problems.

2.4 Selection of materials in engineering design

In a design where we have already decided on the shape of the cross-section and the length (so that it will fulfil the basic design requirements, for example so that an aircraft wing gives the required lift) we still have the flexibility to choose the cross-sectional thickness and the material from which it is made. We could choose a lightweight (low-density) material, even though it may have inferior properties (lower strength or stiffness), but if we increase the thickness of the structure, we can reduce the stress it experiences.

This is fine in concept but by how much would we need to increase the thickness to achieve the desired response? Would we still achieve a weight saving? Would all the extra material required make it more expensive than our original choice? It sounds as though we would have to calculate the stresses, dimensions and costs for many different designs, with many different materials, until we find the best option.

We will see in this section that this is not the case. The method described in the following examples enables a simple comparison of materials to be made for a given application, with a minimum of effort.

Material selection for a stiff lightweight beam

We commonly want to design objects or structures to be as stiff as possible (to minimize the deflection under an applied load) and to achieve this with minimum weight. Examples of this are the deflection of tennis rackets, bridges and aircraft wings. The benefits of mass reduction are clear. In the case of an aircraft wing, this mass reduction could be used to increase either the payload or the fuel load, so that the aircraft can become more cost-effective, more manoeuvrable, fly further before refuelling or reduce emissions.

The following example is for the end deflection of a cantilever beam (one which is fixed at only one end). Figure 2.16 illustrates this case. Equations defining the deflection, δ, of a beam in terms of the applied load, P, the length of the beam, L, the stiffness of the beam, E, and the second moment of area of the beam, I, can be found in standard textbooks on mechanics for different modes of loading. In the case of the end loading shown in Figure 2.16, the maximum (end) deflection is given by equation (2.8):

$$\delta = \frac{PL^3}{3EI} \tag{2.8}$$

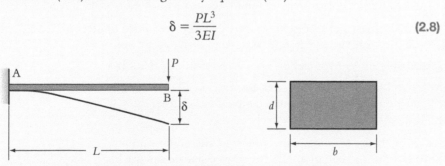

Figure 2.16 Schematic (left) of the end deflection of a cantilever beam and (right) the cross-section of the beam

For a beam with a rectangular cross-section, the second moment of area, I, is defined in equation (2.9), where b is the breadth or width of the beam and d is the depth:

$$I = \frac{bd^3}{12} \tag{2.9}$$

Combining equations (2.8) and (2.9) gives

$$\delta = \frac{4PL^3}{Ebd^3} \qquad (2.10)$$

Designs of this type require that the beam can sustain a certain force without deflecting more than a given amount, δ. The values of the force and deflection will of course vary from design to design. We will assume that in this design the length and breadth of the beam are fixed and that we can change the depth, d, (or thickness) of our design along with the material (which will give us different values for E, the stiffness, in equation (2.10). Clearly, by increasing the thickness and the stiffness of the material used, we will reduce the deflection. We are, however, seeking to achieve a certain deflection (set by the design) but to achieve this using the lightest beam possible. To do this we need to develop an expression for the mass of the beam from its volume multiplied by its density, ρ. The mass of the beam is given by equation (2.11).

$$M = bdL\rho \qquad (2.11)$$

In order to be able to proceed further with our analysis, we must eliminate the free variable (in this example, the depth (d)) from our expressions. This means that we do not have to solve our problem for many different designs. To do this we simply rearrange equation (2.10) in terms of d and substitute this into equation (2.11), giving equation (2.12). This then gives us the mass of the beam that will give us the required deflection for our chosen geometry. Equation (2.12) has been manipulated to give two bracketed terms: on the left-hand side showing the parameters fixed by the design (we can specify the force, deflection and geometry, so this term is constant) and on the right-hand side, the properties of the material:

$$M = \left(\frac{4PL^6 b^2}{\delta}\right)^{\frac{1}{3}} \left(\frac{\rho}{E^{\frac{1}{3}}}\right) \qquad (2.12)$$

It is this right-hand term that is of interest when searching for the best material. If we want to minimize the mass of the beam, then this term should be as small as possible. To choose the best material we simply compare values for this term for different materials. It is important to note that we are not just looking for a material with the highest stiffness and the lowest density. The particular geometry of the problem we have chosen has shown that picking a material with low density will have a greater influence on the mass of the beam than picking a material with high stiffness.

Table 2.8 compares $\frac{\rho}{E^{1/3}}$ values for different materials and indicates that carbon fibre will give us a slightly lighter beam than wood and that these materials will give us much lighter beams than all the others. As the first term in equation (2.12) is a constant, set by our design, it means that since the value of $\frac{\rho}{E^{1/3}}$ for carbon fibre is less than half that for a glass-fibre-reinforced polymer and less than one-fifth that for steel; a carbon fibre beam, which produces the same deflection as glass-fibre-reinforced polymer or steel ones, would weigh less than half or one-fifth of their masses respectively.

To calculate the actual mass of the beam for the material selected, we would simply substitute all the values for the terms into equation (2.12), and to calculate the thickness of the beam, to achieve the stiffness we seek, we would substitute values into equation (2.10). Since the $\frac{\rho}{E^{1/3}}$ term is proportional to the mass, we can derive the relative costs for beams made from the different materials by multiplying by the cost per unit mass. This is also shown in Table 2.8 where it is clear that, for a good balance between weight and cost, wood makes the best choice. The carbon fibre beam is more than 70 times as expensive as the wooden equivalent.

Material	Stiffness (GPa)	Density (kg m^{-3})	$\rho/E^{1/3}$	Cost per kg	Relative cost
Nylon	3	1100	763	7.5	5723
Wood (parallel to grain)	11	600	270	1.25	338
Aluminium alloy	70	2700	655	2	1310
Glass-fibre-reinforced polymer (GFRP)	40	2000	585	5	2925
Titanium	110	4500	941	50	47,050
Carbon-fibre-reinforced polymer (CFRP)	200	1500	256	100	25,600
Steel	210	7850	1321	0.5	661

Table 2.8 Comparison of materials for a light, stiff and low-cost beam

The final selection of the best material would depend on the application. Carbon-fibre-reinforced polymers have superseded wood for light, stiff (and more expensive) tennis rackets. Wood was initially used for plane wings but does not weather well (it can warp and catch fire easily). Aluminium replaced wood for modern, large planes and it is now being replaced with more expensive composites. The lighter composites offset their high materials cost through improved fuel efficiency over the lifetime of the aircraft. Carbon fibre composites are used widely in high-performance military aircraft where cost is not such a concern.

Material selection for a high-speed compact disc

We often need to design structures to be strong and light. Examples include dent-resistant car body panels and bumpers, lightweight pressure vessels (the bodies of planes and spacecraft) and spinning discs or blades such as those in a fan. In the design of spinning objects we would like them to spin as fast as possible, but are worried about the stresses generated due to the centrifugal force that might cause the disc or blade to fail. There is a need to find a balance between a material with low density, so that the stresses are low, and high strength, so that it can withstand high stresses before yielding or failing. The following example is for the selection of a material for a compact disc (CD) which spins at high speed, but it is relevant to any spinning disc and is very similar to the case for rotating fan blades.

In order to read data from CD drives more rapidly, there is a need to increase the speed at which CD drives operate. A 52× drive spins at 10,500 RPM (revolutions per minute) when reading the outer tracks, but needs to spin at 27,500 RPM to read the inner tracks at the same linear velocity. Most drives do not often reach these speeds and hence the data is read at a lower rate towards the centre of the disc (we will see why shortly). CDs are currently made from 120 mm diameter, 1.2 mm thick polycarbonate discs with a film of aluminium applied to the surface onto which the data is written. They could be made from any material, as long as an aluminium film can be applied to the surface. The radius of the disc should remain the same so that no modifications to the equipment for spinning and reading the disc are required.

The maximum radial tensile stress acting on a spinning disc is given by equation (2.13) where ρ is the density of the material, ω the angular velocity of the disc (in radians per second) and r the radius of the disc. It is evident that the maximum stress is independent of the thickness of the disc. For a polycarbonate disc, with a density of $1200\,\mathrm{kg\,m^{-3}}$, equation (2.13) indicates that the stress generated is roughly 9 MPa when spun at 10,500 RPM, increasing to 60 MPa when spun at 27,500 RPM:

$$\sigma_{max} = 0.42\,\rho\omega^2 r^2 \tag{2.13}$$

Reading the entire disc at high speed will generate stresses that will exceed the yield stress of polycarbonate (which is about 55 MPa) resulting in permanent deformation and could exceed the tensile stress of the disc (about 60 MPa), in which case the disc will shatter. Catastrophic disc failures at speeds above 48× have been recorded that carry the risk of damaging more vital (and expensive) parts of the computer or causing injury to the user.

To help understand the problem, it is useful to rearrange equation (2.13) to make the angular velocity the subject (since this is what we want to maximize) and split the equation into design terms that are fixed and materials terms that we can change. This yields equation (2.14):

$$\omega = \left(\frac{1}{0.42\,r^2}\right)^{\frac{1}{2}}\left(\frac{\sigma_{max}}{\rho}\right)^{\frac{1}{2}} \tag{2.14}$$

from which we can see that, to maximize the spinning speed, we need to maximize the term $(\sigma_{max}/\rho)^{1/2}$. If we assume that the maximum stress allowed is the yield stress of the material (or the failure stress for brittle materials) we can calculate this term for suitable materials and compare them. This information is tabulated in Table 2.9.

The mass of the disc is simply the volume multiplied by the density. If we assume that the disc thickness is the same, irrespective of the material used (it could be slightly thinner for stiffer materials but we would need to run a separate stiffness-based analysis to determine the exact changes in thickness that could be made), then the mass will simply vary with the density. The cost of the material for the disc will be proportional to the cost per kg multiplied by the density (giving cost per $\mathrm{m^3}$). This is also shown in Table 2.9.

Material	Yield strength (MPa)	Density (kg m^{-3})	$(\sigma_y/\rho)^{1/2}$	Cost per kg	Relative cost
Mild steel	220	7850	0.17	0.5	3925
Titanium alloy	830	4500	0.43	50	225,000
Aluminium alloy	400	2700	0.38	2	5400
GFRP	300*	2000	0.39	5	10,000
Polycarbonate	55	1200	0.21	5	6000
CFRP	650*	1500	0.66	100	150,000

* failure strength

Table 2.9 Comparison of materials, mass and cost for a compact disc

The table shows that CFRP would give us the CD capable of being spun the fastest. Comparing the values for $(\sigma_{max}/\rho)^{1/2}$ indicates we could spin it three times faster than polycarbonate (although the stress given for CFRP is the tensile stress – there is no plastic deformation in CFRP – and spinning above this speed would lead to fracture). CFRP is, of course, too expensive for the application (25 times more expensive than polycarbonate for the material alone) and it is not easy to manufacture CDs quickly and cheaply enough in CFRP. CFRP is, however, starting to replace titanium (the second best option) for fan blades in jet engines (essentially the same problem, but an application where the cost is not such an issue). Aluminium is a promising alternative to polycarbonate, and lower in cost, but a disc of similar size would be more than twice as heavy. This would require twice the energy to spin it at the same rate, which might require expensive upgrades to the motor, drive unit and bearings. A change in material is, however, needed if discs are to spin faster!

Material selection for small and light springs

We are often concerned with how we can store and recover energy, for example using a spring. There are many different types of spring, for example coil or helical springs, leaf springs (shown in Figure 2.17) and cantilever springs. While they can work in compression or tension and some springs are more efficient than others, the principle of energy storage is the same and is governed by the spring material's ability to store elastic energy.

Figure 2.17 Images showing (left) helical or coil springs and (right) a leaf spring

The elastic energy stored in a unit volume of material, W_v, is described mathematically in equation (2.15):

$$W_v = \frac{1}{2}\frac{\sigma_y^2}{E} \qquad (2.15)$$

where σ_y is the yield stress for ductile materials, or the fracture strength for a brittle material, and E is the stiffness.

To select a spring that stores the largest amount of energy for a given volume (or size), or the smallest spring to store a given amount of energy, a material with the largest value of $\frac{\sigma_y^2}{E}$ should be chosen. This equation can be further developed, to give equation (2.16), expressing the energy stored per unit mass (Wm in $J\,kg^{-1}$) by dividing the energy stored per unit volume (in $J\,m^{-3}$) by the density of the spring material (in $kg\,m^{-3}$):

$$W_m = \frac{1}{2}\frac{\sigma_y^2}{E\rho} \qquad (2.16)$$

Materials with high values of $\frac{\sigma_y^2}{\rho E}$ will produce lightweight springs. Values for both these terms are presented in Table 2.10, which reveals a number of interesting points. It shows that by changing the strength of the steel used (through its composition and through processing) the maximum stored energy increases (since the yield stress increases and the stiffness remains unchanged). High-strength (spring) steels are a good and common choice for small springs (surpassed only by rubber for energy storage per unit volume, but the use of rubber can be limited by the magnitude of the load that can be applied before it fails). Titanium springs are good and corrosion resistant but are expensive. For light springs, steel is no longer so attractive and titanium (used as fasteners in F1 cars), GFRP and CFRP (used in truck and high-performance car leaf springs) and polymers (used in cheap toys) become better choices. Ceramics could be used as springs; in fact, glass springs are widely used in scientific instruments, but they can only be used safely in compression owing to their tendency to fail at low stresses in tension if they contain defects or if they are damaged in use.

Material	Yield strength (MPa)	Young's modulus (GPa)	Density (kg m^{-3})	σ_y^2/E	$\sigma_y^2/\rho E$ ($\times$ 1000)
Mild steel	220	210	7850	230	0.03
Low alloy steel	690	210	7850	2267	0.29
Spring steel	1300	210	7850	8047	1.03
CFRP	650*	200	1500	2113	1.41
GFRP	300*	40	2000	2250	1.13
Titanium alloy	830	110	4500	6263	1.39
Aluminium alloy	400	70	2700	2286	0.85
Rubber	30*	0.05	850	18,000	21.18
High-density polyethylene	30	0.7	970	1286	1.33
Nylon	45	3	1100	675	0.61
Wood (parallel to grain)	55*	11	600	275	0.46

* failure strength.

Table 2.10 Comparison of materials for small and light springs

The examples presented give an overview of the approach taken to select a material for a given application where constraints such as cost and weight may be a concern. The methodology is, however, applicable to any number of different design problems.

More generally, we need to analyse the problem by obtaining expressions that describe the response we wish to achieve from our design (the bending or stretching) under a given environment (load, temperature). In most cases we then need to obtain a second expression for our constraint, for example the mass, as a function of the geometry and density of the material chosen. Using this simple analysis method (where we do not have the complete freedom to change the shape of our design) we must decide which one of the dimensions can be varied, so that we might achieve the desired response for any material selected.

To enable a comparison of materials without having to make hundreds of individual calculations, we need to substitute and rearrange our expressions to make the constraint the subject of the equation, eliminating the free variable in the process. By separating terms that are fixed by the design and hence will not vary with the material selected (geometry, force, temperature, magnitude of deflection, extension) and terms that will vary with the material chosen (strength, stiffness, density, cost) we can substitute data for different materials and rank them accordingly.

This method is a useful means of narrowing the field of candidate materials, considering the most important property required to achieve functionality. The final choice of material may require consideration of other important materials properties (and hence repetition of this process), the material's performance in its environment (its susceptibility to corrosion, for example), and how easily it can be manufactured to the desired shape.

Learning summary

By the end of this section you will have learnt:

✔ The material selection process for an engineering design requires a mathematical analysis of the problem, in combination with the use of relevant material property data;

✔ The generic method for this material selection process;

✔ That the method used is ideal for narrowing the field of candidate materials and that the final choice may require consideration of other factors such as cost and ease of manufacture.

2.5 Materials processing

Materials are selected for a particular application in a component as a result of an attribute (or combination of their attributes) of the type that has been discussed previously, such as strength, density, thermal conductivity or cost.

There is a very wide range of processes that take a material and convert it into a component of the required shape. The selection of a suitable manufacturing process depends to some degree upon the attributes of the material. However, we must also be aware that certain properties of materials can be changed (sometimes quite markedly) by the processing of the material itself. Sometimes, this change in properties is a byproduct of the process; for example, in bending a metal sheet to form a shape, the properties of the metal in the region of deformation will be markedly changed. In other cases, the process is conducted primarily to change the material properties; this is the basis of many processes where materials are heat treated.

Many of these changes in material properties are often controlled by changes in the material at the scale of the microstructure or at the atomic scale. It is thus very important for the engineer to understand the basic mechanisms that control these changes, since these will directly influence the attributes of the material.

When a component is being designed, the material and the processing route need to be considered at the same time. The properties of the component will be governed by the design and the material properties; however, the design and the choice of materials will constrain the choice of manufacturing route for the component. As such, the design, material and manufacture need to be considered concurrently as part of the design process; designs where material and manufacture are considered following the main mechanical design will tend to require more design iterations and will tend to be suboptimal designs. Thus, for optimal designs, we need to understand the capabilities of various manufacturing processes before we can opt to use a given material–manufacturing route solution.

Casting

Casting is a process where a material is melted to form a liquid, poured into a mould and then solidified such that the material takes the shape of the mould. Casting almost always refers to the production of metallic components, which may range in size from a few millimetres (e.g. electrical connectors) to tens of metres (e.g. a propeller for a ship). The solidified component needs to be removed from the mould; with expendable moulds, this is achieved by breaking the mould away from the component, and with reusable moulds, this is achieved by pulling apart a multipart mould and ejecting the component. Casting is generally used for making parts of complex shape that would be difficult or uneconomical to make by other methods, such as cutting from solid material. In its most advanced form, it is employed to make single-crystal turbine blades from nickel-based alloys for use in aeroengine turbines.

To achieve solidification, heat needs to be removed from the material, and this heat will be extracted through the walls of the mould; the rate of heat extraction will depend upon the thermal conductivity and heat capacity of the mould material, the size of the mould and how easily heat is transferred across the interface between the material and the mould. On solidification of metals, there is volume change, normally a contraction (see Underpinning Principles 2); this volume change needs to be taken account of when designing a casting so that the mould remains full during solidification and no cavities are formed in the structure. When a metal alloy is melted and then resolidified, the resolidification may be a complex process, leading to a non-uniform structure in the metal. For details, see Underpinning Principles 4.

Underpinning Principles 4: Phase diagrams and the solidification of alloys

Phase diagrams

In the casting of pure metals, solidification takes place at a given temperature (the melting point of the metal) and the microstructure consists of a single solid, normally in a polycrystalline form.

However, in practice, most metals that are cast are not pure metals, but are alloys containing more than one element. In this section, the concept of a phase diagram will be introduced.

Phase diagrams are based around the concept of solubility of components in each other. We are used to considering solubility of certain materials in water, such as salt and sugar. If the sugar–water system is considered, sugar and water are defined as the components. If pure water is taken and some sugar added, the sugar will dissolve in the water to form a solution known as syrup. This solution is known as a phase. As the amount of sugar is increased, all the sugar will dissolve, but at a certain point, no more will go into the solution and as further sugar is added it will not dissolve, but instead remain as sugar. At this point, there are now two phases present, pure sugar and syrup (with the syrup having the maximum amount of sugar that can possibly be dissolved in it at that temperature). The solubility limit of the sugar in water is dependent upon temperature.

So we can see that the phases present in a given system depend upon the temperature and the proportions of the components. Such information is commonly presented in the form of a diagram, known as a phase diagram. Figure UP 4.1 shows a phase diagram of the sugar–water system. The overall composition of the system is defined along the x-axis; the fraction of one of the components is defined (in this case in weight percentage terms) and thus the fraction of the other component is also defined (since they must both add up to 100 per cent).

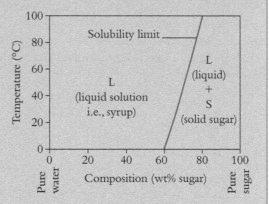

Figure UP 4.1 Schematic phase diagram of the sugar–water binary system

The sugar–water phase diagram (Figure UP 4.1) shows only two regions over the temperature range plotted. A specific combination of temperature and bulk composition defines a point on the phase diagram that will lie in one of these regions.

To the left is the region marked L; this is a region where there is a single phase. In a single-phase region, the phase present will be as indicated and it will have the composition of the bulk.

To the right is the region marked L + S; this is a region where two phases exist, namely solid sugar and syrup.

System analysis in two-phase regions

As we have seen, there are regions on phase diagrams where two phases coexist in the stable condition, known as two-phase regions. Figure UP 4.2 shows a section of such a diagram with two single-phase regions, the α-phase region and the β-phase region. Conventionally, solid phases are labelled with Greek letters or, if they are compounds, they are sometimes given the chemical formula of the compound; liquid phases are normally labelled 'L'. The names for the phases will mean different things on different diagrams, so do not be put off by this. Remember that they are just names, and that the way that we operate with them is always the same.

In Figure UP 4.2, the two single-phase regions are separated by a two-phase region, marked '$\alpha + \beta$'. Two-phase regions always contain the two single phases that surround the two-phase region when you move from any point within the two-phase region along a line of constant temperature (in Figure UP 4.2, the two single-phase regions are separated by a two-phase region, marked $\alpha + \beta$). To determine the nature of the two phases that are present in such a field, simply take a point within the field and travel along a line of constant temperature (a horizontal line) to both the left and the right. The phases present in your two-phase field will be those of the two single-phase fields that you encounter in this way.

In Figure UP 4.2, the point of interest is defined by a specific alloy composition C_0 at a specific temperature T. This point is defined by the intersection of the two dashed lines. It can be seen that this point lies within the two-phase region, marked $\alpha + \beta$, indicating that both of these phases will be present in the stable (equilibrium) condition.

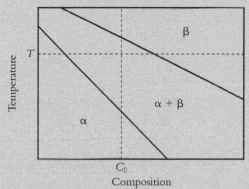

Figure UP 4.2 Generic section of a phase diagram with two single-phase regions (α and β) marked, along with a two-phase region ($\alpha + \beta$). The position of interest on the phase diagram, as defined by the temperature T and the composition C_0, lies within the two-phase region, $\alpha + \beta$

Determination of the compositions of phases present

To determine the compositions of the phases present in a two-phase region, a horizontal line (a line of constant temperature) is drawn from the point of interest to both of the surrounding single-phase boundaries. This line is known as the tie-line.

Where the tie-line intersects with the phase boundaries defines the compositions of the two phases present. In Figure UP 4.3, the compositions of the phases α and β at the temperature T are C_α and C_β respectively.

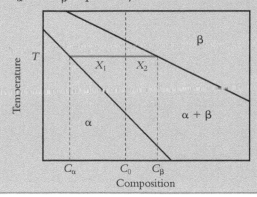

Figure UP 4.3 Section of phase diagram shown in Figure UP 4.2 with tie-line added

Determination of the proportions of the phases present

The overall composition of the material remains constant even though the material is now made up of two separate phases with their compositions C_α and C_β defined. The proportions of the phases present are such that the overall composition is preserved.

The tie-line construction is again employed to calculate the proportions of the phases present as follows:

The tie-line is divided into two sections, with the division taking place at the overall composition. In Figure UP 4.3, the two tie-line lengths are marked X_1 and X_2, where $X_1 = C_0 - C_\alpha$ and $X_2 = C_\beta - C_0$.

The fraction of the two phases (F_α and F_β) are defined as follows:

$$F_\alpha = \frac{X_2}{X_1 + X_2} \text{ and } F_\beta = \frac{X_1}{X_1 + X_2}$$

These are simply the ratios of the tie-line lengths to the left and right of the bulk composition as a function of the total line length. Note, however, that when evaluating the proportion of the phase with the single phase field to the *left* of the bulk composition, the ratio of line lengths used is that of the fraction of the tie-line to the *right* of the bulk composition. This construction is known as the Lever Rule.

Development of microstructure on passing through a two-phase region

Referring to Figure UP 4.4, the changes in phase compositions and proportions for an alloy of bulk composition C_0 as it is cooled from the β region to the $\alpha + \beta$ region can be understood. At T_β, the alloy is in the single β-phase region, with the composition C_0. As the temperature is dropped to T_1, the material has entered the two-phase $\alpha + \beta$ region. The compositions of the α and β phases are $C_\alpha^{T_1}$ and $C_\beta^{T_1}$ respectively. It can be seen that there is a high fraction of β phase at this temperature, given by

$$F_\beta = \frac{C_0 - C_\alpha^{T_1}}{C_\beta^{T_1} - C_\alpha^{T_1}}$$

(not surprising, given that the temperature is only just below that at which the alloy is in the single β-phase field). The fraction of the α phase is correspondingly given by

$$F_\alpha = \frac{C_\beta^{T_1} - C_0}{C_\beta^{T_1} - C_\alpha^{T_1}}$$

As the temperature is further reduced to T_2 and T_3, the composition of the α phase changes towards C_0 and its proportion increases. At the same time, the composition of the β phase changes from $C_\beta^{T_1}$ to $C_\beta^{T_3}$ and its proportion decreases.

As the temperature is further reduced to T_α, the alloy enters the single-phase α field, and as such only α with a composition C_0 is present.

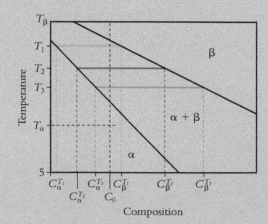

Figure UP 4.4 **Change of phase composition and proportion with temperature for an alloy of bulk composition C_0**

The eutectic

A binary phase diagram containing a eutectic is shown in Figure UP 4.5. The two components of the system are A and B, and the composition of an alloy is defined by the fraction of each of these (traditionally written as A − x wt%B, which indicates an alloy with x wt%B and (100 − x) wt% A).

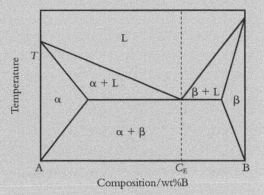

Figure UP 4.5 **Schematic diagram of a binary phase diagram with a eutectic**

There are three single-phase regions on the diagram. These are as follows:

α – a solid phase made up of A with a small amount of B dissolved in it to form a solid solution. The solubility of B into α depends upon the temperature.

β – a solid phase made up of B with a small amount of A dissolved in it to form a solid solution. The solubility of A into β depends upon the temperature.

L – a liquid phase which, at high temperatures, exists across the composition range from A to B.

Each of the single-phase regions is separated from other single-phase regions by a two-phase region. These operate in exactly the same way as before.

However, the diagram also contains a feature known as the eutectic. At the eutectic composition, marked C_E, the liquid can freeze to form a solid at a specific temperature without passing through a two-phase liquid–solid region. Apart from the pure components, A and B, it is the only composition for which this is true.

However, on freezing of a liquid of composition C_E, it does not freeze to form a single solid phase of the same composition, but it passes into a two-phase α + β region. The compositions of the two phases are found (as always), by drawing a tie-line to the boundaries of the single-phase regions. It can be seen that, in this case, the composition of the α and β phases formed are quite different. This requires movement of atoms so that one region (α) can become rich in A and another region (β) can become rich in B.

This motion of atoms occurs by diffusion (see Underpinning Principles 6, 'Diffusion'). Rearrangement and partitioning of atoms takes time, since the atoms need to move over macroscopic distances. To minimize diffusion distances, the alloy does not form into two discrete blocks of α and β as it solidifies from the liquid. Instead, it splits up into many more regions so that the distances that atoms have to move (diffuse) to be able to be in the right region thermodynamically are minimized. However, interfaces are regions of high energy, and a solution with many interfaces is far more unstable thermodynamically than one with less.

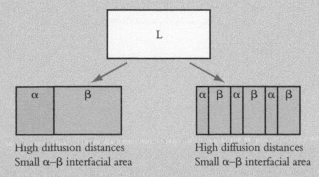

Figure UP 4.6 **Development of a eutectic microstructure on growth of an α–β lamellar solid from a liquid. In the α–β solid on the left, the diffusion distances are high, which makes this structure difficult to form kinetically. In the α–β solid on the right, the diffusion distances are much lower and thus a structure of this type is formed in practice**

There are many casting processes, and these tend to be classified by whether the mould is expendable or permanent, and whether the molten material is fed into the mould by gravity or under applied pressure. If a permanent mould is to be used, it needs to be designed such that the component will be able to be removed following solidification.

Casting processes

Sand casting

In sand casting, the mould is made from sand mixed with various binders so that it is able to hold a shape. A pattern is made from a material that is easy to work (such as wood); the pattern broadly defines the shape of the outer surface of the cast component although it may have features which allow for the addition of cores. The pattern is used to define the shape of a cavity in a (normally) two-part sand box. The sand box is filled and packed with sand around the pattern. The two halves of the sand box can be separated along the parting plane (often referred to as the parting line). Cores may be added following the formation of the primary shape if internal features are required; in Figure 2.18, a hollow cylinder is being cast, with the internal bore being created by the use of a cylindrical core. Note that features are required in the pattern to hold the core parts.

The liquid metal is poured into the cavity as defined by the shape of the mould and the cores. As the liquid metal fills the cavity, air needs to be displaced and to stop the air from preventing filling of the cavity, vents are provided. As the casting solidifies, the metal contracts and risers are provided in the system to keep the casting fed with liquid metal in order to prevent voids from being formed in the cast component. The casting generally solidifies slowly owing to the low thermal conductivity of the mould; the sand box is then opened and the sand removed (including the cores) to leave the cast component.

The production of moulds can be done manually, whereupon the capital costs of the equipment are low. However, manual production of moulds is time consuming, making the cost of labour (and other overheads) for the process high. If a higher rate of production is required (such as for the production of cylinder blocks for internal combustion engines), mechanized mould production is employed; the higher capital costs are offset by higher production rates.

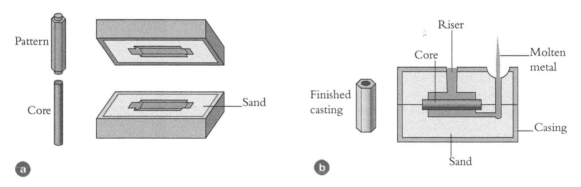

Figure 2.18 **Schematic diagram of the sand casting process**

Investment casting

Investment casting is an expendable mould process. It is used for components as diverse as jewellery and turbochargers, and it forms the basis of one of the most sophisticated casting processes operated in the casting industry today, namely casting of single-crystal internally cooled turbine blades.

A schematic diagram of the investment casting process is shown in Figure 2.19. Similarly to sand casting, a pattern is required. However, this pattern needs to be made of a wax material as

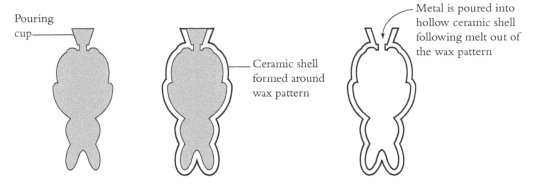

Figure 2.19 Schematic diagram of the processes in investment casting

it will not be removed as a whole once the mould has been formed around it; instead it will be melted out of the mould. The wax pattern is commonly injection moulded (see section on 'Injection moulding'), thus requiring a mould for the wax pattern to be made.

The mould is built up around the wax pattern by dipping the pattern into a slurry of fine ceramic particles suspended in water. Upon removal of the pattern from the slurry, the slurry on the wax is dried and forms a layer of ceramic powder around the pattern. This dipping-drying process is repeated may times to build up a thick layer; the outer layers are built up more quickly by the addition of coarser powder to the layer. The wax pattern is then removed from the mould by heating, and then the mould is fired at high temperatures, which causes the ceramic particles to bond together (called sintering – see section on 'Sintering'). The mould can now be filled with metal, and the component is released from the mould by breaking the mould away.

The mould in the investment casting process can be made of a highly temperature-resistant ceramic, and this investment casting can be used with high-melting-point materials such as steel and nickel alloys, as well as with low-melting-point materials such as copper and silver. As the pattern is removed from the mould by melting and the mould is removed from the component by fracture, the intricacy in the form of the castings produced in this way is almost limitless. Highly detailed surfaces can be cast, and the surface roughness is controlled by the nature of the first ceramic layers to be built around the pattern; use of a fine ceramic slurry yields very smooth surfaces.

A modern internally cooled turbine blade has an internal labyrinthine channel within it to allow it to be internally cooled. This channel has to be cast in during the investment casting process, and this is achieved by placing a ceramic core in the mould. The wax pattern is injection moulded around the ceramic core and the core is thus incorporated into the final mould upon removal of the wax. The mould is filled with liquid metal inside a furnace, and, by careful imposition of temperature gradients during cooling, along with seeding, the multiple nucleation of crystals can be avoided, yielding a blade which is a single crystal (i.e. with no grain boundaries throughout the component). This expensive process is employed since the single-crystal product is required to resist creep under the high loads and temperatures observed in service.

Permanent mould casting processes

In permanent mould casting, the mould is referred to as a die. The die is made up of multiple parts (two as a minimum) which can be mechanically assembled to leave a cavity, and mechanically disassembled following casting to allow removal of the cast component. The die system may have ejector pins embedded in one of the sections of the die which push the component away from the die following solidification.

The moulds are normally made of a high-melting-point material so that the molten metal which is to be fed into it does not cause rapid degradation of the mould itself; cast iron, steel and graphite are common mould materials. The moulds are normally expensive, but this is offset by the fact that the mould can be reused many times. To avoid rapid mould degradation, permanent mould casting is normally limited to casting of alloys of aluminium, zinc, magnesium and lead.

The two main forms of permanent mould casting are gravity die casting and pressure die casting (often referred to simply as die casting). In gravity die casting, the molten metal is simply poured into the mould to fill the cavity; gas pressure or vacuum may be used to aid mould filling. In pressure die casting (Figure 2.20), liquid metal is forced into the die under pressure (typically between 10 MPa and 170 MPa) from a ram which ensures good mould filling, and good contact between the component and die during solidification (resulting in high solidification rates). In both cases, air leaves the mould through gaps between the die sections.

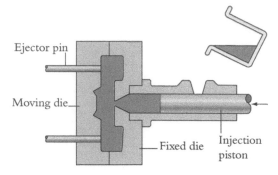

Figure 2.20 Schematic diagram of pressure die casting process

Both forms of casting result in high-quality surfaces (replicating the surface of the mould well) and good dimensional accuracy. Die casting is able to produce components with fine surface detail with component thicknesses of between 0.75 mm and 12 mm. The production rates in permanent mould casting are high (especially for die casting) and this offsets the high cost of both the capital equipment and the tooling.

Other processes

Many other variants on these processes exist, such as squeeze casting (a hybrid process between forging and casting), centrifugal casting (where rotation of the mould is used to cause mould filling) and continuous casting (for casting of raw material on a continuous basis). Whatever the specific details of the casting process, all processes are controlled by the flow of the liquid metal and the rate of heat extraction from the material through the mould.

Castability

The castability of a metal is the ease with which that metal can be cast. Generally, the following increase the castability of a metal:

- low viscosity and surface tension: these allow fine detail and complex shapes to be cast
- low solidification contraction: this reduces the tendency to form voids and cracks in the casting
- low solubility for gases to avoid porosity

In addition, the following increase castability in permanent mould processes:

- low melting temperature: this enhances mould life in permanent mould processes
- low thermal capacity and high conductivity: this promotes high production rates

Aluminium has one of the larger solidification contractions at 7.1 per cent, with steels exhibiting solidification contractions typically between 2.5 per cent and 4 per cent. In contrast, grey cast iron exhibits a solidification expansion of 2.5 per cent.

Casting defects and design of castings

Casting is a process which depends on fluid flow of the liquid metal and heat flow in solidification. Knowledge of these together with the solidification properties of the metal allow moulds to be designed by sophisticated computer modelling.

Materials and processing

Common casting defects:

- Misrun: the liquid did not fill the cavity before freezing.
- Cold shut: two flows within the casting come together but do not mix properly as they were partially solidified upon contact.
- Washout: in sand casting, turbulent flow has caused a part of the sand mould to wash away into the liquid metal, leading to sand inclusions in the casting and loss of integrity of the component shape.
- Porosity: a high solubility for gas in the liquid results in the formation of gas pores (bubbles) in the casting on solidification when the gas solubility is significantly reduced.
- Shrinkage cavities and cracking: owing to the high solidification shrinkage, the large temperature range over which freezing occurs and the high coefficient of thermal expansion.

Although the design of castings is now performed by modelling, certain basic rules still apply.

- The casting should be evenly distributed around the parting plane(s).
- The pattern must be removable from mould or, in permanent mould casting, the mould from casting. As part of this, a draft angle should be incorporated to allow ease of removal of the cast component.
- Shrinkage must be allowed for in mould design. An allowance needs to be made on final dimensions for contraction after solidification (patterns are thus made using shrink rules).
- Avoid rapid changes in section since these cause turbulence.
- Avoid changes in section since this causes solidification times to extend in thicker sections, possibly leading to shrinkage cavities – see Figure 2.21(a).
- Avoid enclosed sections solidifying last. The last material to solidify is usually associated with the formation of voids. This can be achieved by the use of chills in the mould which enhance the cooling rate in that area – see Figure 2.21(b).

Figure 2.21 Simple design changes to avoid shrinkage cavities forming in cast components: (a) avoidance of section thickness change; (b) use of a chill to avoid the thickest section solidifying last

Learning summary

By the end of this section you will have learnt:

✔ Casting is the formation of shaped components involving filling a shaped mould with a liquid and solidifying it;

✔ Casting is normally only employed for metallic components;

✔ Many casting methods are available;

✔ Casting may be a slow manual process or a fast automated process;

✔ The main phenomena which control casting are fluid flow, heat transfer and solidification behaviour of the liquid;

✔ Certain rules need to be followed to reduce the possibility of casting defects.

Moulding

Moulding has some similarities to casting in that a mould is used to confer the shape of a component to be manufactured. Moulding refers to the shaping of a viscous material (as opposed to casting, which deals with a liquid). Moulding is a term generally applied to the processing of polymers and glasses.

Glasses and thermoplastic polymers have attributes which make them suitable for moulding. In these cases, the material can be made viscous by heating; the temperatures tend to be relatively low for the most thermoplastic polymers, and somewhat higher for glasses. The viscous material needs to be forced into the shape of the mould, and a number of methods can be used to do this. Once the material has taken up the shape of the cavity, heat is extracted from the part, causing it to increase in viscosity (to the point where it can be considered a solid), whereupon it can be extracted from the mould.

Thermosetting polymers can also be moulded, although in this case, the forming of the solid component is not achieved by the extraction of heat from the part (as described above) but by a chemical reaction at an elevated temperature.

Moulding processes

Press moulding

In this process, a gob of viscous material (generally made viscous by heating elsewhere) is placed into the lower die surface and an upper die is brought down, causing the material to flow. This is commonly used to form glassware, etc. Figure 2.22 shows a schematic diagram of a press-moulding operation.

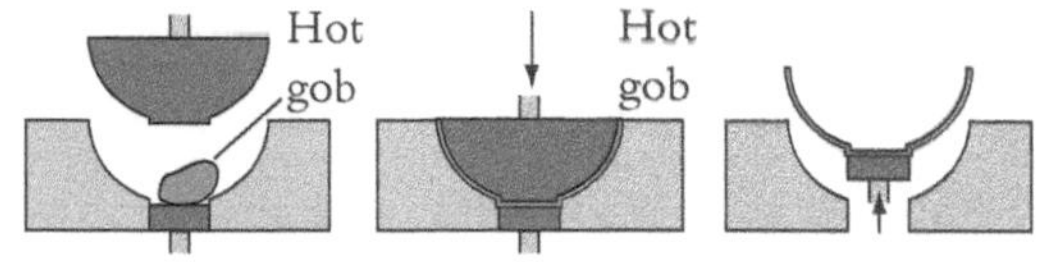

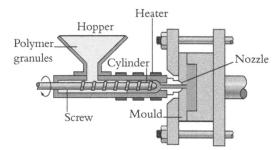

Figure 2.22 Press moulding of a viscous gob of polymer or glass to form a shaped object

Injection moulding

Similar to die casting, a viscous polymer is forced into a split cavity. The polymer is fed into a screw in the form of pellets, and this polymer is heated by (i) external heaters on the cylinder; (ii) rotation of the screw which causes deformation heating. The screw can also act as a plunger in the cylinder, and when the screw is translated along the cylinder, the viscous polymer is injected into the mould cavity. The cavity is normally made of metal and is generally water cooled to aid heat extraction from the part. The mould is held closed for a period after injection of the viscous polymer for the part to have heat extracted from it and thus to gain strength. The cavity is then opened, and ejector pins force the part out before a new cycle starts. The cavity needs to be constructed in such a way that it can be disassembled (as part of the automatic process) to allow the component to be removed. Figure 2.23 shows a schematic diagram of the apparatus required for injection moulding.

Figure 2.23 Apparatus for injection moulding of polymeric components

Extrusion

Polymer components can be moulded by extrusion. In a similar way to that of polymer injection moulding, the polymer is again fed into a screw in the form of granules. The polymer is made viscous by the action of the screw and also by external heating. In contrast to injection moulding, this is a continuous process which does not involve translation of the screw in the barrel. The viscous material is forced through a shaped orifice, and the component cross-section takes up the shape of the orifice cross-section, forming a prismatic product. In the case of polymer extrusion, the product is commonly fed into a waterbath as it leaves the extrusion orifice to remove heat, thus conferring strength to the product, enabling it to be handled. Figure 2.24 shows a schematic diagram of a polymer extrusion apparatus.

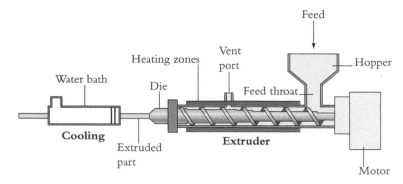

Figure 2.24 Apparatus for extrusion of polymeric components

Materials and processing

Blow moulding

Since moulding deals with a viscous solid, the solid can be inflated with pressurized gas to take up the shape of a cavity. This would be the most common way to form a polymeric bottle. A parison is formed, normally either by injection moulding or by extrusion. The parison is then inflated to form the component.

Figure 2.25 shows a schematic diagram of an extrusion–blow-moulding operation. A tube is formed by extrusion which is clamped into a split mould which seals the lower end. The upper portion is then inflated with gas, stretching out the polymer to the shape of the cavity. Although thickness variations are quite coarse across the component, this is a very cost-effective process.

Glass containers are also commonly formed by press-and-blow methods. Here a viscous gob of glass is pressed to a shape in a mould. It is then taken to a second mould, whereupon it is blown out to the shape of the second cavity to form the container.

In addition to three-dimensional components, polymer film (such as is commonly used in packaging, etc.) is generally made by a continuous blow moulding operation. Here, there is no mould as such, but instead the extruded tube is inflated out to form a balloon, the size of which is limited by a sizing basket. The balloon travels continuously upwards until it has cooled enough for the film to be collapsed without welding together. The bubble is then collapsed, the edges trimmed off both sides (to make two layers of polymer film out of the tube) and the film wound up.

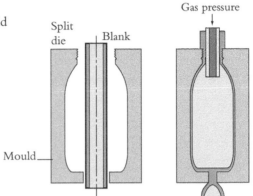

Figure 2.25 Schematic diagram of a extrusion–blow-moulding operation

Rotational moulding

Rotational moulding is commonly used to form large hollow components from thermosetting polymers. A two-part mould is taken and opened (see Figure 2.27). Polymer granules are fed into the mould and the mould heated from the exterior to allow the polymer to become viscous. The mould is then spun (often with a relatively complex motion) to ensure that the polymer coats all of the internal walls. The mould walls are cooled, and the mould opened to allow extraction of the component. The cost of the capital equipment is relatively low as no pressure or force is required to form the components. The dimensional accuracy of such components is generally poor, but the costs are low.

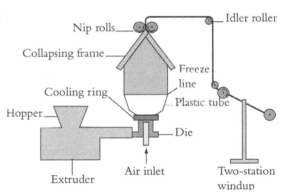

Figure 2.26 Schematic diagram of a blow moulding plant for production of polymer film

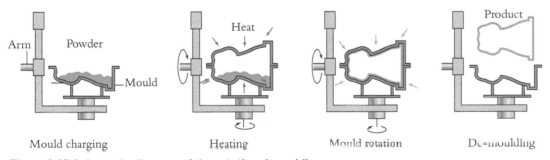

Figure 2.27 Schematic diagram of the rotational moulding process

Moulding of thermosetting polymers

Unlike thermoplastics, the change from solid to viscous melt and back again cannot simply and repeatably be made by heating and cooling of thermosetting polymers. For thermosetting polymers, the application of heat to an unpolymerized precursor initiates the polymerization reaction, whereupon the material solidifies. The application of heat to the polymerized solid does not, however, result in reduction in viscosity but instead to charring and burning of the polymer (at high enough temperatures).

In moulding operations involving thermosetting polymers, the polymer itself is heated in a shaped die (cf. moulding of thermoplastics where the heat is *extracted* from the viscous polymer via the die). The unpolymerized material fills the cavity and takes its shape, and also polymerizes in the cavity under the action of the heat. There are a number of methods for moulding thermosetting resins that are similar but differ in their specific details. One of the methods (compression moulding) is shown schematically in Figure 2.28.

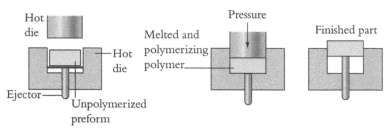

Figure 2.28 Schematic diagram of compression moulding operation for thermosetting polymers

Fibre forming

Fibres of polymers are formed by a range of methods (dependent upon the properties of the polymer). Many can be extruded as a melt with low viscosity through a plate with many holes in it (called a spinneret). The fibres are then cooled to impart strength. They can be drawn to result in molecular orientation and thus to increase strength and stiffness.

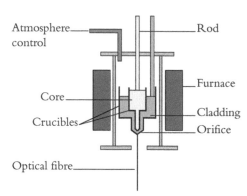

Figure 2.29 Drawing of glass fibre using the double-crucible method

Glass fibre is most commonly employed as optical fibre, employed for rapid data transmission. To function, the glass fibre needs to keep the light in the fibre and not lose it from the surfaces; to promote this, the fibre is made of a core glass (with high refractive index) and a sheath glass (with a lower refractive index). A fibre of this type can be formed in a number of ways. One way is known as double-crucible drawing; two crucibles (containing high-purity charges of the two types of glass) are placed concentrically. A fibre is drawn from the inner crucible and glass from the outer crucible is then drawn along with it to form the sheath.

Changes in properties of thermoplastics during deformation

Polymers are made up of chain-like molecules (see Underpinning Principles 1), and during drawing (stretching) of thermoplastic polymers, these chains may become aligned in the draw direction.

Drawing of polymers in this way increases both the strength and the modulus of the material in the draw direction, but reduces ductility in the draw direction. However, these effects can be reversed by annealing following drawing; here, the polymer is heated, allowing chains to slip, decreasing alignment. Sometimes annealing can occur while a polymer is being drawn if the draw temperature is high; in this case, the effect of chain alignment on mechanical properties following slip will not be observed.

The drawing of thermoplastic polymers results in a directionality in mechanical properties; this directionality is known as anisotropy.

Deformation processing

Deformation processing takes a solid metal as its feedstock and deforms the solid to shape in a variety of ways. Deformation processing is limited to metals, as a good level of ductility is required to make materials suitable for this process. The metals can be processed at a range of temperatures. The yield strength of all metals decreases as the temperature is increased; as such, deformation (taking a metal above its yield point) at elevated temperatures requires less force and processing equipment can be less robust and thus less expensive. The disadvantage with elevated temperature processing is that dealing with hot metal is more difficult than dealing with cold metal.

In addition to the processing characteristics that are governed by the metal temperature, the final microstructure (and therefore properties) of the material are governed by the metal temperature during processing.

Underpinning Principles 5: Annealing of metals

As we have seen in Underpinning Principles 2, metals are made up of atoms that are packed in regular fashion; this ordered material is described as being crystalline. A bulk crystalline material is normally made up of small regions of these crystals, randomly oriented with respect to each other to make up the whole body. Each of the regions is termed a grain. In the regions between grains, the atoms do not pack together so well; these regions are called grain boundaries. The crystalline bonding in the grain boundaries is not perfect, so grain boundaries are regions of high energy within the structure. Dislocations (see Underpinning Principles 3) are also regions where the bonding is not perfect, and these two are high-energy features within the structure.

Thermodynamically, materials would like to move towards their lowest energy state and, microstructurally, this implies removal of dislocations and removal of grain boundaries. For most metals at room temperature this will not occur, since these processes require movement of atoms by diffusion, which is too slow at these temperatures. However, at elevated temperatures, atomic movement by diffusion can be significant (see Underpinning Principles 3).

Cold-worked structures (materials with high dislocation densities) have a strong driving force to reduce their internal energies by reducing the density of dislocations. The rearrangement of the atoms means that the dislocations can be annihilated as a completely new grain structure forms (recrystallization). When metals recrystallize, their grain size is reduced and the dislocation density decreases.

There are three main microstructural processes that may occur during annealing, which are illustrated schematically in Figure UP 5.1.

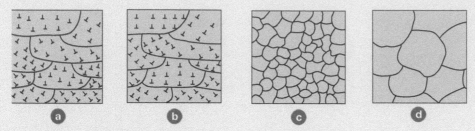

Figure UP 5.1 **Illustration of processes occurring in annealing of a cold-worked metal: (a) cold-worked; (b) after recovery; (c) after crystallization; (d) after grain growth**

Recovery

Recovery is a relatively low-temperature process. Dislocations move to form a polygonalized structure, driven by the reduction in strain energy associated with this structure. However, dislocation density is not significantly reduced and mechanical properties remain virtually unchanged. The only significant change is that residual stress in the component is reduced or eliminated.

Recrystallization

Recrystallization takes place at higher temperatures. This process (recrystallization) involves nucleation and growth of new grains within the structure, and as these new grains are formed, dislocations are eliminated. Following recrystallization, the material will have a lower strength due to removal of work strengthening, but this will be compensated for by grain-size strengthening. The ductility increases significantly following recrystallization. A rough estimate of the temperature at which recrystallization occurs is $0.4\,T_m$ (where T_m is the melting point of the metal in Kelvin).

Grain growth

Following recrystallization, grain growth may occur where grain boundaries are eliminated to reduce further the energy stored in the structure. As a result of grain growth, both strength (see Underpinning Principles 3) and toughness are reduced.

Deformation processes

Rolling

In rolling, a material is taken and passed between rotating rolls. It is commonly employed for taking material from ingot to plate or sheet, but can also be employed to produce shaped material, such as tube. Figure 2.30 shows a schematic diagram of a rolling process to reduce the thickness of a metal. The reduction in thickness per pass is limited by machine capacity and material properties; higher yield stresses and higher reductions require higher machine power. To achieve high reductions, material is commonly passed through a rolling mill a number of times, with each pass resulting in a small reduction in thickness.

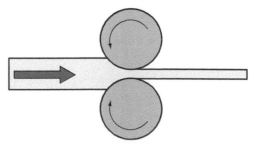

Figure 2.30 **Schematic diagram of the rolling process**

Materials and processing

Forging

Forging is the deformation of metal between dies (and is akin to press moulding). Here, dies are pressed against the material to change its shape. The dies can be shaped or plane. Forging can be used simply to change the shape of material stock or to form it into a shaped component. Figure 2.31 shows a schematic diagram of a closed die forging operation to produce a shaped component.

Extrusion

We have seen in the section on moulding, exclusion is a process for extrusion of thermoplastic polymers into shaped components with constant cross-section. The process for extrusion of metals is very similar. However, unlike the extrusion of polymeric components, the process is not continuous, since here the feedstock material is a block of metal rather than thermoplastic pellets. The metal is forced through a shaped die by the application of a force to a billet of material by a ram which forces it through the die. Variants exist for the process, such as direct and indirect extrusion. Figure 2.32 shows a schematic diagram for a direct extrusion process.

Drawing

Sheet metal is commonly drawn to form components. Drawing involves change of shape and, commonly, change of thickness of the sheet as part of the process. Beverage cans are typically made by a complex drawing process from sheet material. Figure 2.33 shows a drawing process to produce a cup from a sheet of material. The pressure pad and punch both exert forces on the sheet, causing it to deform in a controlled manner.

Effect of deformation on material characteristics

Grain shape and dislocation density

When a metal is deformed, its grains change shape. This is illustrated schematically in Figure 2.34. This may lead to anisotropy, i.e. the properties of the material are different in different directions. However, whether this deformed grain shape survives into the product depends largely upon the temperature at which the material is being deformed. At high temperatures, recrystallization may occur (see the next section).

Deformation of metals results in increase of dislocation density in the structure (see Underpinning Principles 3). However, whether these dislocations survive into the product depends largely upon the temperature at which the material is being deformed. At high temperatures, dislocation annihilation may occur (see the next section).

Hot and cold deformation

In Underpinning Principles 5 (Annealing of Metals), we can see that there are various stages of the annealing process. In the recrystallization process, the dislocations are annihilated, as a new, fine and equiaxed grain structure grows through the material. This recrystallization stage is of great significance in deformation processes. In Underpinning Principles 5, we define the recrystallization temperature as $\sim 0.4\ T_m$; however, we define some different limits when we refer to recrystallization in deformation processing.

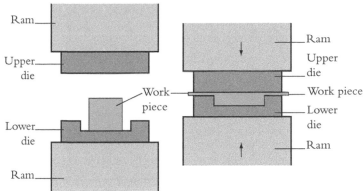

Figure 2.31 Schematic diagram of a closed die forging operation to produce a shaped component

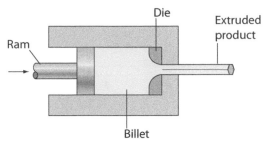

Figure 2.32 Schematic diagram of a direct extrusion process for metals

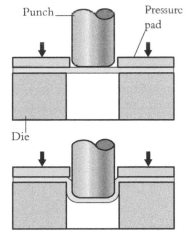

Figure 2.33 Schematic diagram of the drawing process

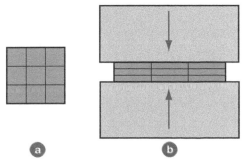

Figure 2.34 Schematic diagram of the change of shape of grains during a deformation (forging) process

If a process is described as 'hot' (such as hot rolling, hot extrusion, etc.), then this indicates that annealing during the process is significant. A process is termed 'hot' when the temperature is greater than $0.55\,T_m$ (where T_m is the melting point of the metal on the absolute (Kelvin) temperature scale). At this temperature, recrystallization is very fast, and we can be sure that a material will fully recrystallize in a short period under these conditions.

The annealing during the deformation process results in the dislocations that were formed being annihilated as the material recrystallizes. The recrystallization also removes the anisotropy of the grain structure, and the final grain structure is equiaxed (similar in all directions) and fine in scale. This is said to be a refined microstructure. Owing to the continual recrystallization during the processing, the amount of deformation in the process itself is almost unlimited, and ductile fracture will not occur.

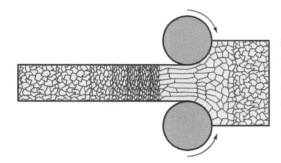

Figure 2.35 Schematic diagram of the development of grain shape during hot rolling. Grain elongation is observed as the material is deformed, but then the material recrystallizes to a very fine grain structure; with time, some grain growth is also observed (here, as the material moves away from the rollers)

Because of the annihilation of dislocations during the recrystallization process, the final material exhibits high ductility and low yield strengths.

If a process is described as 'cold' (such as cold rolling, cold extrusion, etc.), then this indicates that annealing during the process is insignificant. A process is termed 'cold' when the temperature is less than $0.35\,T_m$. Below this temperature, we can be sure that no recrystallization will occur.

In this case, the dislocations generated by the deformation will be retained in the final product. As such, the product will have high yield strength and low ductility (due to dislocation strengthening). Moreover, during the processing, the amount of deformation before fracture is limited because damage is continuously accumulated.

The microstructural anisotropy (i.e. the elongated grain structure) as seen in the previous section will also be preserved in products formed by cold deformation processes.

The intermediate temperatures ($0.35\,T_m < T < 0.55\,T_m$) are normally referred to as 'warm'; in this interval and in the timescales of the deformation process, recrystallization may occur to some extent, but may not be complete.

Residual stress

Deformation (especially cold deformation) can lead to residual stresses in a component. These are often useful. Residual stresses cannot be uniform through a component since they must balance in the end.

- Rolling may induce a compressive residual stress in a plate.
- If the surface of such a plate is machined, then the plate must distort to redress the stress and movement balance.
- Residual stresses also affect the ability of a component to carry load. The stresses are summative and yielding may occur before expected.
- Components subjected to fatigue loading may be made resistant to fatigue by application of a residual compressive stress in the surface. Shot peening can be used to achieve this (cold deformation on a micro scale).
- Residual stresses can be eliminated by annealing or by further deformation. Both may give shape change and require a machining allowance.

Learning summary

By the end of this section you will have learnt:

✔ Metals are commonly deformed to form shaped components;

✔ A wide range of specific methods are available;

✔ Deformation can be conducted hot or cold;

✔ Deformation produces an increase in dislocation density in the metals;

✔ The final properties of the material will depend critically upon the temperature at which the metal is when it is deformed;

✔ If the material being deformed is 'hot' (> 0.55 Tm), the dislocations will be removed by in-process annealing, resulting in a fine recrystallized structure;

✔ If the material being deformed is 'cold' (< 0.35 Tm), the dislocations will be not be removed and the increased dislocation density will result in a stronger but less ductile material;

✔ Deformation of metals to shape may result in residual stresses in the components.

Powder processing

The last of the main classes of primary shaping process is that of powder processing. Here, material in the form of a powder is taken and mixed with a binding agent, causing the particles to bond. The powder can be formed into shapes in a number of ways but the particles remain as individual particles, and thus such components have limited mechanical strength. At this stage in the process, the component is referred to as a *green body*. The green body is heated (a process called sintering) to bond the individual particles into a three-dimensional structure. The bonds form in the solid state by a process called diffusion (see Underpinning Principles 6).

The use of a powdered substance for the raw material for the process allows the material to be in a state where it can easily flow without being hot. Thus, powder processing is a very suitable forming technique for materials with very high melting points (where casting would be difficult), and for very strong, low-ductility materials (where deformation processing would not be suitable). Since it is a process where the components are formed close to net shape, then it can be a much more economically favourable option than machining components from stock.

Underpinning Principles 6: Diffusion

Diffusion is the process by which there is long-range movement of atoms within a structure. Atoms will move down a concentration gradient (or more accurately down an *activity* gradient). The driving force for their movement is to reduce the overall energy of the system. This is normally observed in reduction of stresses or strains, and removal of high-energy features in a structure such as dislocations or grain boundaries.

Diffusion is the physical movement of atoms and as such takes time. A schematic diagram of a diffusional movement is shown in Figure UP 6.1 for both substitutional and interstitial atoms. In both cases, the atoms marked in green are changing places in the structure, but in doing so require other atoms to be

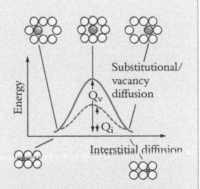

Figure UP 6.1 Activation energy barrier associated with diffusion of both substitutional and interstitial atoms

displaced from their preferred position. This displacement requires energy, and the energy to make this displacement is known as the activation energy, Q. In systems where Q is large, diffusion is restricted, whereas diffusion is easier in systems with low values of Q.

The energy that an atom may need to displace its neighbours to allow it to move past them comes from the thermal energy of the atom, associated with its temperature. Atoms at higher temperatures are more likely to be able to make the jump in position than the same atoms at lower temperature.

A rough estimate can be made as to how far, x, an atom might diffuse in a time, t, as follows:

$$x \simeq \sqrt{Dt}$$

where D is a term known as the diffusion coefficient. The diffusion coefficient is strongly dependent upon temperature as follows:

$$D = D_0 \exp\left(-\frac{Q}{RT}\right)$$

where D_0 is a constant for the system, Q is the activation energy for the process, R is the gas constant ($8.314\,\mathrm{J\,mol^{-1}\,K^{-1}}$) and T is the absolute temperature (the temperature measured on the Kelvin scale).

The diffusion coefficients for many diffusional processes in metals rise rapidly with temperature.

Calculate the self-diffusion distance for aluminium for 15 min at room temperature (20°C), 200°C and 400°C (remembering to use units of Kelvin in these calculations). Present these numbers both in terms of metres and also in terms of numbers of atomic diameters of aluminium (the atomic diameter of aluminium is approximately 250 pm (i.e. 250×10^{-12} m)). Comment on these in relation to the annealing process.

For aluminium alloys, $D_0 = 1 \times 10^{-5}\,\mathrm{m^2\,s^{-1}}$ and $Q = 135\,\mathrm{kJ\,mol^{-1}}$.

If the diffusion coefficients D are known, the diffusion distance x can be estimated from the equation

$$x \sim \sqrt{Dt}$$

knowing that 15 minutes = 900 seconds.

Using the equation

$$D = D_0 \exp\left(-\frac{Q}{RT}\right)$$

The diffusion distance at 200°C is given by

$$D = 1 \times 10^{-5} \exp\left(-\frac{135,000}{8.314 \times (200 + 273)}\right) = 1.23 \times 10^{-20}\,\mathrm{m^2\,s^{-1}}$$

The diffusion distance is thus

$$x \sim \sqrt{Dt} = \sqrt{1.23 \times 10^{-20} \times 900} = 3.3 \times 10^{-9}\,\mathrm{m} \approx 13 \text{ atomic diameters}$$

At 400°C, the diffusion coefficient is $3.32 \times 10^{-16}\,\mathrm{m^2\,s^{-1}}$ (almost 30,000 times larger than at 200°C) and thus $x \sim 5.5 \times 10^{-7}\,\mathrm{m}$ (~2200 atomic diameters).

Annealing involves rearrangement of atomic structure; at 200°C, the rearrangement is limited since the atoms won't diffuse far enough. However, at 400°C, the diffusion distances are much larger and significant annealing will occur in the time interval indicated.

Materials and forming processes

Powder processing is used for three main classes of material, namely claywares, engineering ceramics and metals. Claywares are generally processed by different routes from the other two materials.

Claywares

Clay is commonly used to form a wide variety of components, such as bricks, tiles, tableware and sanitaryware. Clay is a naturally abundant raw material and is generally composed of oxides and hydroxides of aluminium and silicon. When mixed with the right amount of water, it can form a solid mass that can easily be deformed, but will hold its shape. As such, this material can be formed into shape by a number of the processes that are commonly used for moulding, such as pressing or extrusion.

Moreover, if the water content of the clay is raised, the material can be formed into a liquid, known as a *slip*. As we saw in the section on casting of metals, liquids can be employed to make intricate components with fine detail. However, the process must incorporate a mechanism by which the liquid is returned to a solid. In the casting of metals, this is achieved by removal of heat, but in the case of slips, the material is returned to solid form by removal of water. Figure 2.36 schematically shows the formation of a thin-walled structure by slip casting. The slip is poured into a highly porous mould; the mould absorbs water from the slip close to it, and in doing so reduces the water content of that slip so that it forms a solid shell. The shell will thicken with time as more water is drawn into the mould; once the desired thickness is achieved, the mould is inverted and the remaining slip removed.

Both of these processes result in green bodies that have adequate strength for handling. Different green bodies can be joined together by the use of a slip as an adhesive agent.

The green bodies will be dried (often in a heated oven) before sintering.

Metals and engineering ceramics

Many high-melting-point metals (such as steels) and engineering ceramics (such as alumina, silicon nitride, etc.) are commonly processed by powder processing. Tungsten wires for the filaments of incandescent light bulbs, steel connecting rods for internal combustion engines and zirconia discs for 'quarter-turn' tap valves are examples of components commonly made by powder processing. For these types of material, the powder particles are mixed with a small amount of temporary binding agent, chosen such that it does not react with the powders during further processing; for metals, soaps are commonly employed as these also provide lubrication during processing.

The powders and binding agents are pressed into a shaped die under high pressures, resulting in high-density green components (see Figure 2.37); higher pressures result in higher densities (and less shrinkage during the sintering stage). Also, pressing operations will be designed to minimize differences in porosity throughout the green body; differential porosity will lead to uneven shrinkage in sintering and thus to distortion. As in other moulding operations, the die needs to be designed to allow easy removal of the green component.

Slip poured into mould

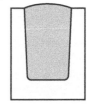

Draining mould

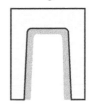

Figure 2.36 Slip casting of a thin-walled structure into a porous mould

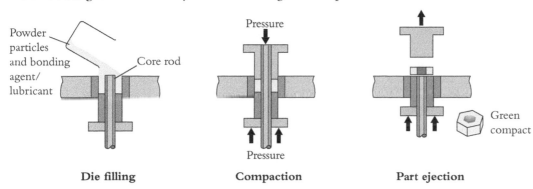

Figure 2.37 Schematic diagram of a powder compaction process

Sintering

Green parts are robust enough to be handled, but are still composed of individual particles that are not bonded to each other. To allow the individual particles to bond to each other to form a three-dimensional network, the green parts are raised to an elevated temperature to allow diffusion to occur (see Underpinning Principless 6). In the initial stages of heating, the cycle is controlled to allow the removal of any binder phases (such as water or soap), before being raised to the sintering temperature.

In claywares, some of the components fuse at elevated temperatures to form a liquid which flows around the other solid particles. On cooling of the part, this liquid (which is a small fraction of the volume of the component) solidifies and bonds the whole structure together. This process is known as liquid phase sintering.

For metals and engineering ceramics, no liquid phase forms during the sintering process. On heating, the diffusion rate increases, allowing atoms to move toward the point of contact between two particles and form a neck between them; the two particles thus bond and form part of a three-dimensional solid network (see Figure 2.38). The movement of atoms is driven by a reduction in energy of the system; surfaces are regions of high energy in a system (the bonds between atoms are not well satisfied at a surface), and the amount of surface can be reduced by forming necks between particles. As sintering proceeds, the product will further reduce the amount of surface area, and porosity will shrink and be eliminated.

Sintering can take place either in a batch furnace, where components are loaded onto shelves or trays, or green components can be placed on a conveyor belt which moves slowly through a long furnace. Sintering may need to take place in protective atmospheres to avoid oxidation of components of the product during the sinter cycle.

Compacted product   Partly sintered product

Figure 2.38 Schematic diagram of atomic diffusion during neck formation in sintering

Machining and cutting

Unlike the processes discussed so far, machining is used to form a shape out of a larger body by material removal. There are many processes that fit under this heading, but all involve a tool that is moved relative to the work, which results in material removal from the work. The most common of these processes are drilling, milling, turning and grinding.

In the drilling process, the work is normally held still while the tool is rotated and cuts a cylindrical hole in the work. In the milling process, the work is generally moved slowly with respect to a rapidly turning tool which removes a layer from the surface of the work. In the turning process, the work is normally rotated while the tool is slowly moved with respect to the work to remove a circumferential layer. The grinding process is similar to the milling process: the milling tool is replaced by an abrasive wheel which rotates at high speed, the circumference of which is moved against the work and removes a thin layer from the surface. The surface of the grinding wheel can be shaped to form a profile on a component. Schematic diagrams of these machining processes are shown in Figure 2.39.

In the cutting processes, a tool is employed to remove material, but here, neither the tool nor workpiece needs to be rotated. Cutting can take place by a variety of methods, some using a conventional tool (such as sawing and punching) while others use other methods to remove material (such as a laser, waterjet or electrical discharge). Cutting processes are commonly used to remove material in sheet metal work.

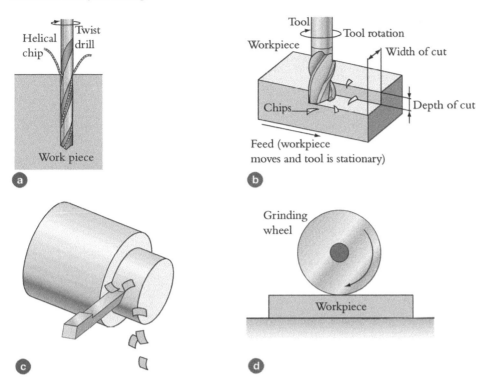

Figure 2.39 Schematic diagram of the main machining processes: (a) drilling; (b) milling; (c) turning; (d) grinding

---

Learning summary

By the end of this section you will have learnt:

✔ As a material shaping processes, machining and cutting involves material removal from a larger body to form a component;

✔ Machining is commonly employed for metals, less for polymers (which can be easily moulded to shape) and ceramics (which are very hard to machine);

✔ It involves motion between a tool and the workpiece;

✔ Cutting is commonly used for metals and polymers.

Heat treatment of metals

Metals are commonly heat treated to change their properties. This section will concentrate on two very common heat treatments which can be applied to certain metals.

Heat treatment of cold-worked structures

As we saw in Underpinning Principles 3 and 5 and the section 'Yield and tensile strength', annealing is commonly applied to cold-worked structures to change properties.

Heat treatment of steels

Steels are iron–carbon-based alloys in which the carbon content typically lies between 0.05 and 1.0 wt%C. Other alloying elements such as Mn, Ni and Cr are often added for a variety of purposes. To understand the microstructures observed in steel (and thus the mechanical properties of the material) requires a basic understanding of the relevant section of the iron-carbon phase diagram (see Underpinning Principles 4 for backgrounds on phase diagrams).

Figure 2.40 shows the relevant part of the phase diagram. This diagram is not unlike the diagram contained in Underpinning Principles 4 for eutectic solidification. However, there are some significant differences.

In this part of the phase diagram, there are no liquid phases, only solid phases. The three single phases are as follows:

- Ferrite. The symbol on the phase diagram is α. Ferrite has a body-centred cubic crystal structure with carbon as an interstitial element in the iron lattice (see UP 2.11). The solubility of carbon in ferrite is very low. Ferrite is a soft, relatively ductile phase.

- Austenite. The symbol on the phase diagram is γ. Austenite has a face-centred cubic crystal structure with carbon as an interstitial element in the iron lattice. The solubility of carbon in austenite is significantly higher than in ferrite. Austenite is a soft, relatively ductile phase.

- Cementite. The symbol on the phase diagram is Fe_3C, which is the chemical formula of this ceramic phase. Cementite contains 6.67 wt%C and is a hard, brittle phase.

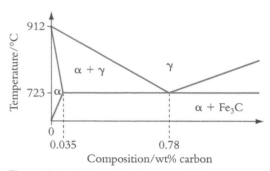

Figure 2.40 Schematic diagram of the low-temperature low-carbon section of the iron–carbon phase diagram

Many of the useful properties of steel are derived from the fact that, at elevated temperature, when the material is held in the austenite phase field, all the carbon dissolves in the metal. This process is known as *austenitizing*.

The eutectoid reaction

Processing of many steels is centred on appropriate control of the eutectoid reaction. The eutectoid reaction is similar to that of the eutectic reaction (see Underpinning Principles 4), except that the eutectoid involves a transformation from a solid phase (rather than a liquid phase) on cooling. However, the basic principles of the two reactions are the same.

In the case of the Fe–C system, the eutectoid composition is ~Fe–0.78 wt%C (see Figure 2.40). If an alloy of the eutectoid composition is heated above 723°C, a structure containing only austenite grains results. When the austenite cools back through 723°C, the eutectoid reaction begins.

The two phases that form in the eutectoid reaction have very different compositions, one with a high carbon concentration and the other with a low carbon concentration. As such (and remembering that this is a solid–solid transformation), carbon atoms must diffuse during the reaction to achieve this partitioning of carbon between the two phases. This redistribution of atoms is easiest both if the temperature is high and if the diffusion distances are short, which is the case when the α and Fe_3C form as thin plates, or lamellae (see Underpinning Principles 4).

In the case of steels, the lamellar structure of α and Fe_3C is known as *pearlite*. The pearlite is composed of soft and ductile ferrite interspersed with plates of hard, brittle cementite. The 'grain size' (plate width) of the ferrite is small and so pearlite is a strong material. The pearlitic structure becomes finer if it is cooled more rapidly.

Microstructure of steels

Many heat treatments for steels raise the steel into the austenite regime, and then cool it again relatively slowly back into the low-temperature ferrite–cementite regime. The slow cooling allows the carbon to diffuse and the pearlite microstructure to form. Typically a general engineering steel will have a composition (denoted C_0) somewhere between 0.035 wt%C and 0.78 wt%C. When this steel is cooled from the austenite regime, it will first enter the austenite + ferrite phase field. As we saw in Underpinning Principles 4, the compositions and proportions of the two phases at any temperature can be derived by drawing a tie-line construction and by use of the Lever rule.

As the alloy is cooled to fractionally above the eutectoid temperature of 723°C, the alloy will still consist of ferrite and austenite. However, at this point, the austenite that remains will have been enriched in carbon so that this is now eutectoid austenite. Also, the ferrite that has formed by this stage is referred to as *primary ferrite*, and tends to form as blocky regions within the microstructure. As can be seen by reference to Underpinning Principles 4 and the phase diagram in Figure UP 4.4, higher carbon levels in the range being considered will result in a higher fraction of eutectoid austenite in this two-phase microstructure.

As this mixture of eutectoid austenite and primary ferrite is cooled below the eutectoid temperature, the remaining austenite (because it has eutectoid composition) will all transform into pearlite (remembering that pearlite is itself not a phase but a mixture of ferrite and cementite). The primary ferrite will remain basically unchanged as it cools through the eutectoid. Since the eutectoid austenite transforms to pearlite, higher carbon levels in the range being considered will result in a higher fraction of pearlite in the final microstructure. As pearlite is the microstructural feature which confers strength to such steels, both the yield strength and the ultimate tensile strength are observed to rise as the carbon level increases, with a commensurate reduction in the ductility and toughness.

Martensite formation in steel

The formation of pearlite as steels are cooled below the eutectoid temperature depends upon diffusion of carbon (see Underpinning Principles 6). However, if the high-temperature austenite is cooled rapidly enough (we call such a rapid cool a *quench*), the carbon does not have enough time at elevated temperature to diffuse, and thus it gets trapped in position. The ferrite cannot grow in its normal manner from the austenite (owing to the high carbon levels which are not permitted in ferrite); however, as the steel is cooled below a certain temperature, the driving force to form ferrite from the unstable austenite becomes too great, and the ferrite forms with the carbon trapped in it. The transformation of austenite to ferrite occurs typically at temperatures in the vicinity of 200–300°C where there is not enough diffusion to allow the austenite structure to rearrange itself into a ferritic structure. As such, the transformation occurs in a diffusionless manner, which involves distortion and shearing of the crystal structures from one to the other. The ferrite-type structure thus formed is highly distorted by the high concentration of carbon trapped in it; this distorted ferrite structure is known as *martensite*. The high level of distortion in the crystal structure severely impedes the motion of dislocations in the structure. Remembering that motion of dislocations is the mechanism of plastic flow in metal (see Underpinning Principles 3), this distortion of the crystal structure results in a steel with a high yield strength; however, along with this goes brittleness which makes martensite of little use as an engineering steel. Higher carbon levels in the steel result in materials with higher yield strength, but also more brittle materials.

Hardenability of steels

In the previous section it was explained that if a steel was quenched rapidly enough from the austenite, martensite will form; however, if the quench rate is too low, then a mixture of ferrite and pearlite will form.

The rate at which the steel needs to be cooled to form martensite depends upon the alloy chemistry. If the steel has a high alloy content (such as high carbon, high manganese, etc.), then the rate at which it has to be quenched to form martensite is quite low – it is easy for martensite to form even with a low quench rate, and such a steel is said to have a high *hardenability* (this is not to be confused with the term hardness, which measures the ability of a material to resist plastic deformation). Conversely, if the steel has low levels of alloying, we need a very high quench rate to form martensite – the steel is said to have a low hardenability.

Tempering of martensite

In *martensite*, the high level of carbon in the ferritic steel structure can come out of the ferrite (and form very small particles of iron carbide) if the carbon is given enough thermal energy to *diffuse* (see Underpinning Principles 6). This is done by raising the temperature of the martensite in a process known as *tempering*; the tempering temperature must not be high enough to reaustenitize the steel. During tempering of martensite, the carbon comes out of the ferrite lattice, leaving the latter less strained and more amenable to dislocation motion.

Accordingly, the yield and tensile strength is reduced, but the toughness and ductility rise substantially. The effect of the tempering increases with increasing tempering temperature and time. It must also be remembered that tempering (with its changes in mechanical properties) can inadvertently occur during service at elevated temperature.

Precipitation strengthening

Another heat treatment process which has a strong influence on the mechanical properties of materials is that of precipitation strengthening. Alloys need to be specifically designed to be suitable for the precipitation strengthening process. By manipulation of the heat treatment cycle, an array of small, second-phase particles (precipitates) can be generated throughout the material. These precipitates can act as dislocation pinning points, and by restricting the motion of dislocations, the yield strength of the material is seen to rise (see Underpinning Principles 3). It is found that the increase in yield strength is proportional to the inverse of the distance between adjacent precipitates.

However, if the precipitates are too small, they do not act as an effective barrier to dislocation motion, and are simply cut by the dislocations. Since the volume fraction of precipitates is governed by the composition of the system, the growth of precipitate size to avoid cutting results in an increase in the distance between adjacent pinning points, and thus a reduced increase in yield strength. As such, the heat treatment of such materials for precipitation strengthening is a balance which aims to control the size and distribution of the precipitates.

Aluminium with 4 wt% copper (Al–4 wt%Cu) is the basis of a series of heat treatable, precipitation-strengthened aluminium alloys; the phase diagram of the aluminium copper system is shown in Figure 2.41. (See Underpinning Principles 4.)

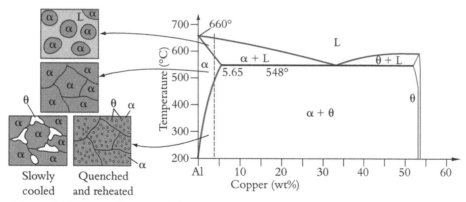

Figure 2.41 Phase diagram of the aluminium–copper system

It can be seen that, at low temperatures, the phases present in the alloy will be α and θ. The θ phase is known as an intermetallic phase, in that it is a compound formed of two metals. It has a higher yield strength and is less ductile than the α phase, and it acts as an effective barrier to dislocation motion.

To precipitation-strengthen the alloy, the θ phase must be dispersed throughout the α phase as small precipitates. As such, the first step is to take all the θ phase into solution (known as *solutionizing*). This requires that the temperature is raised such that the material enters the single-phase α region. It can be seen from Figure 2.41 that there is a limited range of compositions over which this operation is possible; for a copper content greater than 5.65 wt%, the alloy will enter the ($\alpha + L$) phase field as it leaves the ($\alpha + \theta$) phase field, and the formation of a liquid phase will result in unacceptable distortion of parts during the heat treatment. In reality, to give a reasonable range of temperatures over which the solutionizing step can take place, the copper content of the alloy must remain below around 4 wt%. However, as the copper content is reduced, the amount of θ phase in the final alloy is also reduced, and thus the strengthening effect of the precipitation sequence is reduced.

Following solutionizing, the alloy must be cooled again into the ($\alpha + \theta$) phase field to form the precipitates. If it is cooled slowly, the copper has time to diffuse over relatively long distances, and large θ phase regions are formed. These regions have distances between adjacent precipitates which are large and are thus not effective barriers to dislocation motion; as such, the strengthening effect of such a microstructure is small. To avoid this rapid growth, the alloy must be quenched from the single-phase α region so that the copper is trapped in the α structure at low temperature. This structure is unstable; if the temperature is raised enough to allow diffusion, the copper will come out of solution in the α phase and form the θ phase. However, since the temperature of this precipitation heat treatment is low, the diffusion distances are small, and thus a fine dispersion of the θ phase forms. Again, the growth of these precipitates (size and thus distance between them) can be controlled by the temperature and time of the precipitation step, and thus the properties of the alloy can be tailored through the heat treatment. As with martensitic steels, it must be remembered that the precipitation sequence can inadvertently continue if the material is used at elevated temperature; this would normally result in growth of the individual θ phase precipitates and the corresponding increase in distance between adjacent precipitates, and thus a decrease in strength of the alloy.

Learning summary

By the end of this section you will have learnt:

✔ Some properties of metals can be substantially changed by heat treatment;

✔ The changes depend upon the alloy and the heat treatment;

✔ An understanding of the alloy phase diagram is required to understand many heat treatment schedules;

✔ The strength and toughness of many steels can be widely varied; structures such as ferrite-pearlite and tempered martensite can be formed by heat treatment;

✔ Alloys (such as Al - 4wt% Cu) can be strengthened by controlling the formation of small precipitates in the metal via a controlled heat treatment schedule.

Materials joining

Components are often joined together to make a larger or more complex structure or assembly. In some cases, the materials to be joined may be similar (or the same) and the joining is employed to make a more complex structure which would have been difficult or costly to make in one piece. In other cases, the materials to be joined may be dissimilar, having been chosen for different parts of a structure owing to their different attributes; in these cases, it would often be impractical (or impossible) to make the structure in a single operation, and joining is thus necessary.

There is a wide variety of joining technologies available. Many of these technologies (such as mechanical fastening) do not significantly alter the properties of the materials being joined, but some do (such as fusion welding). This section will briefly outline some of the major joining technologies, before concentrating on welding of metals and the microstructural changes (and thus changes in properties) that result from the welding process.

Adhesive bonding

In adhesive bonding, a thin layer of adhesive is placed between two surfaces to be joined. The adhesive is normally a polymeric material. The two surfaces are pressed together and the adhesive sets (either by evaporation of a solvent or by cooling). Modern adhesives can be very strong. The adhesive layer bonds to the two surfaces by a combination of chemical and mechanical bonding. Adhesive bonding is generally a low-temperature process and can be automated. It is commonly employed for bonding of polymeric components, and is ideal for bonding of dissimilar classes of materials (such as a join between a metal and a glass). However, it can be used for bonding metals, and it is commonly used in this respect in the fabrication of automobile structures.

Mechanical fastening

Mechanical fastening takes a variety of forms, where components are held together by constraint imposed by the fastener itself. A representative selection of forms of fastener is shown in Figure 2.42. Some mechanical fasteners allow easy disassembly of the structure. Threaded fasteners (such as nuts and bolts), rivets, staples and screws are normally made of metal while snap-through fasteners are normally made of a polymer. Mechanical fasteners can be used to join both similar and dissimilar materials.

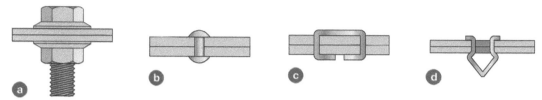

Figure 2.42 **Forms of mechanical fasteners: (a) threaded; (b) rivet; (c) staple; (d) snap-through**

Soldering and brazing

Soldering and brazing are terms used to describe a process where two metallic surfaces are bonded together with a third metal. The solder or braze is usually of considerably lower melting point than the materials that are being joined, but the filler material does metallurgically interact with the materials being joined; as such, a metallurgical bond is formed. Soldering is a lower-temperature process than brazing, but brazing forms a stronger bond than soldering.

Traditionally, soldering has been employed to join electronic components to printed circuit boards with solder materials containing low-melting-point lead. Such solders typically melt at around 200°C. Environmental concerns around the use of lead solders has resulted in the recent development of lead-free materials.

Braze tends to melt at higher temperatures (typically 450°C and above). The degree of metallurgical interaction with the materials being joined is more significant than in soldering owing to the higher temperatures employed. Steel, copper and brass can be easily brazed, but more reactive materials (such as aluminium and titanium) are difficult to braze.

Welding

Many metallic structures (from power-station pipework to mountain-bike frames) are fabricated from parts that are joined together by welds. In welding, two metals are joined together by the addition of a filler material between them; however, in contrast to soldering and brazing, not only is the filler material molten, but there is also significant melting back of the two metal surfaces to be joined. Figure 2.43 shows a schematic diagram of a welding process.

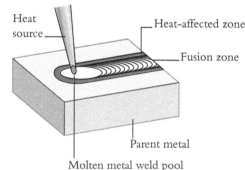

Figure 2.43 **Schematic diagram of a welding process**

Various sources of heat can be used in welding (such as a flame, plasma, laser beam or electron beam). However, the most common form of welding employs an electric arc which is struck between an electrode (part of the welding apparatus) and the materials to be joined. The heat from the arc melts a filler metal which runs into the gap to form a molten metal pool. Although the pool loses heat to the surrounding cold metal, this is replaced by energy from the arc. Some of the parent metal is melted back; as the arc moves on, the molten metal that is left behind solidifies rapidly, fusing the two plates together.

The part of the weld (including the new filler metal) that has been molten is termed the fusion zone. The area surrounding the fusion zone (part of the parent metal) will have been heated to very high temperatures through conduction of heat from the molten metal in the fusion zone, and although it has not melted, it will have been metallurgically changed by this short, high-temperature heat treatment. This region where this heat treatment has been significant is known as the heat-affected zone (HAZ).

Metallurgical changes during welding

The fusion zone

The fusion zone is essentially a casting and tends to have a directional grain structure. A weld bead is typically up to 3 mm wide and 3 mm thick. If a bigger weld is required, this is achieved by laying a number of weld beads on top of each other. As welds are castings, they can have the same problems as experienced in castings, such as porosity in the structure, and shrinkage of the weld metal as it solidifies and cools.

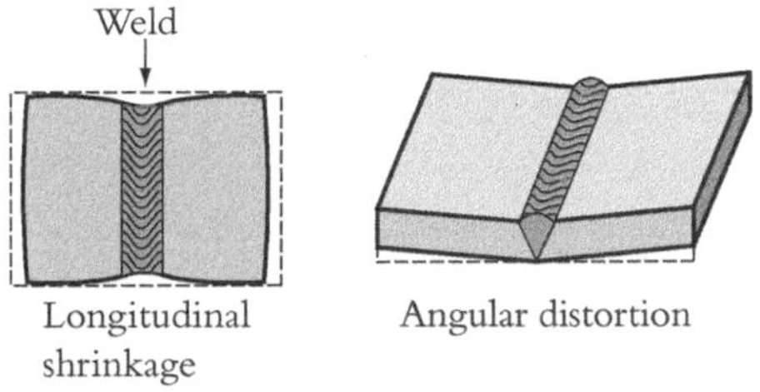

The shrinkage as the weld solidifies and cools will cause distortion of the welded structure (see Figure 2.44), which then results in stresses being generated in the structure, even with no applied external loads – these are known as residual stresses. The contraction will tend to be more significant where the weld material is thicker (i.e. at the top of the weld) and thus angular distortion may be observed in the structure. Also, the shrinkage may lead to solidification cracking along the centreline of the weld bead, as this is the last material to solidify.

Figure 2.44 Distortions in a welded structure owing to shrinkage during solidification and cooling of the weld metal

The HAZ

The temperatures in the HAZ will vary with distance from the boundary with the fusion zone, with regions closest to the fusion zone being close to the melting point of the material. The cooling rates in the HAZ following the passage of the arc are very high. All the effects of heat treatment of metals may thus be observed in the HAZ; some of these effects may be unwanted and lead to undesirable mechanical properties. Two examples where there are significant microstructural changes will be considered in the following sections.

The HAZ in welding of cold-worked metals

The thermal cycle experienced in the HAZ will allow any of the annealing processes discussed in Underpinning Principles 5 to operate. However, the effect of this will be most significant in the annealing of cold-worked metal, where the annealing processes are rapid and the effects of the annealing process are most marked.

In the section 'Hot and cold deformation', it was seen that cold deformation resulted in a microstructure with elongated grains and a high dislocation density. The high dislocation density results in a high strength but low ductility in the material. It was also seen in the same section, that if the deformation process was conducted at a sufficiently elevated temperature, then the material may exhibit recovery, recrystallization and grain growth; the main effect of these was the significant reduction in dislocation density and the corresponding reduction in strength and increase in ductility.

Similar processes of recovery, recrystallization and grain growth (see Underpinning Principles 5) may occur in the HAZ of a weld in a cold-worked structure. Figure 2.45 shows schematically the changes that might be expected in the microstructure, and the corresponding changes in the strength of the material. It can be seen that grain growth has taken place closest to the weld metal (where the HAZ temperature is highest) with only recrystallization occurring further away from the weld as the HAZ temperature is reduced.

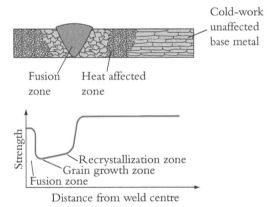

Figure 2.45 Schematic diagram of microstructural changes observed following welding of a cold-rolled sheet, along with an indication of variations of strength throughout the weldment

The HAZ in welding of steels

During welding of normal (ferritic) steels, parts of the heat-affected zone will be heated into the austenite regime. Under the conditions of rapid cooling observed in a weldment, there may be a tendency to form undesirable brittle martensite. The tendency depends upon the *hardenability* of the steel (see the section 'Hardenability of steels').

High hardenability is associated with steels with high concentrations of alloying elements. Steels with high hardenability are difficult to weld as they have a higher tendency to form undesirable martensite than steels with lower hardenability. Thus, low hardenability is associated with high *weldability*, and vice versa. The weldability of a steel is expressed quantitatively as its *carbon equivalent value* (CEV). The CEV is calculated by taking the weight percentages (wt%) of the relevant elements in the steel and summing them as follows:

$$CEV = \frac{wt\% \, C + wt\% \, Mn}{6} + \frac{(wt\% \, Cr + wt\% \, Mo + wt\%V)}{5} + \frac{(wt\% \, Ni + wt\% \, Cu)}{15}$$

where the elements of interest are indicated by their chemical symbols. As a rule of thumb, a material with a CEV > 0.4 wt% will form brittle martensite upon welding.

In certain cases, there may be a need to weld steels with CEV > 0.4 wt%. There are two main approaches to avoiding the presence of brittle martensite in the final structure.

The first approach seeks to avoid the formation of martensite during welding. As discussed in section 'Hardenability of steels', martensite is known to form in a certain steel when the quench rate is high enough. Martensite formation in that steel can be avoided if the quench rate can be reduced. The HAZ is quenched primarily by heat conduction into the cold metal surrounding the weld. The rate of heat extraction (and thus cooling) can be reduced by increasing the general temperature of the metal surrounding the weld; this process is known as preheating.

The second approach seeks to reduce the brittleness of the martensite that forms in the HAZ. The weld is performed normally, and martensite forms in the HAZ. The whole weld area is heated following the welding process to a temperature below the eutectoid; the martensite will then be tempered, with its brittleness being reduced. This process is known as a post-weld heat treatment.

Learning summary

By the end of this section you will have learnt:

✔ Materials and components are commonly joined to make complex structures;

✔ Adhesive bonding is commonly used for joining a wide variety of materials, and can be used to join dissimilar materials;

✔ Mechanical fasteners are commonly used for joining a wide variety of materials, and can be used to join dissimilar materials;

✔ Soldering and brazing are used to join metals. Soldering is ideally used in the electronics industry;

✔ Welding of metals involves joining with a molten metal; the components to be joined also melt close to the weld zone;

✔ Welding can produce very strong joints, but also can affect detrimentally the microstructure and properties of the components being welded;

✔ Welding often results in residual stresses;

✔ Welding may anneal the components being joined close to the weld zone (in the HAZ); this could reduce the strength of a cold worked metal considerably;

✔ In steels, brittle martensite can sometimes form in the HAZ; pre-heating before welding and PWHT can be used to avoid the problems associated with this.

2.6 Failure of materials

One of the most important aspects of designing with materials is picking the right material for the job and ensuring, as well as possible, that it will not fail in service. A number of high-profile engineering disasters have had poor design to blame for failure – for example, forgetting to account for wind loading or the weight of the building itself, or the improper use of materials. In addition to poor engineering practice, extraordinary unforeseen events such as freak weather, explosions, fire and vehicle impact have also led to disaster. In many cases – the two space shuttle disasters being prime examples – the failure of apparently minor components (an 'O' ring and foam insulation) has led to a catastrophic chain of events.

In this final section, we will look at the ways in which a material might fail in service. This will cover the service conditions that are relevant to the failure process and how to design and select materials to prevent failure. We shall look, in turn, at failure by stress overload, brittle fracture, fatigue and creep.

Failure by stress overload

Definition and service conditions

Failure by stress overload occurs when the service stress in a component exceeds the tensile strength of the material from which it is made. Exceeding the tensile strength will result in fracture of the component, resulting in its inevitable loss of functionality. It is possible that exceeding the yield stress of the material, causing permanent distortion of the component, may prevent it from being able to function, but this, and the resultant safety of the structure, are highly dependent upon the component and the application.

An appreciation of the service stresses and the incorporation of safety margins should rule out the component encountering stresses above the yield point, let alone the tensile strength. Stress overload, however, occurs most often under unpredictable conditions. Examples might be where very high forces are encountered due to freak natural events, such as high winds or under the impact of large waves, or where the component is exposed to higher forces than normally expected due to the failure of other components in the structure (perhaps as a result of mechanical, thermal or electrical failure).

Example of stress overload

In 1966 a Boeing 707 disintegrated and crashed near Mount Fuji, Japan, shortly after departure from Tokyo International Airport, killing all passengers. While flying into the wind, approaching Mount Fuji from the downwind side, the aircraft encountered severe clear-air turbulence, causing a sudden structural failure (detachment of the vertical stabilizer) that initiated the in-flight break-up of the aircraft. Destabilization of the aircraft led to high stresses on other components and they in turn failed. Although some stress cracking was found in the vertical stabilizer bolt holes, it was determined, by subsequent testing, that this did not contribute to this accident and that the probable cause was that the aircraft suddenly encountered abnormally severe turbulence which imposed a gust load considerably in excess of the design limit.

Designing against stress overload

Freak conditions aside, it is relatively simple to avoid stress overload. An appreciation of the maximum forces that could act on a particular component in a structure (even if others fail!) should enable the dimensions to be specified so that the stresses acting are below the yield stress, let alone the tensile stress. The incorporation of a safety factor should ensure that even unexpectedly high forces produce no permanent shape change, let alone failure.

The following example first illustrates the selection of an appropriate thickness of steel for a pressure vessel (in this case to avoid yielding, but the process is the same for failure) and then shows the process for selecting the best material for this application.

Selection of materials to avoid stress overload

We commonly view pressure vessels as storage for liquids or gases, or as reactor vessels, but aircraft, spacecraft and submarine superstructures are essentially vessels resisting a pressure difference. In all cases, we would rather that our vessel did not explode or implode (depending on the pressure difference from outside to inside). If they are going to move, we might also want them to be as light as possible. Often we just want them to be as cheap as possible. We shall look at a simple, spherical, thin-walled pressure vessel in which a pressure difference, P, causes a stress in the walls, σ, which for a wall thickness, t, and sphere radius r, shown in Figure 2.46, can be described by equation (2.17).

$$\sigma = \frac{Pr}{2t} \qquad (2.17)$$

If we wish to find the thickness of the vessel (for a given radius and pressure) to prevent it from yielding, we can simply rearrange equation (2.17) and substitute a value for the yield stress of the material used to make the vessel. For example, a submersible diving 4500 m below the ocean surface experiences a pressure difference of approximately 45 MPa (the pressure being greater on the outside). In the case of Alvin (a manned submersible used for research on hydrothermal vents, and to explore the wreck of *RMS Titanic*), the sphere containing the (three) crew is 2.1 m in diameter and made from a high-strength steel (with a yield strength of 1000 MPa). Equation (2.17) indicates that to avoid yielding (in this instance in compression) the thickness of the sphere wall must be at least 23.6 mm (in reality, a safety factor would also be incorporated).

If we wish to minimize the mass of the vessel (it does need to be raised and lowered into the water) we need to be able to express its mass in terms of the geometry. If we just consider the personnel sphere, the volume of material used can be approximated to the surface area, $4\pi r^2$, multiplied by the thickness, t. Its mass is hence the volume multiplied by the density, ρ, and is given in equation (2.18):

$$M = 4\pi r^2 t\rho \qquad (2.18)$$

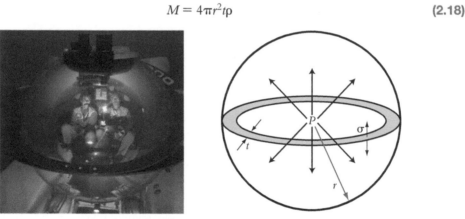

Figure 2.46 A submersible 'pressure vessel' (left) and (right) a schematic of a thin-walled spherical pressure vessel

As in previous examples we must eliminate the free variable, the thickness (since it can be any value to cope with the design stress) by rearranging equation (2.17) and substituting it into equation (2.18). Since the stress in the walls must be below that for permanent yielding (we could just as easily use the tensile stress as the limit but would rather avoid distortion of our vessel every time it is used), equation (2.19) gives us the mass in terms of the geometry, pressure and properties of the material used.

$$M = 2\pi r^3 P\left(\frac{\rho}{\sigma_y}\right) \qquad (2.19)$$

Equation (2.19) indicates that a light vessel requires the smallest value of $\dfrac{\rho}{\sigma_y}$. The cost of the material to make the vessel (not including the manufacturing cost!) is proportional to this term multiplied by the cost of the material per kg. Table 2.11 compares data for suitable materials. CFRP makes the lightest pressure vessel, with titanium and aluminium, glass-fibre-reinforced polymer (GFRP) close behind (these materials are ideal for aircraft and spacecraft – in fact in a recent refit, Alvin's steel hull was replaced by an expensive titanium one). Concrete, polyethylene and mild steel make heavy vessels. If we are concerned about cost, concrete is best (which is why it is used for water towers and nuclear reactors). Pressure vessel steel (an alloy steel) offers the best compromise between cost and weight for many applications.

Material	Yield strength (MPa)	Density (kg m^{-3})	ρ/σ_y	Cost per kg	Relative cost
Mild steel	220	7850	35.7	0.5	17.8
Alloy steel	1000	7850	7.9	1	7.9
Titanium alloy	830	4500	5.4	50	270
Reinforced concrete	200	2500	12.5	0.25	3.13
Aluminium alloy	400	2700	6.8	2	13.6
GFRP	300*	2000	6.7	5	33.3
Polyethylene	30	970	32.3	1	32.3
CFRP	650*	1500	2.3	100	230

* failure strength

Table 2.11 Data for the selection of light and cheap pressure vessels

Failure by fast fracture

Definition and service conditions

Fast fracture is caused by the growth of pre-existing cracks in a material. Internal and surface flaws or cracks act as points of stress concentration. When load is applied, the resultant stresses at the cracks, if high enough, can cause them to grow or propagate rapidly (at the speed of sound in that material) and so cause failure of the material. The stress at which this happens is highly dependent upon the material type; brittle materials such as ceramics fail at low stresses. It should be noted that cracks will only propagate in tension or shear, rather than in compression (which will cause the crack to close rather than open).

Example of fast fracture

During the Second World War, the USA built Liberty ships (cargo ships that were cheap and quick to build). Although their original design life was only for five years, early Liberty ships suffered hull and deck cracks and there were nearly 1500 instances of significant brittle fractures. During the Second World War, 19 ships broke in half without warning. Catastrophic failures started from cracks which formed at the square corners of hatches that coincided with welded seams. It was also discovered that the grade of steel used became more brittle than predicted when exposed to cold sea temperatures in the North Atlantic. Under increased service stresses, due to stress concentration at both the sharp corners and the welds, overloading of the ships and severe storms at sea, the cracks propagated, leading to fast fracture. The predominantly welded (as opposed to riveted) hull construction then allowed cracks to run for large distances unimpeded and the ships literally snapped in half.

Designing against fast fracture

A material's ability to resist fracture, if it contains a crack, is defined by a property called the fracture toughness. Materials with low toughness (described earlier as the energy absorbed during failure) also tend to have low values for the fracture toughness. The relationship between the stress required for a crack to propagate (resulting in fracture), the fracture toughness, K_{IC}, and an internal crack of length $2a$, or surface crack of length a, is given by:

$$\sigma_{max} = \frac{K_{IC}}{Y\sqrt{\pi a}} \quad (2.20)$$

This expression contains a dimensionless factor Y that enables a wide range of crack geometries to be considered. Y is equal to unity for a centre-line crack, and 1.1 for a surface crack. The subscript I arises because of the different ways of loading a material to enable a crack to propagate. It refers to loading via mode I, the opening mode, where a tensile stress is applied normal to the plane of the crack. Figure 2.47 shows the geometries of the crack and the loading.

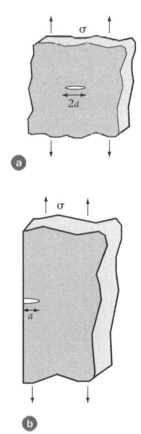

Figure 2.47 Schematic of the loading and crack geometries for (a) an internal crack of length 2a (Y = 1) and (b) a surface crack of length a (Y = 1.1)

A material with a low value of fracture toughness will undergo fast fracture at stresses below those for yielding. For example, a material with a fracture toughness of 5 MPa m$^{1/2}$ (typical of a ceramic), containing a 3 mm long central crack ($Y = 1$, $2a = 3$ mm) will undergo fast fracture at stresses above 73 MPa (well below the tensile stress). Ceramics contain crack defects from the processes used to form and shape them. As a result, fast fracture, rather than yielding, takes place under tensile loading, limiting their use, in tension, in engineering applications. To enable ceramics to be used in tension to their full potential, crack defects must be eliminated during or after processing – this is neither easy nor cheap!

Materials with high fracture toughness will have to experience high stresses or have very large cracks present, before they fail by fast fracture. For example, a material with a fracture toughness of 100 MPa m$^{1/2}$ (typical of a metal), containing a 3 mm long central crack ($Y = 1$, $2a = 3$ mm), would fail by fast fracture at stresses above 1457 MPa. This stress is likely to be higher than that for yielding and hence yielding will occur in preference to fast fracture. Long 'cracks' can be introduced into a structure that has been fabricated by welding (for example a pressure vessel). Improper welding can result in unbonded regions that act like cracks, which, if undetected, could result in disaster.

Design against fracture often involves the calculation of a maximum crack size that can be tolerated in a material before fast fracture will occur. In demanding engineering applications where high stresses are encountered, for example when designing and manufacturing aircraft and pressure vessels, the critical crack size may be small. For example, if an aluminium alloy with a fracture toughness of 30 MPa m$^{1/2}$ is to be subjected to a service stress (let us say including a safety factor) of 300 MPa (below the yield stress of 400 MPa), we can use equation (2.20) to determine a maximum surface defect ($Y = 1.1$, crack length = a) of 2.6 mm or an internal crack ($Y = 1$, crack length = $2a$) of 6.4 mm. Components are inspected, when new and intermittently in service, using non-destructive methods (which include X-ray radiography, ultrasonic inspection and the use of fluorescent dyes) to measure the size and quantity of any defects. Any parts with cracks approaching the critical size must be repaired or scrapped.

Selection of materials to avoid fast fracture

Pressure vessels, from the simplest aerosol can to the biggest boiler, are designed, for safety reasons, to yield or leak before they fracture (i.e. explode, creating a shower of fragments). Small pressure vessels are usually designed to allow yielding at a pressure still too low to propagate any crack that the vessel may contain. The distortion caused by yielding is easy to detect and the pressure can be released safely. With large pressure vessels this may not be possible; instead, safe design is achieved by ensuring that the smallest crack that will propagate has a length greater than the thickness of the vessel wall. The leak is easily detected, and it releases pressure gradually and thus safely (as long as the contents are not poisonous).

We shall look at this 'leak-before-break' case. The stress in the wall of a thin-walled spherical pressure vessel of radius, r, is shown in equation (2.21). Designing against stress overload ensures that the wall thickness, t, is chosen so that, at the maximum working pressure, P, the stress is no greater than the yield stress, σ_y:

$$\sigma_y = \frac{Pr}{2t} \qquad (2.21)$$

If there are cracks present in the wall of the pressure vessel, we can ensure that fast fracture does not occur by arranging that any cracks, just large enough to penetrate both the inner and the outer surface of the vessel ($2a = t$), are still stable. The leaking of gas through this crack will then prevent any further build-up in pressure. A schematic of this condition is shown in Figure 2.48, along with a power station 'reactor' vessel, and is defined in equation (2.22):

$$\sigma_{max} = \frac{K_{IC}}{Y\sqrt{\pi\left(\dfrac{t}{2}\right)}} \qquad (2.22)$$

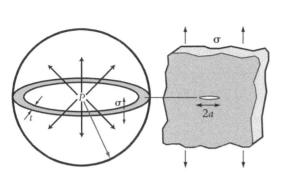

Figure 2.48 Schematic diagram of the forces in a spherical pressure vessel and the crack geometry and (right) a power station reactor 'pressure' vessel

In order to work out the maximum pressure that can be carried safely by the vessel, we need to eliminate the free variable, the thickness, t, and to do this we rearrange equation (2.21) and substitute it into equation (2.22). Knowing that the maximum stress encountered will be the yield stress, and rearranging in terms of the pressure, gives:

$$P = \left(\frac{4}{Y^2 \pi R}\right)\left(\frac{K_{IC}^2}{\sigma_y}\right) \tag{2.23}$$

This equation has been split into design terms (in the left-hand bracket) and materials terms (in the right-hand bracket). The maximum pressure is carried most safely by the material with the largest value of $\frac{K_{IC}^2}{\sigma_y}$.

Table 2.12 compares $\frac{K_{IC}^2}{\sigma_y}$ data for different materials. It should be noted that high values for $\frac{K_{IC}^2}{\sigma_y}$ can be obtained by choosing a material with a low yield stress, but this makes the vessel very thick (and potentially heavy) in order to withstand the pressure, and hence a high value for the yield stress is also desirable. The selection process shows that tough alloy steels are a good choice (steels are used for most pressure vessels) but copper is good for small vessels where corrosion may be a problem (for brewing and distilling alcohol). Polyethylene is only good for lightly stressed vessels (such as fizzy drinks bottles). Materials with high σ_y, such as titanium and CFRP, are good for lightweight pressure vessels.

Material	Yield strength (MPa)	Fracture toughness K_{IC} (MPa m$^{1/2}$)	K_{IC}^2/σ_y
Structural steel	220	50	11.4
Alloy steel	1000	170	28.9
Titanium alloy	830	100	12.0
Reinforced concrete	200	10	0.5
Aluminium alloy	400	30	2.3
Copper	450	90	18.0
GFRP	300*	50	8.3
Polyethylene	30	2	0.1
CFRP	650*	40	2.5

Table 2.12 Data for the selection of leak-before-break pressure vessels

Failure by fatigue

Definition and service conditions

Fatigue is the progressive and localized structural damage that occurs when a material is subjected to cyclic loading. When a material undergoes repeated stress cycles (for example (+) tensile, then (−) compressive) new cracks can be initiated and both new and existing cracks can grow. The importance of this is that small cracks that are below the critical size for fast fracture can grow, by fatigue, to exceed this limit and catastrophic failure (by fast fracture) will follow. Depending on the material, shape and stress experienced, failure may require thousands, millions, billions or trillions of stress cycles.

Figure 2.49 shows a typical stress cycle for fatigue (stress against time), showing the mean stress and the stress range, where the stress imposed can be a result of mechanical or thermal loading. The example in the figure is a regular stress cycle which might be encountered in a vibrating system such as an oscillating aircraft wing. The loading experienced, for example, by a ship at sea, will be much less regular. In most cases of fatigue the maximum stress is below the yield stress and must be tensile to allow crack opening and advancement of the crack. Often the stress cycles about a non-zero mean stress (as is shown in Figure 2.49).

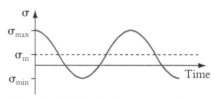

Figure 2.49 A typical stress cycle (stress against time) for fatigue

Example of metal fatigue

Metal fatigue in conjunction with fast fracture has been a major cause of aircraft failure, and was deemed to be the cause of the explosive decompression, at altitude, of a de Havilland Comet passenger jet in 1954, killing all those on board. After forensic investigation of the wreck and tests on a model, aircraft investigators told a public enquiry that the sharp corners around the plane's window openings (actually the forward antenna window in the roof) acted as initiation sites for cracks that grew under fatigue loading (caused by the repeated pressurization and depressurization of the aircraft cabin) until they were large enough to cause fast fracture of the fuselage. The sharp-cornered window cut outs created stress concentrations that were much larger than expected, enabling fatigue cracks to start at the rivet holes around the windows. As a result, all aircraft windows were immediately redesigned with rounded corners to reduce the stress concentration.

Design against fatigue failure

The careful manufacture of small engineering components (for example, gears, axles, connecting rods) can ensure that they are free from cracks. In this case, fatigue is controlled by the initiation of a crack, commonly in regions of stress concentration (such as at the sharp corners at the join of the head and shank of a bolt).

If the maximum stress is below the yield stress (which tends to be the case in vibrating systems), the number of cycles to failure, N_f and the cyclic stress, $\Delta\sigma$, follow an equation of the form:

$$\Delta\sigma N_f^m = C \qquad (2.24)$$

where m and C are constants.

The fatigue life of a material (derived from experiments) can be characterized by an S–N curve, a graph of the magnitude of the cyclical stress amplitude ($\Delta\sigma$ or S) about a zero mean stress, plotted, on a logarithmic scale, against the number of cycles to failure (N_f or N). Figure 2.50 shows schematic S–N curves typical of steel and aluminium. The curve for steel follows the dependence described in equation (2.24) at high stress amplitudes but indicates the existence of a minimum stress amplitude (S_{limit} – more commonly called the fatigue limit) below which crack growth does not occur and hence fatigue does not take place. This is not the case for all materials and the curve for aluminium demonstrates this case, showing that, for any given stress amplitude, there will be a finite number of stress cycles before failure. These plots can be used as a guide to specify a safe stress amplitude to withstand a given number of cycles (or identify the life for a given stress amplitude).

Materials and processing

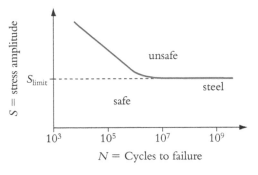

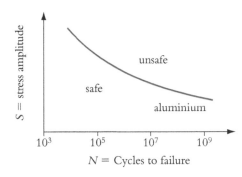

Figure 2.50 Schematic *S–N* curves typical of (left) steel and (right) aluminium

Large structures, particularly those containing welds, will contain cracks. In this case we need to know the size of the cracks that are present (we can use non-destructive inspection to measure the size of these cracks) and then be able to predict how long it will take for these cracks to grow to reach the critical length for fast fracture.

For the greatest part of the fatigue process the relationship between the rate of crack growth $\left(\dfrac{da}{dN}\right)$ and the cyclic stress intensity factor, ΔK, is described by Paris' law, shown in equation (2.25):

$$\frac{da}{dN} = A(\Delta K)^n \tag{2.25}$$

where A and n are constants that vary with material. The cyclic stress intensity factor, ΔK, varies with the applied stress and the crack geometry and is described by:

$$\Delta K = \Delta\sigma\sqrt{\pi a} \tag{2.26}$$

It should be noted that only the tensile part of the stress cycle will contribute to crack opening and growth and hence should be used to determine ΔK.

Equations (2.25) and (2.26) indicate that the crack growth rate increases as the stress amplitude and crack length increase. Since the crack length is not constant during the fatigue process, the Paris equation (equation (2.25)) must be integrated to calculate the number of cycles to failure. If we assume that $n = 4$ (it sits within the range 1 to 6 for most materials) and that the cyclic stress varies from zero to some tensile stress σ_{max}, then we can calculate the number of cycles it takes to grow a crack from size a to a_f (if we assume that it would cause fast fracture when it reached this size, then the number of cycles is the number to failure, N_f).

Expanding the Paris equation gives:

$$\frac{da}{dN} = A\left(\sigma_{max}\sqrt{\pi a}\right)^4 \tag{2.27}$$

Separating the variables and setting the limits gives:

$$\int_0^{N_f} dN = \frac{1}{A(\sigma^4{}_{max}\pi^2)}\int_a^{a_f}\frac{1}{a^2}da \tag{2.28}$$

Integrating between these limits gives:

$$N_f = \frac{1}{A(\sigma^4{}_{max}\pi^2)}\left(\frac{1}{a} - \frac{1}{a_f}\right) \tag{2.29}$$

We can use this approach to determine the safe life for a crack in an aluminium aircraft component. If a component is subjected to an alternating stress of range 160 MPa ($\pm$ 80 MPa) about a mean stress of 50 MPa, then the maximum stress, σ_{max}, is 130 MPa, the minimum is -30 MPa. If the fracture toughness of the alloy is 35 MPa m$^{1/2}$, then by using:

$$\sigma_{max} = \frac{K_{IC}}{1.1\sqrt{\pi a}} \tag{2.30}$$

we can calculate the critical crack length for fast fracture (for a surface crack) to be 19 mm.

If the largest surface crack found by non-destructive testing is 2 mm deep ($a = 2$ mm), we can determine the number of cycles that the wing can survive before this crack would grow to reach the critical crack length (19 mm).

Using equation (2.29), which was derived from equation (2.27) assuming that $n = 4$, substituting data for the initial and final crack lengths, the maximum tensile stress and assuming $A = 5 \times 10^{-12}$ m (MPa m$^{1/2}$)$^{-4}$, gives:

$$N_f = \frac{1}{5 \times 10^{-12}(130^4\pi^2)}\left(\frac{1}{2 \times 10^{-3}} - \frac{1}{19 \times 10^{-3}}\right) \qquad (2.31)$$

Note that the stress should be kept in MPa and the crack length expressed in metres to have units that are compatible with A. The number of cycles to failure is 31,741. The frequency of the oscillation will determine the actual life in hours (or more properly flying hours) before failure.

To design against fatigue failure, we would ideally like to keep the stress amplitude below the fatigue limit (if the material has one) or select a material that will not suffer from metal fatigue during the life of the product (this is also not always possible). If not, then we can design for a fixed life and replace the part with a new one before it reaches this limit (we know that planes have a certain number of flying hours before they are retired). In the case of the example given above, we would use the predicted life to define the time period between inspections and measurement of the length of the (growing) crack. Once the crack exceeds a specified length (one well below the 19 mm required for fast fracture!) the part would be repaired or scrapped.

Failure by creep

Creep is the term used to describe the tendency of a material to move or to deform permanently to relieve stresses. Creep is a slow, continuous 'time-dependent' deformation, under long-term exposure to constant load, at stresses that are below the yield strength of the material. The accumulation of creep strain causes damage and ultimately failure of the material. Creep is more severe in materials when the temperature is raised to above 40 per cent of the melting point (in Kelvin).

The temperature range in which creep deformation may occur differs in various materials. For example, tungsten requires a temperature above 1000°C before creep deformation can occur, while plastics and low-melting-temperature metals, including many solders, creep at room temperature. Creep deformation is thus important not only in systems where high temperatures are endured, such as steam turbine power plants, jet engines and heat exchangers, but also in the design of many everyday objects made from polymers.

Examples of creep failure

The collapse of the World Trade Center was credited, in part, to creep. The aircraft impacts dislodged some of the fireproofing from the structural steel members supporting the towers, increasing their exposure to the heat of the fires caused by the burning jet fuel (it is estimated that the burning jet fuel could have easily caused sustained temperatures in excess of 800 °C). Although the steel would have reached temperatures that were well below the melting point, they would have been high enough to weaken the core columns so that they underwent plastic deformation and creep from the weight of the floors above. The resulting failure of the 2–3 floor system at the site of impact would have sent the 20–25 floors above free-falling onto the 80–85-floor structure below. The enormous energy released by this collapse would have been too large to be absorbed by the structure below, causing the explosive buckling, floor by floor, of the towers.

The failure of tungsten light bulb filaments can also occur by creep. When hot (i.e when the bulb is switched on), the filament coil progressively sags between its supports owing to creep deformation caused by the weight of the filament itself. If too much deformation occurs, the adjacent turns of the coil touch one another, causing an electrical short and local overheating, leading to rapid failure of the filament.

Design against creep

Figure 2.51 shows schematically how materials (metals, ceramics and polymers) deform with time under a constant tensile load (creep will also occur in compression or shear) as a function of temperature. Initially, the strain rate (the slope of the plot) slows with increasing time. This is known as primary creep. The strain rate eventually reaches a minimum and becomes near-constant. This is known as secondary or steady-state creep. In tertiary creep, the strain rate exponentially increases with strain, damage starts to occur followed by failure. The creep rate can also be seen to increase with increasing temperature.

At high applied stresses, creep in metals is controlled by the movement of dislocations. At high temperatures, atomic diffusion helps 'unlock' dislocations that have been stopped by obstacles (such as precipitates) in the material. Diffusion allows the dislocations to climb over the obstacles, to move through the material, resulting in plastic deformation. The creep rate (in the steady state region) can be described by:

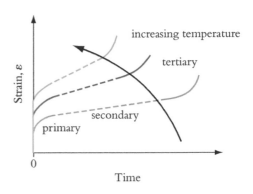

Figure 2.51 Schematic illustration of the change in creep strain as a function of time at different temperatures

$$\frac{d\varepsilon}{dt} = \dot\varepsilon = A\sigma^n \exp\left(\frac{-Q}{RT}\right) \qquad (2.32)$$

where A and n are constants that vary with material type, T is the temperature (in Kelvin), R is the gas constant ($8.31\,\mathrm{J\,mol^{-1}\,K^{-1}}$) and Q is the activation energy for the creep process (with units of $\mathrm{J\,mol^{-1}}$) and is also a material parameter. The strain rate (or creep rate), as described by this equation, increases with both temperature and stress. At a constant temperature, the creep rate can be given by:

$$\dot\varepsilon = B\sigma^n \qquad (2.33)$$

where B is also a constant. For example, if $Q = 160\,\mathrm{kJ\,mol^{-1}}$, $n = 5$ and $A = 7.2 \times 10^{-10}\,\mathrm{MPa^{-5}\,s^{-1}}$, for an applied stress of 100 MPa at 600°C (873 K) the creep rate is $1.9 \times 10^{-9}\,\mathrm{s^{-1}}$. (Note: the stress should be in MPa to be compatible with the units for A.) If the creep strain allowed is 0.01 (1 per cent), then this will be achieved in 5.26×10^6 s or 1461 h. Doubling the stress will increase the strain rate 32 times, while increasing the temperature by approximately 165°C (for this example) will achieve the same effect.

At lower applied stresses, creep deformation of metals occurs by elongation of the grains through the diffusion of atoms through the grains. Since the rate of diffusion increases with temperature, so does the creep rate. The creep rate via this mechanism can be described by equation (2.34), where C is a constant. It can be seen that the creep rate scales inversely to the square of the grain size, d, (i.e. increasing the grain size by a factor of 100 will decrease the creep rate by 10,000).

$$\frac{d\varepsilon}{dt} = \dot\varepsilon = \frac{C\sigma}{d^2} \exp\left(\frac{-Q}{RT}\right) \qquad (2.34)$$

On account of the high strength of ceramics and their small grain size (a result of their manufacture from fine powders), creep in ceramics occurs mainly by this process. Creep in polymers occurs at temperatures above the glass transition temperature where the polymer behaves like a rubber (if it has many chemical crosslinks) or a viscous liquid (if it does not). Above the glass transition temperature, creep occurs by viscous flow of the polymer chains past each other and follows the same linear dependence upon the applied stress (i.e. $n = 1$) as for diffusional flow as shown:

$$\frac{d\varepsilon}{dt} = \dot\varepsilon = C'\sigma\exp\left(\frac{-Q}{RT}\right) \qquad (2.35)$$

where C' is a constant and Q is the activation energy for viscous flow.

As a consequence of creep deformation, several processes can occur. At a constant stress or load (if the material or component is free to deform), strain accumulates with time. Components will elongate in the stress direction and, depending on the magnitude of the applied stress and its duration, the deformation may become so large that a component can no longer perform its function (for example, creep of a turbine blade will cause the blade to contact the casing, resulting in seizure and failure of the blade). Alternatively, if the component is constrained and cannot move, then the stress will relax with time. An example of this is the stress in a tightened bolt. The creep strain will reduce the elastic strain and hence the bolt must be repeatedly tightened.

If there is no limiting strain (from the design) that will cause failure, there is a limit to the amount of creep deformation that can occur in a material before it fails as a result of internal damage. This results in the concept of a creep life that we must ensure is longer than the service life. The creep life is determined from empirical relationships obtained by fitting simple equations to experimental data. An example of this is shown in Figure 2.52, where the creep life or time to failure (in hours) can be determined for a given applied stress and is shown at various temperatures (where $T_3 > T_2 > T_1$).

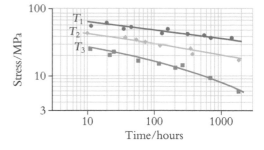

Figure 2.52 Typical plot of creep life (or time to failure) as a function of applied stress at a number of temperatures ($T_3 > T_2 > T_1$)

To minimize creep, materials with high melting points should be chosen as diffusion will be slower at a given temperature. Materials with a large resistance to dislocation motion (ceramics) and those with many fine, hard precipitates (that are stable at the service temperature) are also favoured. Materials with large grains (or even one single grain) are more resistant to creep, since the diffusion paths for grain deformation are increased. More extensively crosslinked polymers (for example, thermosets such as epoxies) have higher glass transition temperatures and are more resistant to creep. Increasing the percentage crystallinity of a (thermoplastic) polymer (reducing the fraction of amorphous material) also increases the creep resistance. The creep resistance of a polymer can also be improved by filling the polymer with glass or silica powders or, better still, continuous glass and carbon fibres which bear all the load and eliminate creep.

Creep of a turbine blade

In a rotating turbine blade operating at high temperature, strain accumulates with time due to creep. The centrifugal load gives rise to an elastic stress and drives creep deformation in the radial direction. Extensive deformation will cause the blade to foul against the casing in which it is housed. Figure 2.53 shows the typical clearance between blades and the casing in a turbine engine.

The clearance between the casing and the blade tip defines the maximum permissible strain. For a gap of 0.5 mm, the maximum strain on a 50 mm long blade is 1.0×10^{-2}. The elastic stretching of the blade is given by equation (2.36) where the stiffness of the material at the operating temperature would be used. For a nickel alloy turbine blade, this is approximately 160 GPa at 900°C:

Figure 2.53 Typical clearance between the blades and the casing in a turbine engine

$$\varepsilon_{\text{elastic}} = \frac{\sigma}{E} \qquad (2.36)$$

The stress in a rotating blade is proportional to the square of its angular velocity and for a turbine blade spinning at 16,000 RPM (1676 rad s^{-1}) the average stress is typically 200 MPa. The elastic strain is 1.25×10^3 and hence the maximum permissible creep strain is 8.75×10^{-3}.

The creep rate of the nickel alloy at a given temperature (in this case 900°C or 1173 K) can be described by an equation of the form:

$$\dot{\varepsilon} = B\sigma^n \qquad (2.37)$$

where, at 1173 K, $B = 3.8 \times 10^{-22}$ MPa^{-5}s^{-1} and $n = 5$. Substituting these values, along with that for the stress (in MPa), gives a strain rate of 1.216×10^{-10}.

Since the creep strain is the product of the creep strain rate and the time period over which creep takes place, the blade will creep by the maximum permissible strain (8.75×10^{-3}) after nearly 72×10^{6} or 20,000 h in service.

Learning summary

By the end of this section you will have learnt:

✔ A number of ways in which materials can fail in service and the relavance of these failure methods to engineering situations;

✔ The relevant equations that govern the failure process along with typical data for engineering materials;

✔ How to use these equations to design against material failure and to select suitable materials to avoid failure.

Fluid mechanics

Stephen Pickering

UNIT OVERVIEW

- Introductory concepts
- Fluids at rest – hydrostatics
- Fluids in motion
- Fluids in motion – linear momentum

Fluids (the generic name given to both liquids and gases) are all around us – the air we breathe, the water in lakes, rivers and the sea, the liquids we drink. They are a familiar part of everyday life and the way in which fluids behave is understood by everyday experiences. For instance, liquids have a defined volume but take the shape of the container they are in; a kilogram of water generally has a fixed volume but it can be confined inside a bottle or spread out over a large area if spilled on the floor. Some fluids flow more easily than others: water is easy to pour, but syrup moves more slowly. Fluids that are moving create forces on solid objects: the umbrella blows inside out on a windy day! In engineering it is important to understand the behaviour of fluids in a scientific way, to be able to quantify the effects so that calculations can be undertaken to design systems involving fluids and to predict their performance. For instance, if water needs to be pumped 25 metres uphill at a rate of 10 litres per second, how big does the pump have to be and how much energy is needed to pump the water? Could the system be designed to be more sustainable and use less energy? If an oil storage tank needs to be built, how thick will the walls need to be? How much force does the oil exert on the walls? In this chapter the behaviour of fluids will be explained in a mathematical way and tools will be introduced to enable design calculations to be performed. These will be related to some of the everyday experiences that we have of fluids to help understanding.

The chapter is divided into five main sections. Following the introduction in Section 3.1, Section 3.2 looks at fluid statics. The concept of pressure is described, showing how to calculate the forces created by the pressure of a fluid on a solid surface and how pressure can be measured using a manometer. The concept of buoyancy and why some things float and others sink is explained. In the remaining part of the chapter, moving fluids are considered, starting in Section 3.3 with the simplified case of what are known as ideal fluids (fluids in which viscosity is ignored). Section 3.4 moves on to real (viscous) fluids and the ways of estimating the amount of energy dissipated to friction when these fluids flow. Finally, in Section 3.5 the forces exerted by moving fluids are explained in terms of changes in momentum.

3.1 Introductory concepts

What is a fluid?

A fluid is a substance that has the ability to flow, i.e. to take up the shape of its container. This is in contrast to solids, which generally have a fixed rigid shape unless acted upon by large forces. Liquids, gases, vapours and plasmas are fluids, and even some plastic solids (such as bitumen) can be treated as fluids. The differences between solids and fluids relate to what is happening on a molecular scale. In a solid the atoms or molecules are in fixed positions relative to each other and, although the atoms may be vibrating, they do not move past each other and so the solid retains its shape. By contrast a fluid consists of atoms or molecules that are in continuous random motion. They are not held in fixed positions relative to each other but are able to move freely and they continually collide with each other and their surroundings. As the molecules are free to move, if a force is applied to a fluid, the molecules move past each other and the fluid deforms and changes shape. By contrast, when a force is applied to a solid, the molecules resist moving past each other and the solid will retain its shape unless the forces are large enough to cause it to deform or break. This distinction is expressed more exactly by stating that a fluid that is not moving cannot support a shear stress and any applied shear will lead to deformation of the fluid. Once the shear stress is removed, there is no residual stress within the fluid, no matter how much the molecules within the fluid are displaced from their original locations, as shown in Figure 3.1.

The difference between a liquid and a gas again relates to what is happening on a molecular scale. In a liquid the forces of attraction (known as cohesion) between the molecules prevent the molecules from moving too far away from each other and so the liquid retains a definite volume. Therefore, a fixed amount of liquid has a certain volume associated with it and it takes a great deal of applied force to squash a liquid into a smaller volume. If a volume of liquid is placed into a larger container, there will be a free surface. In a gas the movement of the molecules is much greater, so that the forces of attraction between the molecules are overcome and the molecules can move away from each other only to be confined by the surfaces of the container they are in. Therefore, if a fixed amount of gas is placed into a container, it will expand until it completely fills the container; so there is no free surface in a gas. Liquids turn into gases when they are given more (kinetic) energy to make the molecules move faster. The amount of kinetic energy in the molecules is what results in the property known as temperature.

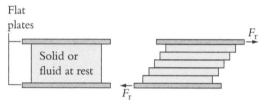

Figure 3.1 If a shear force (F_T) is applied, a solid will deform slightly, whereas a fluid will continue to deform indefinitely as the layers of fluid keep sliding over each other

> A stationary fluid cannot withstand a shear stress but deforms under the action of a shear.

Continuum

All matter is made up of atoms or molecules, but these are so small that the volumes of fluids involved in engineering calculations usually contain sufficient elementary particles that the fluid can be considered to have homogeneous properties, i.e. it has continuous properties rather than being made up of individual particles. For example, a 1 mm diameter spherical drop of water contains approximately 2.9×10^{18} water molecules. On a very, very small scale, individual atoms and molecules can be distinguished and a fluid is not homogeneous. However, in all the calculations in this book, a fluid will be treated as a continuum.

A similar argument can be used when considering the random motion of a fluid. With a very, very small sample, there may be, say, fewer than 20 molecules inside the boundaries of a volume at one time and more or less at a different time, due to random motion. The timescales involved with this random motion are very short and so we assume that no matter how small the time step we are considering, the properties do not change with time.

> Fluids can be treated as continuous in time and space.

Fluid mechanics

Mass and weight

The mass is the amount of a substance and its weight is the force produced by a mass when acted upon by gravity:

$$\text{Weight} = \text{mass} \times \text{acceleration due to gravity}$$

$$W = mg \tag{3.1}$$

In the SI system of units, weight is measured in newtons (N), mass in kilograms (kg) and acceleration in m s^{-2}. The acceleration due to gravity on earth is generally taken as 9.81 m s^{-2}.

Mass is independent of where a substance is, so a certain amount of a substance will have the same mass whether it is on the earth, orbiting in space or on the moon. The weight depends on the local gravity, so while a mass of 1 kg may have a weight of about 10 N on the earth, it would only have about a sixth of this weight on the moon.

Density

Density is the amount of mass in a given volume and so relates to the weight of a given volume of a fluid. Density typically uses the greek symbol ρ (rho, pronounced *row* as in *row a boat*) and the equation for density is

$$\rho = \frac{m}{V} \tag{3.2}$$

The SI unit for density is kg m^{-3}. The density of a fluid can be found either from the mass of a known volume or from the volume of a known mass, and this gives the mean density.

In engineering the terms relative density or specific gravity (SG) are sometimes used when referring to liquids. For example, a liquid may be said to have a specific gravity of 0.96. The SG (or relative density) of a fluid is the ratio of the density of the fluid to the density of water at standard temperature and pressure (usually taken to be 1000 kg m^{-3}).

$$SG = \frac{\rho_{\text{liquid}}}{\rho_{\text{water}}} \tag{3.3}$$

So a fluid with a specific gravity or relative density of 0.83 has a density of 830 kg m^{-3}.

The specific gravity of olive oil is 0.85. When a container of 2 kg mass and 20 litres internal volume is completely filled with the oil, what is the combined mass of the container and oil?

We need to know the mass of the oil inside the container and can calculate this from the known volume and density. Two small calculations are required to obtain the volume and density in SI units:

$$\rho_{\text{oil}} = SG_{\text{oil}} \times 1000 = 850 \text{ kg m}^{-3}$$

$$20 \text{ litres} = 20 \times 10^{-3} \text{ m}^3$$

We can now calculate the mass using equation (3.2):

$$\rho = \frac{m}{V}, \quad \text{so} \quad m = \rho V = 850 \times 20 \times 10^{-3} = 17 \text{ kg}$$

Consequently, the combined mass is $17 + 2 = 19$ kg.

Pressure

Pressure is caused by the atoms and molecules in a fluid colliding with each other and with the surfaces of a container. Consider a box containing a gas. The molecules of the gas are in constant random motion, and sometimes they collide with each other. When this happens, an equal and opposite force is exerted on each molecule and there is no net force on the box, but there is an internal pressure within the fluid. However, when one of the molecules hits one of the walls of the box (as shown in Figure 3.2), it bounces off, and the change in momentum of the molecule creates a force on the wall perpendicular to the wall. The velocity of the molecule parallel to the wall does not change during the impact (assuming that it is a perfectly elastic collision), but the velocity perpendicular to the wall retains the same magnitude but is reversed in direction. For a single molecular impact, this would be a small force, but there are many molecules and so there will be many collisions, creating what appears to be a uniform distribution of force over the surface.

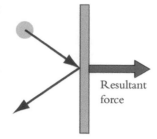

Resultant force

Figure 3.2

Pressure is a scalar quantity as it does not have a direction associated with it – the pressure within a fluid acts in every direction as the molecules bounce off each other. However, wherever a fluid is in contact with a surface, the force exerted on the surface is perpendicular to the surface, and so it is a vector quantity as it has direction. For an element of the wall, where all the molecules impinging on it have the same pressure, the pressure is defined as the force acting on the element divided by the area of the element:

$$\frac{\delta F}{\delta A} = p \tag{3.4}$$

To find the total force acting on a surface, sum all the elemental forces. Thus, $F = \Sigma \delta F$. In the mathematical limit, as the elemental areas become infinitesimally small, this may be expressed as the integral

$$F = \int p \, dA \tag{3.5}$$

If the pressure is constant, this reduces to the well-known formula *force equals pressure times area*:

$$F = pA \tag{3.6}$$

If the pressure is not uniform, the integration can only be performed if the variation in pressure over the area is known. (Equation (3.5) can be integrated numerically if there is experimental data available, or analytically if a formula for p as a function of A can be derived.)

As pressure acts in all directions, a pressure should never be represented as an arrow on a diagram! However, the force created by a pressure acting on a surface does have direction and can be shown by an arrow.

Pressure is force per unit area and has the units of $N \, m^{-2}$. The SI unit for pressure is the pascal (Pa), and $1 \, Pa = 1 \, N \, m^{-2}$. However, another unit commonly used for pressure in engineering is the bar. One bar is $100\,000$ Pa or 10^5 Pa, and it is a useful engineering unit because it approximates atmospheric pressure. Standard atmospheric pressure is $1.013\,25$ bar. A millibar, 1 mbar, is equal to 100 Pa and is also a unit in common usage.

Absolute pressure and gauge pressure

If there are no molecules/atoms (complete vacuum), then the pressure will be zero. Also, if there is no molecular movement (absolute zero temperature, 0 K), then the pressure will be zero. Pressures above this zero condition are referred to as **absolute** pressures. As the atmosphere is all around, in engineering it is often the pressure present *in addition* to atmospheric pressure that is of interest, and this gives rise to the concept of **gauge** pressure. Most instruments for measuring pressure are so constructed that they indicate gauge pressure (i.e. the pressure *relative* to atmospheric pressure). Gauge pressure is related to atmospheric pressure thus:

> Gauge pressure = absolute pressure − atmospheric pressure
> or
> Absolute pressure = gauge pressure + atmospheric pressure

Fluid mechanics

Thus, if the absolute pressure (pressure relative to a vacuum) is 3.5 bar and atmospheric pressure is 1.0 bar, then the gauge pressure will be 2.5 bar (or, commonly, the pressure is 2.5 bar gauge). Although gauge pressures can be very useful, it is important to note that atmospheric pressure changes daily and is also significantly affected by elevation. For instance, at an altitude of 1000 m, atmospheric pressure is only 0.9 bar.

A sealed container is completely filled with a gas at an absolute pressure of 3370 mbar. Air pressure outside the container is 1.1 bar. What is the gauge pressure of the gas? The container is then taken to the top of a mountain where air pressure is 0.9 bar. If the container does not deform and the ambient temperature does not change, what is the gauge pressure of the gas now? Why would the answer be different if the ambient temperature changed?

First, express both pressures in the same units:

$$3370 \text{ mbar is } 3.37 \text{ bar}$$

Gauge pressure = absolute pressure − atmospheric pressure:

$$P_{\text{gauge, ground}} = P_{\text{abs, ground}} - P_{\text{at, ground}}$$

$$P_{\text{gauge, ground}} = 3.37 - 1.1 = 2.27 \text{ bar}$$

At ground level the gauge pressure of the gas is 2.27 bar. This means that it has a pressure of 2.27 bar more than atmospheric.

When the container is taken to the top of the mountain, the pressure of the gas inside it remains the same. This is because the container is sealed and so there is the same mass of gas inside. The container does not deform and so has the same volume. If the temperature does not change either, then the gas inside the container will be in exactly the same state as it was at the ground, and its absolute pressure is therefore still 3.37 bar.

Atmospheric pressure has changed, however, and so there is a new gauge pressure:

$$P_{\text{gauge, mountain}} = P_{\text{abs, mountain}} - P_{\text{at, mountain}}$$

$$P_{\text{gauge, mountain}} = 3.37 - 0.9 = 2.47 \text{ bar}$$

If the temperature of the gas changes, then its pressure will change too. The change in pressure can be calculated using the perfect gas law (see later in this section).

The perfect gas equation of state

(The relationship between pressure, volume and temperature in gases.)

A gas can be thought of as behaving in the same way as a collection of rigid, perfectly elastic spheres in constant motion that interact with each other and with the walls of the container. Using this concept it is possible to deduce equations for behaviour, and this idealized representation (known as a perfect gas) results in the **perfect gas equation of state**, which holds true over a wide range of pressures and temperatures commonly experienced in engineering. At low pressures and high temperatures the approximation is not so good. Another name for a **perfect** gas is an **ideal** gas.

The perfect gas equation is derived by considering Boyle's law and Charles's law. Boyle's law states that, when temperature is constant, the volume occupied by a gas is inversely proportional to its pressure, and Charles's law states that, when pressure is constant, the volume occupied by a gas is proportional to its absolute temperature (in kelvin). Together, these give

$$\frac{pV}{T} = \text{constant} \tag{3.7}$$

The constant depends on the quantity of gas, and Avogadro's hypothesis states that equal volumes of all gases at a given temperature and pressure contain the same number of molecules. Therefore, at a particular temperature and pressure, a certain number of atoms or molecules of *any* gas will occupy the same volume. For engineering calculations the quantity of atoms or molecules usually referred to is the **kilomole**, usually written as kmol. A kilomole of molecules always occupies the same volume at a particular temperature and pressure, and this volume is 22.41 m^3 at the reference temperature and pressure (0 °C and 1.013 25 bar; these conditions are known as standard temperature and pressure, or STP). Thus, for 1 kmol of gas, the constant in equation (3.7) is 8314 J kmol^{-1} K^{-1}, and it is known as the molar or universal gas constant and given the symbol $\widetilde{R}$.

Thus, in general, for n kmol of gas, equation (3.7) becomes

$$pV = n\widetilde{R}T \tag{3.8}$$

The mass of 1 kmol of a substance depends on the **molar mass**, and this is determined from the atomic weights of the elements in the molecules. For example, the molar mass of hydrogen gas (H_2), in which there are two atoms per molecule, is 2.0 kg kmol^{-1}. For oxygen gas (O_2) the molar mass is 32 kg kmol^{-1}. The number of moles of a quantity of a substance, n, is its mass divided by the molar mass. So

$$n = \frac{m}{\widetilde{m}}$$

Substituting this relationship into equation (3.8) gives:

$$pV = mRT \tag{3.9}$$

where

$$R = \frac{\widetilde{R}}{\widetilde{m}}$$

This is known as the specific gas constant and is dependent on the particular gas under consideration. For air, which is approximately 79 per cent nitrogen and 21 per cent oxygen, the average molar mass is 29 kg kmol^{-1}, and so the specific gas constant is 287.1 J kg^{-1} K^{-1}. (It is worth noting that many people think that use of the word **specific** in the term **specific gas constant** means that the constant is specific to a particular gas, which it is. However, it is really used in its other meaning, where specific quantities are quantities per unit mass; and the specific gas constant has units J kg^{-1} K^{-1} and so refers to a kilogram of a particular gas.)

Dividing equation (3.9) by the mass of gas gives

$$pv = RT \tag{3.10}$$

where v (lower case) is the specific volume, or the volume of 1 kg of gas. It has the units m^3 kg^{-1}. It is the reciprocal of density ρ (which has the units kg m^{-3}). Therefore, an alternative form of the perfect gas equation is

$$p = \rho RT \tag{3.11}$$

Therefore, the density of a gas can conveniently be determined from pressure and temperature as

$$\rho = \frac{p}{RT} \tag{3.12}$$

In using the various forms of the perfect gas equation, or the equation of state for a gas, as it is otherwise termed, it is important to remember to use the correct units: temperature must be expressed as absolute temperature in kelvin, and pressure must be expressed as absolute pressure in N m^{-2} or Pa.

What is the specific gas constant for oxygen gas?

Oxygen is normally present in the atmosphere as molecules of two atoms (O_2). The atomic mass of oxygen is $16 \, \text{kg} \, \text{kmol}^{-1}$, so the molar mass of molecular gaseous oxygen is $32 \, \text{kg} \, \text{kmol}^{-1}$.

$$R_{\text{oxygen}} = \frac{\widetilde{R}}{\widetilde{m}_{\text{oxygen}}} = \frac{8314}{32} = 259.8 \, \text{J} \, \text{kg}^{-1} \, \text{K}^{-1}$$

A helium-filled weather balloon is to expand to a sphere of 20 m diameter at a height of 30 km where the absolute pressure is 1200 Pa and the temperature is $-47 \, °\text{C}$. If there is to be no stress in the fabric of the balloon, what volume of helium must be added at ground level where the absolute pressure is 101.3 kPa and the temperature is $15 \, °\text{C}$?

There are two key points to note here: first, if there is no stress in the fabric of the balloon, then the pressure inside the balloon is equal to the pressure outside. Secondly, there is the same quantity (or mass) of hydrogen in the balloon at ground level as at 30 km.

Apply

$$pV = mRT$$

The parameters m and R are constant, so

$$\left(\frac{pV}{T}\right)_{\text{ground}} = \left(\frac{pV}{T}\right)_{30 \, \text{km}}$$

and

$$V_{\text{ground}} = \left(\frac{p_{30 \, \text{km}}}{p_{\text{ground}}}\right)\left(\frac{T_{\text{ground}}}{T_{30 \, \text{km}}}\right)V_{30 \, \text{km}}$$

$$= \left(\frac{1200}{101 \, 300}\right)\left(\frac{273.15 + 15}{273.15 - 47}\right)\left(\frac{\pi 20^3}{6}\right)$$

$$= 63.2 \, \text{m}^3$$

The perfect gas equation only applies to gases at reasonable pressures and temperatures found in engineering. It does not apply to liquids, which for most cases can be assumed to be incompressible, so the density does not vary significantly with pressure or temperature.

Remember temperatures must be in kelvin in the perfect gas equation.
To convert from °C to K, add 273.15.

Compressibility

In general, a pressure increase in a substance causes a reduction in the volume occupied by a given mass; thus the density increases. For solids and liquids the reduction in volume is very small but it is important in some cases. For gases, even a small change in pressure can cause a significant change in volume and therefore density.

The equation that relates the change in volume to the change in pressure is given below, where K is the bulk modulus of elasticity:

$$\delta p = K\frac{\delta V}{V} \tag{3.13}$$

K does vary slightly with pressure, but, over ranges where it can be regarded as constant, we can write

$$\Delta p = K\frac{\Delta V}{V}$$

For solids, K is of the order of 10^{11} Pa, and for liquids it is of the order of 10^9 Pa (increasing with pressure). For example, K for water is 2×10^9 Pa, and therefore a pressure increase of 100 bar will only cause a volume reduction of 0.5 per cent:

$$\frac{\Delta V}{V} = \frac{\Delta p}{K} = \frac{100 \times 10^5}{2 \times 10^9} = 0.005$$

It is for this reason that liquids are generally regarded as incompressible and will be treated as such in this chapter.

For gases, K is approximately equal to the pressure of the gas, provided that the pressure is not excessively high. Thus, for a gas at 1 bar, K is of the order of 1 bar, and a 0.1 per cent volumetric strain only requires a pressure change of approximately 0.1 per cent of 1 bar (100 Pa). Gases can only really be treated as incompressible if the pressure changes involved are very small. However, in this chapter only fluids where the behaviour can be considered to be incompressible are dealt with. In the chapter of this book describing thermodynamics, the behaviour of gases subject to large pressure changes is considered.

Surface tension

Surface tension arises from the forces between the molecules of a liquid (cohesive forces). At the surface of a liquid these forces of attraction act like a 'skin' holding the liquid together. This is the reason that small drops of liquid tend to assume a spherical shape, e.g. raindrops and drops of dew. Surface tension is defined as the tensile force acting per unit length at the interface of a liquid. It therefore has the dimensional formula

$$\left[\frac{\text{Force}}{\text{Length}}\right] = [MLT^{-2}L^{-1}] = [MLT^{-2}]$$

with units of $N\,m^{-1}$ and the symbol γ.

There are also forces of adhesion between liquid molecules and the molecules of a solid boundary. If these forces of adhesion exceed the forces of cohesion within the fluid, then the liquid tends to spread out over (or wet) the solid surface. Alternatively, if the surface tension forces are greater than the forces of adhesion with a solid, then the liquid does not wet the surface but tends to form discrete droplets. Water will wet clean glass but mercury will not. Water will not wet a greasy surface.

In engineering, surface tension is not usually a significant force except where the dimensions of the volume of water are small compared with the area of the surface with which it is in contact, and it is the reason for the phenomenon of capillary action. From experience it is known that water will tend to be drawn along a piece of string and this is due to surface tension. Surface tension is only of significance in this chapter when considering the measurement of pressure using a manometer tube and will be dealt with in that section.

Key points from Section 3.1
- A fluid is shaped by external forces (i.e. a fluid takes up the shape of its container).
- A fluid at rest cannot support a shear stress.
- In fluid mechanics, all fluids are treated as having local properties that are the same as the bulk properties and do not vary with time (continuum).
- Pressure in a fluid is associated with molecular motion.
- When pressure is constant over an area, $F = pA$.
- Absolute pressure (relative to vacuum) and gauge pressure (relative to atmospheric pressure) are both used in engineering.
- The perfect gas equation of state can be used to obtain gas properties.
- Liquids can usually be treated as incompressible but gases cannot.

3.2 Fluids at rest – hydrostatics

Hydrostatics is the study of fluids at rest, but can also concern fluids in uniform motion such that all the fluid has the same velocity throughout, i.e. there is no motion of one part of the fluid relative to another part or to a surface. For example, liquid inside a road tanker moving along may be analysed as a hydrostatics problem if there is no relative motion within the fluid.

Pressure

As described in the introduction, pressure results from the impact of the molecules in a fluid with each other and with the walls of a container. At a solid surface, pressure exerts a force perpendicular to the surface and the force can be determined by integrating the pressure:

$$F = \int p \, dA$$

If the pressure is uniform over a surface, the equation then becomes the familiar

$$F = pA$$

Within a fluid the pressure acts equally in all directions; in the absence of any shear forces owing to movement of the fluid, this is known as *Pascal's law* and can be proved as shown below for information.

Pascal's law

In the absence of shear forces the pressure at a point in a fluid acts equally in all directions.

A surface submerged in a fluid experiences a force perpendicular to the surface (provided that there are no shear forces) as a result of the pressure; but a pressure itself acts in all directions.

Proof of Pascal's Law

This can be shown as follows.

Let an infinitesimal triangular prism of fluid, surrounding any point S, have thickness δy (perpendicular to the paper). Let pressures p_1, p_2, p_3 act on faces AC, CB, AB respectively.

If no shear forces act on the prism, then the only forces acting are:

(i) normal forces F_x, F_z and F_n on faces AC, CB and AB respectively, due to pressures p_1, p_2 and p_3. This type of force is called a surface force.

(ii) forces that act on the mass as a whole, such as gravity forces. These are called body forces.

F_x, F_z and F_n must be normal to their respective faces, otherwise there would be components of force along the faces and shear forces would exist. A force applied normal to a surface will not cause sliding.

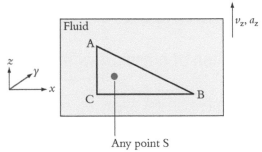

Any point S

Figure 3.3

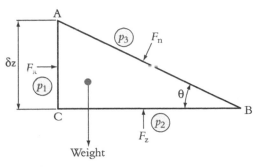

Weight

Figure 3.4

Resolve forces in the z-direction and apply Newton's second law of motion:

Net force in z-direction = mass of element $\times\ a_z$

where a_z represents the acceleration (zero if the fluid is at rest).

Thus

$$F_z - F_n \cos\theta - [\rho g(\tfrac{1}{2}\delta x\ \delta y\ \delta z)] = \rho(\tfrac{1}{2}\delta x\ \delta y\ \delta z)a_z$$

where ρ represents fluid density and the bracketed quantity [ ] represents the body force (weight).

But

$$\cos\theta = \frac{\delta x}{\text{AB}}, \quad F_z = p_2\ \delta x\ \delta y, \quad F_n = p_3(\text{AB})\delta y$$

Therefore

$$p_2\ \delta x\ \delta y - p_3(\text{AB})\delta y\frac{\delta x}{\text{AB}} = \tfrac{1}{2}\rho\ \delta x\ \delta y\ \delta z(a_z + g)$$

and so

$$p_2 - p_3 = \tfrac{1}{2}\rho\ \mathrm{d}z(a_z + g) \tag{3.14}$$

As the size of the prism tends to zero, then δz tends to zero so that in the limit

$$p_2 = p_3$$

Similarly, resolving in the x-direction, in the limit as the size of the prism tends to zero, we have

$$p_1 = p_3$$
$$\therefore \qquad p_1 = p_2 = p_3 \tag{3.15}$$

Since θ can take any value, the pressure at a point is independent of direction, provided that there are no shear forces in the fluid. If shear forces are involved, equations (a) and (b) are invalid. However, in most cases of real fluid flow, the shear stresses are very small compared with the pressures and so Pascal's law can still be applied.

Since pressure has magnitude, but pressure at a point is independent of direction, pressure must be a scalar. However, force has a magnitude and direction and is thus a vector. Arrows should never be used to indicate pressure, since these imply that direction is relevant.

Variation of pressure with elevation

Within a stationary fluid, gravity acts downwards, and this causes a vertical pressure gradient within the fluid. The variation in pressure with depth can be determined by considering the vertical forces acting on an element of fluid at rest, as shown in Figure 3.5.

Let the fluid pressure be p at height z, and $p + \delta p$ at height $z + \delta z$. Force F_z, due to pressure p acts upwards to support the element. Force $F_z + \delta F_z$ due to pressure $p + \mathrm{d}p$ acts downwards owing to the weight of fluid above. Now

$$F_z = p\ \delta x\ \delta y$$
$$F_z + \delta F_z = (p + \delta p)\ \delta x\ \delta y$$
$$\therefore \delta F_z = \delta p(\delta x\ \delta y)$$

Resolve forces in the z-direction:

$$F_z - (F_z + \delta F_z) - g\ \delta m = 0$$

but

$$\delta m = \rho\ \delta x\ \delta y\ \delta z$$
$$\therefore \qquad -\delta p(\delta x\ \delta y) = g\rho(\delta x\ \delta y\ \delta z)$$

Fluid mechanics

or

$$\frac{\delta p}{\delta z} = -\rho g$$

As

$$\delta z \to 0, \quad \frac{\delta p}{\delta z} \to \frac{\delta p}{\delta z}$$

$$\therefore \quad \frac{dp}{dz} = -\rho g$$

$$\Delta p = -\rho g \Delta z \qquad\qquad (3.16)$$

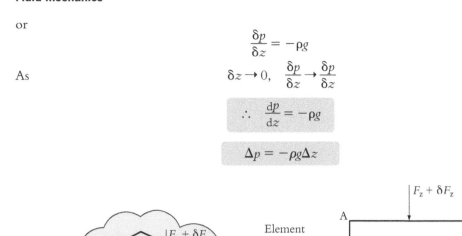

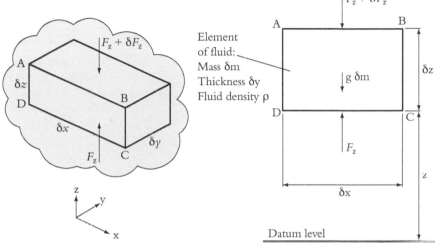

Figure 3.5 **Vertical forces acting on a fluid element at rest**

This equation applies for any fluid. The minus sign indicates that, as z increases, p decreases. For example, pressure increases as a diver descends.

Figure 3.6 shows the forces acting in the x-direction.

Resolve forces in the x-direction:

$$F_{x_1} = F_{x_2}$$

since the element is at rest.

Thus, the pressure on face AD is equal to that on face BC:

$$\therefore \qquad \frac{dp}{dx} = 0, \text{ and similarly } \frac{dp}{dy} = 0$$

> For a fluid at rest, p does not vary with x or y and must be the same everywhere on a horizontal plane. However, p does vary with z (on account of gravity) from one horizontal plane to another.

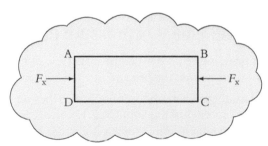

Figure 3.6 **Horizontal forces acting on fluid element**

For an incompressible fluid, e.g. a liquid, density (ρ) is constant and equation (3.16) can be integrated to give

$$\int_1^2 dp = -\int_1^2 \rho g \, dz = -\rho g \int_1^2 dz$$

$$\therefore \qquad p_2 - p_1 = -\rho g(z_2 - z_1) \qquad\qquad (3.17)$$

Or, considering h as being positive in the downwards direction, then

$$\therefore \qquad p_2 - p_1 = \rho g(h_2 - h_1)$$

Note that we conventionally use the notation where z represents elevation above a datum and h represents depth below a datum (usually the surface of a liquid).

Liquid columns

It is useful to consider a simple liquid column as illustrated in Figure 3.7.

Applying equation (3.17) gives

$$p_2 - p_1 = 2\rho g(z_2 - z_1)$$
$$p_2 - p_1 = \rho g h$$

It is often useful to think of this as

> Pressure at base of column = pressure at top + $\rho g h$

In fact, for any column, the pressure at the base of a column depends only on the vertical height of the column and not on the shape. This may not be immediately obvious, but it must be true as the pressure on a horizontal plane in a static fluid is constant. Consider all the vessels shown below. The pressure at the base of each is identical, even though the volume of fluid differs. In some cases, some of the weight of the fluid is supported by the container sides as well as the base.

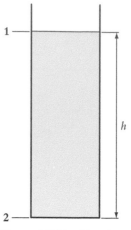

Figure 3.7 Liquid column

Figure 3.8

Atmospheric pressure of 1.01 bar acts on the surface of a lake 10 m deep. Find the pressure at the bottom of the lake.

Take $\rho = 1000 \text{ kg m}^{-3}$.

$p_{\text{bottom}} = P_{\text{top}} + \rho g h$
$p_{\text{bottom}} = 1.01 \times 10^5 + (1000 \times 9.81 \times 10)$
$\quad = 1.991 \times 10^5 \text{ Pa} = 1.991 \text{ bar}$

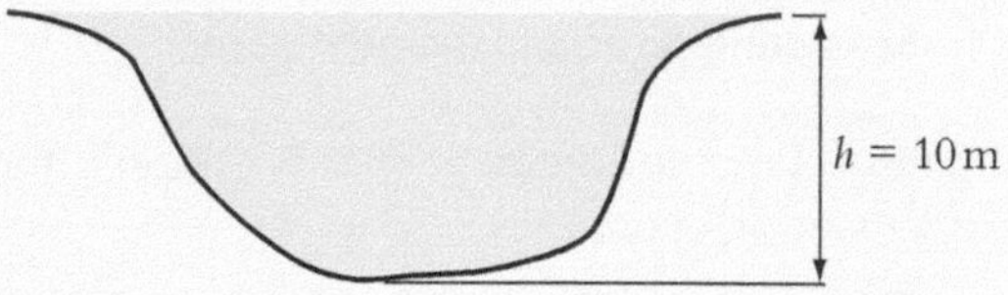

$h = 10\text{m}$

Figure 3.9

Measurement of pressure using manometers

A manometer is a device that makes use of the fact that the height of a static liquid column can be used to measure the pressure in a fluid. There are several different types, discussed briefly below.

Manometers have the advantage that they are simple and do not need calibrating, provided that the properties of the liquids used in them are known. They are thus ideal for use in a laboratory. However, their limited operating range and lack of easy interface to electronic control systems means that they are commonly being replaced with electronic pressure transducers.

Piezometer

This is the simplest type of manometer. It is used for measuring the pressure in a liquid.

With reference to Figure 3.10, let the absolute pressure of a liquid in a pipe be p_1 at level 1, where p_1 is greater than atmospheric pressure p_a. The pressure in the pipe forces the liquid up the piezometer tube until the height of the column of liquid balances the pressure difference between the pipe and the atmosphere. The greater the value of p_1, the greater is the height of the liquid column. For a liquid, the density, ρ, can be assumed to be constant.

$$\Delta p = -\rho g \Delta z \quad \text{or} \quad p_1 - p_2 = -\rho g(z_1 - z_2)$$

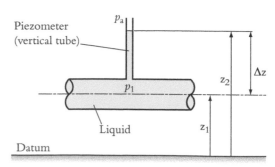

Figure 3.10 Piezometer

Fluid mechanics

but

$$p_2 = p_a$$

$$\therefore \qquad p_1 - p_a = p_{1\text{gauge}} = \rho g(z_2 - z_1)$$

Thus, the piezometer reading Δz is a measure of gauge pressure.

U-tube manometer

This can be used for measuring the pressure of a gas, vapour or liquid and is suitable for higher pressures than the piezometer. A U-tube manometer is illustrated in Figure 3.11.

Fluid 2 (a high-density liquid) is usually water, mercury or an oil (such as paraffin). It prevents escape of the pressurized fluid and does not mix with it.

Let Fluid 1 have constant density ρ_1 and absolute pressure p_1 at level B. Fluid 2 has constant density ρ_2.

Consider the horizontal plane AA′ which passes through the common interface at A. Fluid 2 is continuous beneath AA′ and is at rest. The points A and A′ are on the same horizontal plane within Fluid 2 and so the pressure must be the same as pressure is constant on a horizontal plane in a continuous fluid. Consequently,

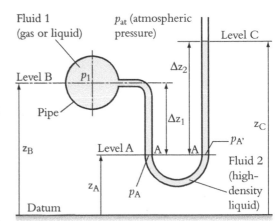

Figure 3.11 U-tube manometer

$$p_A = p_{A'} \qquad\qquad (3.18)$$

In the left-hand limb of the manometer, for fluid 1, taking ρ_1 as constant and using

$$p_{\text{bottom}} = p_{\text{top}} + \rho g h$$

gives

$$p_A = p_1 + \rho_1 g(z_B - z_A) = \rho_1 g \Delta z_1$$

and so

$$p_A = p_1 + \rho_1 g \Delta z_1 \qquad\qquad (3.19)$$

In the right-hand limb of the manometer, for Fluid 2

$$p_{A'} = p_a + \rho_2 g(z_C - z_A)$$

so

$$p_{A'} = p_a + \rho_2 g \Delta z_2 \qquad\qquad (3.20)$$

Combining equations (3.18), (3.19) and (3.20) yields

$$p_1 + \rho_1 g \Delta z_1 = p_a + \rho_2 g \Delta z_2$$

$$\therefore \qquad p_1 - p_a = p_{1\text{gauge}} = g(\rho_2 \Delta z_2 - \rho_1 \Delta z_1) \qquad\qquad (3.21)$$

If fluid 1 is a gas at fairly low pressure (1 or 2 bar), then $\rho_1 \ll \rho_2$, and hence

$$p_{1\text{gauge}} \cong \rho_2 g \Delta z_2$$

For example, if fluid 1 is air at about 1 bar, then $\rho_1 \cong 1.2\ \text{kg m}^{-3}$, and if fluid 2 is water, then $\rho_2 \cong 1000\ \text{kg m}^{-3}$. The error involved in using the approximation ignoring the density of the air in equation (3.21) would be approximately 0.12 per cent; but if $p_1 \cong 10$ bar, the error $\cong 1.2$ per cent.

Differential manometer (also often referred to as a U-tube)

This manometer can be used for finding the difference between two unknown pressures, for example at different sections in a pipe or across a flow meter as shown in Figure 3.12.

The fluid in the pipe can be any gas, vapour or liquid, provided that it does not mix or react chemically with the manometer liquid, which is usually mercury, water or a light oil. Later, it is shown that the manometer reading, related to $p_1 - p_2$, can be used to determine volume flow rate in a pipe. For the determination of small pressure differences in liquids an inverted U–tube, with the liquid columns separated by a low-density fluid, such as air, can be used for greater sensitivity.

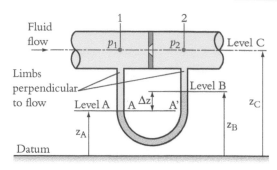

Figure 3.12 **Differential manometer**

Since the manometer liquid is at rest, the pressure is the same at A and A′ on the horizontal plane AA′. Consider a case where there is water in the pipe and mercury is the manometer fluid.

In the left-hand limb:

Water

$$p_A = p_1 + \rho_w g(z_C - z_A) \tag{3.22}$$

In the right-hand limb:

Water

$$p_B = p_2 + \rho_w g(z_C - z_B) \tag{3.23}$$

Mercury

$$p_{A'} = p_B + \rho_m g \Delta z \tag{3.24}$$

As $p_A = p_{A'}$, equate equations (3.22) and (3.24) and substitute for p_B from equation (3.23):

$$p_1 + \rho_w g(z_C - z_A) = [p_2 + \rho_w g(z_C - z_B)] + \rho_m g \Delta z$$

$$p_1 - p_2 = \rho_w g(z_C - z_B - z_C - z_A) + \rho_m g \Delta z$$

$$\Delta p = p_1 - p_a = (\rho_m - \rho_w)g \Delta z \tag{3.25}$$

Note that pressure, or pressure difference, can be expressed as the pressure exerted by a vertical column of water of given height or 'head'. For example, a pressure of 3000 Pa can be described as 0.31 m of water $\left(\dfrac{3000}{\rho_w g} = \dfrac{3000}{1000 \times 9.81}\right)$.

If the pipe is not horizontal, as illustrated in Figure 3.13, the difference in levels between stations 1 and 2 must be allowed for as follows:

$$p_1 - p_2 = (\rho_m - \rho_w)g \Delta z + \rho_w g(z_{C_2} - z_{C_1})$$

Note that the extra term in this equation accounts for the difference in height between the two pressure tappings in the pipe.

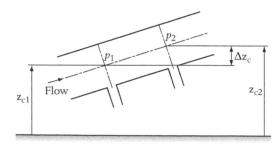

Figure 3.13 **Non-horizontal U-tube manometer**

Fluid mechanics

A differential manometer contains water (density $1000\,\text{kg m}^{-3}$). The manometer is connected between two points in a horizontal pipe carrying air (density $1.2\,\text{kg m}^{-3}$). When the difference in levels is 54 mm, what is the pressure difference between the two points?

For the left-hand limb in the manometer:

$$p_{XX} = p_1 + \rho_{air}g(h + y)$$

For the right-hand limb in the manometer:

$$p_{XX} = p_2 + \rho_{air}gy + \rho_{water}gh$$

Equating the two pressures:

$$p_1 + \rho_{air}gh = p_2 + \rho_{water}gh$$
$$p_1 - p_2 = gh(\rho_{water} - \rho_{air})$$
$$= 9.81 \times 0.054(1000 - 1.2)$$
$$= 529.1\,\text{Pa}$$

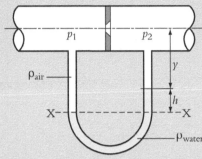

Figure 3.14

If the density of the air had been neglected, then the pressure difference would have been calculated as

$$p_1 - p_2 = 9.81 \times 0.054 \times 1000 = 529.7\,\text{Pa}$$

An error of only 0.1 per cent.

Inclined-tube manometer

This device, illustrated in Figure 3.15, provides greater sensitivity in pressure measurement. Resulting from the inclination of the tube, the displacement, L, of the liquid is greater than Δz_2, which would be the displacement if the tube were vertical.

The reservoir has a much larger cross-sectional area (A_1) than the right-hand limb (area A_2). Thus, the displacement Δz_1 is much smaller than Δz_2 and can often be neglected. This is important, since the instrument can be calibrated to enable pressure to be determined simply from the reading, L, the distance that the liquid travels along the right-hand limb of the manometer when pressure difference is applied. Only the right-hand limb then needs to be observed, unlike normal U-tube manometers. With no air flow, $p_1 = p_2$ and the paraffin levels lie on the zero line.

At level A:

$$p_A = p_{A'}$$

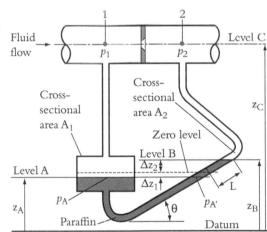

Figure 3.15 Inclined-tube manometer

Use $\Delta p = \rho g h$ for pressure changes with height in both the air and the paraffin. Since pressure differences are small, ρ_{air} is approximately constant and p depends on the vertical height of liquid.

Left-hand side:

$$p_A = p_1 + \rho_{air}g(z_C - z_A) \qquad (3.26)$$

Right-hand side:

$$p_{A'} = p_2 + \rho_{air}g(z_C - z_A) + \rho_p g(z_B - z_A) \qquad (3.27)$$

where ρ_p represents the density of paraffin.

Equate equations (3.26) and (3.27):

$$p_1 + \rho_{air}g(z_C - z_A) = p_2 + \rho_{air}g(z_C - z_B) + \rho_p g(z_B - z_A)$$
$$\therefore \qquad p_1 - p_2 = (\rho_p - \rho_{air})g(z_B - z_A)$$

However, $z_B - z_A = \Delta z_1 + \Delta z_2$, and, by trigonometry, $\Delta z_2 = L \sin\theta$. Also, since the volume of paraffin in the manometer is constant, the volume of liquid pushed out of the left-hand limb is equal to the volume of liquid moving up the right-hand limb, so $A_1\Delta z_1 = A_2 L$, and hence

$$p_1 - p_2 = (\rho_p - \rho_{air})gL\left(\frac{A_2}{A_1} + \sin\theta\right)$$ (3.28)

Now, ρ_{air} can usually be neglected, since $\rho_{air} \ll \rho_p$, and so

$$p_1 - p_2 = \rho_p gL\left(\frac{A_2}{A_1} + \sin\theta\right)$$ (3.29)

Thus, L is proportional to $p_1 - p_2$, and the measurement of L alone can give a direct measurement of the differential pressure. This also applies in the case of a normal vertical differential manometer, since putting $\frac{A_2}{A_1} = 1$ and $\theta = 90°$ into equation (3.28) gives

$$p_1 - p_2 = 2(\rho_p - \rho_{air})gL_{vertical}$$

where $2L$ is equivalent to the Δz of equation (3.25).

If $\frac{A_2}{A_1} = 0.01$, say, and $\theta = 15°$, then using equation (3.28),

$$p_1 - p_2 = 0.269(\rho_p - \rho_{air})gL_{inclined}$$

and, for a given $p_1 - p_2$, $L_{inclined} > L_{vertical}$, thus giving improved sensitivity in an inclined-tube manometer.

A manometer consists of two tubes, A and B, with vertical axes and uniform cross-sectional areas of 500 mm^2 and 800 mm^2 respectively, connected by a U-tube, C, of cross-sectional area 70 mm^2 throughout. Tube A contains a liquid of relative density 0.8, and tube B contains one of relative density 0.9. The surface of separation between the two liquids is in the vertical side of C connected to tube A. Determine the additional pressure that, when applied to tube B, will cause the surface of separation to rise 60 mm in tube C from level 1 to level 2.

- Use continuity (to recognize that the volume of liquid pushed from tube B enters tube A).
- Evaluate p_1 *(original position)* for left-hand and right-hand limbs.
- Evaluate p_2 *(new position)* for left-hand and right-hand limbs.
- Equate ($p_1 - p_2$) for left-hand and right-hand limbs to give additional pressure.

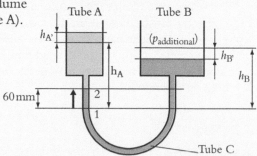

If the interface rises by 60 mm, then the surface of A rises and the surface of B falls. As no extra liquid is added or removed, then

Figure 3.16

$$A_C h_C = A_A h_{A'}, = A_B h_{B'}$$
$$\therefore \quad (70 \times 10^{-6})(60 \times 10^{-3}) = (500 \times 10^{-6})h_{A'}$$
$$\Rightarrow h_{A'} = 8.4 \text{ mm}$$

and

$$(70 \times 10^{-6})(60 \times 10^{-3}) = (800 \times 10^{-6})h_{B'}$$
$$\Rightarrow h_{B'} = 5.3 \text{ mm}$$

In the original position

$$p_1 = p_A + \rho_A g h_A \quad \text{and} \quad p_1 = p_B + \rho_B g h_B$$

In the new position

$$p_2 = p_A + \rho_A g (h_A + 8.4 \times 10^{-3} - 60 \times 10^{-3})$$

and

$$p_2 = p_B + p_{additional} + \rho_B g (h_B - 5.3 \times 10^{-3} - 60 \times 10^{-3})$$

$\therefore$

$$(p_2 - p_1) = [p_A + \rho_A g (h_A - 51.6 \times 10^{-3})] - [p_A + \rho_A g h_A]$$

$$= -51.6 \times 10^{-3} \rho_A g$$

and

$$(p_2 - p_1) = [p_B + p_{additional} + \rho_B g (h_B - 65.3 \times 10^{-3})] - [p_B + \rho_B g h_B]$$

$$= p_{additional} - 65.3 \times 10^{-3} \rho_B g$$

$\therefore$

$$p_{additional} = 65.3 \times 10^{-3} \rho_B g - 51.6 \times 10^{-3} \rho_A g = 171.6 \, \text{Pa}$$

$$\rho_A = 800 \, \text{kg m}^{-3}, \rho_B = 900 \, \text{kg m}^{-3}, g = 9.81 \, \text{m s}^{-2}$$

Surface tension and manometers

The forces of attraction between the molecules in a liquid and the forces of adhesion between the molecules in a liquid and those on a solid surface in contact with the liquid can cause undesirable effects in manometers. First, there may be a meniscus at the interface between the fluid and the manometer tube, i.e. the interface may not be flat, and this makes it difficult to define where the level of the liquid is in the tube.

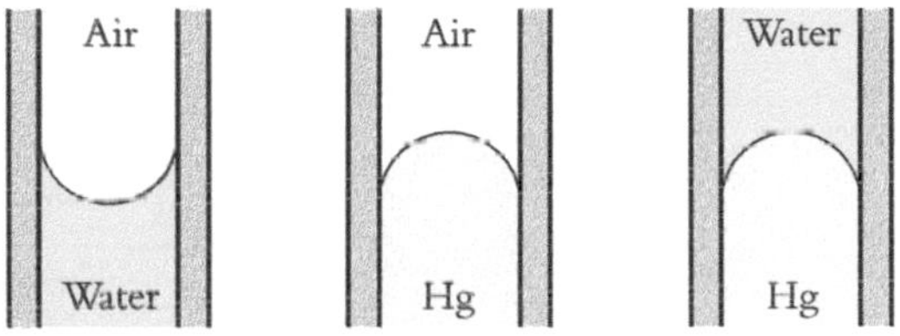

The contact angle between the liquid and the tube depends on the liquids involved and on the tube material.

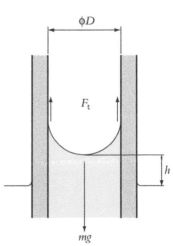

Figure 3.17 **Illustrations of the meniscus**

A second problem is that of capillary rise, where liquid rises up narrow tubes owing to the forces of attraction between the molecules of the fluid and the tube. If the meniscus is approximated to a hemisphere, then the capillary rise can be estimated.

Surface tension is force per unit length and usually has the symbol γ (gamma). The amount of capillary rise can be determined by equating the surface tension force with the weight of the fluid.

In Figure 3.18, the upwards force (due to surface tension) is

$$F_1 = \gamma \pi D$$

The downwards force (due to the weight of water) is

$$mg = \frac{\pi D^2}{4} h \rho g$$

Equating:

$$\gamma \pi D = \frac{\pi D^2}{4} h \rho g$$

$\therefore$

$$h = \frac{4 \gamma}{\rho g D} \tag{3.30}$$

Figure 3.18 **Forces acting on a meniscus**

Examination of equation (3.30) shows that the capillary rise is clearly less if the diameter of the tube is larger. The equation only really applies for tubes less than 3 mm in diameter and will tend to overestimate the effect. When applied to larger tubes, the result is less accurate.

If $\gamma = 0.073\,\text{N}\,\text{m}^{-1}$ (water), $\rho = 1000\,\text{kg}\,\text{m}^{-3}$ (water) and $D = 3\,\text{mm}$, then $h = 9.9\,\text{mm}$, which is significant. In a U-tube manometer the capillary rise should be the same on both sides and the effect cancels. Wetting agents may be added to manometer fluids to reduce the effect.

Where a meniscus is present, the manometer reading should always be taken at the position of the meniscus in the middle of the tube, whether the meniscus is convex or concave.

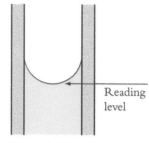

Figure 3.19

Submerged surfaces

As the pressure in a static liquid increases with depth below the surface, the pressure will exert a force on any surface submerged in the liquid. Some practical examples of these forces are: the force on the wall of a dam holding back the water in a reservoir; the force on a wall of an oil storage tank; and the force on a glass observation window in a submarine or in an aquarium. The surfaces may be flat or curved and have any orientation (vertical, horizontal or inclined).

By convention, h is used to refer to the depth below the surface of a liquid and z is used to refer to the elevation above some datum level, as shown below.

The pressure at the surface is $p_0 = p_a$ then, from equation (3.17), the pressure at a depth below the surface can be found relative to p_0 as:

$$p_1 - p_0 = \rho g(h_1 - h_0) = p_1 - p_a = p_{1\text{gauge}} \qquad (3.31)$$

The force on a surface resulting from pressure is given by equation (3.5):

$$F = \int p\,\mathrm{d}A.$$

If the pressure varies over an area, then an integral over the area will need to be performed, an example of which is given below.

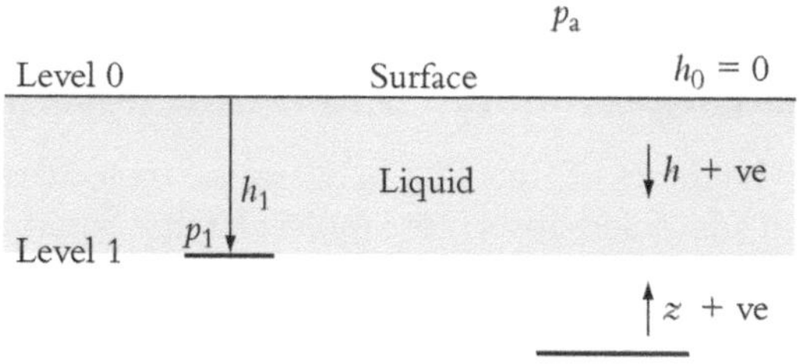

Figure 3.20

Integral over an area

Consider the rectangle ABCD illustrated in Figure 3.21, and let an element of width w and thickness $\mathrm{d}y$ lie at a perpendicular distance y from OO. The area of the element $= w\,\mathrm{d}y = \mathrm{d}A$. The total area is obtained by summing (integrating) all the elemental areas:

$$A = \int_A \mathrm{d}A = \int_{y_1}^{y_2} w\,\mathrm{d}y$$

and if w is constant

$$A = w\int_{y_1}^{y_2} \mathrm{d}y = w[y]_{y_1}^{y_2} = w(y_2 - y_1) = wb \qquad (3.32)$$

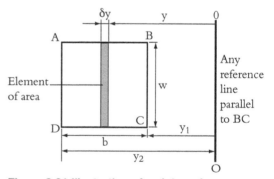

Figure 3.21 Illustration of an integral over an area

This worked example shows the method for calculating the area of a non-rectangular flat surface by integration, where the width varies and an expression for width in terms of height is derived. This is a useful technique for solving problems involving forces on submerged surfaces.

Find the area of a triangle by integration.

The area of a small element at distance h from the apex and of width w and thickness δh

$$dA = w\,dh$$

The area of the triangle is the summation of all the elemental areas:

$$A = \int_A dA = \int_{h1}^{h2} w\,dh = \int_0^a w\,dh$$

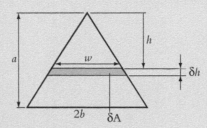

Figure 3.22

Find the width, w, as a function of h. This is a linear relationship, so $w = k_1 h + k_2$.

Boundary conditions:

when $h = 0$, $w = 0$, $\therefore k_2 = 0$

when $h = a$, $w = 2b$, $\therefore k_1 = \dfrac{2b}{a}$

$\therefore$ $\qquad\qquad w = \dfrac{2b}{a}h$

Sum all the small areas to obtain the area of the triangle:

$$\text{Area} = \int_{\text{area}} w\,dh = \int_0^a \frac{2b}{a}h\,dh$$

$$= \frac{2b}{a}\left[\frac{h^2}{2}\right]_0^a = \frac{2b}{a}\left[\frac{a^2}{2}\right] = ab$$

$$= \tfrac{1}{2}(\text{base} \times \text{height})$$

Centroid (centre of area)

Another useful concept to understand is that of the centre of area or **centroid**. Just as the centre of gravity is the point about which a body will balance, and it is the point at which all the mass of the body can be considered to act, so the centroid is the point about which a lamina (a very thin flat plate) would balance and where the centre of mass acts.

Take a lamina of area A and any shape, as illustrated in Figure 3.23, and consider an element parallel to OO at a perpendicular distance y from OO (where OO is any convenient reference line).

The area of an element of area $= \delta A = w\,\delta y$.

The product $y\,dA$ is called the first moment of area dA about axis OO.

The first moment of area A about OO $= \int y\,dA$ evaluated over area A.

Suppose that the centroid is at a perpendicular distance y_C from OO. Then, considering the whole area A to 'act' at C, the total first moment of area is Ay_C. Thus:

$$Ay_C = \int_A y\,dA \quad \text{or} \quad y_C = \frac{1}{A}\int_A y\,dA \qquad (3.33)$$

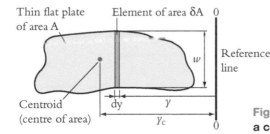

Figure 3.23 Illustration of a centroid of area

Determine the location of the centroid of a plane rectangular lamina.

Take moments for area about OO:

$$\delta M = y\,\delta A = y(b\,\delta y)$$

$$M = b\int_0^d y\,\mathrm{d}y = b\left[\frac{y^2}{2}\right]_0^d = \frac{bd^2}{2}$$

but

$$M = (\text{area})(y_C) = bd\,y_C$$

so

$$y_C = \frac{1}{bd}\frac{bd^2}{2} = \frac{d}{2}$$

and similarly

$$x_C = \frac{b}{2}$$

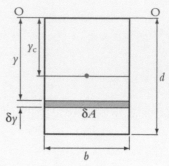

Figure 3.24

Horizontal submerged surfaces

For a submerged surface it is usual to want to know the single total force that is equivalent to the sum of all the elemental forces due to pressure, and the point at which that equivalent force acts (the centre of pressure). To find the total force, all the elemental forces are summed (integrated). The point of action can be found by taking moments for the total force and the distributed force as illustrated below.

For a horizontal surface submerged in a stationary fluid, the pressure over the surface is the same at all points because the pressure is constant on a horizontal plane. The total force is equal to the pressure multiplied by the total area. The point of action of the force is at the centroid (centre of area) of the surface. A fuller mathematical treatment is given in the following section, but for a horizontal submerged surface the key points are:

> Force = pressure × area
> Centre of pressure is at the centre of area (centroid)

Calculation of force and centre of pressure:

Figure 3.25 shows a horizontal submerged surface.

Consider a horizontal lamina at depth h beneath a liquid surface. The absolute pressure at any point on the submerged lamina is $p_a + \rho gh$ and the gauge pressure (or net pressure) is simply ρgh. δF is the force acting on the element on account of hydrostatic pressure. Since the liquid is at rest, the shear forces are zero and δF must act normal to the lamina.

The force on the element = $\delta F = (\rho gh)\,\delta A$.

The total force due to hydrostatic pressure on area A is

$$F_p = \int_A (\rho gh)\mathrm{d}A = (\rho gh)A \qquad (3.34)$$

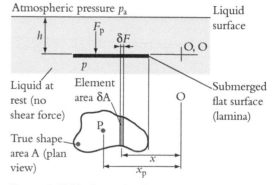

Figure 3.25 Horizontal submerged surface

since h is constant for a horizontal lamina, it follows that

> $$F_p = (\text{uniform pressure}) \times (\text{area})$$

F_p can be considered to act entirely at point P, which is called the centre of pressure. If the perpendicular distance of point P from axis OO at which the force F_p acts is x_p, then $F_p x_p$ = moment about OO of the total force acting on the lamina.

Fluid mechanics

However, the moment due to the elemental force δF about OO is

$$\delta Mo = \delta Fx = [(\rho gh)\delta A]x$$

The total moment of all the elemental forces about OO can be found by integrating to give the sum of the moments of all elements:

$$Mo = \int_A (\rho gh)x \, \mathrm{d}A = (\rho gh)\int_A x \, \mathrm{d}A$$

$\therefore$
$$Mo = (\rho gh)Ax_C \qquad (3.35)$$

Since the integral is the first moment of area about OO, it follows that x_C is the perpendicular distance from OO to the centroid of area A. As $M_O = F_p x_p$, substituting for F_p from equation (3.34) gives

$$M_O = F_p x_p = (\rho gh)Ax_p \qquad (3.36)$$

Equating equations (3.35) and (3.36) shows that the centre of pressure acts at the centroid:

$$x_p = x_C$$

so it can be concluded that

For a horizontal lamina, the centre of pressure P acts at the centroid.

Vertical submerged surfaces

For a vertical surface, the equivalent total force and its line of action can be found using a similar procedure. To find the total force, the elemental forces can be integrated over the area of the vertical surface. To find the line of action of the force, integrate the moments of all the elemental forces about a point and equate this to the moment of the total force about the same point.

A vertical submerged surface is shown in Figure 3.26. As the fluid is static, there are no shear forces and so forces arising due to hydrostatic pressure act perpendicular to the surface of the lamina. The pressure acting on the surface will be expressed as gauge pressure (pressure relative to the local atmospheric pressure). This is appropriate as almost all practical problems involving forces on surfaces (e.g. the forces on the wall of a dam or the force on a sides of a tank) are where there is atmospheric pressure acting on the other side of the surface and so the resultant or net force on the surface is that due to the pressure acting in the fluid in addition to atmospheric pressure.

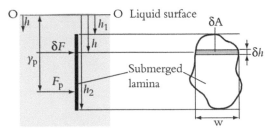

Figure 3.26 Vertical submerged surface

The lamina is split into small elements of width w (w is a variable), height δh and area δA as shown. At depth h the net pressure force will be the gauge pressure $= \rho gh$, and so the force acting on an element is

$$\delta F = \rho gh \, \delta A = \rho ghw \, \delta h$$

The total force on the element is found by summing these elemental forces:

$$F_p = \int_{area} \rho gh \, \mathrm{d}A = \int_{h_2}^{h_1} \rho ghw \, \mathrm{d}h$$

The total force, F_p, acts at P, the centre of pressure, which can be found by taking moments about a suitable axis (usually the surface of the fluid) as the moment of the point force must equal the sum of the moments due to the distributed forces. The moment of the force on the small element is

$$\delta M_\infty = \delta Fh = (\rho gh \, \delta A)h = \rho gh^2 \, \delta A = \rho gh^2 w \, \delta h$$

155

Therefore

$$M_\infty = F_p y_p = \int_{\text{area}} \rho g h^2 \, \mathrm{d}A = \int_{h_2}^{h_1} \rho g h^2 w \, \mathrm{d}h$$

and

$$y_p = \frac{M_\infty}{F_p} \tag{3.37}$$

An important feature of vertical surfaces (and any surface that is not horizontal) is that the centre of pressure is *not* at the centroid of the shape. This is because the pressure is not uniform over the surface as it increases with depth, so there is a greater force per unit area over parts of the surface at greater depths. Consequently, the centre of area is not the centre at which the pressure force appears to act, and it can be shown that the centre of pressure is always lower than the centroid of the surface.

Numerical examples explain more clearly how this can be applied.

The wall of a reservoir contains a vertical gate, 2 m wide and 3 m high. The upper edge of the gate is level with the water surface. Taking $\rho = 1000 \text{ kg m}^{-3}$ and $g = 9.81 \text{ m s}^{-2}$, determine the net force on the gate and the depth of the centre of pressure.

Find the force acting on a small element:

$$\delta F_p = \rho g h(w \, \delta h)$$
$$= \rho g h(2 \, \delta h)$$

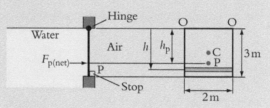

Integrate to find the total hydrostatic force:

$$\therefore \quad F_p = 2\rho g \int_0^3 h \, \mathrm{d}h = 2\rho g \left[\frac{h^2}{2} \right]_0^3 = 9\rho g$$

Figure 3.27

$$= 88.3 \text{ kN}$$

To find the line of action of the force, find the moment of force on the small element:

$$\delta M_\infty = \delta F_p h = \rho g h(2 \, \delta h)h$$

$$\therefore \qquad M_\infty = 2\rho g \int_0^3 h^2 \, \mathrm{d}h = 2\rho g \left[\frac{h^2}{2} \right]_0^3 = 18\rho g$$

The moment due to the distributed force must equal the moment due to the equivalent point force, and therefore

$$F_p h_p = M_\infty \Rightarrow h_p = \frac{18\rho g}{9\rho g} = 2 \text{ m}$$

It is worth noting that the centre of pressure is that due to the gauge pressure on the gate. And this is the position of the true net force on the gate. There is also atmospheric pressure acting on the surface of the water and on the outer surface of the gate, and the forces on either side of the gate due to atmospheric pressure cancel out. Were the outer surface of the gate exposed to a vacuum, then the absolute pressure on the gate surface would have to be considered. This would generate a larger total force and the centre of pressure would also be in a different position.

A triangular aperture in a reservoir wall is sealed by a vertical gate. The gate is hinged along the bottom horizontal edge and is kept closed by a horizontal force F_N applied to the apex A. Find the required value of F_N and the depth to the centre of pressure on the gate.

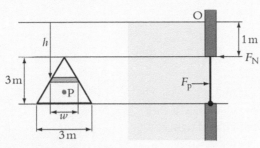

Figure 3.28

Consider a small element at depth h. The hydrostatic force acting on that element = pressure $\times$ area:

$$\delta F = (\rho g h)(w\,\delta h)$$

where w and h are both variables.

Find w as a function of h.

Variable w varies linearly with h, and

$\therefore \qquad w = k_1 h + k_2$

when $h = 1$, $\quad w = 0$ so $0 = k_1 + k_2$,

when $h = 4$, $\quad w = 3$ so $3 = 4k_1 + k_2$

$\therefore \qquad 3 = 3k_1$

so $\qquad k_1 = 1, k_2 = -1$

$\therefore \qquad w = h - 1$

and

$$\delta F = (\rho g h)(h - 1)\,\delta h$$

Integrate

$$F_{\text{p}} = \int_1^4 \rho g h(h - 1)\,\mathrm{d}h = \rho g \int_1^4 (h^2 - h)\,\mathrm{d}h = \rho g \left[\frac{h^3}{3} - \frac{h^2}{2}\right]_1^4$$

$$= 13.5\rho g = 132.4\ \text{kN}$$

Find the location of P, the centre of pressure. Take the moments of force on a small element about O:

$$\delta M = \delta F h = (\rho g h)(h - 1)\,\delta h(h) = \rho g(h^3 - h^2)\,\delta h$$

$$M = \rho g \int_1^4 (h^3 - h^2)\mathrm{d}h = \rho g\left[\frac{h^4}{4} - \frac{h^3}{3}\right]_1^4 = 42\tfrac{3}{4}\,\rho g$$

$$F_{\text{p}} h_{\text{p}} = M \Rightarrow h_{\text{p}} = \frac{42\tfrac{3}{4}\rho g}{13.5\rho g} = 3.167\ \text{m}$$

The force required to keep the gate shut is F_{N}. Take the moments about the gate hinge:

$F_{\text{p}}(4 - h_{\text{p}}) = F_{\text{N}}(3)$

$\therefore \qquad F_{\text{N}} = \dfrac{132.4(4 - 3.167)}{3}$

$\qquad\quad = 36.8\ \text{kN}$

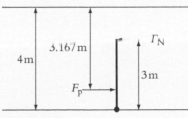

Figure 3.29

Inclined submerged surfaces (direct approach by integration)

An inclined submerged surface is shown in Figure 3.30.

For an inclined surface (as with a vertical surface) the fluid pressure is not uniform over the area A, as it varies with depth h and the centre of pressure does not lie at the centroid.

In analysing an inclined surface it is usually most convenient to take distances along the inclined surface from an axis OO which lies at the liquid surface. Thus, in the diagram, y_C is the perpendicular distance from OO to the centroid C of the surface, y_p is the perpendicular distance to the centre of pressure P, and so on. The force on an inclined surface can be calculated in the same way as that for a vertical surface by considering an element of the surface of width w and thickness δy (area $\delta A = w\,\delta y$) and then integrating over the surface to find the total force and then integrating the moment of each element about OO over the surface to find the centre of pressure.

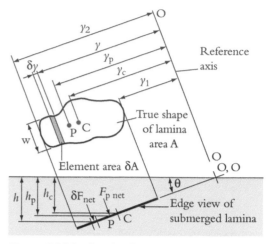

Figure 3.30 **Inclined submerged surface**

Let the element be at depth h and in a liquid at rest.

The net force on the element due to gauge pressure is

$$\delta F_{net} = \rho g h\, \delta A$$

This force, and all others due to fluid pressure, must act perpendicularly to the submerged surface, since shear forces are zero in a fluid at rest.

The total net force on the whole area is found by integraton:

$$F_{p\,net} = \int_A \rho g h\, dA$$

However, h is not constant as it varies along the inclined plate, $h = y\sin\theta$. Therefore, the total net force is

$$F_{p\,net} = \rho g \sin\theta \int_A y\, dA \qquad (3.38)$$

The integral in equation (3.38) is the first moment of area (Ay_C) of the surface about OO (see equation (3.33)). Note that y_C depends only on the shape of the area and is the same for all values of θ. Thus

$$F_{p\,net} = \rho g A (y_C \sin\theta) = \rho g A h_C \qquad (3.39)$$

where h_C is the depth to the centroid. This equation is valid irrespective of the shape of area A and for any angle θ. (Note that, as OO is at the liquid surface, when u 5 0 the surface lies horizontally at the surface of the liquid and so there are no pressure forces due to the liquid acting on it.)

To find the distance to the centre of pressure, P, take the moments of the force on an element of area about *any* convenient axis, such as OO. The moment due to δF_{net} about OO is

$$\delta M_o = (\delta F_{net})y = (\rho g h\, \delta A)y$$

Therefore, the total moment about OO is

$$M_o = \int_A \rho g h y\, \delta A \qquad (3.40)$$

The total moment about OO owing to the net total force can also be written in terms of the distance from OO to the centre of pressure:

$$M_o = F_{p\,net}\, y_p$$

Hence, y_p can be found:

$$y_p = \frac{M_o}{F_{p\,net}} \qquad (3.41)$$

Fluid mechanics

This gives the value of y_p corresponding to gauge pressure only because $F_{p\,net}$ is used.

As $h = y\sin\theta$, equation (3.40) becomes

$$M_o = \rho g\sin\theta \int_A y^2 \, dA \qquad (3.42)$$

where the integral defines the second moment of area A about OO, sometimes given the symbol I_{OO}.

From equations (3.38), (3.41) and (3.42) we have

$$y_p = \frac{\int_A y^2 \, dA}{\int_A y \, dA} = \frac{\int_{y1}^{y2} wy^2 \, dy}{\int_{y1}^{y2} wy \, dy} \qquad (3.43)$$

In general, w does not cancel, since w is a function of y for a gate of variable width.

It is worth reiterating that the magnitude of the total force F_p is equal to the pressure at the centroid of the area multiplied by the area (equation (3.39)). However, the centre of pressure is not at the centroid, but below it for any surface that is not horizontal. Considering this another way, for a horizontal surface the centre of pressure acts at the centroid; but as the surface is inclined, the centre of pressure moves further down the surface away from the centroid until it is furthest from the centroid when the surface is vertical.

Curved surfaces can also be considered by integration. The forces in the horizontal and vertical directions can be determined by resolving the pressure forces on a curved surface into the horizontal and vertical components. However, an alternative approach is given below that involves less mathematics and is more intuitive.

Curved and inclined surfaces (an alternative approach)

Components of $F_{p\,net}$

Consider first a thin, flat, submerged, rectangular plate of any width w and length L and inclined at any angle θ to a liquid surface, as shown in Figure 3.31. C denotes the centroid of the plate (lies at the centre of area), and P the centre of pressure.

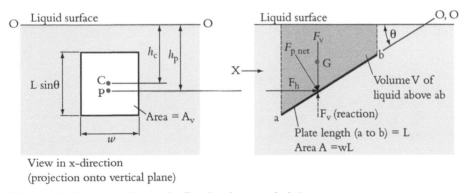

View in x-direction
(projection onto vertical plane)

Figure 3.31 Forces acting on inclined, submerged plate

The resultant net pressure force $F_{p\,net}$ acts at the centre of pressure, P, and, since the liquid is at rest, $F_{p\,net}$ must be perpendicular to the plate (as there is no shear force). Now, $F_{p\,net}$ can be resolved into vertical and horizontal components given by

$$F_v = F_{p\,net}\cos\theta \quad \text{and} \quad F_h = F_{p\,net}\sin\theta$$

The components of F_h and F_v along the plate must cancel as the resultant force is perpendicular to the gate.

Vertical component F_v

For equilibrium, the vertical component F_v must be balanced by an equal and opposite reaction F_v. This reaction must support the weight of all the liquid lying directly above the plate (between the dotted lines). If the volume of this liquid is V, then

$$\text{Weight of liquid} = mg = (\rho V)g = F_v \tag{3.44}$$

> The vertical force on the plate is equal to the weight of water above the plate.

The weight of the liquid acts at the centre of gravity G of the volume V. Thus, the line of action of F_v must pass through G. Equation (3.44) applies to any shape of flat, submerged surface and also to curved as well as flat surfaces.

Horizontal component, F_h

For a flat, submerged surface of any shape at any θ, the horizontal component of force (the resolved component from equation (3.39)) is

$$F_h = \rho g A h_C \sin\theta \tag{3.45}$$

Consider an end view of the inclined plate on a vertical plane (direction X, illustrated in Figure 3.31). The view is the vertical projection of the plate and forms an imaginary vertical surface of width w, height L $\sin\theta$ and area $A_v = wL \sin\theta$ (h_C and h_p are the same as for the inclined surface). The net (horizontal) force on the vertical surface, using equation (3.45), is

$$\begin{aligned}
F_h &= \rho g A_v h_C \\
&= \rho g(wL \sin\theta)h_C \\
&= \rho g A h_C \sin\theta
\end{aligned}$$

which is identical to equation (3.30). Therefore, F_h can be found by considering the horizontal force on the projected area A_v of the inclined plate.

The same also applies for the horizontal force on curved surfaces and can be explained by considering a submerged body of uniform width w and in a state of equilibrium, as shown in Figure 3.32. One end of the body is curved and the other end is flat and vertical. Whether the body is viewed in direction X or in direction X' exactly the same rectangular shape is seen. The net force on the vertical end is F_{h2} and must be horizontal. F_{h2} acts at the centre of pressure P$_2$. The horizontal component of the net force on the curved end is F_{h1} acting through P$_1$ (which does not lie on the curved surface).

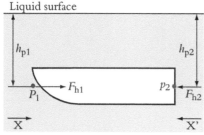
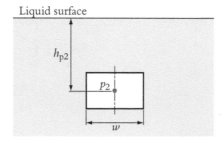

Figure 3.32 Submerged body

For equilibrium

$$F_{h_1} = F_{h_2}$$

otherwise the body would move in one direction or the other, and

$$h_{p_1} = h_{p_2}$$

otherwise the body would rotate.

Fluid mechanics

To summarize, the horizontal force on any curved or inclined surface can be determined in the following way:

> To find F_h and h_p for any curved or inclined flat surface, find F_h and h_p for the projection of that surface onto a vertical plane.

Surfaces can be flat or curved (concave or convex) with the fluid above or below them. In all cases the vertical component of the hydrostatic force is equal to the weight of water above the gate (*or the weight of the water that would be there if the gate were not*), as shown in Figure 3.33. The line of action of the vertical force passes through the centre of gravity of the fluid above the gate.

The horizontal component of force is the same as that on the vertical projection of the vertical gate.

Two other comments need to be made about curved surfaces:

- On the surface of a curved plate, the lines of action of the horizontal and vertical components of force will not meet *on the surface* of the plate, and the point of intersection of the forces is not referred to as the centre of pressure as it is not on the surface of the plate.

- When the surface is split into small elements, the force from each element is locally perpendicular to the surface of the plate. This means that, for a plate made in the shape of a circular arc, the resultant force will pass through the centre of the arc.

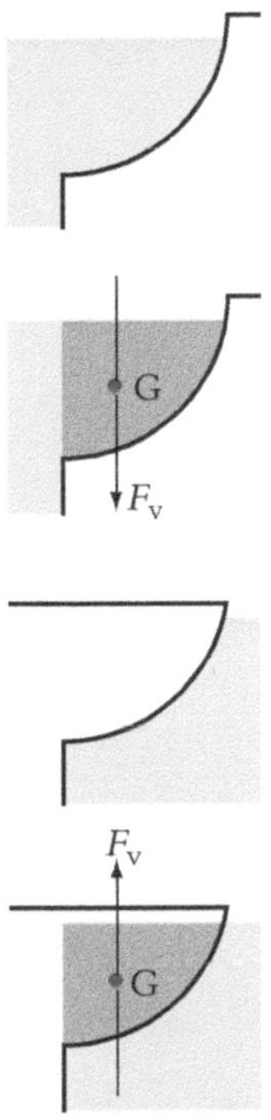

Figure 3.33 Illustration of vertical force acting on curved gates

---

A rectangular sluice gate, 1 m wide and 3 m long, is submerged in water with the upper edge hinged to a wall at a depth of 1 m below the water surface. The gate is inclined at 30° to the horizontal. Taking ρ as $1000\ kg\ m^{-3}$ and g as $9.81\ m\ s^{-2}$, find the net force on the gate that is due to gauge pressure, the depth of the centre of pressure and the normal force F that must be applied at the lower edge of the gate to open it. Neglect friction at the hinge.

F_p acts perpendicular to the gate.

To find the weight of water above the gate, consider the shape of the volume of water to be made up of a cuboid and triangular prism.

$F_V = $ (weight of water above the gate) $ = \rho V g$

$F_V = [(3\cos 30)(1) + \frac{1}{2}(3\cos 30)(3\sin 30)]\ (1)(\rho g)$
$\quad = 44.60\ kN$

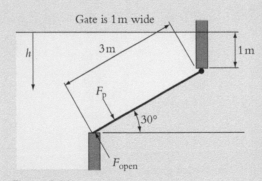

Figure 3.34

The horizontal force is found by considering the vertical projection of the gate.

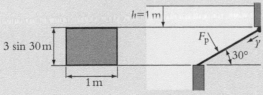

Figure 3.35

$$dF_h = \rho ghw \, dh$$

$$F_h = \rho g w \int_1^{13\sin 30} h \, dh = 1\rho g \left[\frac{h^2}{2} \right]_1^{2.5} = 2.625\rho g = 25.75 \text{ kN}$$

The total force on the gate is found by resolving the vertical and horizontal forces

$$F_p = \sqrt{F_v^2 + F_h^2}$$
$$= \sqrt{44.6^2 + 25.75^2}$$
$$= 51.5 \text{ kN}$$

The line of action of the horizontal force is found by considering the force on the vertical projection of the gate (as shown in the previous worked example) and taking moments about the water surface:

$$dM = \rho ghw \, dhh$$

$$M = \rho g w \int_1^{2.5} h^2 \, dh = 1\rho g \left[\frac{h^3}{3} \right]_1^{2.5}$$

$$= 4.875 \, \rho g$$

$$M = F_h h_p$$

$$h_p = \frac{4.875}{2.625} = 1.86 \text{ m}$$

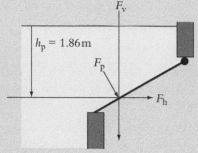

Figure 3.36

To find the force required to open the gate, take the moments for all the forces about the hinge:

$$F_p y_p = F_{open} \times 3$$

$$\sin 30 = \frac{h_p - 1}{y_p}$$

$$\therefore \qquad y_p = 1.72 \text{ m}$$

$$\Rightarrow F_{open} = \frac{51.5 \times 10^3 \times 1.72}{3}$$

$$= 29.5 \text{ kN}$$

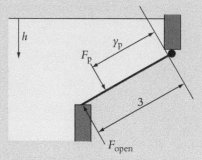

Figure 3.37

A sluice gate has a width 2 m and a radius of 6 m. The gate pivot is level with the water surface. Find the horizontal and vertical components of force on the gate and the line of action of both. Also find the resultant force.

Find the horizontal component of the force by considering the projected vertical gate:

$$F_h = \rho g A_v h_C = 1000 \times 9.81 \times (6 \times 2) \times 3 = 353.2 \text{ kN}$$

or

$$\delta F_h = \rho g h (2\delta h)$$

$$\therefore \qquad F_h = 2\rho g \int_0^6 h \, \delta h = 2\rho g \left[\frac{h^2}{2} \right]_0^6 = 36\rho g = 353.2 \text{ kN}$$

Find the line of action of the horizontal force by taking the moments about S for the horizontal component only:

$$\delta M_{S,h} = \rho g h (2\delta h) h$$

$$\therefore \quad M_{S,h} = 2\rho g \int_0^6 h^2 \, \delta h = 2\rho g \left[\frac{h^3}{3}\right]_0^6 = 144\rho g$$

$$F_h y_p = M_{S,h}$$

$$\therefore \quad y_p = \frac{144\rho g}{36\rho g} = 4 \text{ m}$$

$$F_v = \text{weight of water displaced}$$

$$= \frac{\pi r^2 w}{4}\rho g = 18\pi\rho g$$

$$= 554.7 \text{ kN}$$

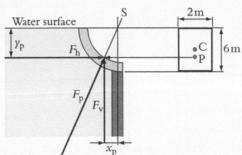

Water surface

Figure 3.38

The resultant force due to hydrostatic pressure must act through S.

Take the moments about S for hydrostatic pressure:

$$F_h y_p = F_v x_p$$

$$\therefore \quad x_p = \frac{353.2 \times 10^3 \times 4}{554.7 \times 10^3} = 2.55 \text{ m}$$

The resultant force

$$F_p = \sqrt{F_h^2 = F_v^2}$$
$$= \sqrt{(353.2 \times 10^3)^2 + (554.7 \times 10^3)^2}$$
$$= 657.6 \text{ kN}$$

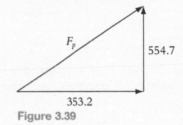

Figure 3.39

Floating bodies – buoyancy

As pressure in a fluid at rest increases with depth, any body immersed in the fluid will have a greater pressure exerted over its lower surfaces than over its upper surfaces. Consequently, there will be a net upwards force on the body and this is known as the **buoyancy force**. If this force equals the weight of the body then it will float and the body is said to be buoyant. This buoyancy force is expressed by the two laws of buoyancy discovered by Archimedes in the third century BC:

1. A body immersed in a fluid experiences a vertical buoyant force equal to the weight of the fluid it displaces.

2. A floating body displaces its own weight in the fluid in which it floats.

The buoyancy force has no horizontal component, because the horizontal thrust in any direction is the same as on a vertical projection of the surface of the body perpendicular to that direction, and the thrusts on the two faces of such a vertical projection balance exactly.

The buoyancy law can be proved as shown below.

Figure 3.39 shows a totally immersed body PQRS floating beneath free surface MN,

The upward thrust on the lower surface PSR of the body corresponds to the weight of fluid, real or imaginary, vertically above that surface, i.e. the volume PSRNM. The downward thrust on surface PQR equals the weight of fluid PQRNM. Therefore the resultant upward force is

$$\text{weight}_{PSRNM} - \text{weight}_{PQRNM} = \text{weight}_{PQRS}$$

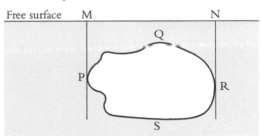

Figure 3.40 Immersed floating body

So the upward force is equal to the weight of the volume of water displaced by the body.

Note that there is no requirement for uniform density in the body.

Since the fluid is in equilibrium, we can imagine the body removed and its place occupied by an equal volume of the fluid. This extra fluid is in equilibrium under the action of its own weight and the thrusts exerted by the surrounding fluid. The resultant of these thrusts (buoyancy force) must therefore be equal and opposite to the weight of the fluid taking the place of the body (or displaced by the body) and must also pass through the centre of gravity of that volume of fluid. This point, corresponding to the centroid of the volume of the fluid (displaced), is called the centre of buoyancy and does not, in general, correspond to the centre of gravity of the body. The position of the centre of buoyancy depends on the shape of the volume considered.

Similar considerations apply to a partly immersed body as shown in Figure 3.40. In this case the buoyancy force corresponds to the weight of fluid within the volume PQR.

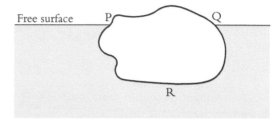

Figure 3.41 **Partially immersed body**

For a floating body to be in equilibrium, the buoyancy force (equal to the weight of the displaced fluid) must equal the weight of the body. This is known as Archimedes' principle:

$$F_B = W$$
$$\text{Bouyancy force} = \rho V g$$

where ρ = fluid density and V = immersed volume.

If the buoyancy force exceeds the weight of the body, then the body rises (bubbles, for example), and if the weight exceeds the buoyancy force, then the body sinks (a stone in water, for example).

$$F_B = W \Rightarrow \text{body floats}$$
$$F_B > W \Rightarrow \text{body rises}$$
$$F_B > W \Rightarrow \text{body sinks}$$

Stability

Buoyancy considerations alone will not determine whether or not a body will float in a chosen orientation. For example, a pencil does not float vertically (point down), although the same proportion of the body is submerged as when the pencil floats horizontally. Stability is related to relative positions of the centre of buoyancy and the centre of gravity of a body but is not covered in this chapter. The relevant theory and equations can be found in most fluid mechanics textbooks, (see References on page 212).

A cube of 35 mm sides is floating in water of 1000 kg m^{-3} density.

(a) What is the density of the cube material if the depth of submersion is 27 mm?

(b) A hemispherical scoop of 15 mm diameter is removed from the top face (volume of a sphere = $\frac{4}{3}\pi r^3$).

What is the new depth of immersion?

(a) When the cube is floating in a stable position, the buoyancy force is equal to the weight of the cube:

Weight of cube $W = mg = \rho_{cube} V_{cube} g = (0.035)^3 \rho_{cube} g = 4.21 \times 10^{-4} \rho_{cube}$

However, the buoyancy force is equal to the weight of displaced fluid:

$$F_B = \rho_{water} V_{displaced} g = 1000(0.035^2 \times 0.027)g = 0.324\,\text{N}$$

where 0.027 is the depth of immersion.

Equate $\qquad W = F_B$

$$4.21 \times 10^{-4}\,\rho_{cube} = 0.324$$

$$\rho_{cube} = 770\,\text{kg/m}^3$$

Figure 3.42

(b) Find the new weight of the cube:

Equate $\quad W = 0.324 - \frac{1}{2}(\frac{4}{3}\pi 0.0075^3)\rho_{cube}g = 0.317\,\text{N}$

When the depth of immersion is h, the buoyancy force is

$$F_B = (0.035^2 h)\rho_{water}g$$

Equate $\qquad F_B = W$

$$12.017h = 0.317$$

$$h = 26.4\,\text{mm}$$

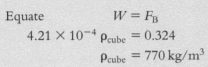

Figure 3.43

A helium balloon has a mass of 300 kg (balloonist + balloon structure); an additional 400 kg of sand ballast is also carried.

(a) At ground level the air pressure is 1.01 bar and the temperature is 300 K. Calculate the volume of helium that must be put into the balloon so that it can rise?

R for air is 287.1 J kg^{-1}K^{-1}, R for helium is 2077 J kg^{-1}K^{-1}.

(b) The balloon rises to 2000 m where the air pressure is 0.795 bar and the temperature is 275 K. To what volume must the balloon expand?

(c) The balloon is now completely full and cannot expand any more. Helium must be vented if it rises any higher to prevent it bursting. The balloon needs to rise further to get over some mountains. How much sand ballast must be released if the balloon is to rise to 3000 m where the air pressure is 0.7012 bar and the temperature is 269 K?

(d) Show that the buoyancy of the balloon is related only to the mass of helium with which it is filled, provided that the balloon is in equilibrium with the surrounding air.

(a) The total mass of the balloon plus ballast is 700 kg. This must be balanced by buoyancy from the helium. The helium displaces air, but has mass that must be balanced.

Total weight of balloon + weight of helium = weight of air displaced

$$m_b g + V_b \rho_{he} g = V_b \rho_{air} g$$

$$m_b = V_b(\rho_{air} - \rho_{he})$$

From $pV = mRT$ and $\rho = \dfrac{m}{V} = \dfrac{p}{RT}$

$$m_b = \frac{pV_b}{T}\left(\frac{1}{R_{air}} - \frac{1}{R_{he}}\right) \qquad (3.46)$$

At ground level, $p = 1.01$ bar and the temperature is 300 K

$$700 = \frac{1.01 \times 10^5 V_b}{300}\left(\frac{1}{287.1} - \frac{1}{2077}\right)$$

$V_b = 692 \text{ m}^3 = $ volume of helium that must be put into the balloon at ground level.

(b) The balloon rises to 2000 m where air pressure is 0.795 bar and the temperature is 275 K (assume that the balloon is in equilibrium with surrounding air).

From equation (3.46)

$$700 = \frac{0.795 \times 10^5 V_b}{275}\left(\frac{1}{287.1} - \frac{1}{2077}\right)$$

$V_b = 806.6 \text{ m}^3 = $ volume to which the balloon expands at 2000 m.

(c) The balloon rises to 3000 m where the air pressure is 0.7012 bar and the temperature is 269 K, but the volume remains at 806.6 m³. Therefore, helium must be vented (assume that the balloon is in equilibrium with surrounding air).

From equation (3.46)

$$m_b = \frac{0.7012 \times 10^5 \times 806.6}{269}\left(\frac{1}{287.1} - \frac{1}{2077}\right)$$

$$m_b = 631.2 \text{ kg}$$

So the balloonist must jettison $700 - 631.2 = 68.8$ kg of ballast.

(d) From equation (3.46)

$$m_b = \frac{pV_b}{T}\left(\frac{1}{R_{air}} - \frac{1}{R_{he}}\right)$$

From $pV = mRT$

$$\frac{pV_b}{T} = m_{he}R_{he}$$

$$m_b = m_{he}\left(\frac{R_{he}}{R_{air}} - 1\right)$$

So the mass that can be lifted is determined solely by the mass of helium in the balloon.

Key points from Section 3.2

- In a fluid at rest, pressure acts equally in all directions.
- Where a fluid is in contact with a surface, the pressure gives rise to a force acting perpendicular to the surface.
- In a fluid at rest, pressure is constant along a horizontal plane.
- In a fluid at rest, pressure increases with depth according to the relationship $\Delta p = \rho g \Delta h$
- The pressure at the base of a column of fluid of depth h is equal to the pressure at the top $+ \rho g \Delta h$
- Pressures can be measured by manometers.
- Surface tension can affect the readings of manometers.
- On a submerged horizontal surface the pressure is constant and the centre of pressure is also the centre of area (centroid).
- On a submerged vertical surface the pressure increases with depth and the centre of pressure is below the centroid.

- On an inclined flat or curved surface the horizontal force and its line of action is equal to the horizontal force on the vertical projection of the inclined or curved surface.
- On an inclined flat or curved surface, the vertical force is equal to the weight of the volume of water vertically above the surface and its line of action passes through the centre of gravity of that volume.
- A body fully immersed in a fluid experiences a vertical upwards force equal to the weight of the volume of fluid displaced.
- A floating body displaces its own weight in liquid.

Learning summary

At the end of this section you should be able to:

✔ determine the pressure at any depth below the surface of a liquid at rest;

✔ calculate the pressure difference indicated by a manometer;

✔ calculate by integration the magnitude and line of action of the force due to fluid static pressure on a submerged, flat, horizontal or vertical surface;

✔ evaluate the horizontal and vertical components of force on a submerged, inclined, flat or simple curved surface and determine the resultant force and line of action for some simple shapes;

✔ calculate buoyancy forces on submerged and floating objects and determine the conditions for equilibrium.

3.3 Fluids in motion

The third section of this chapter is concerned with fluids in motion. Following the description of some introductory concepts in fluid flow, the basic equations of fluid flow are derived from the laws of physics. The application of the law of conservation of mass is used to derive the continuity equation. Then, by making some simplifying assumptions that a fluid has a constant density and no viscosity, the law of conservation of energy is applied to derive the Euler and Bernoulli equations, which are then applied to a variety of practical fluid flow situations, such as flow-measuring devices. Finally, the steady flow energy equation is derived and the flow of fluids with energy dissipation due to viscosity is considered. This is applied to the flow of fluids in pipes, a very common problem in engineering, and the method of calculating pressure losses in single pipe systems and the energy needed for pumping is explained. Ways of reducing energy loss to make systems more sustainable are described.

Introductory concepts

Ideal fluid (inviscid fluids – no viscosity)

The actual flow pattern within a moving fluid is often complex and difficult to model mathematically. However, the theory can be simplified considerably by the assumption that the fluid is *ideal* and although there are limitations to the theory developed for ideal flow, there are a number of practical applications where ideal flow gives a good approximation to reality and useful calculations can be carried out.

In the context of fluid mechanics, 'ideal' has a specific meaning. An ideal fluid (liquid or gas) is one that has the following properties:

- zero viscosity (viscosity causes friction in a fluid between the fluid and a surface it is flowing over and within the fluid itself between regions moving with different velocity. Assuming zero viscosity neglects these frictional effects and means, for instance that there would be no loss in pressure in an ideal fluid flowing along a pipe);

- incompressible (i.e. does not change volume, no matter what pressure is applied, and hence has constant density. This is a reasonable assumption to apply to liquids, but gases and vapours can usually only be assumed to be incompressible when velocities are relatively low. It will be explained later that gas flows can be assumed to be incompressible if the velocity is less than 30 per cent of the speed of sound in the gas);

- zero surface tension (surface tension acts on the free surface of a liquid and in moving liquids is generally important only when there are very thin films or small droplets less than a few millimetres in thickness or diameter);

- does not change phase (phase changes will not be considered in this chapter).

Steady flow

In general a steady flow process involves no changes with time. Properties within a fluid may vary from place to place in the volume of fluid being considered but are constant at any given place. A hose-pipe spraying a garden exhibits steady flow if the velocity of the fluid flow at any particular place does not change with time. However, there will be changes in velocity from one position to another – e.g. the velocity of water at the nozzle is likely to be higher than that in the supply pipe. Unsteady flow involves changes with time. This would occur while the hose-pipe is being turned on and the flowrate increases from zero to a steady value. In this chapter all fluid flow processes considered will be assumed to be steady flow.

Uniform flow

In uniform flow the properties are the same at all points within the volume being considered at any given instant. For example, if all the fluid in a section of straight pipe is moving with the same velocity and it is at a constant density and pressure it may be regarded as uniform.

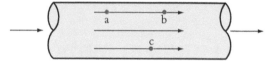

Figure 3.44 Uniform flow

If $v_a = v_b = v_c$ and $p_a = p_b = p_c$, etc., at the same instant, then the flow is uniform. If properties vary from place to place within the control volume at a given instant, then the flow is non-uniform. In practice, due to viscosity, the fluid next to a surface (e.g. at a pipe wall) will be moving more slowly than the bulk of the fluid and so uniform flow rarely occurs. However, uniform flow is a useful approximation that can be made in many situations, such as high-speed flow in a pipe.

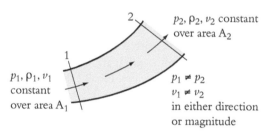

Figure 3.45 One-dimensional flow

One-dimensional flow

In one-dimensional flow it is assumed that all properties are uniform across a plane perpendicular to the flow direction. Properties thus vary in only one direction – usually the direction of flow – as indicated below.

One-dimensional flow also rarely occurs in practice. When a fluid flows along a pipe the velocity is zero at the wall and there is a layer of slower-moving fluid next to the wall owing to the effects of viscosity. The maximum velocity is at the centre of the pipe. Nevertheless, the assumption of one-dimensional flow simplifies analysis and can provide sufficiently accurate results in many situations.

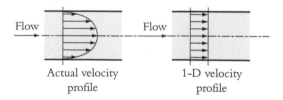

Figure 3.46 Actual velocity profile and equivalent one-dimensional profile

Representing fluid flows

There are a number of different ways of representing the pattern of fluid flow by using lines to show the direction of flow. The different ways of doing this are described below and it is useful to understand what they are and how they differ.

Fluid mechanics

Streamlines

Streamlines provide a picture of the complete flow field at a given instant and they represent the direction of the velocity of the particles. By definition:

> A streamline is a line along which all fluid particles have, at a given instant, velocity vectors that are tangential to the line.

Thus, there is no component of velocity across a streamline. No fluid particle can cross a streamline and *streamlines never cross*. In steady flow there are no changes with time and hence the streamline pattern does not change. If the flow is unsteady, the flow changes with time and so the streamline pattern may also change with time. In a uniform flow, streamlines must be straight and parallel as shown in Figure 3.47(a). If the streamlines are not parallel (Figure 3.47(b)), or are curved (Figure 3.47(c)), then the flow is non-uniform.

a) Uniform flow

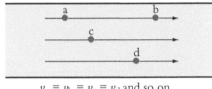

$v_a = v_b = v_c = v_d$ and so on

b) Non-uniform flow

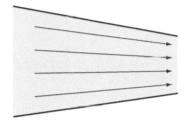

Streamlines are straight but not parallel and the flow is accelerating as the duct converges to a smaller cross-section

c) Non-uniform flow

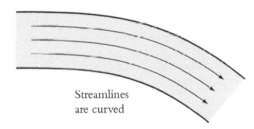

Streamlines are curved

Figure 3.47 Streamline patterns for uniform and non-uniform flow: (a) uniform flow; (b) non-uniform flow; (c) non-uniform flow

If the streamlines are straight but not parallel, the velocity varies in direction and the flow is again non-uniform.

Pathlines

A pathline shows the route taken by a single fluid particle in a time interval. It is equivalent to a time-exposure photograph which traces the movements of visible particles, for example a neutrally buoyant particle – one which has the same density as the fluid and so moves with the fluid. Soap bubbles can be used to show large-scale movement of air.

Pathlines may cross. If the flow is unsteady, as shown below, the irregular pathlines indicate that the speed and direction of particles that leave A vary with time.

Pathline of particle 1, leaving A at time t_1

Pathline of particle 2, leaving A at time t_2

Figure 3.48 Pathlines

Streakline

A streakline joins, at a given time, all particles that have passed through a given point over a given period of time. These are the lines produced, for instance, by a dye line or a smoke stream produced from a continuous supply of dye or smoke. During unsteady flow the streamline pattern (showing the instantaneous directions taken by 'all' particles) can change from moment to moment and will not be the same as a series of pathlines, each of which shows the historical route taken by a different particle over a given or different time period. Streaklines will also differ from streamlines and pathlines. However, under steady flow conditions streamlines, pathlines and streaklines are coincident and do not vary with time.

Consider the flow over an aerofoil shown in Figure 3.49. Assuming steady flow, the velocity v_1 at point 1 will always be the same in magnitude and direction, velocity v_2 at point 2 will always be the same and so on; but $v_1 \neq v_2 \neq v_3$. The line drawn as a tangent to these directions forms streamline S for steady non-uniform flow.

Figure 3.49 Flow over an aerofoil

Under steady conditions, every particle passing through point 1 will proceed in the same direction to point 2, at point 2 it will always have the same velocity v_2, at point 3 it will always have the same velocity v_3 and so on. Therefore, starting at point 1, the route taken by every single particle (the pathline) must be the same as the route taken by a string of particles (the streakline) and these lines must coincide with the streamline.

Continuity equation

The principle of **conservation of mass** is expressed in fluid mechanics by the continuity equation. In the absence of a nuclear reaction, matter is neither created nor destroyed, so considering a fixed volume, the flowrate in minus the flowrate out must equal the rate at which mass accumulates in the volume.

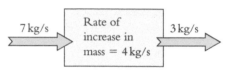

Figure 3.50 Principle of conservation of mass

For example, if 7 kg s^{-1} is flowing into the region and 3 kg s^{-1} is flowing out, then the mass within the region must be increasing by 4 kg s^{-1}, as illustrated in Figure 3.50.

If conditions are *steady* (that is, not changing with time), then the mass within the region does not change and the flow in must equal the flow out. For a steady flow:

Mass flowrate entering = mass flowrate leaving

For more than one inlet or outlet the expression generalizes to:

Sum of mass flowrates entering = sum of mass flowrates leaving

For a slightly more detailed proof, consider Figure 3.51 showing a flow in a duct.

Let the mass of fluid within the duct in the volume *abcd* be *m* at time *t* and $(m + \delta m)$ at time $(t + \delta t)$. During time interval δt, let mass δm_1 enter the control volume at section 1 and mass δm_2 leave at section 2. For the volume, defined by *abcd*, the lines *ab* and *cd* are coincident with the inner surface of the duct wall.

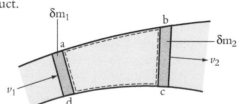

Figure 3.51 Conservation of mass with flow in a duct

By conservation of mass, during the time interval *dt*

Mass flow in − mass flow out = increase in mass within the volume *abcd*

$$\delta m_1 - \delta m_2 = m + \delta m - m = \delta m$$

The rate of change in mass within the volume is

$$\frac{\delta m}{\delta t} = \frac{\delta m_1}{\delta t} - \frac{\delta m_2}{\delta t}$$

As $\delta t \to 0, \dfrac{\delta m}{\delta t} \to \dfrac{dm}{dt} = \dot{m}$

$$\frac{\delta m}{\delta t} = \dot{m}_1 - \dot{m}_2$$

which is the general continuity equation. Here, $\dot{m}_1$ and $\dot{m}_2$ are the inlet and outlet mass flowrates.

Under steady flow conditions, the mass within the volume remains constant, and hence

$$\frac{\mathrm{d}m}{\mathrm{d}t} = 0$$

Thus, $\dot{m}_1 = \dot{m}_2$, indicating that the mass flowrate is constant at any point along the duct for steady flow.

Flow in ducts or pipes

Consider the flow in a pipe and assume one-dimensional flow, i.e. that the velocity is constant across the whole cross-section of the pipe.

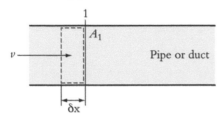

Figure 3.52

Let the inlet area, perpendicular to the flow direction at 1, be A_1. Assume one-dimensional flow. Thus, the density ρ_1 and the velocity v_1 are taken to be uniform over the inlet area A_1.

In time δt the volume of fluid passing point 1 is $A_1 \, \delta x_1$. The mass of this element of fluid is $\delta m_1 = \rho_1 A_1 \, \delta x_1$. Hence

$$\frac{\delta m_1}{\delta t} = \rho_1 A_1 \left(\frac{\delta x_1}{\delta t} \right)$$

Therefore

$$\dot{m}_1 = \rho_1 A_1 v$$

For steady flow, then, anywhere in the pipe the mass flowrate is constant, so

$$\dot{m}_1 = \dot{m}_2$$
$$\rho_1 A_1 v_1 = \rho_2 A_2 v_2$$

This is a very commonly used equation for flows in pipes and ducts. In practice the velocity is not uniform over the cross-section but a mean velocity is used which gives the true mass flowrate when multiplied by the cross-sectional area and mean density of the fluid. It applies to any shape of pipe or duct, not just to circular ones.

An incompressible fluid flows along the pipe shown. If the mean inlet velocity is 10 m s^{-1}, the inlet diameter is 27 mm and the outlet diameter is 43 mm, what is the mean outlet velocity?

As the fluid is incompressible, the density is constant and

$$\dot{V}_1 = \dot{V}_2$$
$$A_1 v_1 = A_2 v_2$$
$$\frac{\pi}{4}(0.027)^2 \times 10 = \frac{\pi}{4}(0.043)^2 v_2$$

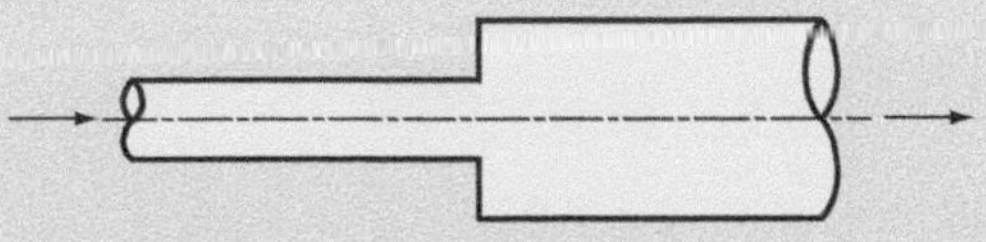

so

$$v_2 = 3.9 \text{ m s}^{-1}$$

Figure 3.53

A 10 litre bucket takes 2 minutes to fill with water (density 1000 kg m^{-3}) from a hosepipe of 20 mm diameter. What is the mass flowrate of water through the hose? What is the mean velocity of water in the hose?

$$\dot{m} = \rho \dot{V}$$

$$\dot{V} = \frac{V}{t} = \frac{10 \times 10^{-3}}{120} = 8.33 \times 10^{-5} \, \text{m}^3 \, \text{s}^{-1}$$

$$\dot{m} = \rho \dot{V} = 1000 \times 8.33 \times 10^{-5} = 0.0833 \, \text{kg s}^{-1}$$

Therefore, the mass flowrate is 0.0833 kg s^{-1}.

$$\dot{V} = Av = \frac{\pi}{4}(0.02)^2 v$$

$$\Rightarrow v = \frac{8.33 \times 10^{-5}}{\frac{\pi}{4}(0.02)^2} = 0.27 \, \text{m s}^{-1}$$

Therefore, the velocity of water in the pipe is 0.27 m s^{-1}.

Fluids in motion – the Euler equation

In a fluid at rest, the pressure varies with elevation. For a fluid in motion, the pressure depends not only on density and elevation but also on:

- acceleration (a pressure difference within the fluid is required to supply the necessary force);
- viscosity (a pressure difference is required to provide the force to overcome resistance from shear stress);
- and on any mechanical work done on or by the fluid (such as the work input from the rotor of a pump or work output if the fluid flows through a turbine to produce mechanical power. Mechanical work interactions such as these are known generally as **shaft work**).

Equations describing all of these effects are quite complex and it is useful to start in a simple way. The Euler equation (**Euler is pronounced 'oiler'**) is a relatively simple expression that allows only for the effect of elevation, density and acceleration on fluid pressure. The following assumptions are made:

(i) steady flow, i.e. properties at a given point, are constant; nothing changes with time (the flow may accelerate as it moves from one position to another, but no acceleration with time takes place);

(ii) frictionless (inviscid) flow – no viscosity, no shear stress;

(iii) zero shaft work (no work input from a pump or fan).

The equation can be derived for flow in one direction, along a streamline, by considering an element of fluid of mass δm moving with velocity v and acceleration a in a direction s (i.e. along a streamline) at a given instant (see Figure 3.54). Streamlines are always parallel to the flow, so there is no motion perpendicular to a streamline. Let the cross-sectional area of the element, perpendicular to the s-direction, be δA and let the length of the element be δs. Fluid pressure varies from p to $p + \delta p$ across the element in the s-direction. Gravity also acts on the element.

The net force on element in the s-direction is

$$\delta F = p\delta A - (p + \delta p)\delta A - g\delta m \cos\theta$$

Elevation z can be related to distance along the streamline s by

$$\cos\theta = \frac{\delta z}{\delta s} \quad \text{and} \quad \delta m = p \, \delta A \, \delta s$$

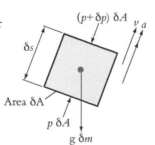
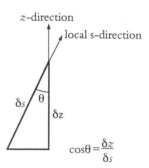

Figure 3.54 Small element of fluid travelling along a streamline

so

$$\delta F = -\delta p \, \delta A - g\rho \, \delta A \, \delta z$$

Applying $F = ma$

$$\delta F = \delta ma$$

$$-\delta p \, \delta A - \rho g \, \delta A \, \delta z = \rho \, \delta A \, \delta s \, a$$

Divide through by $(\delta A \, \delta s)$, and then

$$\frac{1}{\rho}\frac{\delta p}{\delta s} + g\frac{\delta z}{\delta s} + a = 0$$

and in the limit

$$\frac{1}{\rho}\frac{dp}{ds} + g\frac{dz}{ds} + a = 0 \tag{3.47}$$

Acceleration, a, is equal to $\dfrac{dv}{dt}$.

The velocity, v, of the element of fluid will, in general, vary as a function of both time and location. Therefore, $v = f(t, s)$, so a change in velocity, δv, can be expressed in terms of the two variables s and t as

$$\delta v = \frac{\partial v}{\partial s}\delta s + \frac{\partial v}{\partial t}\delta t$$

where $\dfrac{\partial v}{\partial s}$ and $\dfrac{\partial v}{\partial t}$ are the partial derivatives of velocity with respect to distance and time (i.e. $\dfrac{\partial v}{\partial s}$ is the variation in velocity with distance, when other variables (e.g. time) remain constant).

Dividing by δt gives

$$\frac{\delta v}{\delta t} = \frac{\partial v}{\partial s}\frac{\delta s}{\delta t} + \frac{\partial v}{\partial t}$$

which in the limit becomes

$$\frac{dv}{dt} = \frac{\partial v}{\partial s}\frac{ds}{dt} + \frac{\partial v}{\partial t}$$

Under steady conditions, the velocity at any point does not change with time (the velocity of the fluid may change with position but not with time), and consequently $\dfrac{\partial v}{\partial t} = 0$. Also, $\dfrac{ds}{dt} = v$, the velocity, and hence

$$\frac{dv}{dt} = v\frac{dv}{ds}$$

Substituting back into equation (3.47) above yields

$$\frac{1}{\rho}\frac{dp}{ds} + g\frac{dz}{ds} + v\frac{dv}{ds} = 0$$

This is the Euler equation. It only applies for:

- steady flow;
- inviscid flow (no viscosity);
- zero shaft work;
- flow along a streamline.

Euler's equation is often also written as

$$\frac{dp}{\rho} + g \, dz + v \, dv = 0 \tag{3.48}$$

Note that, for a fluid at rest, velocity v is zero, and so the Euler equation becomes

$$\frac{dp}{\rho} + g\,dz = 0$$

$$dp = -\rho g\,dz$$

which integrates to give

$$\Delta p = -\rho g \Delta z$$

which is the familiar equation relating pressure change with depth in a stationary fluid.

The Bernoulli equation

The Euler equation is a differential equation, and, by integrating along a streamline, a practical equation can be derived, known as the Bernoulli equation.

The Bernoulli equation is obtained by integrating the Euler equation (equation (3.48)):

$$\int \frac{dp}{\rho} + g \int dz + \int v\,dv = 0$$

$$\int \frac{dp}{\rho} + gz + \frac{v^2}{2} = \text{constant}$$

If it is assumed that the density is constant (i.e. that the fluid is incompressible), then

$$\frac{p}{\rho} + gz + \frac{v^2}{2} = \text{constant}$$

(3.49)

This is the Bernoulli equation, and it applies only for the following situations:

- inviscid flow; ⎫
  } Ideal flow
- incompressible flow; ⎭
- steady flow;
- zero shaft work;
- flow along a streamline.

Each term in the Bernoulli equation has the units of energy mass^{-1}, and the equation is a special statement of the **law of conservation of energy** that applies to an ideal flow in which there are no frictional losses or energy transfers such as work taking place.

The three terms in the equation relate to three different types of energy that the fluid may have:

- The first term $\frac{p}{\rho}$ is a pressure term and is the energy needed to move the fluid against the local pressure in the fluid. This term is sometimes known as **flow work** or **flow energy**.

- The second term gz is an elevation term and relates to the energy needed to raise the fluid against gravity. It is a potential energy term.

- The third term $\frac{v^2}{2}$ relates to the velocity, and so it represents the kinetic energy in the fluid.

The Bernoulli equation simply states that the energy in a fluid due to pressure, elevation and motion is conserved provided that there is no external energy input from a pump or fan (no shaft work) and that there is no viscosity to dissipate kinetic energy as heat. This is a similar concept to assuming frictionless motion in mechanics.

The Bernoulli equation may also be rearranged into slightly different forms, as shown opposite.

Fluid mechanics

Head form of the Bernoulli equation

If each term in the Bernoulli equation is divided by g, then each term has the units of length (elevation) and is called a **head**. Head represents energy weight^{-1}.

The head form of the Bernoulli equation is

$$\frac{p}{\rho g} + z + \frac{v^2}{2g} = \text{constant}$$

$$(3.50)$$

The use of the term **head** within fluid mechanics is very common. The three terms in the Bernoulli equation are referred to as

$$\frac{p}{\rho g} = H_p = \text{pressure head}$$

$$z = \text{elevation head}$$

$$\frac{v^2}{2g} = H_v = \text{velocity head}$$

The sum of all three heads is referred to as the total head, H_T. Thus

$$H_p + z + H_v = H_T$$

H_T is constant, and so between two locations, 1 and 2, there is no change in total head:

$$H_{T1} = H_{T2}$$

The quantity $(H_p + z)$ is called the piezometric head, H_{PZ}, and $p + \rho gz$ is the piezometric pressure. Piezometric head has no particular significance and is simply a convenient combination.

Pressure form of the Bernoulli equation

If each term in the Bernoulli equation is multiplied by ρ, then each term has the units of pressure. The pressure form of the Bernoulli equation is

$$p + \rho gz + \tfrac{1}{2}\rho v^2 = \text{constant}$$

$$(3.51)$$

The three terms in the Bernoulli equation are referred to as

p	static pressure
ρgz	elevation pressure associated with the height relative to some datum
$\tfrac{1}{2}\rho v^2$	dynamic pressure related to the velocity of the fluid

The sum of all three pressures is referred to as the total pressure p_T. Thus

$$p + \rho gz + \tfrac{1}{2}\rho v^2 = p_T$$

p_T is constant, and so between two locations, 1 and 2, there is no change in total pressure head:

$$p_{T1} = p_{T2}$$

The quantity $p + \rho gz$ is called the **piezometric pressure**.

Applications of the Bernoulli equation

Flow along a pipe

Consider the flow of an ideal, constant density liquid along a uniform diameter, inclined pipe, as illustrated in Figure 3.55. Measured from the pipe centreline at section 1 (height z_1 above the datum), the height of the column of liquid in the piezometer is H_{p1}. At the pipe centre-line:

$$p_{1,\text{abs}} = p_a + \rho g H_{p1}$$

$$p_{1,\text{gauge}} = p_{1,\text{abs}} - p_a = \rho g H_{p1}$$

Thus, the height of the liquid column in the piezometer is equal to the pressure head (based on gauge pressure).

The corresponding piezometric head is

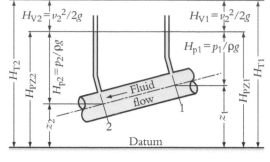

Figure 3.55 **Flow of an ideal fluid along a pipe**

$$H_{\text{PZ1}} = \frac{p_1}{\rho g} + z_1$$

and is represented physically by the distance $(z_1 + H_{p1})$. If the mean velocity at section 1 is v_1, then the velocity head is

$$H_{v1} = \frac{v_1^2}{2g}$$

By definition, the total head

$$H_{\text{T1}} = \frac{p_1}{\rho g} + z_1 + \frac{v_1^2}{2g} = H_{\text{PZ1}} + H_{v1}$$

Similarly, at section 2

$$H_{\text{T2}} = \frac{p_2}{\rho g} + z_2 + \frac{v_2^2}{2g} = H_{\text{PZ2}} + H_{v2}$$

Each streamline has its own values of z, H_p and H_v, but in one-dimensional steady flow it is usual to take the elevation z to the centre-line of the full pipe, i.e. we consider the central streamline. For an **ideal fluid and steady flow**, and irrespective of the angle of inclination of the pipe,

$$H_{\text{T1}} = H_{\text{T2}}$$

that is

$$H_{\text{PZ1}} + H_{v1} = H_{\text{PZ2}} + H_{v2}$$

Also, for steady flow, the **mass flowrate** is constant, so

$$\dot{m}_1 = \dot{m}_2 \Rightarrow \rho_1 A_1 v_1 = \rho_2 A_2 v_2$$

but

$$\rho_1 = \rho_2 \quad \text{and} \quad A_1 = A_2$$

so $v_1 = v_2$ and then $H_{v1} = H_{v2}$, and hence $H_{\text{PZ1}} = H_{\text{PZ2}}$.

Therefore, for the steady, inviscid (**frictionless**) flow of an incompressible fluid at constant velocity (and with zero shaft work) the piezometric head is constant irrespective of the inclination of the pipe. Also, if the pipe is horizontal, then, $z_1 = z_2$ and $p_1 = p_2$.

For pipe and duct flow, the two equations to apply are as follows:

Bernoulli: $\quad p_1 + \rho g z_1 + \frac{1}{2}\rho v_1^2 = p_2 + \rho g z_2 + \frac{1}{2}\rho v_2^2$

Continuity: $\quad \rho_1 A_1 v_1 = \rho_2 A_2 v_2$

A pipe carrying water tapers from a cross-section of $0.3\,\text{m}^2$ at A to $0.15\,\text{m}^2$ at B. At A, the velocity, assumed uniform, is $1.8\,\text{m s}^{-1}$ and the pressure is $117\,\text{kPa}$ gauge. If frictional effects are negligible, determine the pressure at B, which is 6 m above the level of A.

Apply continuity between A and B to find the velocity at B:

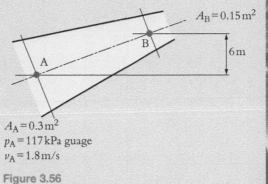

$$\rho v_A A_A = \rho v_B A_B$$

$$\therefore \quad v_B = \frac{v_A A_A}{A_B} = \frac{1.8 \times 0.3}{0.15} = 3.6\,\text{m s}^{-1}$$

Apply Bernoulli between A and B (central streamline) to find the pressure at B:

$A_A = 0.3\,\text{m}^2$
$p_A = 117\,\text{kPa guage}$
$v_A = 1.8\,\text{m/s}$

Figure 3.56

$$p_A + \rho g z_A + \tfrac{1}{2}\rho v_A^2 = p_B + \rho g z_B + \tfrac{1}{2}\rho v_B^2$$

$$P_B = P_A + \rho g(z_A - z_B) + \tfrac{1}{2}\rho(v_A^2 - v_B^2)$$

$$p_B = 117 \times 10^3 + (1000 \times 9.81 \times (-6)) + 500(1.8^2 - 3.6^2)$$

$$= 53.3\,\text{kPa}$$

Therefore, the gauge pressure at B is $53.3\,\text{kPa}$.

Venturimeter

The venturimeter is a well-known fluid flowmeter, widely used in industry. The venturimeter can easily be installed in a pipeline and creates very little overall pressure loss. Essentially the venturimeter is a convergent-divergent pipe. The divergent section with a gradual taper restores the fluid to almost its original state (pressure). Figure 3.57 illustrates a venturimeter.

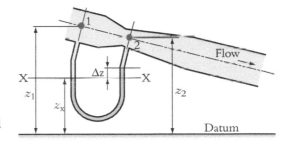

The standard venturimeter has a 30° included angle convergent part and a 7° included angle divergent part (as this is found to minimize the pressure loss) and a diameter ratio $\dfrac{d_1}{d_2} = 2$. As the fluid passes from 1 to

Figure 3.57 Venturimeter

2, the velocity increases owing to the change in flow area. Consequently the pressure falls and this differential pressure can be measured. As the flow travels through the divergent section, the flow area is increased to the original value and the flow velocity is reduced and the pressure increases again. As the inlet and outlet areas are the same, there is no overall change in velocity and, if there are no frictional losses, then there will be no change in pressure. In practice there are always some frictional losses and these result in a small overall loss of pressure across the device.

Consider the steady flow of an ideal fluid (inviscid, incompressible) along a venturimeter and apply the Bernoulli equation to the central streamline:

$$\frac{p_1}{\rho g} + z_1 + \frac{v_1^2}{2g} = \frac{p_2}{\rho g} + z_2 + \frac{v_2^2}{2g}$$

Rearranging yields

$$\frac{v_1^2 - v_2^2}{2g} = H_{PZ1} - H_{PZ2} = \Delta H_{PZ} \tag{3.52}$$

Apply the continuity equation between 1 and 2, assuming that v_1 is uniform over A_1 and that v_2 is uniform over A_2:

$$\rho_1 A_1 v_1 = \rho_2 A_2 v_2$$

As the fluid is incompressible, it follows that $\rho_1 = \rho_2$, so

$$v_1 = \frac{A_2}{A_1} v_2$$

and hence

$$v_2^2 - v_1^2 = v_2^2\left(1 - \left(\frac{A_2}{A_1}\right)^2\right)$$

Substituting this into equation (3.52) gives

$$v_2^2 = \frac{2g\Delta H_{PZ}}{\left(1 - \left(\frac{A_2}{A_1}\right)^2\right)} \qquad (3.53)$$

This is the velocity that would exist, for the given pressure difference, if the flow were frictionless and therefore represents the maximum velocity at the throat of the venturi.

The volume flowrate

$$\dot{V} = \frac{\dot{m}}{\rho} = v_2 A_2 \ (= v_1 A_1)$$

The volume flowrate calculated will be the *ideal* volume flowrate because it is based on the assumption of an ideal fluid:

$$\dot{V}_{ideal} = A_2 v_2 = \sqrt{\frac{2g\Delta H_{PZ}}{\left(1 - \left(\frac{A_2}{A_1}\right)^2\right)}}$$

The real volume flowrate will be less than the ideal value, as any real fluid will have a viscosity. This means that some of the mechanical energy in the fluid is dissipated as heat, so the fluid slows down and the flowrate will be lower. It is usual to account for this difference by mulitplying the ideal volume flowrate by a discharge coefficient, C_d:

$$\dot{V}_{real} = C_d \dot{V}_{ideal} = C_d A_2 \sqrt{\frac{2g\Delta H_{PZ}}{\left(1 - \left(\frac{A_2}{A_1}\right)^2\right)}}$$

This applies to horizontal, vertical or inclined venturimeters, since ΔH_{PZ} allows for the appropriate elevations z_1 and z_2.

C_d is called the coefficient of discharge or discharge coefficient, defined by

$$C_d = \frac{\dot{m}_{real}}{\dot{m}_{ideal}} = \frac{\rho \dot{V}_{real}}{\rho \dot{V}_{ideal}} = \frac{\dot{V}_{real}}{\dot{V}_{ideal}}$$

where ρ is constant for liquids and low speed gas flows. The coefficient C_d is determined experimentally, by individual calibration of the venturimeter with a particular fluid, and is typically found to be between 0.95 and 0.99 for low-viscosity fluids such as water and air. It accounts for the effects of friction and non-uniformity of velocity profile through the device and depends on $\frac{d_2}{d_1}$, the flowrate, viscosity, surface roughness and the positions of manometer tappings. For the venturimeter, Bernoulli's equation gives a good approximation to the real flowrate, since C_d is so close to 1.0.

Water flows along a horizontal venturimeter which has inlet and throat areas of 180 cm² and 45 cm² respectively. A mercury-under-water manometer connected to the inlet and throat sections shows a difference in mercury levels of 10 cm. The coefficient of discharge is 0.96. Calculate the pressure difference between inlet and throat and the mass flowrate of the water ($\rho_w = 1000\ \mathrm{kg\,m^{-3}}$ and $\rho_m = 13{,}600\ \mathrm{kg\,m^{-3}}$).

$$\Delta p = \Delta z_m(\rho_m - \rho)g = 0.1(13\,600 - 1000)9.81$$

$$= 12.36\text{ kPa}$$

$$\Delta H_{PZ} = \left(\frac{p_1}{\rho g} + z_1\right) - \left(\frac{p_2}{\rho g} + z_2\right)$$

and, as $z_1 = z_2$,

$$\Delta H_{PZ} = \frac{\Delta p}{\rho g} = \frac{12.36 \times 10^3}{1000 \times 9.81}$$

$$= 1.26\text{ m}$$

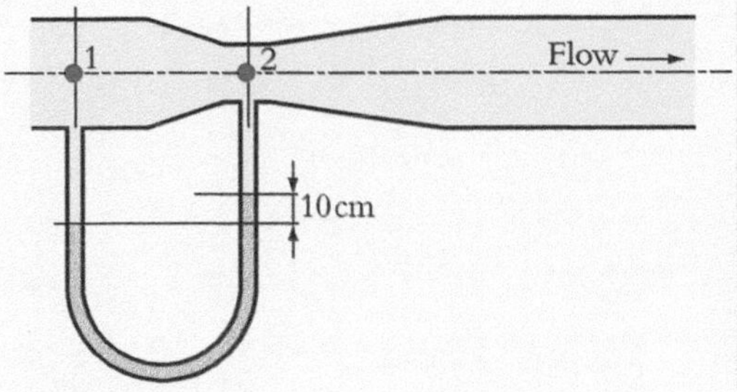

Figure 3.58

$$\dot{V}_{real} = C_d\dot{V}_{ideal} = C_dA_2\sqrt{\frac{2g\Delta H_{PZ}}{\left(1 - \left(\frac{A_2}{A_1}\right)^2\right)}}$$

so

$$\dot{V}_{real} = 0.96 \times 45 \times 10^{-4}\sqrt{\frac{2 \times 9.81 \times 1.26}{\left(1 - \left(\frac{45}{180}\right)^2\right)}}$$

$$= 0.02218\text{ m}^3/\text{s}$$

and

$$\dot{m}_{water} = \rho\dot{V} = 1000 \times 0.02218 = 22.18\text{ kg s}^{-1}$$

Therefore, the mass flowrate of water is 22.2 kg s^{-1}.

A vertical venturimeter carries a liquid of relative density 0.8 and has inlet and throat diameters of 150 mm and 75 mm respectively. The pressure connection at the throat is 150 mm above that at the inlet. If the actual rate of flow is 40 litres s^{-1} and C_d is 0.96, calculate:

a) the pressure difference between inlet and throat and

b) the difference of levels in a vertical U-tube mercury manometer connected between inlet and throat ($\rho m = 13{,}600$ kg m^{-3}).

For the venturimeter

$$\dot{V}_{real} = C_d\dot{V}_{ideal} = C_dA_2\sqrt{\frac{2g(\Delta H_{PZ})}{\left(1 - \left(\frac{A_2}{A_1}\right)^2\right)}}$$

and hence
$$40 \times 10^{-3} = 0.96\left(\frac{\pi}{4}0.075^2\right)\sqrt{\frac{2 \times 9.81(\Delta H_{PZ})}{\left(1 - \left(\frac{75}{150}\right)^4\right)}}$$

and

$$\Delta H_{PZ} = 4.25\text{ m}$$

$$\Delta H_{PZ} = \left(\frac{p_1}{\rho g} + z_1\right) - \left(\frac{p_2}{\rho g} + z_2\right)$$

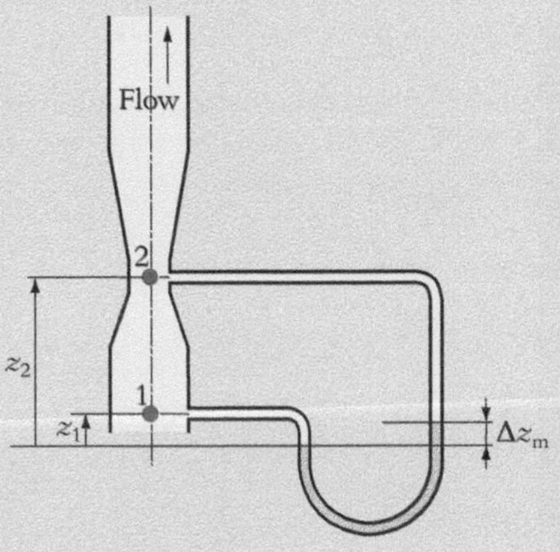

Figure 3.59

$$\therefore \qquad p_1 - p_2 = \rho g(\Delta H_{PZ} + (z_2 - z_1))$$
$$= 800 \times 9.81(4.25 + 0.15)$$
$$= 34.53 \text{ kPa}$$

For the manometer

$$p_1 + z_1\rho g = p_2 + \rho g(z_2 - \Delta z_m) + \rho_m g \Delta z_m$$

$$\frac{p_1}{\rho g} + z_1 = \frac{p_2}{\rho g} + z_2 - \Delta z_m + \frac{\rho_m}{\rho}\Delta z_m$$

$$\Delta H_{PZ} = \Delta z_m\left(\frac{\rho_m}{\rho} - 1\right)$$

$$\therefore \qquad \Delta z_m = \frac{4.25}{\left(\dfrac{13\,600}{800} - 1\right)} = 0.266 \text{ m}$$

Therefore, the mass flowrate of water is 22.2 kg s^{-1}.

Nozzle flow meter

A nozzle flow meter (illustrated in Figure 3.60) is essentially a venturimeter with the divergent section omitted. The nozzle (convergent part) can be formed from a plate or cast and is simpler and cheaper to make than a complete venturi. It is also shorter and can be installed easily in between two flanges. Pressure tappings 1 and 2 are provided where piezometer tubes or the limbs of a differential manometer can be connected. This enables the pressure difference $(p_1 - p_2)$ or the difference in piezometric head (ΔH_{PZ}) to be determined.

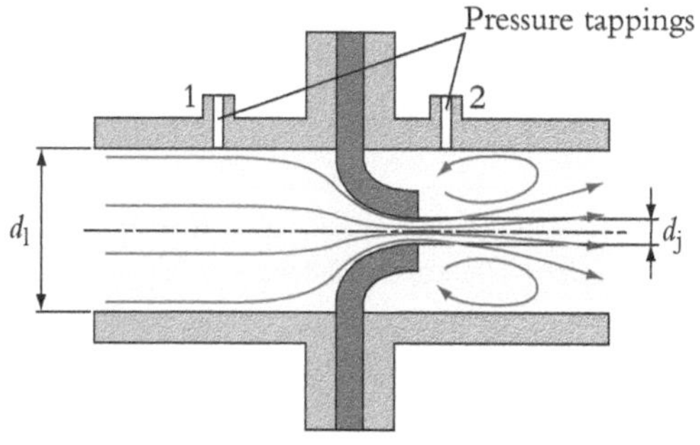

Figure 3.60 **Nozzle flow meter**

Immediately after the nozzle exit there is a jet of fluid separating from the nozzle and this causes a recirculation region in which, in real fluids with viscosity, much of the kinetic energy of the fluid at the nozzle exit is dissipated by friction. This means that there is little recovery of pressure after the nozzle and so there is a significant overall pressure loss. This is in contrast to the venturimeter, where the gradual divergence after the throat avoids flow separation and recirculation and helps to minimize energy dissipation and overall pressure losses.

The same analysis is used as for the venturimeter and Bernoulli's equation is applied at sections 1 and 2. The inlet area is taken to be A_1 (as before) and the throat area, A_2, is replaced by the nozzle outlet area, designated A_j, where $A_j = \frac{\pi}{4}d_j^2$.

Thus, the real flowrate through the nozzle flow meter is

$$\dot{V}_{real} = C_d A_j \sqrt{\frac{2g(\Delta H_{PZ})}{\left(1 - \left(\dfrac{A_j}{A_1}\right)^2\right)}}$$

C_d usually lies between 0.95 and 0.98, its value depending upon the positions of tappings 1 and 2, the nozzle shape, the ratio $\dfrac{d_j}{d_1}$ and the flow conditions.

Fluid mechanics

Orifice plate meter

This consists essentially of a flat round plate containing a central sharp-edged hole and is placed concentrically across the flow. It is simpler and cheaper to make than the nozzle meter but has a much higher overall pressure loss as there is more flow separation and recirculation downstream of the orifice plate. It tends to be less accurate than the venturimeter (in the order of ± 2 per cent error compared with ±0.75 per cent). A disadvantage is that the orifice plate is much more susceptible to wear than the nozzle meter. Any erosion of the sharp edge by the fluid, or possible build-up by deposits, affects the value of discharge coefficient and hence the calibration. As the edge becomes rounded C_d increases.

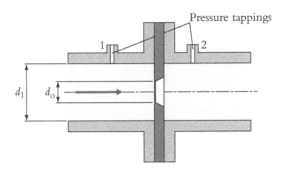

Figure 3.61 Orifice plate meter

As in previous cases, pressure tappings 1 and 2 enable $(p_1 - p_2)$ or ΔH_{PZ} to be determined. The analysis is basically the same as for the nozzle flow meter and the flowrate is given by the equation:

$$\dot{V}_{real} = C_d A_o \sqrt{\frac{2g(\Delta H_{PZ})}{\left(1 - \left(\frac{A_o}{A_1}\right)^2\right)}}$$

where A_1 is the upstream pipe diameter and A_o is the diameter of the hole in the orifice plate; C_d is the **discharge coefficient** and depends on the positions and type of the pressure tappings and on whether the flow is laminar or turbulent. It is determined experimentally from readings of $\dot{m}$ and ΔH_{PZ}, and appropriate values can be found in the standards for orifice plates. Values of C_d for orifice plates are in the range 0.6–0.65 for $0.2 < \dfrac{d_o}{d_1} < 0.75$ and high flowrates. Orifice plates are extensively used in industry because they are cheap and easy to install. There is a British Standard covering their calibration and installation (BS 1042).

The discharge coefficient (C_d) is much lower than for the venturimeter or nozzle flow meter. The main reason for this is because of the way the flow separates from the orifice plate. As the orifice has a sharp-edged hole and is quite thin, the flow continues to converge downstream of the hole for a short distance, becoming parallel at about position 2, as shown in Figure 3.62. Consequently the minimum flow area is not at the orifice itself but at 2. This is called the **vena contracta** and has an effective flow area A_j.

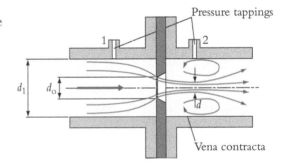

Figure 3.62 The streamlines in an orifice plate meter

The area of the vena contracta, A_j, is less than the orifice area, A_o. The two areas are related by a coefficient of contraction C_c, where

$$C_c = \frac{A_2}{A_o} = \frac{A_j}{A_o} = \left(\frac{d_j}{d_o}\right)^2 \qquad (3.54)$$

$$A_2 = C_c A_o$$

The coefficient of contraction varies with the ratio of the orifice diameter to the pipe diameter and the flowrate and is typically in the order of 0.6 for a thin orifice plate with a sharp edge for turbulent flow.

This phenomenon of the vena contracta commonly occurs wherever a flow passes through a thin sharp-edged hole and it effectively reduces the area of the hole, typically to around 60 per cent of the actual diameter, depending on the upstream flow conditions.

An orifice plate meter situated in a horizontal pipe of 100 mm diameter has an orifice of diameter 60 mm and $C_d = 0.6$. The pressure difference across the orifice is 0.1 bar. Calculate the mass flowrate assuming that the fluid is paraffin of density $755\ \mathrm{kg\,m^{-3}}$.

Orifice plate equation

$$\dot{V}_{\mathrm{real}} = CA_o \sqrt{\frac{2g(\Delta H_{\mathrm{PZ}})}{1 - \left(\dfrac{A_0}{A_1}\right)^2}}$$

$$\Delta p = p_1 - p_2 = 0.1\ \mathrm{bar}$$

$$\therefore\ \Delta H_{\mathrm{PZ}} = \left(\frac{p_1}{\rho g} + z\right) - \left(\frac{p_2}{\rho g} + z\right) = \frac{\Delta p}{\rho g}$$

$$\Delta H_{\mathrm{PZ}} = \frac{\Delta P}{\rho g} = \frac{0.1 \times 10^5}{755\,g} = 1.35\ \mathrm{m}$$

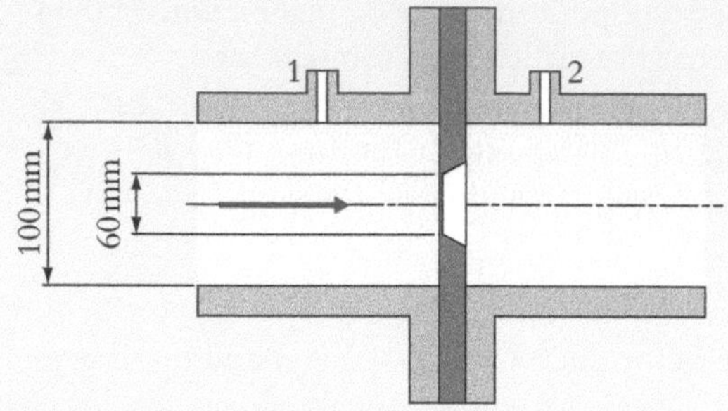

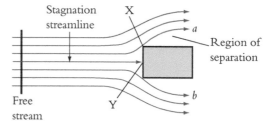

Figure 3.63

$$\dot{V}_{\mathrm{real}} = 0.6\left(\frac{\pi}{4}0.06^2\right)\sqrt{\frac{2g(1.35)}{1 - \left(\dfrac{0.06}{0.1}\right)^2}}$$

$$= 9.36 \times 10^{-3}\ \mathrm{m^3\,s^{-1}}$$

$$\dot{m} = \rho\dot{V} = 755 \times 9.36 \times 10^{-3} = 7.066\ \mathrm{kg\,s^{-1}}$$

Pitot-static probe

Pitot-static probes are used to give velocity information at a point within a flowfield. They are widely used in experimental fluids and also in many industrial applications.

Consider a stream of fluid flowing towards a fixed body, as illustrated in Figure 3.64.

The fluid is deflected by the object and flows around it as indicated by the streamlines. Some fluid will flow along the vertical face of the body and will follow streamline a and some streamline b. Depending on whether the front of the body is round or bluff (as in Figure 3.64) the flow may separate from the body at a separation point (X in Figure 3.64). Sharp corners always cause separation, although flow may still separate even if the body is smooth and rounded. Fluid flowing along the central streamline hits the body at Y and is brought suddenly to rest. Point Y is referred to as the stagnation point as the fluid stagnates there.

Figure 3.64 Flow round a fixed body

Apply Bernoulli's equation to the central streamline between a point upstream where the flow is undisturbed by the body and point Y. Let the upstream pressure, velocity and elevation be p, v and z respectively:

$$\frac{p}{\rho} + v^2 + gz = \frac{p_Y}{\rho} + v_Y^2 + gz_Y$$

However, $v_Y = 0$, and, if the streamline is horizontal, then $z = z_Y$, and so

$$p_Y = p + \tfrac{1}{2}\rho v^2$$

p_Y is called the stagnation pressure and is often denoted by p^+. p is the static pressure and is the dynamic pressure. If p^+ and p can be measured, then the velocity of flow can be deduced:

$$p^+ - p = \tfrac{1}{2}\rho v^2$$

Fluid mechanics

so

$$v = \sqrt{\frac{2}{\rho}(p^+ - p)}$$

The static pressure p can be measured using a piezometer attached to a static pressure tapping and the stagnation pressure, p^+, with a small tube known as a pitot tube facing upsteam directly into the flow, as shown in Figure 3.65.

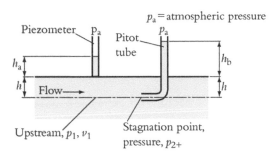

Some of the liquid flowing along the horizontal pipe rises into the piezometer and some into the pitot tube to levels which depend on p_1 and p_2^+ respectively. When equilibrium is reached, the fluid in the tubes is at rest. While fluid in the pipe flows past the piezometer opening, fluid flowing directly towards the pitot tube opening is brought suddenly to rest at 2 and hence exerts the stagnation pressure p_2^+.

Figure 3.65 Pitot tube and piezometer

The static pressure at 1 is

$$p_1 = p_a + \rho g(h_a + h)$$

The stagnation pressure at 2 is

$$p_2^+ = p_a + \rho g(h_b - h)$$

Therefore

$$p_2^+ - p_1 = \rho g(h_b - h_a)$$

However, as $z_1 = z_2$ and velocity v_1 is constant as the pipe is of constant diameter, it follows that

$$p_2^+ - p_1 = \tfrac{1}{2}\rho v_1^2$$

and so

$$\rho g(h_b - h_a) = \tfrac{1}{2}\rho v_1^2$$

Thus

$$v_1 = \sqrt{2g(h_b - h_a)}$$

This is a useful result because v_1 can be determined from measurements of h_b and h_a.

The two tubes (piezometer and pitot) can be combined in a single instrument called a pitot-static probe. This consists of two concentric tubes, as shown in Figure 3.65.

The inner tube is subject to the stagnation pressure while the radial holes in the outer tube are static pressure tappings. Thus, the annular space between the two tubes is subject to a pressure that, by careful design of the probe, is very close to the static pressure in the upstream region. The rounded ellipsoidal front of the probe prevents separation of the flow and helps to ensure that the flow over the static pressure holes is fairly uniform. These holes must be located at a sufficient distance from the front, around which the fluid accelerates and the pressure decreases. The instrument must also be aligned accurately with the flow direction (see inset graph on Figure 3.66). If the probe is not directly facing the oncoming flow, then the stagnation pressure measured is lower and the static pressure can be higher, resulting in a lower pressure difference and hence under estimate of the flow velocity. This can be a limitation if the predominant flow direction is not known in advance. Pitot-static probes can be used for fluids and gases where density can be assumed constant.

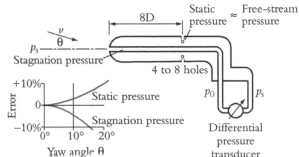

Figure 3.66 Pitot-static probe (reproduced from *Fluid Mechanics* by Frank M. White)

A pitot-static probe is attached to the wing of a light aircraft flying at an altitude of 2000 m, where the air temperature is 2 °C and the pressure is 0.8 bar. The pressure difference measured by the probe is 3.3 kPa. Take the gas constant for air to be 287.1 J kg^{-1} K^{-1}. What is the speed of the plane in miles per hour?

For a pitot-static probe

$$p^+ - p_s = \tfrac{1}{2}\rho v^2$$

$$\therefore \quad v^2 = \frac{2}{\rho}(3.3 \times 10)$$

$$\rho = \frac{p}{RT} = \frac{0.8 \times 10^5}{287.1 \times 275} = 1.013 \text{ kg m}^{-1}$$

Velocity $= 80.7 \text{ m s}^{-1} = 181 \text{ miles h}^{-1}$

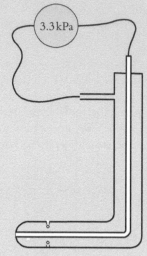

Figure 3.67

Free surface flow

There are many applications for the Bernoulli equation in free surface flows. Consider the exit flow through a hole in the wall of a tank as illustrated in Figure 3.68. All streamlines must start from the free surface (otherwise there would be a situation with fluid being removed from within the flow and not being replenished).

At the water surface the pressure is atmospheric and, if the tank is large, the velocity is negligible. At 2, the pressure is also atmospheric. Applying the Bernoulli equation between 1 and 2 gives

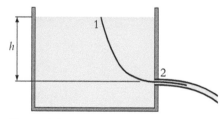

Figure 3.68

$$\rho g z_1 = \rho g z_2 + \tfrac{1}{2}\rho v_2^2$$

which simplifies to

$$v_2 = \sqrt{2g(z_1 - z_2)} = \sqrt{2gh}$$

This kind of analysis is only appropriate where the drop in the free surface is negligible and steady conditions exist.

The volume flowrate through the hole would depend on the size and shape of the hole. If the hole had a smooth rounded intake (similar to the orifice in a nozzle flow meter) the discharge coefficient would be close to unity. A hole with a sharp edge may behave like an orifice plate and then the discharge coefficient may be about 0.6.

Applicability of the Bernoulli equation

The Bernoulli equation can be used in flows where the fluid can be considered to be incompressible and where the pressure losses due to viscosity in the fluid are small compared to the changes in static pressure, elevation pressure and/or dynamic pressure.

In devices where there are significant changes in velocity, such as the flow meters considered earlier or in the discharge of fluids from a large reservoir through a small hole, then the Bernoulli equation can be applied with good accuracy. However, in situations such as flows in long horizontal pipes of constant diameter (e.g. pipelines), the pressure changes along the pipe are dominated by frictional effects due to the viscosity of the fluid. These situations will be considered later in this chapter.

Liquids can be considered incompressible under most practical situations and so the Bernoulli equation can be applied. It is generally found that in gas flows where the velocity of the gas is less than 30 per cent of the sonic velocity (the speed of sound), then compressibility effects are generally negligible and the Bernoulli equation can be applied with good accuracy. However, at higher velocities the Bernoulli equation can give large errors, for example in the discharge of compressed air from pipes. As guide, the sonic velocity a in a gas is given by the equation $a = \sqrt{\gamma RT}$, where R is the specific gas constant, T is the absolute temperature and γ is the ratio of the specific heat capacity at constant pressure to the specific heat capacity at constant volume for a gas. For air, γ is approximately 1.4.

The study of compressible flow in gases is beyond the scope of this book and readers are advised to consult a more advanced fluid mechanics text.

3.4 Viscous (real) fluids

Viscosity

A fluid offers resistance to motion due to its viscosity or 'internal friction'. The greater the resistance to flow, the greater the viscosity. For example, water flows (can be poured) more readily than some oils and we say that oil has a higher viscosity.

Viscosity arises from two effects: in liquids, intermolecular forces act as drag between layers of fluid moving at different velocities, and this acts like friction; in gases, it is the mixing of molecules as they move, in continuous random motion, between regions of faster and slower moving fluid that causes friction due to momentum transfer. Slower layers tend to retard faster layers, hence resistance.

> The higher the viscosity of a fluid, the greater is the resistance to motion between fluid layers, and, for a given applied shear stress, the lower is the rate of shear deformations between layers.

Viscosity can be defined in terms of the forces generated at rate of shear or velocity gradient. Consider the straight and parallel flow of a fluid over a fixed, horizontal surface as shown in Figure 3.69. Fluid in direct contact with the surface has zero velocity because surface irregularities trap molecules of the fluid. A short distance from the surface, the fluid has a relatively low velocity, but in the free-stream region, the velocity is v_f.

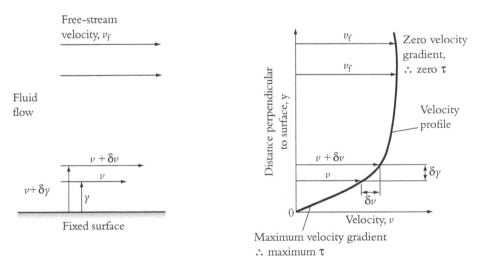

Figure 3.69 **Velocity gradient near a fixed surface**

The velocity of flow increases continuously from zero at the fixed surface to v_{f_1} in the main stream and is usually represented by a smooth curve, known as the **velocity profile**. At a distance y, let the velocity be v and at a distance $y + \delta y$, let the velocity be $v + \delta v$. The ratio $\dfrac{\delta v}{\delta y}$ is the average velocity gradient (rate of change of velocity with distance) over the distance δy, but, as δy tends to zero,

$\dfrac{\delta v}{\delta y} \rightarrow$ the value of the differential $\dfrac{dv}{dy}$ at a point such as point A;

$\dfrac{dv}{dy}$ at point A is the true **velocity gradient**, or rate of shear, at point A.

For most fluids used in engineering, it is found that the shear stress τ is directly proportional to the velocity gradient when straight and parallel flow is involved. Thus:

$$\tau \propto \frac{dv}{dy} \quad \text{or} \quad \tau = \text{constant} \left(\frac{dv}{dy} \right)$$

The constant of proportionality is called the **dynamic viscosity** or often just the **viscosity** of the fluid, and is denoted by μ. Hence

$$\tau = \mu \frac{dv}{dy}$$

This is **Newton's law of viscosity**, and fluids that obey it are known as **Newtonian fluids**. The equation is limited to straight and parallel (laminar) flow. Only if the flow is of this form does dv represent the time rate of sliding of one layer over another. If angular velocity is involved, the velocity gradient is not necessarily equal to the rate of shear.

The equation indicates that, in the main stream where $\dfrac{dv}{dy} = 0, \tau = 0$, and, at the fixed surface where $\dfrac{dv}{dy}$ has its maximum value, τ is a maximum.

The value of μ depends on the type of fluid, on molecular motion between fluid layers, and on inter-molecular (cohesive) forces. However, for a fluid exhibiting Newtonian properties, μ is independent of velocity gradient and shear stress but depends considerably on fluid temperature and, to a very small extent, on fluid pressure. For **liquids** μ decreases as temperature increases due to reduced cohesive forces between molecules. For **gases**, μ increases as temperature increases due to increasing momentum of molecules leading to greater momentum exchange.

In some fluids the viscosity varies with the rate of shear and these are known as **non-Newtonian** fluids. Figure 3.70 shows the relationship between $\dfrac{dv}{dy}$ and τ for Newtonian and non-Newtonian fluids. For a Newtonian fluid there is a linear relationship and the gradient of the line, μ, is constant for each fluid at a given temperature. For fluids, subject to a given τ and a given temperature, μ varies inversely with the velocity gradient.

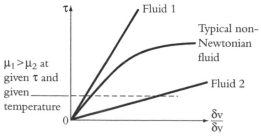

Figure 3.70 **Relationship between velocity gradient and shear stress for Newtonian and non-Newtonian fluids**

Fluid mechanics

There are different kinds of non-Newtonian fluid in which the viscosity varies in different ways with the applied shear stress. In some fluids (a common example is tomato ketchup) the viscosity can be made to decrease when a high rate of shear is applied for a period of time; so if you shake the bottle the ketchup flows more easily. Conversely, some fluids increase in viscosity when the rate of shear increases.

The units of μ can be found by considering the dimensions of the quantities involved.

$$[\mu] \equiv \left[\frac{\tau}{\left(\dfrac{dv}{dy}\right)}\right] \equiv \left[\frac{N}{m^2}\frac{s}{m}m\right] \equiv [Pa\ s]$$

$$\equiv \left[\frac{kg\ m}{s^2}\frac{1}{m^2}\frac{s}{m}m\right] \equiv \left[\frac{kg}{m\ s}\right]$$

Thus, the unit of dynamic viscosity is the **pascal second** (Pa s) or $kg\ m\ s^{-1}$.

(Another commonly used unit of viscosity is the poise, which was derived from the c.g.s. system of units (centimetre, gram, second): 1 poise = 0.1 Pa s.)

Kinematic viscosity, denoted by ν, is defined as the ratio μ/ρ, where ρ is the fluid density.

The **unit of kinematic viscosity** is $m^2\ s^{-1}$.

Viscosity and lubrication

Many commonly occurring fluids in engineering applications conform to Newton's law of viscosity (e.g. water, gases and many oils), such that

$$\tau = \mu\frac{dv}{dy}$$

These fluids are Newtonian. In general the velocity gradient is not known, but where thin films exist (such as lubricating flows) the velocity gradient can be approximated as linear and an estimate of shear stress obtained. In these lubrication cases, the frictional drag due to the viscosity in the fluid is often a significant force and can be estimated as follows:

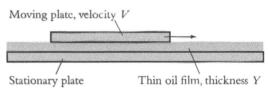

Moving plate, velocity V

Stationary plate   Thin oil film, thickness Y

Figure 3.71

In this case

$$\tau = \mu\frac{\partial v}{\partial y} \approx \mu\frac{V}{Y}$$

and the drag force F on the plate is thus equal to the viscous shear stress τ multiplied by the area of the plate A subject to the stress: $F = \tau A$.

For a cylindrical journal bearing

$$\tau \approx \mu\frac{r\omega}{t}$$

Drag torque $T = \tau(2\pi rL)r$

Power dissipated $= T\omega = \mu\left(\dfrac{2\pi r^3 L}{t}\right)\omega_2$

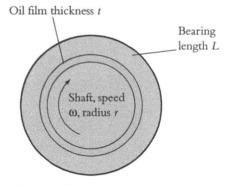

Oil film thickness t

Bearing length L

Shaft, speed ω, radius r

Figure 3.72

While a lubricant is necessary to reduce wear and friction between two surfaces, there is a drag force due to the viscosity of the lubricant. This represents an energy loss in the system, so, for a more sustainable system, it is desirable wherever possible to reduce the frictional drag due to oil viscosity. This can be achieved by using an oil of lower viscosity, although the lubricating effect must be maintained, or by having a thicker lubricating film. In many situations, such as hydrodynamic bearings, for instance, the thickness of the film of a lubricant is not fixed but determined by the loads between the two surfaces and the details of the oil flow in the gap. The details of lubrication theory are beyond the scope of this book.

A hydraulic ram 200 mm in diameter and 1.2 m long moves wholly within a concentric cylinder 200.2 mm in diameter, and the annular clearance is filled with oil of relative density 0.85 and kinematic viscosity 400 mm^2 s^{-1}. What is the viscous force resisting the motion when the ram moves at 120 mm s^{-1}?

$$\tau = \mu\frac{\partial u}{\partial y}$$

$$\mu = \nu\rho = 400 \times 10^{-6} \times 850 = 0.34 \text{ kg m}^{-1}\text{s}^{-1}$$

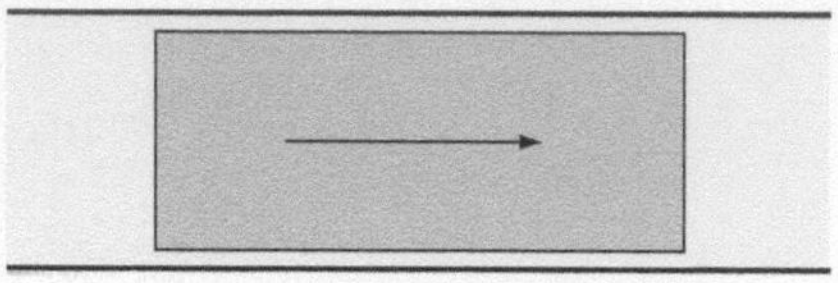

$$\frac{\partial u}{\partial y} \approx \frac{u}{y}$$

so

$$\tau = \mu\frac{u}{y} = 0.34\left(\frac{120 \times 10^{-3}}{0.1 \times 10^{-3}}\right) = 408 \text{ N m}^{-2}$$

$$\tau = \frac{F}{A}$$

$$\therefore F = 408\pi \times 0.2 \times 1.2 = 307.6 \text{ N}$$

Therefore, the drag force on the ram is 308 N.

Figure 3.73

Laminar and turbulent flow

Experimental visualization shows that there are essentially two types of flow. For flow at relatively low velocities, it is found that the fluid pathlines tend to be smooth and parallel. At high velocities pathlines tend to be very irregular and cross over one another. For example, the flow from a tap at low velocity is steady, the surface of the water is smooth and there may be no perceptible changes in the flow with time. If the flowrate is increased, the surface breaks up and is rough and although the bulk flow may be steady, there are very noticeable unsteady local disturbances in the flow.

Osbourne Reynolds (1842–1912) carried out a series of simple experiments to illustrate and distinguish these two types of flow. His experimental test facility is shown below in Figure 3.74.

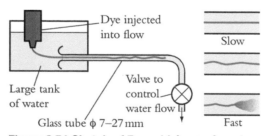

Figure 3.74 Sketch of Reynolds' experiment

As the valve was opened and the flowrate increased, Reynolds observed different flow patterns. At low flowrates the dye filament was smooth and unbroken along the length of the pipe. At higher flowrates the filment was initially smooth but became wavy at a certain distance along the pipe. As the flowrate was increased further, the filament fluctuations increased until there came a point where the dye filament suddenly mixed across the tube. Further increase in flowrate did not change the flow pattern, but the point at which mixing occurred moved closer to the inlet. These two distinct types of flow are known as laminar and turbulent.

Fluid mechanics

Laminar flow

Laminar flow occurs at low flowrates. The fluid particles move in smooth parallel lines and the fluid can be considered to be moving in layers or laminae. Because the fluid stays in distinct layers there is no mixing across the streamlines (except for a small amount due to molecular diffusion).

Turbulent flow

At high flowrates, the paths taken by individual particles are not straight but very disorderly. A substantial amount of mixing takes place within the flow and consequently the dye line in Reynolds' experiment dispersed very rapidly when the flow became turbulent. Although the bulk flow may be steady, locally it is very unsteady with many small-scale fluctuations.

Transition

Between fully laminar and fully turbulent flow there is a transition region and the flow may oscillate between laminar and turbulent.

At very low rates the flow may be entirely laminar (smooth) across the whole cross-section of the pipe. At higher flowrates turbulent conditions prevail across the pipe except for a very thin region of laminar flow adjacent to the wall, where the presence of the wall prevents the turbulent fluctuations taking place. Most flows in engineering situations are turbulent. However, viscous, laminar flows are the usual type in lubrication applications and in the flow through porous media.

Turbulence occurs where there is shear in a flow and different layers of fluid move at different velocities. Fluid particles touching solid surfaces do not move but stick to the surface. This is a fundamental principle in fluid mechanics and is known as the **no-slip** condition at the wall. It occurs because of the adhesive forces between the fluid and solid molecules and also because the surface of a solid is not smooth, but rough, on a molecular scale, and so the fluid particles become trapped in the crevices. When a fluid moves over a surface, there is sliding between the layers of fluid against the wall and the fluid further away from the wall and this region of slower moving fluid near to the wall is known as a **boundary layer**. In the boundary layer the velocity increases as the distance from the wall increases and results in a velocity gradient. Viscosity in a fluid causes shear stresses within the boundary layer as viscosity acts like friction between the layers of fluid moving at different velocity. At low velocity, the fluid particles slide smoothly over each other and the flow is laminar. However, as the velocity (and momentum) of the flow increases, the layers of fluid do not slide smoothly over each other but the flow becomes unstable, and a rolling, swirling motion occurs between the layers of fluid. This swirling motion causes what are known as **eddies**. These are localized unsteady disturbances and cause mixing to take place between the streamlines. They can be observed in everyday life, for instance when pouring milk into black coffee or in the smoke coming out of a chimney. It is evident that the turbulence causes mixing as the milk mixes well with the coffee. The localized rapid changes in velocity cause small pressure fluctuations in the fluid and these can be heard, for instance in the rushing of the wind past your ears or when someone blows into a microphone. The size of the turbulent eddies depends on the scale of the flow. In the atmosphere, when there is a wind blowing, the boundary layer can be tens of metres thick and the turbulent eddies are very large, particularly near to buildings, which provide large disturbances to the air flow. The gusts experienced on a windy day are large, unsteady turbulent eddies swirling along in the air.

Turbulence is inhibited by viscosity in the fluid, and it is found that more viscous fluids need to move at higher velocities before turbulence commences. This can also be seen when a thick cream is poured into a cup of coffee and it is noticed that the turbulent eddies are much reduced or non-existent; so the tendency for a flow to become turbulent depends on the viscosity of the fluid and the velocity of the flow which can be characterized by its momentum. A simple numerical parameter is used to distinguish between laminar and turbulent flows; it is called the Reynolds number (Re) and is the ratio of the dynamic forces in the fluid (characterized by momentum) to the viscous forces (characterized by the viscous shear stress). It is defined by the equation:

$$Re = \frac{\rho v l}{\mu}$$

where ρ is the fluid density, V is the fluid velocity, l is a characteristic linear dimension in the flow and μ is the fluid viscosity.

The derivation of the number is explained below.

In laminar flow, the most influential factor is the magnitude of the viscous forces. Newton's law of viscosity states that the shear stress is

$$\tau = \mu \frac{\partial u}{\partial y}$$

If some characteristic velocity and distance are chosen, then the viscous shear stress is proportional to $\frac{(\mu v)}{L}$, where v is the characteristic velocity and L is the characteristic length. As force is equal to stress times area (area $\propto L^2$), it follows that

$$\text{Viscous forces} \propto \mu \left(\frac{v}{L}\right) L^2 = \mu v L$$

In turbulent flow, viscous effects are less significant, but inertial effects (mixing, momentum exchange, etc.) are. Inertial forces due to accelerations can be represented by $F = ma$:

$$m \propto \rho L^3$$

$$a = \frac{dv}{dt}$$

so

$$a \propto \frac{v}{t} \quad \text{and} \quad t = \frac{L}{v}$$

$$\therefore \qquad a \propto \frac{v^2}{L}$$

and hence

$$\text{Inertia forces} \propto \rho L^3 \left(\frac{v^2}{L}\right) = \rho L^2 v^2$$

The ratio of inertial forces to viscous forces is called the Reynolds number and is abbreviated Re. Because Re is a ratio of forces, it has no units and is referred to as a non-dimensional or dimensionless group. In fluid mechanics there are many such non-dimensional groups.

$$\text{Reynolds number} = \frac{\text{inertia forces}}{\text{viscous forces}} = \frac{\rho L^2 v^2}{\mu v L}$$

$$\text{Reynolds number } (Re) = \frac{\rho v L}{\mu}$$

If the flowrate is low, the flow is laminar, viscous forces dominate and Re is small. If the flowrate is high, inertia forces dominate and Re is large.

The choice of the velocity v and characteristic dimension l is important and defined for different flow situations. For flows in pipes or ducts (known as internal flows) the velocity is the mean velocity in the pipe and the dimension is the pipe diameter or some equivalent diameter for non-circular ducts. However, for flows over the outside of bodies or surfaces (known as external flows) the velocity used is the free stream velocity (the velocity far away from the object) and the dimension is a characteristic diameter for flows over objects or a distance along a surface in the flow direction for flows over a surface.

Fluid mechanics

Non-dimensional groups are very important in fluid mechanics because it is found that when comparing two similar flow situations, if the appropriate non-dimensional numbers are the same then the flow behaviour will be the same, regardless of the actual dimensions and fluids involved.

For example, consider the flow along a circular cross-section pipe. The Re is

$$Re = \frac{\rho v d}{\mu}$$

where d = diameter of the pipe, and v = mean velocity = $\frac{\dot{V}}{A}$

Atmospheric air flowing at $10\ \text{m s}^{-1}$ in a 10 mm diameter pipe has an Re

$$Re_{\text{air}} = \frac{\rho v d}{\mu} = \frac{1.2 \times 10 \times 0.01}{1.8 \times 10^{-5}} = 13{,}300$$

Water flowing in a 1 m diameter pipe at $0.133\ \text{m s}^{-1}$ has an Re

$$Re = \frac{1000 \times 0.133 \times 1}{10^{-3}} = 13{,}300$$

These two flows will behave in the same manner because they have the same Reynolds number.

For flow in a smooth circular pipe, under normal engineering conditions the following approximate values can be assumed:

$$Re < 2000 \text{ laminar flow}$$

$$2000 < Re < 4000 \text{ transitional flow}$$

$$Re > 4000 \text{ turbulent flow}$$

These figures are only an approximate guide and will be significantly affected by surface roughness.

Water of $1000\ \text{kg m}^{-3}$ density and 0.001 viscosity flows steadily along a pipe of 3 mm internal diameter. 200 ml of water is collected in a measuring beaker in 43 s. What is the Re of the flow? Is the flow laminar, transitional or turbulent?

First, calculate the volume flowrate:

$$\dot{V} = \frac{V}{t} = \frac{200 \times 10^{-6}}{43} = 4.65 \times 10^{-6}\ \text{m}^3\ \text{s}$$

Then calculate the mean flow velocity:

$$\dot{V} = vA = v\left(\frac{\pi}{4}d^2\right)$$

∴

$$v = \frac{4.65 \times 10^{-6}}{\left[\frac{\pi}{4}0.003^2\right]} = 0.66\ \text{m s}^{-1}$$

Finally, calculate the Re:

$$Re = \frac{\rho v d}{\mu} = \frac{1000 \times 0.66 \times 0.003}{0.001} = 1980$$

As the Re is approaching 2000, the flow is likely to be transitional.

Steady flow energy equation

So far, ideal flows have been considered in which there are no viscous frictional losses and the Bernoulli equation has been derived and applied to some situations where frictional losses are not significant. For flows in long pipes, viscous frictional losses are very important and the next section examines how to consider these types of flow.

In a real viscous fluid, the effect of viscosity is to degrade mechanical energy into heat, in the same way that mechanical friction, for example in an object sliding along a surface, degrades mechanical energy into heat. The Bernoulli equation is a statement of the **law of conservation of energy** but only accounts for mechanical forms of energy. To take into account the effects of viscosity in which some mechanical energy is converted into heat, a fuller expression of the energy conservation law must be made. In this section the steady flow energy equation (SFEE) will be derived. This is a fuller mathematical statement of the law of conservation of energy and includes thermal energy terms as well as mechanical energy terms but only applies to steady flows. The equation can be derived by considering an open system or control volume that undergoes a steady flow process and a brief introduction to the concept of a control volume is needed first.

Control volumes

A *control volume* is an imaginary boundary drawn to facilitate analysis of a flow. It is drawn at places where the fluid flow conditions are known, or where a good approximation can be made; for instance, where the fluid velocity, pressure and temperature is known. The region within the control volume is analysed in terms of the energy and mass flows entering or leaving the control volume and it is not necessary to understand the detail of the flows inside the control volume.

Control volumes may be likened to **free body diagrams** in solid mechanics, in that the forces on an object can be analysed without having to understand the details of the stresses within the object.

Derivation of the SFEE

A control volume is shown in Figure 3.75 through which a flow moves. Fluid enters at section 1, at mean height z_1 above the datum, with velocity v_1 and pressure p_1. Fluid leaves at section 2, at mean height z_2 (in general, $z_1 \neq z_2$) above the datum, with velocity v_2 and pressure p_2.

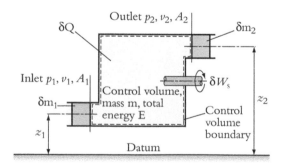

Figure 3.75 Open system undergoing a steady flow process

During the process heat δQ may be transferred to or from the control volume in time δt and shaft work δW_s may be done on or by the fluid. (For shaft work, either the shaft is attached to a rotor and the fluid causes the rotor to turn (this is a turbine and gives a work output) or the shaft may drive the rotor to give an increase in fluid pressure (this is then a pump, fan or compressor and work is input to the control volume). At this stage the mechanical details are not of interest as the concern is only with overall changes in energy.

Fluid mechanics

Steady flow exists, therefore all properties are constant with time at any given location. It is further assumed that all properties are uniform over the whole of the inlet and outlet areas A_1 and A_2. Mass flowrate, heat transfer and shaft work rates are steady and do not change with time. For **steady flow** conditions the mass m of fluid within the control volume and the total energy E of the fluid within the control volume must be constant with time. E includes all possible forms of energy possessed by a fluid in motion:

- internal energy (thermal energy);
- kinetic energy;
- potential energy.

Let mass δm_1 enter the control volume in time δt, and mass δm_2 leave in the same time interval. Let the specific energy (energy per unit mass) of the fluid of mass δm_1 be e_1, where

$$e_1 = u_1 + \frac{v_1^2}{2} + gz_1$$

In words this means

Specific energy = specific internal energy + specific kinetic energy + specific potential energy

For uniform (one-dimensional) flow at the inlet and for steady conditions, e_1 will be uniform over the whole of inlet area A_1 and will be constant with time. Thus, the total energy of mass δm_1 is

$$\delta E_1 = e_1 \, \delta m_1$$

and similarly

$$\delta E_2 = e_2 \, \delta m_2$$

As the system is operating steadily, the mass within the control volume does not change, so the mass entering in time δt (δm_1) must be the same as the mass leaving (δm_2). Hence

$$\delta m_1 = \delta m_2$$

Applying conservation of energy to the control volume, the energy entering the control volume due to heat transfer and work must equal the change in energy in the control volume in a given time interval:

$$\delta Q + \delta W = \delta E \tag{3.55}$$

where δQ is the heat transferred to the system in time δt, δW is the total work done on the system and δE is the total change in energy within the control volume and is equal to the difference between the energy in the mass entering the control volume (δE_1) and the energy in the mass leaving the control volume (δE_2):

$$\delta E = \delta E_2 - \delta E_1 = e_2 \, \delta m_2 - e_1 \, \delta m_1 \tag{3.56}$$

The total work δW is

$$\delta W = \delta W_s + \text{net flow work}$$

Flow work is the work needed to push the mass into the control volume against the pressure within the control volume, so the inlet flow work is done by the surroundings on the system to push mass δm_1 into the control volume and outlet flow work is done on the surroundings by the system as mass δm_2 is ejected. An expression for flow work can be obtained as follows.

The work done on δm_1 to push it across section 1 against pressure p_1, assumed uniform over A_1, is

$$\text{Flow work done on } \delta m_1 = p_1 A_1 \, \delta x_1 = p_1 \, \delta V_1$$

where δV_1 is the volume of the element entering. Similarly, flow work is done by the system on δm_2:

Flow work done on $\delta m_2 = (-p_2 A_2)\, \delta x_2 = -p_2\, \delta V_2$

Thus, the net flow work done on the closed system in moving through the control volume is

$$\delta W = \delta W_s + (p_1\, \delta V_1 - p_2\, \delta V_2) \qquad (3.57)$$

Combining equations (3.55), (3.56) and (3.57) gives

$$\delta Q + [\delta W_s + (p_1\, \delta V_1 - p_2\, \delta V_2)] = e_2\, \delta m_2 - e_1\, \delta m_1$$

However, $\delta m_1 = \delta m_2 = \delta m$. Therefore

$$\frac{\delta Q}{\delta m} + \frac{\delta W_s}{\delta m} + (e_2 - e_1) + p_2\frac{\delta V_2}{\delta m} - p_1\frac{\delta V_1}{\delta m}$$

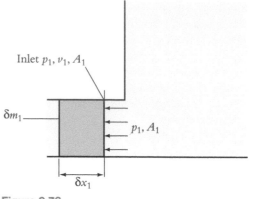

Inlet p_1, v_1, A_1

δm_1

p_1, A_1

δx_1

Figure 3.76

Substituting for the specific energy terms gives

$$q + w_s = \left[u_2 + \frac{p_2}{\rho_2} + gz_2 + \frac{v_2^2}{2} \right] - \left[u_1 + \frac{p_1}{\rho_1} + gz_1 + \frac{v_1^2}{2} \right] \qquad (3.58)$$

where

$$q = \frac{\delta Q}{\delta m}, \quad w_s = \frac{\delta W_s}{\delta m}, \quad \rho_1 = \frac{\delta m}{\delta V_1}, \quad \rho_2 = \frac{\delta m}{\delta V_2}$$

From the definition of enthalpy, the specific enthalpy

$$h = u + pv = u + \frac{p}{\rho}$$

Thus, equation (3.58) can be rewritten as

$$q + w_s = (h_2 - h_1) + g(z_2 - z_1) + \frac{v_2^2 - v_1^2}{2}$$

This is the **steady flow energy equation**. It applies to any fluid, liquid, gas or vapour and to any steady flow process. This equation is commonly used in thermodynamics and may already be familiar.

Extended Bernoulli equation – friction losses

When friction is present, mechanical energy degrades into heat energy and one way of expressing this loss, applicable to incompressible fluids, is to use the concept of friction head loss.

The SFEE (equation 3.58) can be rearranged as

$$q + w_s = (u_2 - u_1) + \left(\frac{p_2}{\rho_2} - \frac{p_1}{\rho_1}\right) + \left(\frac{v_2^2 - v_1^2}{2}\right) + g(z_2 - z_1)$$

or

$$w_s = \left[\frac{p_2}{\rho_2} + gz_2 + \frac{v_2^2}{2}\right] - \left[\frac{p_1}{\rho_1} + gz_1 + \frac{v_1^2}{2}\right] + (u_2 - u_1) - q \qquad (3.59)$$

For an incompressible fluid, $\rho_1 = \rho_2$. Also, the total head is defined as

$$H_T = \frac{p}{\rho g} + z + \frac{v^2}{2g}$$

So equation (3.59) can be written as

$$\frac{w_s}{g} = H_{T2} - H_{T1} + \left[\frac{(u_2 - u_1) - q}{g}\right] \qquad (3.60)$$

Fluid mechanics

Each term in equation (3.60) has the units of elevation or head. The quantity

$$\left[\frac{(u_2 - u_1) - q}{g} \right]$$

is called the **friction head loss** and is denoted by H_f. The friction head loss represents the amount of mechanical energy converted into heat (internal energy). It is the difference in internal energy in a flow between inlet and outlet of the control volume $(u_2 - u_1)$ after allowing for any heat transfer q and so represents the conversion of energy into heat.

Hence

$$\frac{w_s}{g} = H_{T2} - H_{T1} + H_f \tag{3.61}$$

(Note that $\dot{m}gH_f$ therefore gives the power dissipated owing to friction.)

This equation applies to any steady flow process but for incompressible fluids only. A further simplification can be made if the control volume involves no shaft work (i.e. no pumps, compressors, turbines, etc.). In this case:

$$H_{T1} - H_{T2} = H_f \tag{3.62}$$

This is often referred to as the **extended Bernoulli equation (EBE)**. It is applicable to the flow of real (viscous) liquids along pipes or channels. Although the head loss due to friction is expressed in terms of changes in internal energy and heat transfer, it is impractical to determine it from measurements of temperature as, in most cases, the changes in temperature are too small to be measured accurately. Frictional head loss is therefore more easily measured by using equation (3.62) and determining it from changes in total head, expressed in terms of pressure.

Estimation of Hf

The value of H_f represents the energy lost due to friction and depends upon the particular flow system, the fluid involved (viscosity), the velocity of flow, the roughness of surfaces over which flow takes place and the detail of the shape of the flow passages including the existence or otherwise of sharp corners and other geometrical details. In most situations it is not possible to calculate from first principles the friction loss, except by using advanced numerical techniques such as computational fluid dynamics. Therefore, for most practical purposes frictional losses are determined from experimental measurements and methods for applying these to predict losses are given below for incompressible flow in pipes and ducts and for other features in piping systems, such as entrance and exit losses, bends, expansions and contractions.

Incompressible flow in pipes and ducts

For incompressible flow in pipes where there is no shaft work (pump or fan) involved the extended Bernoulli equation (equation (3.62)) applies.

Pipes of uniform diameter

For fully developed flow in round pipes of uniform roughness it was found experimentally that piezometric head falls uniformly along the pipe. One of the key experimentalists was the French engineer Henri Darcy (1803–1858) and the standard formula for friction loss in pipes is named after him. The Darcy equation states that

$$H_f = \frac{4fl}{d} \frac{v^2}{2g} \tag{3.63}$$

(NB Head losses are usually quoted as multiples of $\frac{v^2}{(2g)}$, as this is the velocity head.)

In the Darcy equation, f is called the friction factor and takes values in the range 0.002–0.02, l is the length of pipe, d is the pipe diameter and $\bar{u}$ is the mean flow velocity. The friction factor, f, is dimensionless; its value depends on the pipe roughness and also on the Reynolds number.

For laminar flow it can be shown that $f = \dfrac{16}{Re}$ and is independent of pipe roughness unless the roughness is so great that the pipe diameter can be regarded as varying. For turbulent flow ($Re > 2000$), roughness is important, and it is usually expressed as **relative roughness** $\left(\dfrac{k}{d}\right)$, where d is the internal diameter of the pipe and k is the roughness, the size of the 'bumps' in the wall of the pipe. Friction factors for typical commercial pipes and tubing have been compiled into a chart by Lewis Moody an American engineer (1880–1953) and the Moody chart is the most widely employed method of predicting the friction factor f, although it should only ever be regarded as a good approximation as the roughness of a pipe always has a certain degree of uncertainty. The Moody chart is reproduced in Figure 3.77.

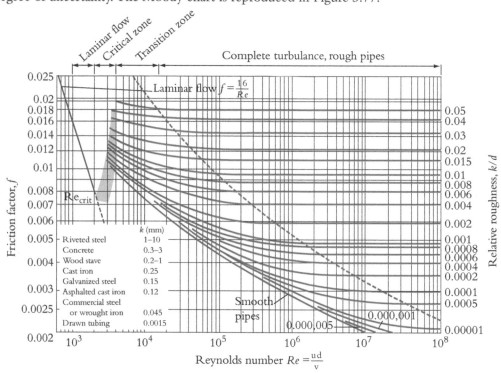

Figure 3.77 Moody chart (reproduced from 'Mechanics of Fluids' by B. S. Massey)

The Moody chart is used to give the friction factor when the Reynolds number and pipe roughness are known. For example, when the Re is 10^5 and the pipe is made of cast iron and has a diameter of 250 mm $\left(\dfrac{k}{d} = \dfrac{0.25}{250} = 0.001\right)$, the friction factor indicated by the Moody chart is 0.0055. (Note that the scales in the Moody chart are logarithmic.)

There are four distinct areas on the Moody chart:
- For laminar flows ($Re < 2000$), the friction factor is independent of pipe roughness.
- For perfectly smooth pipes, the friction factor has its lowest values and it continues to decrease even at very high values of Re. The Blasius formula $f = 0.079(Re)^{-\frac{1}{4}}$ can be used to approximate friction factors for smooth pipes for Re values between 4000 and 10^5.
- In the transition zone, the friction factor decreases as Re increases in the turbulent region.
- Above a certain Re for rough pipes, a constant friction factor is reached that is dependent only on the pipe roughness.

It should be noted that there are two conventions for friction factor. In the Darcy equation (3.63) the friction factor f is known as the Fanning friction factor, and this is the friction factor shown on the Moody chart. There is an alternative form of the Darcy equation, known as the Darcy–Weisbach equation, $H_f = \dfrac{fl}{d}\dfrac{v^2}{2g}$, in which the friction factor f is four times as large as the Fanning friction factor. Moody charts can be drawn either to give the Fanning or Darcy–Weisbach friction factors and engineers must be aware which friction factor they are using in calculations. The Fanning friction factor is often used in the UK, whereas the Darcy–Weisbach friction factor is often used in the USA.

Non-circular ducts or pipes

For non-circular ducts and pipes the Darcy equation can still be applied but the diameter used is an equivalent diameter known as the hydraulic diameter d_h given by the equation

$$d_h = \frac{4 \text{ (pipe cross-sectioned area)}}{\text{perimeter}}$$

Note that for a circular pipe the hydraulic diameter is the same as the pipe diameter.

When the duct or pipe is not completely full, then

$$d_h = \frac{4 \text{ (flow area)}}{\text{wetted perimeter}}$$

Water flows along a horizontal rectangular duct at 1 litre^{-1}. Calculate the difference in pressure between two points a distance of 20 m apart. The cross-section of the pipe is 20 × 40 mm and the friction factor is 0.01.

Calculate the hydraulic diameter and mean velocity of the water in the rectangular duct and then calculate the head loss from the Darcy equation.

$$H_f = \frac{4flv^2}{2gD}$$

$$D = \frac{4A}{P} = \frac{4(0.02 \times 0.04)}{2(0.02 + 0.04)} = 0.0267 \text{ m}$$

$$v = \frac{\dot{V}}{A} = \frac{0.001}{(0.02 + 0.04)} = 1.25 \text{ m s}^{-1}$$

$$H_f = \frac{4 \times 0.005 \times 20 \times 1.25^2}{2 \times 9.81 \times 0.0267} = 1.19 \text{ m}$$

Extended Bernoulli equation: $H_{T1} - H_{T2} = H_f$

$$\left(\frac{p_1}{\rho g} + z_1 + \frac{v_1^2}{2g}\right) - \left(\frac{p_2}{\rho g} + z_2 + \frac{v_2^2}{2g}\right) = H_f$$

$$z_1 + z_2 \quad v_1 + v_2$$

∴

$$p_1 + p_2 = \rho g H_f$$

$$p_1 - p_2 = 1000 \times 9.81 \times 1.19 = 1.17 \times 10^4 \text{ Pa}$$

The pressure loss is 0.117 bar.

Calculate the steady rate at which water (of kinematic viscosity 10^{-6} m^2 s^{-1}) will flow through a cast iron pipe 100 mm in diameter and 120 m long under a head difference of 5 m.

The velocity is unknown, therefore f must be estimated because Re is unknown.

From the Moody chart the roughness of the cast iron pipe is 0.25 mm. Therefore

$$\frac{k}{d} = \frac{0.25}{100} = 0.0025$$

Using the Moody chart (Figure 3.78), and assuming a high Re, the friction factor is 0.0062.

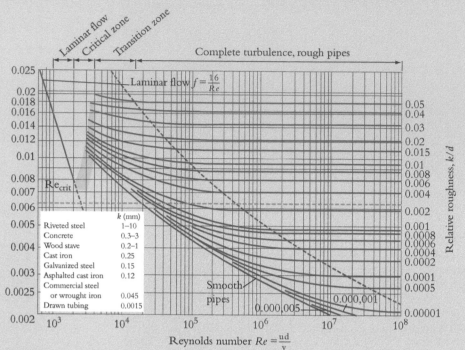

Figure 3.78

Calculate the velocity in the pipe from the Darcy equation:

$$H_f = \frac{4flv^2}{2gD}$$

$$\therefore \quad v = \sqrt{\frac{H_f 2gD}{4fl}}$$

$$v = \sqrt{\frac{5 \times 2 \times 9.81 \times 0.1}{4 \times 0.0062 \times 120}} = 1.81 \text{ m s}^{-1}$$

$$\therefore \quad Re = \frac{vd}{v} = \frac{1.81 \times 0.1}{1 \times 10^{-6}} = 1.8 \times 10^5$$

Then, using the Re, use the Moody chart again to make a better estimate of the friction factor.

At $Re = 180{,}000$ and a relative roughness of 0.0025, the friction factor is 0.0063. In the Darcy equation this gives a new velocity of 1.8 m s^{-1}.

This is sufficiently close to the previous value, so no further iterations are needed.

The flowrate can thus be determined from the velocity:

$$v = 1.8 \text{ m s}^{-1}$$

$$\dot{V} = vA = 1.8 \times \left(\frac{\pi}{4} 0.1^2\right) = 14.1 \text{ litres s}^{-1}$$

Therefore, the flowrate is $14.1 \text{ litres s}^{-1}$.

Entrance, exit and other losses

In a pipe flow system, as well as the losses in the long straight pipes there are head losses due to friction in other parts of the system such as entrances, exits, bends and in components such as

Fluid mechanics

valves. The magnitude of the losses depends on the details of the flow pattern and features such as sharp corners and changes in cross-section, and can result in frictional loss. These other losses are often referred to as **minor** losses, although when a system only involves short lengths of pipe, these minor losses may dominate. It is conventional to express these other losses in terms of velocity head:

$$H_f = K\left(\frac{v^2}{2g}\right)$$

where K is the loss factor and has a value which depends on the geometry and component involved. Loss factors for standard components are available in data books. Manufacturers usually produce data sheets for components such as valves and complicated fittings. When there is a change in diameter in a component, it is important to understand on which velocity the K factor is based. In many cases it is based on the flow velocity downstream of the component.

The loss factors do vary with Re and data may be given, but in many pipe systems the flow is fully turbulent and the loss factors at high Re are generally found to be independent of Re.

A number of standard and illustrative loss factors are tabulated below and further information can be found in the references at the end of this chapter.

Pipe entry loss

When a flow enters a pipe from a larger reservoir the head loss depends critically on the shape of the inlet. When there is an inlet with a sharp corner there is flow separation, the flow reduces in area at the vena contracta and there are frictional losses due to eddies. The value of K is approximately 0.5. For a rounded smooth pipe entry the K value is much lower.

$K \approx 0.5$ for sharp corners

$(K \approx 0.1$ for smooth entry)

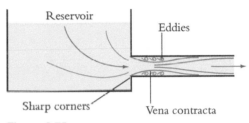

Figure 3.79

$$H_f \approx 0.5\left(\frac{v^2}{2g}\right)$$

v is the mean velocity in the pipe.

Pipe exit loss

Where a pipe exits into a larger reservoir, the velocity reduces to zero and all the dynamic head is lost as friction, so the loss factor $K = 1.0$.

$$H_f = \left(\frac{v^2}{2g}\right)$$

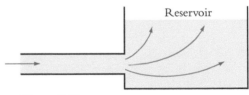

Figure 3.80

v is the velocity in the pipe.

Sudden enlargement

At a sudden enlargement there is flow separation and loss in the eddies:

If $d_1 = d_2$, $K = 0$

If $d_1 \ll d_2$, $K \rightarrow 1.0$

For other values of diameter ratio a fairly accurate expression is

$$K = \left(\frac{A_2}{A_1} - 1\right)^2$$

$$H_f = K\left(\frac{v_2^2}{2g}\right)$$

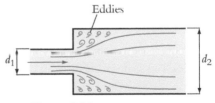

Figure 3.81

Sudden contraction

For a sudden contraction with a sharp corner

$$H_f = K\left(\frac{v_2^2}{2g}\right)$$

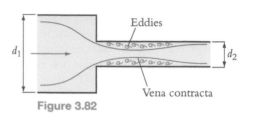

Figure 3.82

For fairly high Re the following are representative. The head loss is based on the downstream velocity:

$\frac{d_2}{d_1}$	0	0.2	0.4	0.6	0.8	1.0
K	0.5	0.45	0.38	0.28	0.14	0

90° elbow (sharp bend)

$K = 0.9$

90° smooth bend

$K = 0.16$–0.35 depending on bend radius

Nozzle

K is very small ≈ 0.05 (based on outlet velocity) and can often be ignored.

Gate valve

Wide open: $K \sim 0.2$

Half open: $K \sim 5$

Figure 3.83

A gate valve is located in a pipe of diameter 50 mm. Water flows at 0.005 m³ s⁻¹ along the pipe. What is the pressure loss due to friction in the valve when it is half open? (Assume K for a half-open gate valve to be 5.)

$$H_f = K\left(\frac{v^2}{2g}\right)$$

$$v = \frac{\dot{V}}{A} = \frac{0.005}{\frac{\pi}{4}(0.05^2)} = 2.55 \text{ m s}^{-1}$$

$$H_f = 5\left(\frac{2.55^2}{2 \times 9.81}\right) = 1.66 \text{ m}$$

$$\Delta p = \rho g H_f = 1000 \times 9.81 \times 1.66 = 16\,300 \text{ Pa}$$

The pressure loss is 0.16 bar.

Application of the steady flow energy equation to include pumps

By including the shaft work term in the steady flow energy equation, the performance of a pump can be determined.

For the pump itself, apply the SFEE between the inlet and outlet flanges. The SFEE is

$$q + w_s = \left[u_2 + \frac{p_2}{\rho_2} + gz_2 + \frac{v_2^2}{2}\right] - \left[u_1 + \frac{p_1}{\rho_1} + gz_1 + \frac{v_1^2}{2}\right]$$

Fluid mechanics

This can then be simplified by assuming the following:

- inlet and outlet pipes are at approximately the same elevation ($z_1 = z_2$);
- inlet and outlet pipes are of the same diameter ($v_1 = v_2$);
- there is no external heat transfer, so $q = 0$ (there is frictional dissipation, and so the fluid will get hotter, but there is no external heat input);
- the flow is incompressible (ρ is constant).

The SFEE then reduces to

$$w_s = \frac{p_2 - p_1}{\rho} + gH_f$$

For a hydraulic pump the pressure is raised by the pump and p_2 is greater than p_1. Hence the work done on the system is positive (according to the sign convention, transfers of heat and work into the system are positive and transfers of heat and work out of the system are negative). The required work input is increased by the friction head loss H_f. The head loss can be estimated for a particular machine operating under given conditions, but frictional effects can be allowed for more conveniently by using the hydraulic efficiency, defined as

$$\eta_{HP} = \frac{w_{s,i}\ (\text{ideal})}{w_s\ (\text{actual})}$$

In this expression, the ideal work input $w_{s,i}$ could be expressed as $\dfrac{(p_2 - p_1)}{\rho}$.

It is found that the efficiencies of pumps η_{HP} lie between 50 per cent (small pump) and 90 per cent (large pump).

Water is pumped from reservoir A to reservoir B. Using the data listed, determine:

(a) the power required to drive the pump;

(b) the total power dissipated in friction;

(c) the pressure at pump inlet.

DATA: Total length of pipe = 10 km
Pipe length from A to pump inlet = 20 m
Internal diameter of pipe = 0.2 m
Mean velocity in pipe = 1 m s^{-1}
H_f for pump = 25 m
Loss coefficients: Pipe entrance (from A) $K = 0.5$
Pipe exit (into B) $K = 1.0$
Valve, $K = 3.0$
$f = 0.005$

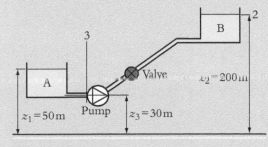

Figure 3.84

(a) Find the total head loss due to friction in the pump and pipework.

Friction in the pipe:

$$H_{f}, \text{pipe} = \frac{4flv^2}{2gd} \quad \text{(Darcy equation)}$$

$$= \frac{4 \times 0.005 \times 10 \times 10^3 \times 1^2}{2 \times 9.81 \times 0.2} = 50.97 \text{ m}$$

$H_{f,\text{pump}}$ is given as 25 m. Pipe losses, entrance, exit and valve:

$$H_f = (k_{\text{entry}} + k_{\text{exit}} + k_{\text{valve}})\left(\frac{v^2}{2g}\right)$$

$$= (0.5 + 1 + 3)\left(\frac{1^2}{2g}\right) = 0.23 \text{ m}$$

$$\therefore \qquad H_{f,\text{total}} = 50.97 + 25 + 0.23 = 76.2 \text{ m}$$

Apply SFEE for between 1 and 2

$$\frac{w_s}{g} = H_{T2} - H_{T1} + H_f$$

gives

$$w_s = g\left[\left(\frac{p_2 - p_1}{\rho g}\right) + \left(\frac{v_2^2 - v_1^2}{2g}\right) + (z_2 - z_1) + H_{f,\text{total}}\right]$$

$v_1 \approx 0$, $v_2 \approx 0$ because the velocity at the surface is negligible.

$p_1 = p_2 = p_{\text{atmospheric}}$, which simplifies to

$$w_s = g[(z_2 - z_1) + H_{f,\text{total}}] \quad \text{and} \quad W_s = \dot{m}w_s$$

Calculate $\dot{m}$:

$$\dot{m} = \rho A_{\text{pipe}} v_{\text{pipe}} = 1000 \times \left(\frac{\pi}{4}0.2^2\right) \times 1 = 31.42 \text{ kg s}^{-1}$$

$$\dot{W}_s = 31.42 \times 9.81[(200 - 50) + 76.2] = 69.72 \text{ kW}$$

(b) Power dissipated in friction $= \dot{m}gH_f$ (because gH_f is the specific energy)
$= 31.42 \times 9.81 \times 76.2 = 23.49 \text{ kW}$.

So the frictional power dissipation is 23.5 kW.

(c) Apply the extended Bernouilli equation from 1 to 3.

$$H_{T1} - H_{T3} = H_{fl \to 3}$$

Friction loss from 1 to 3:

$$H_{f,\text{friction}} = \frac{4flv^2}{2gd} = \frac{4 \times 0.005 \times 20 \times 1^2}{2 \times 9.81 \times 0.2} = 0.102 \text{ m}$$

$$H_{\text{entry}} = k\left(\frac{v^2}{2g}\right) = 0.5\left(\frac{1^2}{2g}\right) = 0.025 \text{ m}$$

$$H_{f,\text{total}} = 0.127 \text{ m}$$

$$H_{T1} - H_{T3} = H_{fl \to 3}$$

$$\left(\frac{p_1 - p_3}{\rho g}\right) + \left(\frac{v_1^2 - v_1^2}{2g}\right) + (z_1 - z_3) = H_f = 0.127 \text{ m}$$

$$p_1 - p_3 = \rho g\left[0.127 - (50 - 30) - \left(\frac{0 - 1^2}{2g}\right)\right]$$

$$= -1.95 \text{ bar}$$

If $p_1 = 1$ bar, then $p_3 = 2.95$ bar or $p_3 = 1.95$ bar gauge.

Therefore, the pressure at the pump inlet is 1.95 bar gauge.

Fluid mechanics

In a pumped pipe flow system, the pump puts energy into the flow to compensate for the losses due to pipe friction and other losses. To make a system more sustainable it is therefore important to minimize the frictional losses and so reduce the pumping power input. For a constant mass flow in a pipe, it can be shown from the continuity equation that the mean velocity is proportional to $\frac{1}{D^2}$, (where D is pipe diameter). Substituting in the Darcy equation (3.63) shows that the head loss due to friction is proportional to $\frac{1}{D^5}$, assuming that the friction factor is constant. Therefore, using larger diameter pipes is an effective way of reducing pumping power. Larger pipes use more material resources and are more expensive to manufacture and so a life cycle environmental and economic analysis is required to determine the optimal pipe size to use.

Key points from Section 3.4

- A steady flow is one that does not change with time.
- A uniform flow is one where the properties do not vary across a plane or within a volume.
- A one-dimensional flow is one where flow properties only vary in one direction.
- An ideal (inviscid) fluid has no viscosity, and is incompressible. No real fluid is ideal but the simplification is valid in some situations.
- For a steady flow, the law of conservation of mass (continuity) means that the flow entering a volume must equal the flow leaving a volume.
- The Euler equation is a differential equation for the flow of an ideal fluid.
- The Bernoulli equation expresses the relationship between pressure, elevation and velocity in an ideal fluid for steady flow along a streamline.
- The Bernoulli equation can be expressed in three ways: in terms of specific energy, pressure or head.
- The Bernoulli equation can be used for real fluids when losses due to friction are negligible and the fluid is incompressible.
- Viscosity in a fluid creates frictional drag in a fluid flow.
- There are two fundamental types of fluid flow: laminar and turbulent. They can be characterized by the Reynolds number of the flow.
- The steady flow energy equation can be applied to the flow of real fluids and leads to the extended Bernoulli equation that can be used to describe the losses resulting from viscous friction in a flow.
- The Moody Chart can be used to estimate the frictional losses in pipe and duct flows.
- The SFEE can be used to determine the performance of a pump needed in a pipe system.

Learning summary

At the end of this Section you should be able to:

✔ calculate the mass flowrate of a steady flow in a pipe or duct;

✔ understand the three forms of the Bernoulli equation;

✔ be able to apply the Bernoulli equation to calculate flows in pipes including the performance of venturi, nozzle and orifice plate flow meters and a pitot-static probe;

✔ calculate the drag forces created by viscosity in thin films between moving surfaces;

✔ calculate the Reynolds number of flows in pipes and ducts and determine whether the flow is likely to be laminar or turbulent;

✔ estimate the pressure losses in flows in pipes due to friction;

✔ calculate the pressures and flows in simple single pipe systems accounting for losses due to friction in pipes and other components of pipe systems;

✔ calculate the performance of a pump needed in a simple pipe flow system.

3.5 Fluids in motion – linear momentum

In the fourth and final part of this chapter the motion of fluids is considered from a different perspective. The forces produced in a fluid when it undergoes an acceleration will be described by applying Newton's second law of motion – force equals rate of change of momentum. Analysis using linear momentum is useful when considering the force produced on an object by fluid flowing over it. For example, there is a force exerted on a solid surface by a jet of fluid impinging on it. There are also the aerodynamic/hydrodynamic forces (lift and drag) on an aircraft wing or hydrofoil, the force on a pipe bend caused by the fluid flowing within it, the thrust from a jet engine, and so on. All these forces are **dynamic** forces and they are associated with a change in the momentum of the fluid. The most convenient way of analysing momentum problems is to use control volume analysis and so this principle will be further explained first.

Control volumes

A **control volume** is an imaginary boundary drawn to facilitate analysis of a flow. It is drawn at places where the fluid flow conditions are known, or where a good approximation can be made. For momentum analysis the region within the control volume is analysed in terms of the mass flows entering or leaving the control volume and other forces such as gravity and pressure acting on it. It is not necessary to understand the detail of the flows inside the control volume.

Example

The thrust produced by a jet engine on an aircraft at rest can be analysed simply in terms of the changes in momentum of the air passing through the engine. So a suitable control volume could be as shown in Figure 3.85.

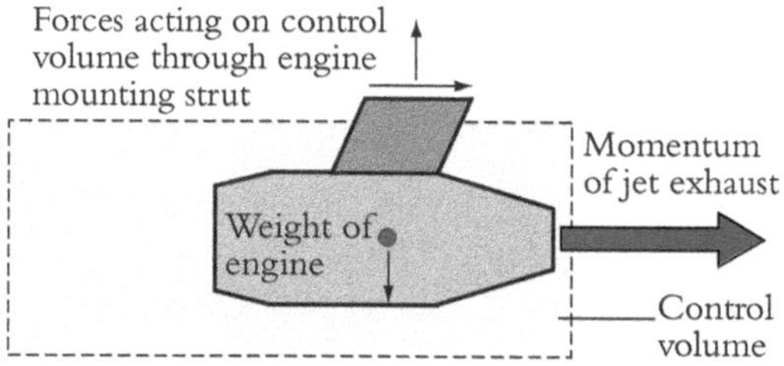

The control volume is drawn sufficiently far in front of the engine that the air velocity entering can be assumed to be at atmospheric pressure and the air velocity is negligible. At exit of the engine the control volume is drawn close to the jet exhaust where the velocity is known and the pressure is atmospheric. The control volume cuts the strut attaching the engine to the aircraft and there will be a force transmitted across the control volume there to oppose the forces on the engine created by the thrust and gravity.

Figure 3.85

The details of all the complex flows within the engine inside the control volume do not need to be known – the thrust produced can be determined simply in terms of the forces and flows crossing the control volume. However, to understand the details of the flows in each part of the jet engine, the combustion chamber or turbine for example, then a more detailed analysis of that component would be required.

Linear momentum equation

The magnitude of the forces due to momentum is determined essentially by Newton's second law of motion. In its most general form, Newton's second law states that:

> The net force acting on a body in any direction is equal to the rate of increase of momentum of the body in that direction.

Since force and momentum are both *vector* quantities, it is essential to specify both magnitudes and directions.

For a complex fluids system a **control volume** can be drawn around the part that is being analysed and Newton's laws of motion applied to that control volume. It is only the external forces that act on the control volume (due to reactions from solid structural components, external pressure of gravity) and the momentum in the flows crossing the control volume boundary that are considered.

Fluid mechanics

To derive a relation by which force may be related to the fluid within a given space, Newton's second law can be applied to the flow in a stream tube. Assume that the flow is smooth and steady and the stream tube remains stationary with respect to the fixed coordinate axes and has a cross-section sufficiently small that the velocity can be considered uniform over both planes AB and CD, as shown in Figure 3.86.

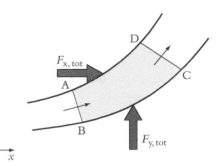

Consider the control volume ABCD around a stream tube. (A stream tube is a region of fluid in which all the fluid flows parallel to the tube and no fluid crosses the sides of the tube.)

Figure 3.86 Flow along a stream tube

The change of momentum in the x-direction is given by

$$\text{Change in } x\text{-momentum} = \dot{m}(v_{x, \text{out}} - v_{x, \text{in}})$$

Thus, the rate of change in momentum in the x-direction, equal to the net force on the control volume, is

$$F_{x, \text{tot}} = \text{total force on streamtube in } x\text{-direction} = \dot{m}(v_{x, \text{out}} - v_{x, \text{in}})$$

In real flow situations, the velocity will change across inlet and outlet planes and so all the forces from all the elemental stream tubes that make up the complete flow need to be summed. In general, therefore,

$$F_{x, \text{tot}} = \Sigma \dot{m} v_{x, \text{out}} - \Sigma \dot{m} v_{x, \text{in}}$$

and, similarly, in the y and z-directions

$$F_{y, \text{tot}} = \Sigma \dot{m} v_{y, \text{out}} - \Sigma \dot{m} v_{y, \text{in}}$$
$$F_{z, \text{tot}} = \Sigma \dot{m} v_{z, \text{out}} - \Sigma \dot{m} v_{z, \text{in}}$$

However, in many cases the inlet and outlet flows will be approximately uniform or a mean flow velocity can be applied and only these cases will be considered in this chapter. Where a mean flow velocity can be applied, the total force on the fluid in the control volume in the x-direction is given by:

$$Fx = \text{total force on CV in } x\text{-direction} = \dot{m}(v_{x, \text{out}} - v_{x, \text{in}})$$

Similarly for the y-direction.

Total force

The **total force** (F) acting on the control volume is made up of forces due to **pressure** at the control volume boundaries and externally applied forces (such as **structural reaction forces** from pipes and **gravitational forces**).

When attempting linear momentum problems, the control volume should be selected so that at the sections where fluid enters or leaves it, the streamlines are straight and parallel so that there are no accelerations perpendicular to them.

The control volume can also be selected so that atmospheric pressure acts on all surfaces where there is no flow. Consequently the net force due to atmospheric pressure is zero (in any direction). It therefore makes the calculations easier if gauge pressure is used throughout in the flows as this effectively compensates for atmospheric pressure.

Applications of the linear momentum equation

Force on fluid as it flows over a vane

A vane is a device for changing the direction of a jet of fluid (see Figure 3.87). Vanes are often used in pumps and turbines to change the direction of a flow of fluid. In the case in Figure 3.87, the fluid is a discrete jet of water open to the atmosphere and so the pressure all around is atmospheric. Draw a control volume some distance away from the vane in a position where you can be sure that you know the velocity and direction of the jet.

Apply the linear momentum equation to the control volume in the x-direction:

The force on the control volume in the x-direction equals the rate of change of momentum in the x-direction.

(N.B. Gravity is in the z-direction)

$F_x =$ total force on the control volume in x-direction $= \dot{m}(v_{x,\,out} - v_{x,\,in})$

$F_x = \dot{m}(v_{out}\cos\theta - v_{in})$

By continuity

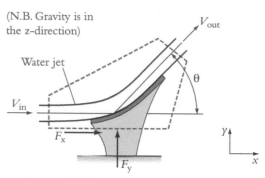

$$\dot{m} = \rho v_{in}A_{in} = \rho v_{out}A_{out}$$
$$F_x = \rho v_{in}A_{in}(v_{out}\cos\theta - v_{in})$$

and similarly in the y-direction:

$F_y =$ total force on the control volume in y-direction $= \dot{m}(v_{y,\,out} - v_{y,\,in})$

Figure 3.87 Flow around a horizontal vane

$$F_y = \rho v_{in}A_{in}(v_{out}\sin\theta - 0)$$

F_x and F_y are the forces that must be exerted *on* the control volume to maintain it in equilibrium. As the vane support is the structure that passes out of the control volume, F_x and F_y are the structural forces that must be exerted by the vane support *on* the vane to hold it in position. Equal and opposite forces are exerted *by* the vane *on* the surroundings. Similarly, the fluid exerts equal and opposite forces to F_x and F_y *on* the vane. Note that in this case, gravity acts in the z-direction, so it has no component in the x and y directions and there is no gravity force acting on the control volume. If gravity were acting in the xy plane, it would be one of the forces acting on the control volume along with the structural force from the vane support. In that case $F_x =$ gravity force in the x-direction + structural force from vane support on control volume in the x-direction. F_x is thus the sum of all forces acting *on* the control volume.

The forces that occur when a jet of water changes direction by passing over an object can easily be investigated in the kitchen by holding a spoon under a running tap. Note that larger forces are produced when the water is deflected by a greater angle – giving a larger change in momentum.

A jet of water flows smoothly on to a curved vane which turns it through 60°. The initial jet is 50 mm in diameter and the velocity (which is uniform) is 36 m s⁻¹. As a result of friction, the velocity of the water leaving the surface is 30 m s⁻¹. Neglecting gravity effects, calculate the force on the vane.

The pressure at 1 and 2 will be atmospheric because the jet is in contact with the atmosphere. Apply the linear momentum equation in the x-direction to find the x-component of the force on the CV:

$$F_x = \dot{m}(v_{x,\,leaving} - v_{x,\,entering})$$
$$F_x = \dot{m}(v_2\cos\theta - v_1)$$
$$\dot{m} = \rho Av$$

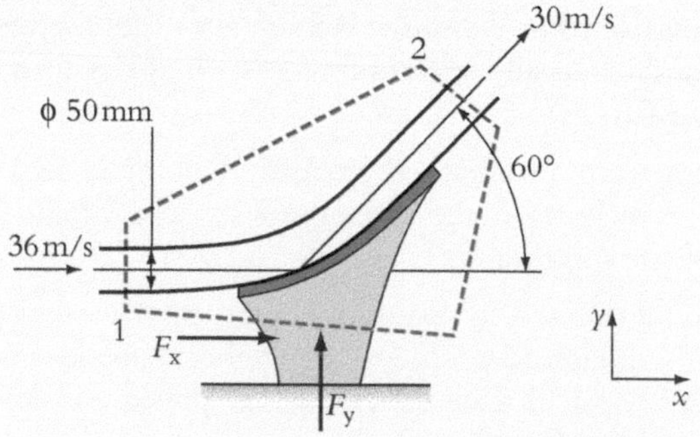

So

Figure 3.88

$$\dot{m} = 1000 \times \left(\frac{\pi}{4}0.05^2\right)36 = 70.69 \text{ kg s}^{-1}$$

Hence

$$F_x = \dot{m}(v_2\cos 60 - v_1)$$
$$= 70.69(30 \times \tfrac{1}{2} - 36) = -1484 \text{ N}$$

and -1484 N acts on the CV in the x-direction.

Apply the linear momentum equation for the fluid in the y-direction:

$$F_y = \dot{m}(v_2\sin60 - 0)$$

$$= 70.69(30\sin60) = 1837\,\text{N}$$

and 1837 N acts on the CV in the y-direction.

Find the resultant force on the CV:

$$F^2_R = 1837^2 + 1484^2$$

$$\therefore \qquad F_R = 2362\,\text{N}$$

$$\tan\varphi = \frac{1837}{1484} \Rightarrow \varphi = 51.1°$$

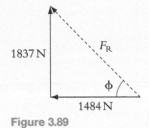

Figure 3.89

So 2362 N acts at 51.1° above horizontal on the CV.

The force on the vane is equal and opposite to this force on the CV.

Force caused by the flow of a fluid around a bend in a pipe

When the flow is confined within a pipe, the static pressure will vary with the location along the pipe. As the pressure differs from atmospheric, the forces due to static pressure must be taken into account. Consider a pipe bend in which not only the direction of flow but also the cross-sectional area is changed, as shown in Figure 3.90.

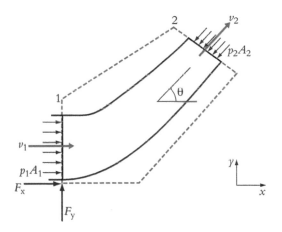

A control volume is selected that includes the pipe between sections 1 and 2. For simplicity the axis of the bend is in the horizontal plane, so there are no changes of elevation and the weights of the pipe and fluid act in a direction perpendicular to the plane and so do not affect the forces in the x and y directions. The conditions at sections 1 and 2 are uniform with the streamlines straight and parallel.

The mean gauge pressure and cross-sectional area at section 1 are p_1 and A_1, so the fluid adjacent to this cross-section exerts a force p_1A_1 on the fluid in the control volume. Similarly, there is a force p_2A_2

Figure 3.90 Flow around a pipe bend

acting at section 2 on the fluid in the control volume. Let forces F_x and F_y be the x and y components of the structural force acting on the control volume to maintain equilibrium. The total force in the x-direction on the control volume equals the increase in x-momentum of the fluid as it passes through the control volume:

$$p_1A_1 - p_2A_2\cos\theta + Fx = \dot{m}(v_2\cos\theta - v_1)$$

so F_x can be calculated.

Similarly, the total force in the y-direction acting on the control volume is equal to the increase in y-momentum, and thus F_y can be obtained:

$$-p_2A_2\sin\theta + Fy = \rho\dot{V}(v_2\sin\theta - 0)$$

From the components F_x and F_y, the magnitude and direction of the total force that needs to be exerted on the control volume to maintain equilibrium can readily be calculated. As the pipe structure crosses the control volume boundary, this is the force that must be exerted on the pipe bend through the pipe structure to hold it in position. The force exerted by the fluid on the bend is equal and opposite to this.

Where only one of the pressures p_1 and p_2 is included in the data of the problem, the other may be deduced from the Bernoulli or extended Bernoulli equation if frictional effects are significant. If the bend is located in a vertical plane, the **weight of the water** inside the control volume and also the weight of the pipe structure inside the control volume must also be included.

A 45° reducing pipe bend (in a horizontal plane) tapers from 600 mm diameter at the inlet to 300 mm diameter at the outlet. The pressure at the inlet is 140 kPa and the rate of flow of water through the bend is 0.425 $m^3\ s^{-1}$. Neglecting friction, calculate the net resultant horizontal force exerted by the water on the bend.

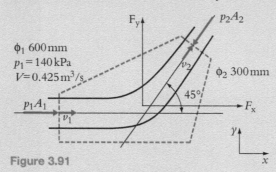

Figure 3.91

Find v_1, v_2.

Assume water density is constant (1000 kgm^{-3}).

Volume flowrate $\dot{V} = A_1 v_1 = A_2 v_2$

$$\therefore\ 0.425 = \left(\frac{\pi}{4}0.6^2\right)v_1 \Rightarrow v_1 = 1.5\ m\ s^{-1}, v_2 = 6.0\ m\ s^{-1}$$

Find the pressure at 2 using the Bernouilli equation:

$$p_1 + \rho g z_1 + \frac{\rho v_1^2}{2} = p_2 + \rho g z_2 + \frac{\rho v_2^2}{2}$$

$z_1 = z_2$ pipe bend in horizontal plane

$\therefore$
$$p_2 = p_1 + \tfrac{1}{2}\rho(v_1^2 - v_2^2) = 140 \times 10^3 + 500(1.5^2 - 6^2)$$

$$p_2 = 123.1\ \text{kPa}$$

Apply the linear momentum equation in the x-direction for the fluid:

$$F_x + p_1 A_1 - p_2 A_2 \cos 45 = \rho \dot{V}(v_2 \cos 45 - v_1)$$

$$F_x = 1000 \times 0.425(6 \cos 45 - 1.5) - (140 \times 10^3\left(\frac{\pi}{4}0.6^2\right)) - 123.1 \times 10^3\left(\frac{\pi}{4}0.3^2\right)\cos 45$$

and

$$F_x = -32.27\ \text{kN}$$

Apply the linear momentum equation in the y-direction for the fluid:

$$F_y - p_2 A_2 \sin 45 = \rho \dot{V}(v_2 \sin 45 - 0)$$

$$F_y = 1000 \times 0.425(6 \sin 45 - 0) + 123.1 \times 10^3(\frac{\pi}{4}0.3^2)\sin 45$$

$$F_y = 7.96\ \text{kN}$$

$$F_R = \sqrt{32.27^2 + 7.96^2} = 33.24\ \text{kN}$$

$$\tan \varphi = \frac{7.96}{32.27} \Rightarrow \varphi = 13.9°$$

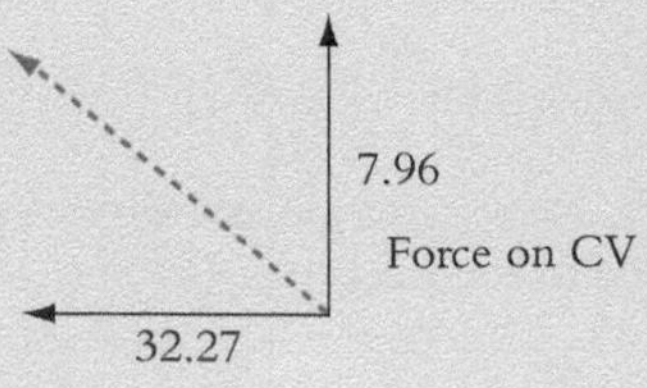

7.96

Force on CV

32.27

Figure 3.92

Force at a nozzle and the reaction of a jet

The forces exerted on nozzles can also be determined from momentum. Consider a horizontal nozzle as shown in Figure 3.93:

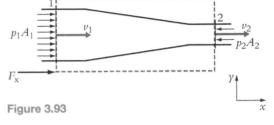

Assume uniform conditions with streamlines straight and parallel at the sections 1 and 2 so that the force exerted in the x-direction on the control volume between planes 1 and 2 is given by

$$p_1A_1 - p_2A_2 + F_x = \rho \dot{V}(v_2 - v_1)$$

Figure 3.93

If the fluid is directed into the atmosphere, then p_2 will be atmospheric pressure (zero gauge pressure), and so the equation becomes

$$p_1A_1 + F_x = \rho \dot{V}(v_2 - v_1)$$

F_x is the force that needs to be exerted on the nozzle to hold it in equilibrium and represents the structural force at the interface between the nozzle and the supply pipe. If the upstream pressure is given, then, assuming that the nozzle is relatively short, frictional losses will be small and the Bernoulli equation can be used to calculate the jet velocity. The continuity equation can be used to calculate v_1. Alternatively, the mass flowrate from the nozzle may be known, in which case the velocities v_1 and v_2 can be calculated from continuity and then the upstream pressure p_1 can be calculated from the Bernoulli equation.

Water flows through the vertical nozzle shown below, which has $D_1 = 150$ mm and $D_2 = 30$ mm. The vertical distance between 1 and 2 is 0.06 m and the velocity at 2 is 20 m s^{-1}. The mass of fluid between 1 and 2 is 2.5 kg. The mass of the nozzle and flange is 2.9 kg. What is the vertical force provided by the flange bolts to hold the nozzle fixed, assuming that there are no frictional losses between 1 and 2?

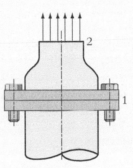

Figure 3.94

Draw a control volume around the nozzle as shown:

$$\dot{m}_1 = \dot{m}_2$$

$$A_1v_1 = A_2v_2 \Rightarrow v_2\left(\frac{d_1}{d_2}\right)^2 v_1$$

$$= \left(\frac{150}{30}\right)^2 v_1 = 25\,v_1$$

$$v_2 = 20 \text{ m s}^{-1}$$

$$\therefore \quad v_1 = 0.8 \text{ m s}^{-1}$$

$$\dot{m} = \rho A_1 v_1$$

$$= 1000 \times \left(\frac{\pi}{4}0.15^2\right)0.8$$

$$= 14.14 \text{ kg s}^{-1}$$

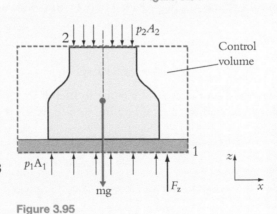

Figure 3.95

$$p_1 + \rho g z_1 + \tfrac{1}{2}\rho v_1^2 = p_2 + \rho g z_2 + \tfrac{1}{2}\rho v_2^2$$

$$p_2 = p_a = 0 \text{ Pa}_{\text{gauge}}$$

$$p_1 = p_a + \rho g(z_2 - z_1) + \tfrac{1}{2}\rho(v_2^2 - v_1^2)$$

$$p_1 = p_a + \rho g(0.06) + \tfrac{1}{2}\rho(20^2 - 0.8^2)$$

$$p_1 = p_a + 588.6 + 199\,680$$

$$= 200{,}269 \text{ Pa}_{\text{gauge}}$$

Apply the linear momentum equation in the y-direction:

$$F_z + p_1 A_1 - p_2 A_2 - mg = \dot{m}(v_{z2} - v_{z1})$$

$$F_z + \left[(200\,269)\frac{\pi}{4}0.15^2 \right] - 5.4g = 14.14(20 - 0.8)$$

$$F_z = 271.49 - 3539.0 + 52.97 = -3214\,\text{N}$$

F_z is the force on the CV. Thus, a downwards force of 3215 N is needed to hold the nozzle in position. This is provided by the tension in the bolts. The water thus exerts an upward force on the nozzle.

The force caused by a jet striking a surface

When a steady jet of fluid strikes a solid surface, the fluid is deflected, moves over the solid surface until the boundaries are reached, and then leaves the surface tangentially. Consider a jet striking a large, plane surface as shown in Figure 3.96.

A suitable control volume is that indicated by dotted lines. Consider the forces in the x- and y-directions.

x-direction

If the x-direction is taken perpendicular to the plane, the fluid, after passing over the surface, will have no component of velocity and therefore no momentum in the x-direction. The rate at which x-momentum enters the control volume is

$$\int \rho_1 v_1 v_{x1}\, dA_1 = \cos\theta \int \rho_1 v_1^2\, dA_1$$

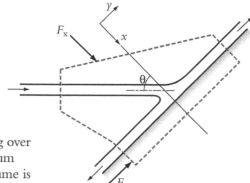

Figure 3.96 **Jet striking a large plane surface**

> Force on control volume in x-direction = rate of increase in x-momentum

Since the pressure is atmospheric, both where the fluid enters the control volume and where it leaves, the force on the control volume can be provided only by the solid surface (effects of gravity being neglected).

Therefore

$$F_{x,\text{on control volume}} = -\cos\theta \int \rho_1 v_1^2\, dA_1$$

F_x is the force that must be exerted on the control volume to keep it in equilibrium and it is equal and opposite to the force exerted by the fluid on the surface which is thus:

$$F_{x,\text{on surface}} = \cos\theta \int \rho_1 v_1^2\, dA_1$$

in the x-direction.

If the jet has uniform density and velocity over its cross-section, the integral reduces to

$$F_x = \rho_1 v_1^2 \cos\theta \int dA_1 = \rho_1 \dot{V}_1 v_1 \cos\theta$$

where

$$\dot{V}_1 = v_1 A_1.$$

y-direction

In the y-direction the jet carries a component of momentum equal to $\sin\theta \int \rho_1 v_1^2\, dA_1$ per unit time. For this component to undergo a change, a force in the y-direction would have to be applied to the fluid. Such a force, being parallel to the surface, would be a shear force exerted by the surface on the fluid. For an ideal fluid moving over a smooth surface, no shear force is possible and so the component $\sin\theta \int \rho_1 v_1^2 \, dA_1$ would be unchanged, and equal to the rate at which y-momentum leaves the control volume. Except when $\theta = 0$, the spreading of the jet over the surface is not symmetrical and the net momentum of the fluid leaving the surface will be $\sin\theta \int \rho_1 v_1^2 \, dA_1$.

For a real fluid the rate at which y-momentum leaves the control volume differs from the rate at which it enters as there will be frictional drag, due to viscosity, on the surface. In general, the force in the y-direction may be calculated if the final velocity of the fluid is known as it leaves the plate in all directions. This, however, requires further experimental data.

When the fluid flows over a curved surface, similar techniques of calculation may be used.

A horizontal jet of water is directed at a curved vane, as shown below. The mean velocity of the water jet is $12 \, \text{m s}^{-1}$, and the volume flowrate is $55 \, \text{m}^3 \, \text{h}^{-1}$. The water jet is deflected such that half the water passes over each side of the vane. The water leaves the vane in the direction shown in the diagram, with 92 per cent of the velocity at which it impacts on the vane. Angle $\theta = 20°$. Calculate the force needed to hold the vane in position.

Assume that the density of the water is $1000 \, \text{kg m}^{-3}$.

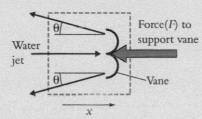

Figure 3.97

The velocity of the water leaving the vane is $0.92 \times 12 = 11.04 \, \text{m s}^{-1}$.

The mass flowrate of water is $\rho \dot{V} = 55 \times \dfrac{1000}{3600} = 15.28 \, \text{kg s}^{-1}$.

Applying the momentum equation to a control volume around the vane:

$$-F = \dot{m}(V_{\text{out}} \cos\theta - V_{\text{in}})$$

$$-F = 15.28(-11.04 \times \cos 20 - 12)$$

$$F = 341.9 \, \text{N}$$

Therefore, the force F needed to hold the vane in position is 342 N. This is equal and opposite to the force of the water on the vane.

Key points from Section 3.5

- The forces exerted by fluids when they change velocity and direction can be evaluated using the momentum equation derived from Newton's second law of motion.
- Control volumes are the easiest way to analyse momentum flow problems.
- The linear momentum equation states that for a steady flow. The force in a particular direction on a control volume is equal to the rate of change in momentum of the fluid flowing in that direction.
- The force on the control volume is the sum of all the forces acting, including external pressure, gravity and structural forces from solid object crossing the control volume boundaries.

3.6 Chapter conclusion

In this chapter the behaviour of fluids has been considered and methods of analysing a number of commonly occurring fluids problems in mechanical engineering have been introduced. Techniques have been described to allow calculations to be carried out for the following situations:

- the variation in pressure with depth in a static fluid;
- the forces produced by a static fluid on a submerged surface;
- the buoyancy force on a submerged object;
- the variation in pressure with velocity and elevation in an ideal (inviscid) fluid (Bernoulli equation) and the application of this to some flow-measuring devices and other simple situations;
- the calculation of the drag forces due to viscosity in some simple flow situations;
- the estimation of the pressure losses in pipes owing to fluid friction and the size of the pump necessary to pump a fluid through a pipe system;
- the determination of the forces resulting from a change in the momentum of a fluid flow.

Additionally, the two fundamental types of fluid flow − laminar and turbulent − have been described, and the use of the dimensionless Reynolds number has been introduced as a means of characterizing whether a flow would be laminar or turbulent.

The chapter has been introductory in nature, and readers are advised to refer to the other fluid mechanics textbooks for a more detailed coverage.

References

1. *Mechanics of Fluids* by B. S. Massey and J. Ward-Smith, 8th Edition, 2008, Taylor and Francis.

2. *Fluid Mechanics* by F. M. White, 6th Edition, 2006, McGraw-Hill.

3. *Fluid Mechanics* by J. F. Douglas, J. M. Gasiorek, J. A. Swaffield and L. B. Jack, 5th Edition, 2005, Pearson

4. *Mechanics of Fluids* by I. H. Shames, 4th Edition, 2003, McGraw-Hill

5. *Introduction to Fluid Mechanics* by R. W. Fox, A. T. McDonald and P. J. Pritchard, 6th Edition, 2004, John Wiley and Sons

Nomenclature

A	m^2	area	p	N/m^2	pressure
C_d		discharge coefficient	q	J/kg	heat transfer per unit mass
C_c		coefficient of contraction	Q	J	heat transfer
c_p	J/kgK	specific heat capacity at constant pressure	R	J/kgK	specific gas constant
c_v	J/kgK	specific heat capacity at constant volume	R	$J/kmol\ K$	Universal gas constant ($8.3145 \times 103\ J/kmol\ K$)
D,d	m	diameter	Re		Reynolds number
e	J/kg	specific energy	T	K	temperature
E	J	energy			(°C, temperature in degrees Celsius = T(K)−273)
F	N	force	$u(U)$	J/kg	specific internal energy (internal energy, J)
f		friction factor	V	$m3$	volume
g	m/s^2	acceleration due to gravity(=9.81 m/s²)	v	m/s	velocity
G		centre of gravity	v	m^3/kg	specific volume (volume per kg)
h	J/kg	enthalpy per unit mass	w	J/kg	specific work
h	m	depth below a liquid surface	W	N	weight
H	m	head	W	J	work
H_{pz}	m	piezometric head ($p/\rho g + z$)	$\dot{W}$		rate of work or power
K	N/m^2	Bulk modulus of elasticity	z	m	height above datum
K		Loss factor	γ	N/m	surface tension
L	m	length	μ	$kg/s(=Ns/m^2)$	dynamic viscosity
m	kg	mass	ν	m^2/s	kinematic viscosity
m	$kg/kmol$	molar mass	ρ	kg/m^3	density (1/v)
M	Nm	moment	τ	N/m^2	shear stress

Unit 4

Thermodynamics

Paul Shayler

UNIT OVERVIEW

- Introduction
- The first law of thermodynamics, conservation of energy, work and heat transfer
- The second law of thermodynamics, heat engines, the Clausius inequality, entropy and irreversibility
- Open systems, mass flow continuity, the steady flow energy equation and enthalpy
- The properties of perfect gases, water and steam
- Types of process and their analyses for work and heat transfer
- Modes of heat transfer and steady-state heat transfer rates
- Cycles, power plant and engines

4.1 Introduction

Thermodynamics is about energy, its conversion from one form to another, the means by which it can be transferred and the efficiency with which it can be harnessed to do work. Although energy exists in many forms, in thermodynamics, greatest attention is given to mechanical and thermal forms that can be manipulated through work and heat transfer. Mechanical forms include kinetic, potential and strain energy, such as that possessed by a weight oscillating on the end of a spring. Thermal forms of energy are evident on the microscopic scale and in the movement and interactions of atoms and molecules; these forms make up the internal energy of matter.

The treatment of laws and principles presented in this chapter is aligned to applications in mechanical engineering. A key to the notation used is given at the end of the chapter. Like many other subjects, thermodynamics has its own language and some key terms and definitions with which to become familiar. These are set out in the following pages. The first and second laws of thermodynamics – two laws of unchallenged importance which stand at the centre of the subject – are introduced and the analysis of work and heat transfer, the properties of working fluids, and the efficiency of cycles producing power are described. The first recorded use of the term thermodynamics was by Lord Kelvin, in a publication in 1849. The term is derived from the Greek *thermos* meaning heat and *dynamics* meaning power. This was a time when more and efficient generation of mechanical power from thermal energy released by burning coal was a pressing need for industry and the thermodynamics of power generation remains an important topic today: the plant of power stations operates on a thermodynamic cycle and thermodynamic machines – steam turbines – drive the alternators which generate the electrical power. More generally, thermodynamics allows the performance of engines to be understood and their efficiency to be analysed. This is required to design the internal combustion engines for cars and the jet engines for aircraft. Knowledge of thermodynamics is used to design the cooling systems for computers and to develop efficient heating systems for homes. How does a refrigerator cool a cold space in a hot room? To what temperature will a brake disc rise during braking? These questions have answers provided by thermodynamics.

Before a problem can be solved, it has to be defined and understood. Producing a conceptual diagram is a useful way of summarizing information which often requires no more than a simple sketch with key words and arrows. The conceptual diagram shown in Figure 4.1 represents a piston-and-cylinder unit which has many important uses. One simple form of this is the pump used to inflate bicycle tyres. The barrel of the pump is a cylinder; the piston is attached to the rod extending from the handle. In a 2005 poll, BBC radio listeners voted the bicycle as the most significant technological innovation since 1800. Today, more than one billion people use bicycles and the bicycle is the principal means of transportation in many parts of the world, so pumping up a bicycle tyre is a task familiar to many of us. But how many of us will have thought about the thermodynamics involved? There is the **work transfer** which occurs as the piston is advanced, the **compression process** raising the air **pressure** in the cylinder and increasing the resistance to the piston movement, the increase in air **temperature** which drives **heat transfer** to the barrel of the pump, making it feel warm, or the **mass transfer** from the pump chamber into the tyre when the tyre valve opens.

A piston-and-cylinder unit is an example of a thermodynamic **system**. This is defined as a region of space containing a quantity of matter, the behaviour of which is under investigation. A system is fully enclosed by a **boundary** which separates it from its **surroundings**, which is everything outside the boundary. The boundary is usually shown as a dashed line in conceptual diagrams; it may be a real boundary, like the surface of a wall, or an imaginary boundary, like the entrance to a pipe, or a combination of sections which are real or imaginary. The examples of conceptual diagrams given in Figure 4.2 show what are likely to be the most appropriate choice of boundary positions in each case, although this might depend on the analysis required. There are two types of system: a **closed system** and an **open system**. No matter can cross the boundary of a **closed system.** When any openings and valves of a piston-and-cylinder unit are closed, trapping the contents, this will be a closed system. The same atoms of matter stay within the boundaries of this type of system and the boundaries are said to be impermeable. In an **open system**, matter can flow into or out of the system across parts of the boundary which are imaginary or permeable. The atoms of matter within the system can be exchanged, and the mass of matter within the system might vary. If the piston-in-cylinder had valves which were open, allowing matter to flow into or out of the cylinder, this would be an open system. In this example, the unit can be changed from a closed system to an open system by opening the valves. As a convenience, the matter contained within or flowing through a system can be referred to as the **working fluid,** regardless of whether it is a gas, liquid or vapour.

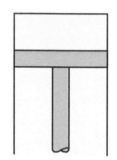

Figure 4.1 A conceptual diagram represents physical components such as a piston-and-cylinder combination in a simplified or symbolic form

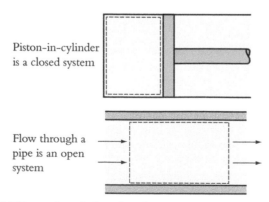

Piston-in-cylinder is a closed system

Flow through a pipe is an open system

Figure 4.2 Examples of closed and open systems, with boundaries shown by the dashed lines

Thermodynamics

The **properties** of a system are characteristics which can be measured or calculated from measurements. A property can be a characteristic defining something particular to the system, such as its volume, or it can be a property of the working fluid. Although a lengthy list of properties can be drawn up, thermodynamics is concerned with a relatively small number. Nine thermodynamic properties, plus velocity and height relative to a datum, form a set of properties commonly examined or used to describe changes when something is done to or by the system. Of these nine, the thermodynamic properties that can be measured are pressure, temperature, volume and mass. Two others can be deduced from values of volume and mass: density, which is the mass per unit volume; and specific volume, which is the inverse of density. The remaining three, internal energy, enthalpy and entropy, which are returned to later in this chapter, are determined from measurements of the others.

A property has a numerical value which depends on the units of measurement or calculation. Although several systems of units have been devised and used in the past, there is now the almost universal adoption of *Le Systeme International d'Unites*, or SI system of units, for engineering. In the SI system, the fundamental units of mass, length, time and absolute temperature are kilograms (kg), metres (m), seconds (s) and degrees kelvin (K) respectively. Although information on these and other properties can be presented in other units, perhaps because of past practice or because they appear to reflect size or some other feature of a problem better, it is important that these are converted to values in SI units for calculations. Length should be in metres not millimetres, mass should be in kilograms not grams, and so on. Values of pressure and temperature are most likely to catch out students because these are particularly likely to be quoted in non-SI units, or in relative not absolute units. Note also that by convention, the units of temperature (kelvin), energy (joule) and force (newton) are not capitalized although their symbols are the capitals K, J and N and the units are named in honour of real scientists: Lord Kelvin, James Joule and Isaac Newton.

Pressure is defined as the force exerted per unit area at a boundary. The boundary can be real or imaginary: a gas filling a container will exert a pressure on the real walls of the container and also on imaginary surfaces within the enclosed volume. The SI unit of pressure is the pascal or Pa, which is numerically equal to a value in newtons per metre squared, $N\,m^{-2}$, but pressure is also commonly stated in the unit of bars. The conversion between these is

$$1\ \text{bar} = 10^5\,N\,m^{-2} = 10^5\,\text{Pa}$$

Pressure can be measured by several methods. For example, a manometer records the height of a water column which the pressure supports. The principle is illustrated in Figure 4.3.

Gas pressure acts at the boundary between the gas and the water; the higher the gas pressure, the higher is the height of the water column this supports. From knowledge of the weight (w) of the water in the column and the cross-sectional area (A) of the column, we obtain the **gauge pressure** (p_g) of the gas:

$$p_g = \frac{w}{A}$$

This is called gauge pressure because it is being measured relative to the pressure of the surrounding air; that is, relative to **ambient pressure** (p_a) which acts on the other end of the water column. For the **absolute pressure** (p) of the gas, these must be added together:

$$p = p_g + p_a$$

Usually, *absolute* pressure values are required for calculations and this is referred to simply as pressure, without the *absolute*.

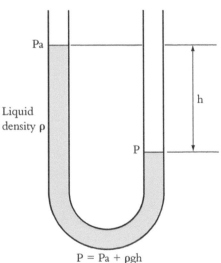

$$P = Pa + \rho g h$$

Figure 4.3 **A manometer measures gauge pressure (that is, relative to atmospheric pressure) in units of the height of liquid supported**

Atmospheric pressure at sea level remains close to the value of standard atmospheric pressure, 1.01325 bar or 101.325 kN m^{-2}, throughout the seasons. As altitude above sea level increases, air pressure falls, to around 0.9 bar (90 kN m^{-2}) at 1000 m above sea level; as water depth below sea level increases, pressure increases, doubling to around 2 bar (200 kN m^{-2}) at 10 m below the surface. Compared with this, pressures in engineering systems vary over a much wider range, from near vacuums of a few thousandths of a bar up to several hundreds of bars.

Values of **temperature** (T) may also be presented in more than one unit, either in degrees celsius or degrees kelvin, but remember that the SI unit of temperature is the kelvin. Although temperature *differences* in kelvin and degrees celsius will have the same numerical value, temperatures used in multiplication or division calculations should always be in degrees kelvin (not following this simple rule is a common source of numerical errors made by students). One early instrument for measuring temperature is the thermometer, which makes use of the observation that a liquid will expand as its temperature is raised to indicate a point on a temperature scale. Mercury thermometers were widely used until recent years but have now been largely replaced by thermocouples. Thermometers and thermocouples give a value for temperature but not an explanation of what temperature *is*. The celsius (and centigrade) scale of temperature was invented as a one-hundred-point scale of temperatures between two reference conditions which were easy to reproduce (the freezing and boiling points of water at a pressure of 1.013 bar). On this scale, temperatures below freezing are negative – at odds with the connection between temperature and energy. The kinetic theory of gases developed in the nineteenth century is more helpful in understanding temperature, relating this to energy stored in the free movement and vibration of gas molecules; at high temperatures, the movement is energetic; at low temperatures it is relatively slow; and at absolute zero it is stilled. On the Kelvin scale, zero degrees kelvin is absolute zero; it is said to be an *absolute* scale of temperature. On the kelvin scale the temperature at which water freezes at a pressure of 1.013 bar is assigned the value 273.15 K. The conversion of temperature values from one scale to the other entails simple addition or subtraction; a *change* of 1 degree is the same in degrees kelvin or celsius, and on the Celsius scale absolute temperature corresponds to −273.15 °C:

$$T \text{ (in degrees kelvin, K)} = T \text{ (in degrees celsius, °C)} + 273.15$$

In principle, the **volume** V (m^3) of a system can be easily calculated from measured dimensions, and **mass** m (kg) can be calculated by dividing measured weight by the gravitational constant ($g = 9.81$ m s^{-2}). From the ratio of these two properties, two others can be deduced: **density** ρ (kg m^{-3}), which is the mass occupying unit volume, and **specific volume** v (m^3 kg^{-1}), which is the volume occupied by unit mass.

The thermodynamic system properties, their symbols and units are summarized in Table 4.1. In the final two columns, the tick indicates whether the property is an intensive or extensive property. An **extensive property** provides information on a total quantity or extent. The mass of matter in a system, and the volume of a system are both extensive properties. The others in the table are **intensive properties**. These provide no indication of size or total quantities. Some properties have extensive and intensive forms which are differentiated by using upper- and lower-case symbols – volume V and specific volume v, for example.

Thermodynamic material properties form a second set of properties which remain constant or change relatively little as system properties change. They include the specific heats and the specific gas constant of a gas, and the thermal conductivity of a solid. They define, for example, the characteristics of a particular working fluid which connect changes in state properties. Material properties can be described without reference to the system under examination. Typically, the values of material properties required for the solution of example or exercise problems are given in the problem description, but there are many sources of data for common solids, gases and liquids which can be drawn upon (for example, Rogers and Mayhew, 1995).

If the values of enough properties of a system can be determined through measurements or calculation to allow the values of all others to be found, the **state of the system** is defined. In many cases of practical importance, for a closed system, knowing the values of only two independent properties is sufficient. For an open system, because matter moves and its kinetic energy and potential energy may change, the variations of velocity and height of matter passing through the system are also required.

Thermodynamics

Property	Symbol	Units	Intensive	Extensive
Pressure	p	N m^{-2} or Pa	✓	
Temperature	T	K	✓	
Volume	V	m^3		✓
Mass	m	kg		✓
Specific volume	$v = \dfrac{V}{m}$	m^3 kg^{-1}	✓	
Density	$\rho = \dfrac{m}{V}$	kg m^{-3}	✓	
Internal energy	U	J		✓
Entropy	S	J K^{-1}		✓
Enthalpy	H	J		✓

Table 4.1 System properties may be intensive or extensive

A state is defined if values of all properties can be defined for all locations, but this will not necessarily be a **state of thermodynamic equilibrium.** This can be described precisely for a closed system as: if a closed system is isolated from its surroundings and its properties are unchanging, the system is said to be in a state of thermodynamic equilibrium. This requires thermal equilibrium (uniform temperature), mechanical equilibrium (uniform pressure) and chemical equilibrium (uniform chemical composition) to be achieved. Within an open system, although property values can vary with location, these can be related to a local state of thermodynamic equilibrium.

When the state of a system or the working fluid passing through a system changes, it is said to undergo a **process**. The initial and final states of a closed system undergoing a process are normally states of thermodynamic equilibrium. Similarly, the inlet and outlet states of matter passing through an open system are assumed to be states of thermodynamic equilibrium. During some processes, the states in between will also be states of thermodynamic equilibrium. This type of process is **reversible**: the changes in the system state can be defined exactly and reversed to restore the initial conditions in the system and the surroundings. In a closed system, for example, during a reversible process, a property will have the same value everywhere within the system at any time. All other processes are **irreversible**. A common cause of irreversibility is the generation of kinetic energy in fluids and gases, perhaps by the rapid movement of a piston, which is subsequently dissipated by friction. All real processes are irreversible to some extent but often the effect is small enough to neglect and the process can be *idealized* as reversible. If the movement of the piston is slow, the generation of kinetic energy is small and the variation of pressure and temperature throughout the cylinder can be neglected and the process can be treated as reversible. Two cases representing reversible and irreversible expansion processes are shown in Figure 4.4.

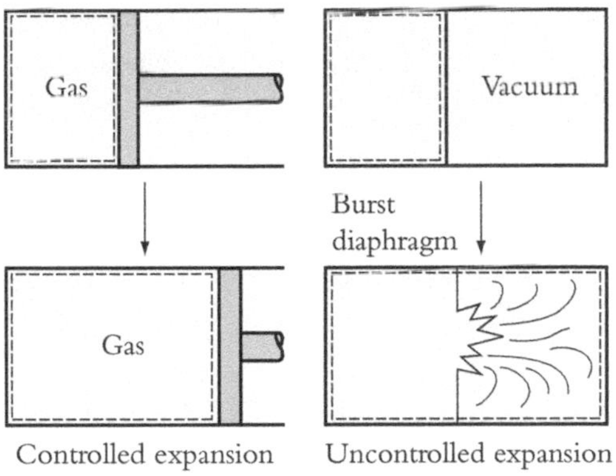

Figure 4.4 Types of expansion. Slowly retracting, the piston will produce a reversible expansion; the bursting of the diaphragm will produce an irreversible expansion

The description of a process can become complicated and, on occasion, difficult to absorb. Sketching a **process or state diagram** is a useful way of summarizing the information in a concise form, showing how the state of matter or a system changes during the process. Remember that for a closed system in a state of thermodynamic equilibrium the values of only two properties are required to define the state. It follows that a reversible process can be plotted on a two-dimensional diagram with axes of any pair of independent properties. Pressure–volume, temperature–volume and temperature–entropy diagrams are three of the most commonly used. Figures 4.5–4.10 illustrate how descriptions of processes can be represented on process diagrams. The descriptions are given in *italics*, and then interpreted before plotting.

A reversible constant-pressure process during which the volume is doubled.

Pressure and volume are the appropriate choices of properties for the axes of the process diagram. If pressure is plotted on the vertical axis, as shown in Figure 4.5, a constant-pressure process appears as a horizontal line. It is usual to identify the initial state as state 1 and the final state as state 2, and to use a suffix to denote the value of a property at the beginning or end of the process, i.e. V_1 is the volume at state 1.

A reversible expansion during which pv = constant.

An expansion is a process during which the volume or specific volume increases, so we can anticipate that v_1 will be smaller than v_2. The description indicates that the product of pressure and specific volume is constant during the expansion, which requires pressure to decrease as specific volume increases and hence for p_1 to be greater than p_2, and because the product is a constant, the process line will appear as a hyperbola on the diagram, as illustrated in Figure 4.6.

A closed system undergoes a reversible constant-volume process during which pressure increases from 1 bar to 3 bar, followed by a reversible constant-pressure compression during which the specific volume decreases from 5 to 2 m³ kg⁻¹.

In this example, illustrated in Figure 4.7, the system undergoes a sequence of two processes. The second begins at state 2 and ends at state 3. Both processes can be plotted on the same process diagram (and in principle, the system might undergo a sequence of several processes all of which could be plotted on the same diagram). The constant-volume process from state 1 to state 2 is a vertical line; the constant-pressure process which follows appears as a horizontal line. This progresses from a larger to a smaller specific volume, which will be a compression process.

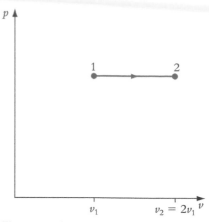

Figure 4.5 **A reversible constant-pressure process during which the volume is doubled**

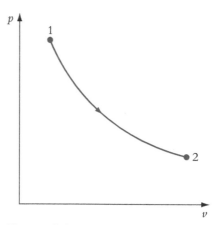

Figure 4.6 **A reversible expansion during which *pv* is constant**

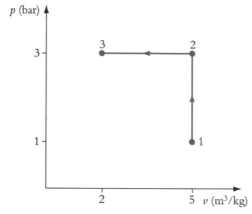

Figure 4.7 **A closed system undergoes a reversible constant-volume process during which pressure increases from 1 bar to 3 bar, followed by a reversible constant-pressure compression during which the specific volume decreases from 5 to 2 m³ kg⁻¹**

A reversible compression process obeys $\dfrac{p}{p_1} = \left(\dfrac{T}{T_1}\right)^{3.5}$. Find temperatures T_2 and T_3 if $p_2 = 3$ bar and $p_3 = 5$ bar given that $p_1 = 2$ bar and $T_1 = 500$ K. Plot the process connecting states 1, 2 and 3 on a state diagram with axes of temperature and pressure.

In this example, calculating T_2 and T_3 aids defining the process line. The power law used to define the relationship between pressure and temperature variations is often used in the description of processes undergone by gases, and the ability to manipulate power law relationships is a valuable skill. To calculate the temperatures, the power law is manipulated to make temperature the subject, $T = T_1\left(\dfrac{p}{p_1}\right)^{1/35}$, and the known values of pressure are substituted to obtain

$T_2 = 561$ K and $T_3 = 650$ K. When plotted, the states 1 to 3 lie on a curve, as shown in Figure 4.8.

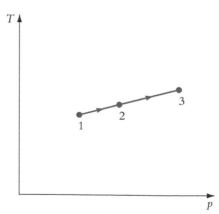

Figure 4.8 **A reversible compression process that obeys $\dfrac{p}{p_1} = \left(\dfrac{T}{T_1}\right)^{3.5}$**

Thermodynamics

A reversible expansion during which pv^n = constant and the pressure falls from 5 to 3 bar, followed by an irreversible compression back to the original specific volume and a final pressure of 4 bar.

All the previous examples describe reversible processes. In this one, the compression process is irreversible, and, although we can plot its initial state (2) and its final state (3), we connect these with a dashed line to indicate that we cannot define the intermediate states. The process diagram is shown in Figure 4.9

The final example is an illustration of a **cycle**. Here is the definition:

If a closed system undergoes a series of processes such that the initial and final states of the system are the same, the system has undergone a cycle.

This will appear on the state diagram as a closed contour, with points denoting the start and end states of each process. For a cycle to be reversible, each of the processes must be reversible. If any one or more of the processes is irreversible, the cycle will be irreversible even though the initial and final states are the same. A cycle comprising a reversible constant-volume reduction in pressure (1 to 2), a reversible constant-pressure expansion (2 to 3) and an irreversible compression (3 to 1) is illustrated in Figure 4.10.

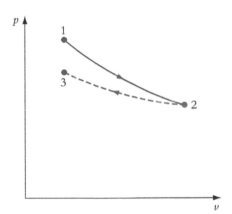

Figure 4.9 A reversible expansion during which pv^n = constant and the pressure falls from 5 to 3 bar, followed by an irreversible compression back to the original specific volume and a final pressure of 4 bar

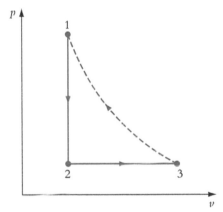

Figure 4.10 A cycle comprised of a reversible constant-volume reduction in pressure (1 to 2), a reversible constant-pressure expansion (2 to 3), and an irreversible compression (3 to 1)

Learning summary

- ✔ You should be familiar with and understand the key terms and definitions given in this section. They are shown in **bold**. The terminology is used throughout the following sections in the presentation of subjects.

- ✔ You should also familiarize yourself with the properties, nomenclature and units introduced here. Some or all of these properties feature strongly in any thermodynamic analysis. Take care to identify the correct units to use in calculations: if in doubt, use SI units and absolute values of pressures and temperatures.

- ✔ Learn to sketch processes on process or state diagrams to aid understanding, and note given values of properties or other defining information on these. This summarizes information in a concise and useful form.

4.2 The first law of thermodynamics, conservation of energy, work and heat transfer

Changes in the state of a closed system during a process are brought about by **work transfer** and **heat transfer**. These are the means by which energy is transferred across the system boundaries. Neither work nor heat is a form of energy; nor are they properties of a system. In mechanics, work is said to be done when a force moves through a distance. In thermodynamics, work can equally be described as 'the energy transfer across the boundary when a system changes its state due to the movement of a part of the boundary under the action of a force'. These statements are consistent and equivalent. Heat can be defined in a similar way, as the energy transfer across the boundary when a system changes its state due to a difference in temperature between the system and its surroundings. 'Heat transfer' is a short-form way of saying 'energy transfer by virtue of a temperature difference'. Because the direction of energy transfer can be into or out of the system, a convention for the positive direction is required. Here, work done *on* the system and heat transfer *to* the system are taken to be positive.

> According to the convention adopted, transfers of energy to the system from the surroundings are positive.

The effect of positive work or heat transfer is to raise the energy of the system and deplete the energy of the surroundings by the same amount. The energy of the system and the surroundings together remains constant. This is the principle of conservation of energy, which holds whether the energy is transferred by heat transfer or by work transfer. **The first law of thermodynamics** is a statement of the same principle, deduced from experiments carried out by James Joule (1818–1899) on closed systems (energy conservation also applies to open systems, which are considered later). The first law is an axiom – known to be true from experience – which has never been disproved.

> When a closed system is taken through a cycle, the sum of the net work transfer and the net heat transfer is zero.

In other words, if the system is returned to its original state, the quantity of energy within the system will be the same; the energy transferred into the system must be equal to the energy transferred out of the system during the cycle.

Work transfer

Work transfer to a closed system can be achieved by any mechanism which moves a force acting on the boundary through a distance. The work done on the system is, according to the mechanics definition:

$$W = \int_{x_1}^{x_2} F \, dx \qquad (4.1)$$

where F is the force and dx is the incremental distance moved in the direction of the force. This does not prescribe either how a force can be applied, or how it can move to do work on a system. It can be achieved by moving a piston forming part of the boundary of a system, or by turning a paddle wheel immersed in a system.

The arrangement of the paddle wheel which stirs the matter in a closed system is illustrated in Figure 4.11. Work transfer occurs as the shaft turns without any change in the volume of the system – it is a constant-volume

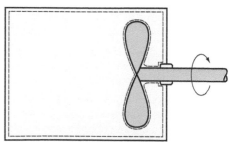

Figure 4.11 **A shaft that rotates against a resisting torque (created by the paddle wheel in the illustration) will produce work transfer across the system boundary without a change in volume**

process – and for this reason it can be called **shaft work**. The boundary of the system lies on the surface of the paddle wheel and moves as the wheel is turned. As the wheel rotates, it exerts a force F on the matter which is equal and opposite to the force resisting motion. In this problem, movement is rotational, not linear, and we can choose to describe work transfer as an applied torque rotated through an angle. Equation (4.1) is transformed by observing that, if the force F acts at a radius r from the axis of the wheel, it moves through a distance $r\,d\theta$ as the wheel rotates through an angle of $d\theta$ radians. Substituting $r\,d\theta$ for dx in equation (4.1) and changing the limits of integration to be consistent with

$$\theta_2 - \theta_1 = \frac{(x_2 - x_1)}{r}$$

gives

$$W = \int_{\theta_1}^{\theta_2} Fr\,d\theta$$

or, if T_o is the torque required to turn the paddle wheel,

$$W = \int_{\theta_1}^{\theta_2} T_o\,d\theta \tag{4.2}$$

The movement of a piston illustrated in Figure 4.12 *does* change the volume of the system – this is necessary for work transfer to occur – and the work done in this case can be called **boundary work**. In most cases, work transfer in closed systems is boundary work and is usually referred to simply as the work done on or by the system. Piston movement to the left is shown to be positive, because this will produce the compression process giving work done on the system.

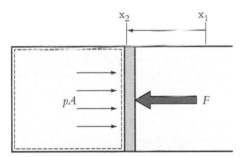

Figure 4.12 In closed systems formed by the piston-in-cylinder, work transfer takes place during a compression or expansion process produced by the movement of the piston. A compression process produces positive work transfer

If the piston has no mass and is free to move, the external force acting on the outer surface of the piston will always be opposed by an equal force exerted by the system on the inner surface of the piston. The piston will move to maintain this equilibrium of forces at all times, and the external force will dictate what pressure will act on the other face of the piston. If the piston area is A and pressure in the system is p, then

$$F = pA$$

and if the external force F causes the piston to advance into the cylinder from an initial position x_1 to any other position x, then substituting for F in equation (4.1) gives

$$W = \int_{x_1}^{x} pA\,dx \tag{4.3}$$

Clearly, as the piston moves further into the cylinder, the volume of the system will contract from V_1 to a value V when the piston is in position x. The changes are related by

$$x - x_1 = -\frac{(V - V_1)}{A}$$

Differentiating this gives $dx = \dfrac{2dV}{A}$, which can be rearranged as $Adx = -dV$ and substituted into equation (4.3) for

$$W = -\int_{v_1}^{v} p\, dV \qquad (4.4)$$

Unlike equations (4.1) and (4.2), there is an important restriction on the validity of using equation (4.4) to calculate the work done during a process:

> If p is taken to be system pressure, the process must be reversible, because otherwise the pressure cannot be defined.

Equations (4.1) and (4.2) are derived by considering how the surroundings act on the system, and there are no restrictions on their validity. Note that if the *local* pressure p, at the piston surface, is known at all times during the process, the force acting on the piston can be calculated and equation (4.4) can be used without qualification. Recall, though, that pressure will only be the same everywhere in the system when this is in a state of thermodynamic equilibrium.

Although energy transferred across the boundary of a closed system by **heat transfer** is indistinguishable from energy transferred by work, the quantity of heat transferred is often inferred by applying the first law to the system rather than by calculation using the heat transfer equivalent to equation (4.4) for work transfer. When applied to a closed system undergoing a cycle, the first law can be expressed as

$$W_{net} + Q_{net} = 0$$

or

$$Q_{net} = -W_{net} \qquad (4.5)$$

where W_{net} $(= \oint dW)$ is the sum of positive and negative work transfers which take place during the cycle, and Q_{net} $(= \oint dQ)$ is the sum of positive and negative heat transfers which take place during the cycle. If the cycle is reversible, the work done can be calculated for each of the processes making up the cycle, and the net heat transfer found from equation (4.5). For example, consider the cycle undergone by the piston-and-cylinder system in Figure 4.13. The cycle is made up of two constant-volume processes and two constant-pressure processes. Denoting the work done during the process with a suffix indicating the start and end points of the process:

$$W_{net} = W_{12} + W_{23} + W_{34} + W_{41}$$

Applying equation (4.4) to each term in turn:

$$W_{net} = 0 + p_2(V_3 - V_2) + 0 + p_4(V_1 - V_4)$$

No work transfer occurs during the two constant-volume processes ($dV = 0$ for these), and for the other two processes, pressure is a constant that can be taken outside the integral. Substituting the pressure and volume values shown in Figure 4.13 gives a value of -400 kJ for W_{net}. The negative value indicates that a net amount of work is done by the system on the surroundings. From equation ((4.5), Q_{net} must be 400 kJ, indicating that 400 kJ of heat transfer has occurred *to* the system *from* the surroundings.

In this example, the system has received a net heat transfer and produced an equivalent amount of work transfer to the surroundings. The system will produce a net amount of work output (to the surroundings) over the cycle every time the cycle is repeated. It behaves as an engine. If the cycle is continually repeated, the engine produces **power**, $\dot{W}$, defined as the rate of work transfer. The unit of power are J s^{-1}, or watts. If the cycle were to be repeated once per second, the power output of the engine would be 400 kJ s^{-1} or 400 kW.

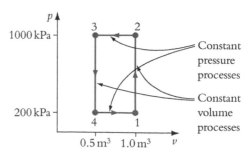

Figure 4.13 **Process diagram for a cycle made up of two constant-volume processes and two constant-pressure processes**

Internal energy

Although over the *cycle* there is no net change of energy in the system, in general it can increase or decrease during the various processes which make up the cycle as work and heat transfers take place. In a thermodynamic system, this will be a change in **internal energy** of the matter in the system. Internal energy U (measured in joules, J) and specific internal energy u (measured in joules per kilogram, J kg^{-1}) are properties. The connection between changes in internal energy and work and heat transfers during a process is stated in a corollary (corollary 1) of the first law:

> The change in internal energy of a closed system is equal to the sum of the heat transferred and the work done during any change of state.

That is, for a process during which the state of the system is changed from state 1 to state 2,

$$W_{12} + Q_{12} = U_2 - U_1 \tag{4.6}$$

Equation (4.6) is valid for any reversible or irreversible process undergone by a closed system, and requires only that the states 1 and 2 are states of thermodynamic equilibrium.

If no work or heat transfers take place during a process, by inspection of equation (4.6) there is no change in internal energy, and

$$U_2 - U_1 = 0$$

This can be stated as a second corollary of the first law:

> The internal energy of a closed system remains unchanged if the system is thermally isolated from its surroundings.

In this statement, **thermally isolated** means no work or heat transfer can take place. This would be the case, for example, if the system boundaries were rigid and fixed like the walls of a rigid container, so that no work transfer could occur, and the surfaces were **thermally insulated** to prevent heat transfers. (Although **isolated** and **insulated** sound similar, work transfer might still be possible if a system was thermally insulated but not isolated.)

The principle of conservation of energy is now readily accepted and the idea that work and heat transfers to a system will raise internal energy may seem obvious. Two hundred years ago, however, the equivalence of energy transfer by work and heat was yet to be established through experiments carried out by James Joule – the discovery is marked by an inscription on Joule's grave – and the existence of internal energy was unproven. The proof is remarkably simple and based on the comparison of two cycles with a common process, as illustrated in Figure 4.14.

The first law says that, for both cycles A–C and B–C, from equation (4.5)

$$W_{net} + Q_{net} = 0$$

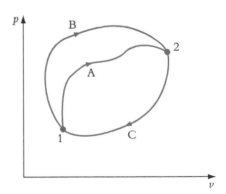

Figure 4.14 **The application of the first law to two cycles with one process common to both allows the existence of internal energy to be proven**

and because process C is common to both cycles, the work and heat transfers to the system during the process are the same for both cycles. It then follows that if the totals for the complete cycles are to be the same, and the contributions of C are the same, the sum of work and heat transfers must also be the same for A and B. In other words, the result is the same regardless of whether process A or B is used to change the state of the system from state 1 to state 2:

$$\int_1^2 (dW + dQ) = \text{constant}$$

Because the result is the same constant for both processes, it must be the difference between the values of a *property* at the two states. The property is **internal energy**, and the result is equation (4.6). Because energy is conserved, if the energy within the system is raised by work and heat transfer, the energy of the surroundings must be depleted by the amount. This principle is depicted in Figure 4.15.

How can different forms of energy be converted and transferred across the boundary of the system?

The energy in the surroundings can be in any of several forms: potential energy, kinetic energy, strain energy, or internal energy. When the internal energy of a closed system is raised, energy is transferred from the surroundings through work and heat transfer. Two examples of ways this can be brought about are illustrated in Figures 4.16 and 4.17 and the following analyses.

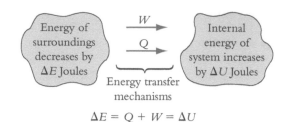

$$\Delta E = Q + W = \Delta U$$

Figure 4.15 Work and heat transfer are the mechanisms by which energy is transferred across the boundary of a closed system

If the weight shown in Figure 4.16 has a mass of 2 kg and descends 1 m, causing the paddle wheel to spin, calculate the gain in internal energy of the system. The temperature of the gas is initially the same as the surroundings and increases as the paddle wheel spins. How much heat transfer to the surroundings must occur before the temperature returns to its initial value?

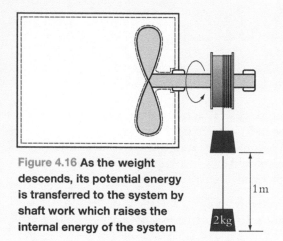

Figure 4.16 As the weight descends, its potential energy is transferred to the system by shaft work which raises the internal energy of the system

The weight gives up potential energy of value mass × gravitational constant × height of decent, mgh (= 19.62 J), as it descends. The paddle wheel transfers this by work transfer to the system, raising internal energy by 19.62 J. The temperature of the gas will also rise and unless the system is thermally insulated, the difference in temperature will produce heat transfer from the system to the surroundings. The gas temperature will return to its initial value when 19.62 J of energy have been transferred back to the surroundings; when this has happened, the temperature difference between the system and the surroundings will be zero and heat transfer will cease.

In Figure 4.17, air trapped in the cylinder by the piston is heated by heat transfer through the cylinder walls. As heat transfer occurs, pressure in the cylinder rises and the piston moves to the right. The position of the piston is dictated by equilibrium of forces acting on the piston:

$$pA = p_a A + F$$

where p is the pressure inside the cylinder, p_a is ambient pressure and F is the force exerted on the piston by the spring. The spring obeys Hooke's law, such that the spring force is

$$F = kx$$

where x is the compression of the spring from its unconstrained or free length and the spring constant k is $100\,\text{kN m}^{-1}$. If the area A of the piston is $0.01\,\text{m}^2$, the pressure in the cylinder rises from 2 to 4 bar, and ambient pressure is 1 bar, calculate the work done on the air in the cylinder and the change in strain energy stored in the spring.

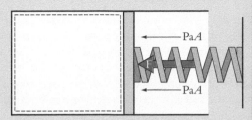

Figure 4.17 Heat transfer to the system produces a pressure rise which in turn causes the movement of the piston to the right. During this expansion process, the system does work on the surrounding air and the spring. As the spring is compressed, the strain energy stored in the spring increases

There are several aspects of this problem to understand. First, the interior walls of the cylinder and the piston (shown dashed in the figure) define the boundaries of a closed system containing the air. As heat transfer occurs and the piston moves to the right, the system volume increases. This is an **expansion** process and work is done **by** the system **on** the surroundings; by our sign convention, it is negative work done on the system. As the piston moves, work is done on the ambient air because this exerts a resistive force $p_a A$ on the outer face of the piston. In addition, work is done on the spring, compressing its length and increasing its strain energy. The strain energy in a spring obeying Hooke's law is equal to the work done compressing (or stretching) the spring; this is a mechanical form of energy. Equating the change in strain energy to the work done compressing the spring from length x_1 to length x_2 yields

$$\Delta SE = \int_{x_1}^{x_2} F\,\mathrm{d}x = \int_{x_1}^{x_2} kx\,\mathrm{d}x = \tfrac{1}{2}k(x_2^2 - x_1^2)$$

The initial and final values of compressed spring length can be found from equilibrium of forces on the piston. Using Hooke's law and rearranging the equation given in the question:

$$x = (p - p_a)\frac{A}{k}$$

Substituting the initial and final cylinder pressures in turn gives

$$x_1 = (2 \times 10^5 - 1 \times 10^5)\frac{0.01}{10^5} = 0.01\,\text{m}$$

and

$$x_2 = (4 \times 10^5 - 1 \times 10^5)\frac{0.01}{10^5} = 0.03\,\text{m}$$

Hence

$$\Delta SE = \tfrac{1}{2}k(x_2^2 - x_1^2) = 0.5 \times 10^5 \times (0.03^2 - 0.01^2) = 40\,\text{J}$$

At the same time as the spring is being compressed, the piston movement is opposed by the force exerted by the surrounding ambient. The work done overcoming this is

$$W = p_a A(x_2 - x_1) = 10^5 \times 0.01 \times 0.02 = 20\,\text{J}$$

The work done **by** the air in the cylinder will therefore be $(40 + 20)$ or 60 J, and hence the work done **on** the air in the cylinder will be -60 J.

Can changes in potential and kinetic energy contained within closed systems be neglected?

Although a closed system is said to undergo non-flow processes, matter in contact with a moving boundary such as the surface of a moving piston will move with it and have velocity. As a result, kinetic energy is imparted to the matter and there can be some non-uniformity of conditions throughout the system. Similarly, there can be a change in the potential energy of matter in a system if its centre of mass is raised as a system boundary is moved. In principle, these changes can be finite and contribute with the change in internal energy ΔU to the energy change so that:

$$W_{12} + Q_{12} = \Delta U + \Delta PE + \Delta KE$$

In practice, the displacement of matter is usually too small for the change in potential energy to be significant and kinetic energy is dissipated into internal energy by friction as conditions of thermodynamic equilibrium are established. Thus, in the analysis of closed systems undergoing reversible processes, generally the changes in potential energy, ΔPE and kinetic energy, ΔKE of matter contained *within the system* can be neglected.

How can a closed system containing a spring be analysed?

This type of closed system might be a shock absorber taking the form of a closed piston–cylinder system containing gas and a spring. Heat transfer will change only internal energy but the movement of the piston can compress the spring and the gas inside the cylinder, changing both strain energy and internal energy. In this case, the problem can be analysed as the combination of thermodynamic and mechanical systems shown in Figure 4.18. The first law is applied to the thermodynamic system, Hooke's law to the spring, and equilibrium of forces on the piston can be used to determine the split of work transfer to these.

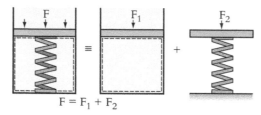

Figure 4.18 This combination of spring within a piston-in-cylinder can be analysed as a mechanical system plus a thermodynamic system

> Work W and heat transfer Q are not properties, but it is sometimes convenient to consider quantities of work and heat transfer per unit mass of matter in the system. These are described as specific work w and specific heat transfer q and have units of $J\,kg^{-1}$.

For example, using specific values, equation (4.6)

$$W_{12} + Q_{12} = U_2 - U_1$$

becomes

$$w_{12} + q_{12} = u_2 - u_1$$

Throughout this section, the suffix of w_{12} and q_{12} denotes the specific work and heat transfer taking place during the process connecting 1 and 2. If a second process 2–3 followed this, the work and heat transfers would be w_{23} and q_{23} respectively. In cases where it is not necessary to distinguish between the values for different processes, the suffix can be dropped. In some of the sections which follow, the suffix has a different meaning. In Section 4.2, for example, a suffix is used to differentiate between heat supplied and heat rejected.

✔ Work and heat transfer are the means by which energy can be transferred across the boundary of a closed system. These are not properties of the system. Our convention is that work or heat transfer to the system will be positive.

✔ The first law of thermodynamics embodies the principle of conservation of energy. When applied to a closed system undergoing a cycle, there is no net transfer of energy into or out of the system over the cycle, nor is there any net change in the energy stored in the system.

✔ When a closed system undergoes a process that changes its state from state 1 to state 2, energy transfer by work or heat transfer will raise or lower the internal energy of the system. Changes in other forms of energy will usually be negligible.

✔ Two important results from the application of the first law of thermodynamics to a closed system are that, when applied to a closed system undergoing a cycle,

$$W_{net} + Q_{net} = 0$$

and, when applied to a closed system undergoing a process 1–2,

$$W_{12} + Q_{12} = U_2 - U_1$$

4.3 The second law of thermodynamics, heat engines, the Clausius inequality, entropy and irreversibility

Although energy is conserved when it is transferred, work transfer is more valuable than heat transfer. Although work can be converted entirely to heat, only some of the heat supplied to a system undergoing a cycle can be converted to work; some of the heat transferred to the system will always be transferred back to the surroundings. The **second law of thermodynamics** recognizes the implications. The Kelvin–Planck statement of the law is this:

> No heat engine can produce a net amount of work output while exchanging heat with a single reservoir only.

Heat engines operate by drawing on a supply of heat from a source, converting some of this to a net work output, and rejecting the balance to a heat sink.

> A heat engine is any device or system designed to convert heat into work output through a cycle of processes; it must have at least one prime mover, one source of heat transfer to the heat engine and one sink of heat transfer from the heat engine.

Prime movers are typically turbines delivering shaft work or reciprocating pistons transferring work to a crankshaft. The source of heat transfer might be the high-temperature products of combustion in a furnace; the sink receiving heat transfer from the engine typically might be a large mass of water such as a lake or sea. A **heat reservoir** is an ideal source or sink which is able to supply or absorb heat without its temperature changing. A representation of a heat engine operating between two reservoirs is shown in Figure 4.19. The convention for the energy transfer to the prime mover to be positive is the same as for the closed systems considered in earlier sections, but in diagrams arrows are used to show the direction of transfer, and in equations Q and W are the magnitudes of energy transfers.

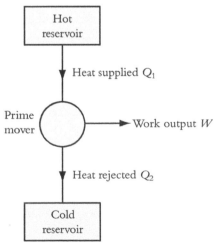

Figure 4.19 A heat engine with a prime mover operating between two reservoirs

Efficiency, the Carnot efficiency and the Carnot principles

The efficiency of a heat engine is defined as

$$\eta = \frac{\text{net work output}}{\text{heat supplied}} = \frac{W}{Q_1} \qquad (4.7)$$

Efficiency is a measure of success in converting heat **supplied** Q_1, which has to be paid for in fuel consumed, into a net quantity of work output W, which is the desired product. To satisfy the second law, the value of efficiency must always to be less than 100 per cent. There must be some heat rejected, Q_1. The first law connects these through energy conservation:

$$Q_1 = W + Q_2$$

Combining the two equations shows that

$$\eta = 1 - \frac{Q_2}{Q_1} \qquad (4.8)$$

The second law says this must be less than 100 per cent, but not by how much. The highest values that can be achieved, and the conditions which apply to this, are described in statements of two **Carnot principles**.

> The first principle is that all reversible heat engines operating on any cycle between the same two reservoirs will have efficiency equal to:
>
> $$\eta_{\text{carnot}} = 1 - \frac{T_2}{T_1} \qquad (4.9)$$
>
> where η_{carnot} is the **Carnot efficiency**, and temperatures T_1 and T_2 are the absolute temperatures, in kelvin, of the heat reservoirs.

The very highest value that can be achieved turns out to be dependent only on the highest temperature at which heat is being supplied and the lowest temperature at which heat is being rejected. The Carnot efficiency is the highest value of efficiency which can be achieved by any heat engine operating between heat reservoirs with upper and lower temperatures T_1 and T_2. This upper limit on efficiency will not be achieved if either the heat engine is irreversible or it operates with more than two heat reservoirs.

> The **second Carnot principle** is that the efficiency of a heat engine operating between two reservoirs will be less than the Carnot efficiency if the heat engine is irreversible.

For a heat engine to be reversible, all the processes making up the cycle must be reversible. If any of the processes are irreversible, the heat engine is irreversible. There is also an efficiency penalty associated with using more than two reservoirs, which can be stated as:

> The efficiency of any heat engine, reversible or irreversible, will be less than the Carnot efficiency if it operates between more than two heat reservoirs.

The efficiency penalty of operating between more than two heat reservoirs has important practical applications in the design of efficient plant for power generation. Efficiency is reduced if some heat is supplied at a temperature below the highest available or rejected at a temperature above the lowest available. The inevitability of a penalty can be illustrated with reference to Figure 4.20.

In the first instance, it is useful to consider the efficiency of two reversible heat engines each operating between two of three heat reservoirs. First, reversible heat engine A operates on its own between the hottest ($T_1 = 1000$ K) and the coldest ($T_2 = 400$ K) reservoirs. The efficiency of engine A will be equal to the Carnot efficiency:

$$\eta_{\text{carnot}} = \left(1 - \frac{T_2}{T_1}\right) = 0.6$$

Thermodynamics

Now a third reservoir with a temperature T_3 between T_1 and T_2 and a second reversible heat engine B, producing part of the required output are introduced. B is reversible and operates between the new reservoir at T_3 and one of the original heat reservoirs. Its efficiency will be equal to the Carnot efficiency calculated using the temperatures of these reservoirs. This must be lower than the Carnot efficiency of heat engine A because the ratio of reservoir temperatures, either $\dfrac{T_3}{T_1}$ or $\dfrac{T_2}{T_3}$ will be higher than the ratio $\dfrac{T_2}{T_1}$. It follows that, if the efficiency of engine B is lower than the efficiency of engine A, the efficiency of the combined two engines must also be lower than the efficiency of A.

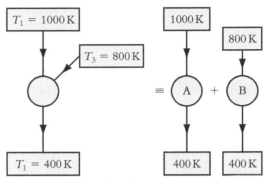

Figure 4.20 **A heat engine operating between more than two reservoirs will always have an efficiency less than the Carnot efficiency**

The next step in the analysis is to recognize that, because both engines contribute to the work output of the unit, they could be replaced by one operating between three reservoirs that will have the same overall efficiency. Thus the reversible heat engine operating between the three reservoirs must have a lower efficiency than a reversible heat engine operating between the hottest and coldest reservoirs. If the temperature of the third reservoir is 800 K, the efficiency of engine B is 0.2 if this is the cold reservoir or 0.5 if it is the hot reservoir. The efficiency of the combined engines η_{A+B} (and the efficiency of the single replacement engine) can be found using the definition of efficiency given in equation (4.7):

$$\eta_{A+B} = \frac{W_A + W_B}{Q_{As} + Q_{Bs}} = \frac{Q_{As}\eta_B + Q_{Bs}\eta_A}{Q_{As} + Q_{Bs}}$$

The value of the efficiency depends on whether the third reservoir is a heat sink or a heat source, and the split of the total heat supplied between the two engines.

The Clausius inequality

Equations (4.8) and (4.9) and the Carnot principles can be used to derive the **Clausius inequality:**

> For any reversible heat engine (or closed system undergoing a reversible cycle), the integral around the cycle of $\dfrac{dQ}{T}$ will be zero; for all irreversible heat engines (or closed systems undergoing an irreversible cycle), this integral will be negative.

This inequality can be expressed mathematically as

$$\oint \frac{dQ}{T} \leqslant 0 \tag{4.10}$$

The result applies to any heat engine or closed system undergoing a cycle, regardless of the number of heat reservoirs involved, and can easily be demonstrated to be true for a heat engine operating between just two heat reservoirs. First, imagine the cycle of the heat engine has only one process during which all the heat supply Q_1 takes place from the hot reservoir at constant temperature T_1, and one process during which all the heat rejection Q_2 takes place to the cold reservoir at constant temperature T_2. If no heat transfer takes place during any other process in the cycle, then

$$\oint \frac{dQ}{T} = \frac{Q_1}{T_1} - \frac{Q_2}{T_2} \tag{4.11}$$

If the cycle is reversible, both expressions (4.8) and (4.9) for efficiency are valid. This requires

$$\frac{Q_2}{Q_1} = \frac{T_2}{T_1}$$

and hence

$$\frac{Q_2}{T_2} = \frac{Q_1}{T_1}$$

Substituting into equation (4.11) gives the result for the first part of the Clausius inequality, that which applies to the operation of a **reversible heat engine**:

$$\oint \frac{dQ}{T} = 0 \qquad (4.12)$$

If the cycle contains irreversible processes, the cycle will be irreversible and the efficiency of the heat engine must be lower than the Carnot efficiency, $\eta < \eta_{Carnot}$. In this case, equations (4.8) and (4.9) now require

$$\frac{Q_1}{T_1} < \frac{Q_2}{T_2}$$

and so

$$\frac{Q_1}{T_1} - \frac{Q_2}{T_2} < 0$$

Substituting this into equation (4.11) gives the result for the second part of the Clausius inequality, that which applies to the operation of an **irreversible heat engine**:

$$\oint \frac{dQ}{T} < 0 \qquad (4.13)$$

Although the Clausius inequality has been derived here through arguments applied to heat engines, the result is true for any closed system undergoing a reversible or irreversible cycle.

The result for the reversible cycle has particular significance, as it led Clausius to conclude that he had discovered a new property, which he named **entropy**.

Entropy (S) and specific entropy (s)

Entropy (S) is an extensive property that has units of joules per degree kelvin ($J\,K^{-1}$). The intensive form is specific entropy (s), or entropy per unit mass, which has units of $J\,kg^{-1}\,K^{-1}$.

As for any other property such as pressure or temperature, the change in the entropy between two states of thermodynamic equilibrium of a closed system must have a value that does not depend on the process connecting these. This observation provides a route to proving entropy exists. The steps are the same as they were for the proof that internal energy exists, and begin with the application of equation (4.13) to two cycles, A–C and B–C, with one process (C) in common, as illustrated in Figure 4.14.

Because the integral of the quantity $\dfrac{dQ}{T}$ between state 2 and state 1 is common to both cycles, it will have the same value for both cycles. It now follows that, if the integration of $\dfrac{dQ}{T}$ around either complete cycle is to be zero, the change between state 1 and state 2 must be the same for both cycles. Finally, because the path connecting states 1 and 2 are different for the two cycles, the change is independent of the path taken and must be the change in value of a property. This leads to the definition of the change in entropy between any two states, 1 and 2:

$$S_2 - S_1 = \int_1^2 \left(\frac{dQ}{T}\right)_{reversible} \qquad (4.14)$$

Note that this is the *change* in entropy; the absolute values of S_1 and S_2 are not defined – for this, the absolute value of entropy at a reference state must be defined. The third law of thermodynamics provides this, defining the entropy of a pure crystalline substance as zero at a temperature of absolute zero, but the analysis of engineering problems usually requires changes in entropy to be determined and any convenient datum for entropy can be used.

Thermodynamics

Equation (4.14) can be thought of as connecting heat transfer to entropy changes during a reversible process undergone by a closed system in a similar way to how work transfer is related to changes in volume:

$$W = -\int_1^2 p\,dV$$

$$Q = \int_1^2 T\,dS$$

What is entropy?

Entropy is a property that, like any other, will have a particular value for a specified state. It has wider significance, however, as a measure of order (or lack of order). Entropy increases as disorder increases, and part of its value in analysing problems is as a measure of the deterioration. In engineering, increases in entropy are often associated with inefficiencies in machine performance or processes which we want to eliminate as far as possible, so minimizing the increase in entropy can be a measure of our success. A simple analogue to aid understanding entropy and the cost of disorder is this: visualize a building in good condition, with all windows and doors in place and no missing roof tiles. This has a structure which is well ordered; everything that makes up the building has its place. In this ordered state, the building has a low value of entropy. Now suppose the building is smashed to rubble, reducing the materials to a disordered state (debris). During the demolition, order has decreased and entropy has increased. What price is there to pay? The cost of the work involved in rebuilding the structure. 'Order' tends to deteriorate into 'disorder', but not vice versa, without work input, so there is a cost.

A remarkable implication of the Clausius inequality is that:

> Entropy is created during an irreversible process.

This can be illustrated by analysing the cycle shown in Figure 4.21. The cycle has a reversible process A, changing state 1 to state 2, during which heat transfer can occur, followed by a second process B, during which no heat transfer occurs. During process A the change in entropy is given by

$$S_2 - S_1 = \int_1^2 \left(\frac{dQ}{T}\right)_{\text{reversible}}$$

During process B, because there is no heat transfer, there can be no entropy transfer from the system and so, around the cycle,

$$\oint \frac{dQ}{T} = \int_1^2 \frac{dQ}{T} + \int_2^1 \frac{dQ}{T} = (S_2 - S_1) + 0 = S_2 - S_1$$

If we now choose to make process B reversible, the cycle is reversible

$$\oint \frac{dQ}{T} = 0$$

and

$$S_1 = S_2$$

If, instead, we make process B irreversible, the cycle is irreversible:

$$\oint \frac{dQ}{T} < 0$$

and

$$S_1 > S_2$$

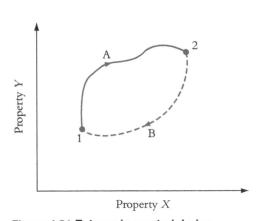

Figure 4.21 **Entropy is created during process B only when this process is irreversible**

Comparing the two cases shows that entropy increased during B when this process was irreversible. The increase is not due to a transfer of entropy from the surroundings; none has been possible because no heat transfer has occurred. The increase must have been generated by the irreversibility of the process.

However, entropy is a property. Does that not mean that $\oint dS = 0$? Well, yes! But, unlike internal energy or mass, which are conserved, the total entropy of the universe is raised each time an irreversible process occurs. We do not have to think on this scale to recognize the implications, only about a system and its surroundings. If an irreversible process creates entropy within the system, and a completed cycle returns the system to its initial state and value of entropy, the entropy generated by irreversibilities must be transferred to the surroundings. This transfer requires there to be heat transfer to the surroundings, and an irreversible cycle must have one or more processes during which heat transfer occurs.

Before leaving this section, here is a note on the question posed in the introduction: How does a refrigerator cool a cold space in a hot room? The answer: It operates as a **reversed** heat engine. The refrigerator operates between a cold reservoir (the interior cold space) and a hot reservoir (the surroundings of the refrigerator) like a heat engine, but the direction of heat transfer is now from the cold reservoir and to the hot reservoir. Instead of producing work output, there has to be work input to the refrigerator. This pumps a refrigerant around a circuit passing between the cold space, where it receives energy, and the surroundings, where it gives up energy. The refrigerant has to be at a lower temperature than the cold space in this part of the circuit and at a higher temperature than the surroundings when in this part of the circuit.

Learning summary

✔ By the end of this section you should understand that the second law of thermodynamics distinguishes between work and heat transfer and recognize that work transfer is the more valuable of these. This does not contradict the first law; energy transferred by one is indistinguishable from energy transferred by the other, and the principle of energy conservation is not violated. There are, however, limits on how efficiently heat can be drawn from a source and converted into work output using a system which operates in a cycle:

> It is impossible to construct a system which will operate in a cycle, extract heat from a reservoir, and do an equivalent amount of work on the surroundings.

✔ As a consequence of the second law, a system operating in a cycle and producing work output must be exchanging heat with at least two reservoirs at different temperatures. If a system is operating in a cycle while exchanging heat with only one reservoir, if a net transfer of work occurs it must be **to** the system. Work can be converted continuously and completely into heat, but heat cannot be converted completely and continuously into work. The efficiency of a heat engine designed to produce a net work output is defined as

$$\eta = \frac{\text{net work output}}{\text{heat supplied}} = \frac{W}{Q_1}$$

✔ Note that it is the heat **supplied** that we are trying to convert to work output, not the net heat transfer.

✔ Efficiency is the measure of success in achieving this. The highest possible efficiency that can be achieved is the Carnot efficiency.

✔ $\eta_{carnot} = 1 - \dfrac{T_2}{T_1}$

✔ The existence of entropy is a corollary of the second law. It is important to remember that entropy is a property, like pressure or temperature, but also that it provides a measure of order and irreversibility. It is not conserved, like mass of energy, and the entropy of the universe is increasing continuously as the result of the myriad irreversible processes taking place: an implication of the Clausius inequality is that entropy is created during an irreversible process; this may result in an increase in entropy of the system and/or of the surroundings.

✔ Entropy can only be transferred across the boundary of a closed system with heat transfer, not work transfer. If no heat transfer takes place, the entropy within a closed system remains constant during reversible processes and increases during irreversible processes.

4.4 Open systems, mass flow continuity, the steady flow energy equation and enthalpy

By definition, matter can cross part or parts of the boundary of an **open system**. This is a fundamental difference from a closed system and there are important differences between the ways the behaviour of the two types of system is analysed.

> The processes undergone in open systems are flow processes.

The **motion** of matter in an open system must be considered. The matter will have kinetic energy and, if its location changes, so might its potential energy. This is different to what happens in a closed system which undergoes non-flow processes and only internal energy changes within the system are significant. For open systems, additional elements of theory are required, as described in this section, and some of the results applied to closed systems in earlier pages will not apply. The main differences in the way steady flow, open system and non-flow, closed system problems are defined and analysed are set out in Table 4.2. The results for open systems are developed in the following.

	Closed system	Open system
Use of suffix to define time or location to which conditions apply	Denotes initial and final states of the system during a discrete process	Denotes locations within the system at which a continuous process starts and ends
Processes	The fluid undergoing the process is retained within the system	The fluid undergoes a process as it passes through the system
State	The fluid state is uniform throughout the system, but varies during a process	The fluid state varies with location through the system. The state of fluid at a particular location is constant
Relationships between properties	Same for both systems	
Mass conservation	The mass of matter in the system remains constant because no matter can cross the boundaries of the system. m = constant	Under steady flow conditions the mass of matter within the system remains constant because the rate at which matter flows into the system is exactly equal to the rate at which matter flows out of the system. $\dot{m}_1 = \dot{m}_2$
Energy conservation (first law)	$Q + W = U_2 - U_1$	Steady flow energy equation
Reversible specific work	$w = -\int_1^2 p \, dv$	$w = \int_1^2 v \, dp + \Delta\text{KE} + \Delta\text{PE}$

Table 4.2 Comparison of closed systems (non-flow processes) and open systems (steady flow processes)

The volume of an open system can change in the most general definition of possibilities, its boundaries can move and the quantities of energy and matter within it can vary with time. However, for many open system problems arising in engineering it is possible to apply assumptions which simplify the analysis. One of the most important is the assumption that steady flow conditions exist, and the analysis of **steady flow problems** is the main topic of this section.

> For steady flow through an open system which has fixed boundaries, the quantities of matter and energy within the system boundaries are each constant and do not change with time.

It follows that the net flows of matter and energy into the system must both be zero. All the sources of energy transfers into and out of the system must be accounted for, including the energy transported by matter flowing into and out of the system as well as work and heat transfers at the boundaries.

Mass conservation and mass flow continuity

The mass m of matter in the volume defined by the boundaries of an open system is given by the integration of density over the volume:

$$m = \int_0^V \rho \, dV$$

Under steady-state conditions this is constant and since matter is neither being created nor destroyed, it follows that the mass of matter entering the volume during a given interval of time must exactly equal the mass of matter leaving during the same interval. This principle of mass conservation leads to the requirement that:

> Under steady flow conditions, there will be mass flow continuity

The **mass flowrates** entering and leaving the open system, in $kg\,s^{-1}$, are equal.

There is no restriction on the number of inlets or outlets of mass flow that a control volume can have, so more generally the **mass flow continuity equation** is

$$\sum_{inlet} \dot{m} = \sum_{outlet} \dot{m} \qquad (4.15)$$

For the open system shown in Figure 4.22, there are two inlets a and b, and one outlet c, and, for this case, mass flow continuity requires

$$\dot{m}_a + \dot{m}_b = \dot{m}_c$$

The mass flowrate across the boundary at an inlet and outlet is related to the cross-sectional area A of the passage, the mass density ρ and the velocity C of the flow normal to the boundary by the **mass flowrate equation**:

$$\dot{m} = \rho A C \qquad (4.16)$$

This equation can be derived by considering the movement of a parcel of matter of mass dm across the boundary shown in Figure 4.23.

The parcel will have a mass $\delta m = \rho A L$ and it will take time $t = \dfrac{L}{C}$ seconds to cross the boundary.

Since mass flowrate is the mass of matter crossing the boundary in unit time, we obtain

$$\dot{m} = \frac{\delta m}{t} = \frac{\rho A L}{\dfrac{L}{C}} = \rho A C$$

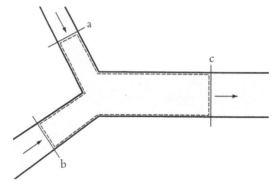

Figure 4.22 **Under steady flow conditions, the combined rates of mass flow into the open system through inlets a and b is equal to the rate of mass flow leaving the system through outlet c**

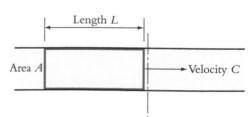

Figure 4.23 **Mass flowrate is the mass of matter passing across the boundary in unit time**

Figure 4.22 illustrates a junction in a high-pressure gas distribution system operating under steady flow conditions. The internal diameters of the pipes a, b and c are 20 cm, 25 cm and 40 cm respectively. The mass flowrate of gas passing through inlet pipe a is 2 kg s^{-1}; at outlet c, the density is 11 kg m^{-3} and the gas flows with a velocity of 3 m s^{-1}. If the density at inlet b is 10 kg m^{-3}, what is the gas velocity in this inlet pipe?

The mass flowrate of gas passing through outlet c will be

$$m_c = \rho_c A_c C_c = 11.0 \times \left(\frac{\pi \times 0.40^2}{4}\right) \times 3.0 = 4.147 \text{ kg s}^{-1}$$

Mass continuity requires

$$m_b = m_c - m_a = 4.147 - 2.0 = 2.147 \text{ kg s}^{-1}$$

Applying the mass flowrate equation to flow through inlet b yields

$$C_b = \frac{m_b}{\rho_b A_b} = \frac{2.147}{10.0 \times \left(\frac{\pi \times 0.25^2}{4}\right)} = 4.374 \text{ m s}^{-1}$$

The steady flow energy equation and enthalpy

Like mass, energy is conserved and in steady flow problems, if the energy contained within the system boundaries does not increase or reduce with time, the energy entering the system during any time interval must be equal to the energy leaving during the same interval. This is the principle of conservation of energy, or the first law of thermodynamics, applied to steady flow through an open system. The equation which expresses this in mathematical form is the **steady flow energy equation** (or **SFEE**). The equation is valid for all reversible and irreversible steady flow processes undergone by a working fluid passing through an open system.

The forms in which energy is transported into or out of the control volume by the movement of matter are illustrated in Figure 4.24. In addition, energy is transferred across the boundaries by work and heat transfer. Part of the work done at the boundaries is done by the working fluid itself, at the inlets and outlets to the system. This is called flow work, because it is work done by matter in motion on matter in front of it on the other side of the system boundary.

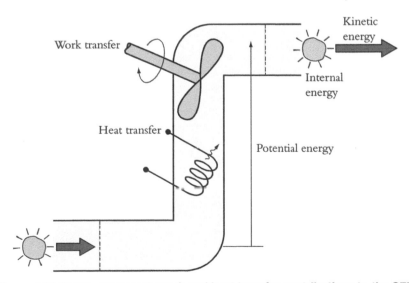

Figure 4.24 **Energy transport, work and heat transfer contributions to the SFEE**

If all the energy inflows and outflows are to balance over a time interval, then

$$\begin{matrix}\text{Heat} \\ \text{transfer}\end{matrix} + \begin{matrix}\text{work} \\ \text{transfer}\end{matrix} + \begin{matrix}\text{flow work} \\ \text{transfer}\end{matrix} = \begin{matrix}\text{energy transported out} \\ \text{of control volume by matter}\end{matrix} - \begin{matrix}\text{energy transported into} \\ \text{control volume by matter}\end{matrix}$$

On the left side of the equality are the sources of energy transfer across the boundary to matter within the control volume. The quantities on the right side are energy being transported across the boundaries by matter. Under steady flow conditions, heat and work transfers take place continuously and these are defined as a rate of heat transfer $\dot{Q}$ or power $\dot{W}$, in J s^{-1} or watts, W. Because we are considering only steady flow conditions, for which the mass flows into and out of the system are equal and also constant, energy transfers can be expressed as values of energy transferred per unit of mass passing through the system. Thus:

> The heat transferred per unit mass (q) is equal to $\dfrac{\dot{Q}}{\dot{m}}$,
>
> the work transferred per unit mass (w) is equal to $\dfrac{\dot{W}}{\dot{m}}$

Using these specific quantities, the SFEE can be written as

$$q + w + p_1 v_1 - p_2 v_2 = \left(u_2 + \frac{C_2^2}{2} + gz_2\right) - \left(u_1 + \frac{C_1^2}{2} + gz_1\right) \qquad (4.17)$$

The right-hand side of the equation expresses the difference between energy transported into and out of the system per unit of mass flow. This is the energy gained by 1 kg of matter as it passes through the system from inlet to outlet. The increase is made up of gains in

specific internal energy $(u_2 - u_1)$, **specific kinetic energy** $\left(\dfrac{C_2^2}{2} + \dfrac{C_1^2}{2}\right)$ and

specific potential energy $(gz_2 - gz_1)$. The specific internal energy is the same form of energy possessed by matter in a closed system; the specific kinetic energy requires the matter to have velocity; and a change in potential energy requires the matter to have changed height above a datum position, where z is taken to be zero.

The final term in the SFEE is the **specific flow work** $(p_1 v_1 - p_2 v_2)$. This is work done on and by matter flowing across the system boundaries. To visualize how flow work is done, imagine the flow of matter as people hurrying towards the same destination. Those at the back of the crowd press forward on those in front, who tend to be assisted forward without so much effort. Figure 4.25 depicts an open system with a parcel of matter A upstream and outside the system boundary and a second parcel, B, inside the downstream boundary of the system. Each exerts a force on the matter downstream of it. The force exerted by A will be $p_1 A_1$, the force exerted by B will be $p_2 A_2$. They have the same mass, and, as B passes through plane 2, A will pass through plane 1.

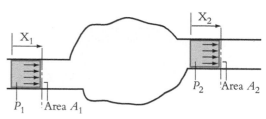

Figure 4.25 **Specific flow work is done at the boundaries of an open system**

As A passes across the boundary at plane 1, it pushes fluid ahead of it and does work $p_1 A_1 x_1$ on fluid inside the system. As B passes across the boundary at plane 2, it does work $p_2 A_2 x_2$ on fluid outside the system. The difference between these is the flow work $(p_1 A_1 x_1 - p_2 A_2 x_2)$ done on the closed system as the transfers across 1 and 2 take place. If A and B each contain mass δm, then the matter in parcel A has a specific volume $v_1 = \dfrac{A_1 x_1}{\delta m}$, and in B $v_2 = \dfrac{A_2 x_2}{\delta m}$.

Hence the work done per unit of mass entering or leaving the system is

$$\text{Specific work flow} = \frac{(p_2 A_2 x_2 - p_2 A_2 x_2)}{\delta m}$$

$$= p_1 v_1 - p_2 v_2$$

Thermodynamics

The calculation of flow work can be repeated each time parcels of matter enter and leave. Each time, the work done represents a transfer of energy across the boundaries of the open system from matter on one side to matter on the other side as a result of flow. This happens only when there is movement of matter, and is a *transfer* of energy between quantities of matter, but it is usual to place the flow work term on the right-hand side of the SFEE:

$$q + w = p_2 v_2 - p_1 v_1 + \left(u_2 + \frac{C_2^2}{2} + gz_2 \right) - \left(u_1 + \frac{C_1^2}{2} + gz_1 \right)$$

or

$$q + w = \left(u_2 + p_2 v_2 + \frac{C_2^2}{2} + gz_2 \right) - \left(u_1 + p_1 v_1 + \frac{C_1^2}{2} + gz_1 \right)$$

The important combination of properties $u + pv$ will *always* appear in the analysis of energy conservation applied to open systems, because matter passing across the boundaries of the system will *always* do flow work and possess internal energy. Because this is a combination of properties, it must be a property itself: at a given state, it will have a particular value. In the intensive form, this is called **specific enthalpy**, h, and has units of joules per kilogram ($J\,kg^{-1}$). (This is a property of the matter, and a quantity of m kilograms of matter has mh joules of **enthalpy** H.) After substituting h for $u + pv$ and regrouping the terms on the right side, the SFEE becomes

$$q + w = \left(h_2 - h_1 + \frac{C_2^2}{2} - \frac{C_1^2}{2} + gz_2 - gz_1 \right) \tag{4.18}$$

If both sides are multiplied by the mass flowrate, then

$$\dot{Q} + \dot{W} = \dot{m} \left(h_2 - h_1 + \frac{C_2^2}{2} - \frac{C_1^2}{2} + gz_2 - gz_1 \right) \tag{4.19}$$

Equations (4.18) and (4.19) will apply to all steady flows through open systems, for any reversible and irreversible process changing the state of matter from state 1 to state 2. For gas flows through open systems at velocities up to tens of metres per second and rising or falling through heights of a few metres, the change in enthalpy is often much larger than the changes in kinetic energy and potential energy. For example, a rise in air temperature from 20 to 30 °C produces a change in the specific enthalpy of air of $10{,}050\,J\,kg^{-1}$; the change in kinetic energy produced by raising velocity from 1 to 10 m s^{-1} is $49.5\,J\,kg^{-1}$, and the change in potential energy produced by raising the height above the datum by 3 m is $29.4\,J\,kg^{-1}$. In many cases such relatively small changes in kinetic and potential energy may be neglected in calculations of heat and work transfer but if, say, the outlet velocity of the flow was to be calculated, the changes could be important.

Air is heated at the rate of 50 kW as it flows at a steady rate downwards through a vertical duct. The mass flowrate of air is 3 kg s^{-1}, and at inlet the air velocity is 25 m s^{-1}. The outlet of the duct is 15 m below. Between the inlet and outlet, the specific enthalpy increases by 16 kJ kg^{-1}. Calculate the air velocity at the exit.

To apply the SFEE in the form given in equation (4.18), the heat input is required as a specific value of heat transfer per unit mass, q:

$$q = \frac{50 \times 10^3}{3} = 16\,667\,J\,kg^{-1}$$

There is no work transfer taking place, so w is zero; the change in specific enthalpy $(h_2 - h_1)$ is $16\,000\,J\,kg^{-1}$, and the change in specific potential energy $g(z_2 - z_1)$ is $147.2\,J\,kg^{-1}$. At the inlet, the specific kinetic energy $\frac{C_1^2}{2}$ is $312.5\,J\,kg^{-1}$. Rearranging (4.17) and substituting values.

$$\frac{C_2^2}{2} = 16\,667 - 16\,000 - 147.2 + 312.5 = 832.3\,J\,kg^{-1}$$

and the air velocity at the outlet is

$$C_2 = 28.8\,m\,s^{-1}$$

What is the connection between the SFEE and the Bernoulli equation with head loss terms?

Although the SFEE can be applied to compressible and incompressible flows through open systems, in fluid mechanics this is more commonly applied as the Bernoulli equation with head loss terms. Historically, fluid mechanics has been the study of incompressible fluids in which only mechanical forms of energy are utilized: if density is constant, internal energy cannot be converted into these other forms and heat transfer raises internal energy. On the other hand, frictional dissipation can convert kinetic energy into internal energy thereby representing a loss of energy which might be harnessed to produce work output. To obtain this form, the SFEE is rearranged from

$$q + w = p_2 v_2 - p_1 v_1 + \left(u_2 + \frac{C_2^2}{2} + g z_2 \right) - \left(u_1 + \frac{C_1^2}{2} - g z_1 \right)$$

For an incompressible fluid, after substituting $v = \dfrac{1}{\rho}$, the result is

$$\frac{p_1}{\rho_1} + \frac{C_1^2}{2} + g z_1 + w = \frac{p_2}{\rho_2} + \frac{C_2^2}{2} + g z_2 + (u_2 - u_1 - q)$$

In fluid mechanics, this would usually be presented as

$$\frac{p_1}{\rho_1} + \frac{C_1^2}{2} + g z_1 + w = \frac{p_2}{\rho_2} + \frac{C_2^2}{2} + g z_2 + g H_{\text{losses}}$$

in which the 'loss' term $g H_{\text{losses}}$ represents the dissipation of mechanically available energy into internal energy.

The SFEE (4.18) or (4.19) can be applied to any steady flow problem and any reversible or irreversible process undergone by the matter flowing through the system. The specific work transfer and power input to an open system can be calculated using the SFEE without qualification. If the flow process is reversible, the dependence of specific work on pressure and specific volume can be expressed explicitly. Equation (4.18) is the starting point for the development. In differential form, and with incremental changes $d(KE)$ in kinetic energy and $d(PE)$ in potential energy per unit mass this becomes:

$$dq + dw = dh + d(KE) + d(PE) \tag{4.20}$$

Taking the definition of specific enthalpy, $h = u + pv$, and differentiating, gives

$$dh = du + d(pv) = du + p\, dv + v\, dp$$

which can be substituted into (4.19) to obtain

$$dq + dw = du + p\, dv + v\, dp + d(KE) + d(PE) \tag{4.21}$$

The next step in the development requires the process undergone by matter passing through the system to be reversible. If this matter is imagined to be sealed inside small parcels with flexible but impervious walls, each parcel can be treated as a miniature closed system being transported in the flow. Since conditions in the open system are reversible, the processes undergone by matter in the moving closed system must also be reversible, and for a reversible process undergone by a closed system, the first law gives

$$dq = du + p\, dv \tag{4.22}$$

Both equations (4.21) and (4.22) can be applied to the *same* matter; the matter is both inside the closed system *and* flowing through the open system. Hence, by equating expressions for dq, and simplifying, we obtain

$$dw = v\, dp + d(KE) + d(PE)$$

and the specific work done between 1 and 2 in a reversible process is

$$w = \int_1^2 v\, dp + \Delta KE + \Delta PE \tag{4.23}$$

Thermodynamics

The rate of work transfer or the power input, in watts, is obtained by multiplying by mass flowrate:

$$\dot{W} = \dot{m} \int_1^2 v\,\mathrm{d}p + \dot{m}(\Delta\mathrm{KE} + \Delta\mathrm{PE})$$

If the changes in kinetic energy and potential energy are small enough to be neglected, (4.23) becomes $w \int_1^2 v\,\mathrm{d}p$. Comparing this to the specific work done on a *closed system* during a reversible process, $w - \int_1^2 v\,\mathrm{d}v$ shows these are not the same; for the same state change 1 to 2, the work done on matter flowing through an open system is greater than the work done on a closed system because of flow work. Section 4.6 contains results for specific work for different types of reversible process when kinetic and potential energy changes can be neglected.

Learning summary

- ✔ Open systems have parts of their boundary which matter can cross. The matter passing through an open system undergoes a flow process. Under steady flow conditions, matter enters and leaves the system at the same mass flowrate and the mass of matter in the system remains constant.

- ✔ The analysis of steady flows through open systems is based on the mass flow continuity equation:

$$\sum_{\text{inlets}} \dot{m} = \sum_{\text{outlets}} \dot{m}$$

  and the SFEE

$$\dot{Q} + \dot{W} = \dot{m}\left(h_2 - h_1 + \frac{C_2^2}{2} - \frac{C_1^2}{2} + gz_2 - gz_1\right)$$

  These equations apply to both reversible and irreversible flow processes.

- ✔ Specific enthalpy h is a property defined as the combination of properties $(u + pv)$. This combination appears as a natural grouping in the steady flow equation, and others, including results which apply to closed system problems. Enthalpy and specific enthalpy have the units of energy (J) or specific energy ($\mathrm{J\,kg^{-1}}$) respectively, but these have no independent physical meaning and enthalpy can be considered to have been invented as a convenience rather than discovered.

- ✔ The specific work done during any, reversible or irreversible, steady flow process can be determined using the SFEE if the remaining terms have known values. In the restricted case of a reversible flow process in which kinetic energy and potential energy changes can be neglected, the specific work can also be determined from

$$w = \int_1^2 v\,\mathrm{d}p$$

- ✔ You must be careful not to confuse this with the corresponding result for specific work done during a reversible process on a closed system:

$$w = -\int_1^2 p\,\mathrm{d}v$$

4.5 The properties of perfect gases and steam

In earlier sections, the matter in or flowing through a system has been described in the most general way as the working fluid, although of course not all gases and fluids behave in the same way or have the same property values. These are distinguished by differences in the values and interdependence of their thermodynamic and transport properties. The properties considered earlier are system properties which change in value when a state change occurs, defining the state of the matter and the system under investigation; these include pressure, temperature and density. Here, we consider how behaviour depends on the nature of the matter and a set of material properties which have constant or weakly varying values. The treatment is selective but provides an introduction to a model of gas behaviour which is widely used in engineering and to the behaviour of water and steam as an important special case of a condensable vapour.

Perfect gases

Many real gases and mixtures of gases (such as air) can be treated as if they are a perfect gas for approximate calculations of properties. No gas is really perfect, but this idealization is very useful in modelling the behaviour of gases over moderate ranges of pressure and temperature. For a gas to be perfect it must meet two requirements, described in this section: (a) gas behaviour obeys the perfect gas equation; and (b) specific heats c_p and c_v are constants.

The perfect gas equation is an **equation of state**; that is, it defines a relationship between the pressure p (Pa or $N\,m^{-2}$), temperature T (K) and density ρ ($kg\,m^{-3}$). This depends only on the characteristics of the gas, and not on the system containing the gas nor on the type of any process undergone. The **perfect gas equation** can be expressed as

$$pV = mRT \qquad (4.24)$$

or, since

$$v = \frac{V}{m}$$

$$pv = RT \qquad (4.25)$$

and, because

$$\rho = \frac{1}{v}$$

$$p = rRT \qquad (4.26)$$

Each gas has a **specific gas constant** R ($J\,kg^{-1}\,K^{-1}$) which is a gas property and a constant. The value of R can be calculated from

$$R = \frac{\tilde{R}}{\tilde{m}}$$

where $\tilde{R}$ is the same for all gases.

> $\tilde{R}$ is the universal gas constant ($\tilde{R} = 8.3145 \times 10^3\,J\,kmol^{-1}\,K^{-1}$). $\tilde{m}$ is the molar mass ($kg\,kmol^{-1}$) of the gas; this is numerically equal to the molecular weight of the gas.

Values of the molar masses and specific gas constants of some common gases are given in Table 4.3. The table includes air, which is a mixture of nitrogen, oxygen and small amounts of other gases, illustrating that a mixture of perfect gases will also behave as a perfect gas.

Gas	Molar mass $\tilde{m}$ (kg kmol^{-1})	Specific gas constant R $R = \tilde{R}/\tilde{m}$ (J kg^{-1} K^{-1})	Specific heat c_p (J kg^{-1} K^{-1})	Specific heat c_v (J kg^{-1} K^{-1})
Air (–)	29	287	1005	718
Argon (Ar)	40	208	520	312
Butane (C_4H_{10})	58	297	1716	1573
Carbon dioxide (CO_2)	44	189	846	657
Carbon monoxide (CO)	28	297	1040	744
Helium (He)	4	208	5193	3116
Hydrogen (H_2)	2	412	14 307	10 183
Methane (CH_4)	16	189	2254	1735
Nitrogen (N_2)	28	297	1039	743
Oxygen (O_2)	32	260	918	658
Propane (C_3H_8)	44	189	1491	1679

Table 4.3 Perfect gas properties of some common gases

A flammable gas mixture with a molar mass of 114 kg kmol^{-1} is burned in a rigid container with a fixed volume. Initially the pressure in the container is 200 kPa and the temperature is 30 °C. After the mixture has burned, the pressure is 400 kPa and the temperature is 900 °C. Calculate the specific gas constant for the products of combustion, assuming the initial gas mixture and the products of combustion behave as perfect gases.

Use suffix (1) for conditions before combustion, and (2) for conditions after combustion. Initially, the specific gas constant will be

$$R_1 = \frac{\tilde{R}}{\tilde{m}_1} = \frac{8.3145 \times 10^3}{114} = 729.3 \, \text{J kg}^{-1}\text{K}^{-1}$$

and because the gas mixture is a perfect gas

$$P_1 V_1 = m_1 R_1 T_1$$

After combustion, because the products of combustion also behave as a perfect gas,

$$P_2 V_2 = m_2 R_2 T_2$$

Although R_2 is unknown because the value of the molar mass has changed during combustion, the volume of the container and the mass of gas do not change, so values of $\frac{V}{m}$ can be equated. Hence

$$\frac{V_1}{m_1} = \frac{V_2}{m_2}$$

and using the perfect gas equation

$$\frac{R_1 T_1}{P_1} = \frac{R_2 T_2}{P_2}$$

Rearranging this, and substituting values, yields

$$R_2 = R_1 \left(\frac{T_1}{T_2}\right)\left(\frac{P_2}{P_1}\right) = 729.3 \left(\frac{273 + 30}{273 + 900}\right)\left(\frac{400 \times 10^3}{200 \times 10^3}\right) = 376.8 \, \text{J kg}^{-1}\text{K}^{-1}$$

Note that the temperatures and pressures have been converted to SI units for the calculation. It is always good practice, and avoids the most common source of calculation errors.

The perfect gas equation defines the value of either p, T or v when the other two property values are specified. If one is held constant, the variation of the other two are linked, as illustrated in Figures 4.26–4.28. This is a *model* of gas behaviour – there are other models which can be used to describe gas behaviour more accurately, but the perfect gas equation is the simplest and most widely used across the range of engineering thermodynamic problems. When applied to air, and using the value of R of 287 J kg^{-1} K^{-1}, the perfect gas equation is sufficiently accurate to use in the majority of applications to engineering problems where pressures are typically between small fractions of a bar and $\sim 10^2$ bar and temperatures are between 200 and 1500 K. For example, at a pressure of 1.013 bar and a temperature of 900 K, the density calculated using the perfect gas equation is 0.3922 kg m^{-3} which is accurate to three significant figures.

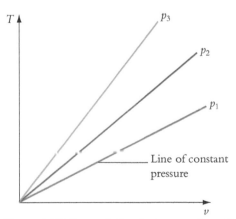

Figure 4.26 The variation in temperature *T* with specific volume *v* at fixed values of pressure *p*, with $p_1 < p_2 < p_3$

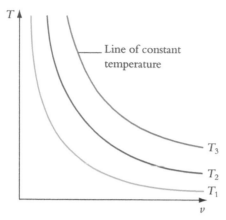

Figure 4.27 **The variation in pressure p with specific volume v at fixed values of temperature, with $T_1 < T_2 < T_3$**

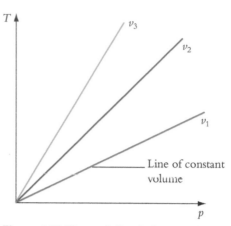

Figure 4.28 **The variation in temperature T with pressure p at fixed values of specific volume, with $v_1 < v_2 < v_3$**

Internal energy, enthalpy and entropy were introduced in earlier sections, internal energy through the study of closed systems, entropy through the study of heat engines and enthalpy through the SFEE, although the significance of these properties extends more widely than this. For example, the specific heat transferred to a closed system during a reversible constant-pressure expansion from 1 to 2 is equal to the change in specific enthalpy of the system. The process is reversible, so specific work is

$$w = -\int_1^2 p \, dv$$

and for a constant-pressure process this becomes

$$w = -p(v_2 - v_1)$$

Applying the first law to the closed system gives

$$q = u_2 - u_1 + p(v_2 - v_1) = (u_2 + pv_2) - (u_1 + pv_1) = h_2 - h_1$$

Internal energy, enthalpy and entropy must be added to the list of key properties such as pressure, temperature and density which change in value when the state of matter changes. There are other properties of matter influencing thermodynamic behaviour which do not change in value nearly so much when the state changes. These include thermal conductivity, viscosity and the specific heats which vary more with temperature than pressure or specific volume but, even so, the variation with temperature is weak and often neglected when the range of temperatures of interest is a few hundred degrees or less.

The specific heats c_v and c_p are properties which relate temperature changes (which can be measured) of a substance to changes in internal energy and enthalpy (which cannot be measured directly). The **specific heat at constant volume**, c_v ($\mathrm{J\,kg^{-1}\,K^{-1}}$), is the change in specific internal energy (u) per degree temperature rise, $c_v = \dfrac{du}{dt}$, during a constant-volume process. The **specific heat at constant pressure**, c_p ($\mathrm{J\,kg^{-1}\,K^{-1}}$), is the change in specific enthalpy (h) per degree temperature rise, $c_p = \dfrac{dh}{dt}$, during a constant-pressure process. Values of the specific heats

of some gases commonly treated as perfect gases are given in Table 4.3.

Because the values of the specific heats of a perfect gas are constants that do not depend on volume, pressure or temperature, the definitions can be rearranged and integrated to give expressions for the changes in specific internal energy and specific enthalpy between any two states 1 and 2:

$$u_2 - u_1 = c_v(T_2 - T_1) \tag{4.27}$$

Thermodynamics

or for the change in internal energy,

$$U_2 - U_1 = mc_v(T_2 - T_1)$$

and for the change in specific enthalpy

$$h_2 - h_1 = c_p(T_2 - T_1) \tag{4.28}$$

or enthalpy

$$H_2 - H_1 = mc_p(T_2 - T_1)$$

A relationship between the specific heats and the specific gas constant can be obtained from the definition of specific enthalpy, $h = u + pv$, and the change between two states 1 and 2:

$$h_2 - h_1 = (u_2 - u_1) + (p_2v_2 - p_1v_1)$$

Substituting from equations (4.28), (4.27) and (4.25) yields

$$c_p(T_2 - T_1) = c_v(T_2 - T_1) + R(T_2 - T_1)$$

and hence, cancelling out $(T_2 - T_1)$ values gives

$$c_p = c_v + R \tag{4.29}$$

The value of c_p is always greater than c_v for a gas. To understand why, consider how values of specific heats might be determined experimentally by heating a closed system containing matter. In these experiments, the joules of heat transfer required to raise the temperature by one degree kelvin is recorded. To determine c_v, the experiment is carried on a closed system with a fixed, constant volume. No work is done on the system because the volume is constant, so applying the first law to the process:

$$q_v = u_2 - u_1$$

and the specific heat at constant volume is

$$c_v = \frac{q_v}{T_2 - T_1}$$

In the experiment to determine c_p, the heat transfer occurs in a constant-pressure process. This can be achieved by allowing the system volume to change, as illustrated in Figure 4.29.

In this case, work is done by the system, transferring energy to the surroundings. Applying the first law to this heating process yields

$$q_p = (u_2 - u_1) + p(v_2 - v_1)$$

and

$$c_p = \frac{q_p}{T_2 - T_1}$$

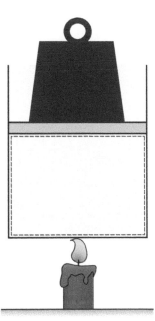

Comparing the results for q_v and q_p shows the increase in internal energy is the same, but, in the constant-pressure case, work transfer returns to the surroundings some of the energy supplied by heat transfer, requiring this to be larger to achieve the same rise in temperature. It follows that c_p must be greater than c_v and the ratio of specific heats $\frac{c_p}{c_v}$, given the symbol γ, is always greater than unity. The value of γ varies between close to 1 for a gas with a large molar mass and $\frac{5}{3}$ for helium which has the lowest molar mass.

Figure 4.29 **An imagined experiment in which a closed system is made to undergo a constant-pressure process in order to determine c_p**

Equation (4.29) and the ratio of specific heats can be combined to find two more useful relationships:

$$c_v = \frac{R}{(\gamma - 1)}$$

and

$$c_p = \frac{\gamma R}{(\gamma - 1)}$$

Calculating the change in entropy of a perfect gas

Calculating the change in the value of any property during a process connecting two states, 1 and 2, is more difficult for some processes than others, but the change must always be the same. The result will be independent of the type of process or whether this is reversible or irreversible. Because the result must always be the same, we can *select* a process to connect the states which we can analyse. This is a very valuable observation. To determine the change in specific entropy of a perfect gas, we can choose to follow a reversible process undergone by a closed system and in *this* case, the incremental changes in specific heat transfer and work done are given by results for a reversible non-flow process:

$$dq = T\,ds$$

$$dw = -p\,dv$$

From the first law

$$dq + dw = du$$

and substituting the results for dq and dw gives a relationship between properties that is independent of the system type or process undergone:

$$T\,ds = du + p\,dv$$

or

$$ds = \frac{du}{T} + \frac{p\,dv}{T}$$

For a perfect gas, $du = c_v\,dT$, and, from the perfect gas equation, $\frac{p}{T} = \frac{R}{v}$, and thus for this case

$$ds = \frac{c_v\,dT}{T} + \frac{R\,dv}{v}$$

or, integrating,

$$s_2 - s_1 = \int_1^2 \frac{c_v\,dT}{T} + \int_1^2 \frac{R\,dv}{v}$$

and hence the change in specific entropy (in $J\,kg^{-1}\,K^{-1}$) between two states 1 and 2 is

$$s_2 - s_1 = c_v \ln\frac{T_2}{T_1} + R \ln\frac{v_2}{v_1} \tag{4.30}$$

Equation (4.30) can be applied to find the entropy change per unit mass either flowing through an open system or enclosed in a closed system. In an open system, $\dot{m}(s_2 - s_1)$ is the increase in the rate of entropy flow between locations 1 and 2. In closed system problems, the increase in entropy $(J\,K^{-1})$ when the system state changes from 1 to 2 is found by multiplying (4.30) by the mass m of matter contained in the system:

$$S_2 - S_1 = mc_v \ln\frac{T_2}{T_1} + mR \ln\frac{v_2}{v_1} \tag{4.31}$$

Two other forms of equations (4.30) and (4.31) can be derived to relate specific entropy changes to changes in temperature and pressure, or pressure and volume. These are included in Table 4.4, in which the relationships for perfect gases are summarized.

Property or equation name	Equation(s)
Perfect gas equation	$pV = mRT$ or $pv = RT$ or $p = \rho RT$
Specific heats and specific gas constant	c_v and c_p are constants $R = c_p - c_v$ and $R = \dfrac{\tilde{R}}{\tilde{m}}$ $\gamma = \dfrac{c_p}{c_v}$ $c_p = \dfrac{\gamma R}{(\gamma - 1)}$ and $c_v = \dfrac{R}{(\gamma - 1)}$
Internal energy and specific internal energy	$U_2 - U_1 = mc_v(T_2 - T_1)$ and $u_2 - u_1 = c_v(T_2 - T_1)$
Enthalpy and specific enthalpy	$H_2 - H_1 = mc_p(T_2 - T_1)$ and $h_2 - h_1 = c_p(T_2 - T_1)$
Entropy and specific entropy	$S_2 - S_1 = mc_v \ln\dfrac{T_2}{T_1} + mR \ln\dfrac{v_2}{v_1}$ or $S_2 - S_1 = mc_v \ln\dfrac{p_2}{p_1} + mc_p \ln\dfrac{v_2}{v_1}$ or $S_2 - S_1 = mc_p \ln\dfrac{T_2}{T_1} - mR \ln\dfrac{p_2}{p_1}$ and $s_2 - s_1 = \dfrac{(S_2 - S_1)}{m}$

Table 4.4 Relationships for perfect gases

Properties of fluids which change phase

In addition to the gases and fluids which remain in the same phase over the range of conditions of interest, there are others (notably water and refrigerants) which may change phase within the range of conditions which arise in particular applications. Although in previous sections the working fluid has been a perfect gas or an incompressible fluid, more generally matter can exist as a solid, liquid, vapour or a gas. A fluid is described as a gas if its temperature exceeds the critical value above which changes of phase by evaporation or condensation do not occur. Fluids also have a critical pressure above which phase changes do not occur, and a fluid at the critical pressure and critical temperature is said to be at the critical state. Fluids undergoing a phase change at constant pressure from liquid to vapour by evaporation absorb energy at constant temperature. When condensing from vapour to liquid, an equal quantity of energy is released at constant temperature. This behaviour is exploited in the operation of thermodynamic plant such as steam for power generation, refrigerators for chilling enclosures and heat pumps for warming enclosures so there are many applications where the change of phase of a working fluid is essential to the efficient performance of a machine or system.

When the matter under investigation changes phase or can exist as a mixture of two phases, additional information is required to define the state of the fluid and the relationship between properties. Several new terms used to describe the behaviour of a condensable vapour are summarized in Table 4.5. The use of these can be illustrated by considering what happens to water being heated at constant pressure. Suppose, for example, that a quantity of tap water at 20°C is sealed in a container with a lid which is free to rise or fall to control the pressure in the container to 1 bar. (This is the same set-up as shown in Figure 4.29, but now the closed system contains water.) The water is heated gradually and when the temperature and volume changes are plotted on a temperature–specific volume diagram, the line of constant pressure will appear as illustrated in Figure 4.30.

The line will start at the initial temperature and specific volume of the water. At this point (A), the water is a **subcooled liquid**. Heating the water raises the temperature and increases the volume of the water until the temperature reaches the **saturation temperature**, which is 99.6 °C at 1 bar. The water is now a **saturated liquid** and the temperature stays constant as the liquid changes phase to vapour by evaporation as further heat addition takes place. The saturated vapour and the remaining saturated liquid coexist in the container as a **saturated mixture** in a state of thermodynamic equilibrium. The **dryness fraction** of the mixture increases from 0, when all the water is saturated liquid, to 1 when all the water is in the saturated vapour state. As the water evaporates, the change in specific volume of the mixture is large. This is due to the specific volume of saturated vapour being much larger than that of saturated liquid at the same pressure and temperature. Eventually, all the liquid will have evaporated to saturated vapour. The energy supplied per unit mass by heating to evaporate the fluid from saturated liquid to saturated vapour is equal to the **specific latent heat of vaporization**. When all the water has become saturated vapour, continued heating will raise the temperature above the saturation value. The vapour is now a **superheated vapour**, with a **degree of superheat** equal to the temperature increase above the saturation temperature.

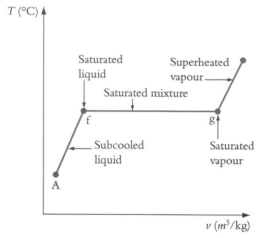

Figure 4.30 A line of constant pressure on a temperature-specific volume diagram for water

SATURATED LIQUID (suffix f)	Liquid on the point of evaporation or the liquid formed during condensation of a vapour.
SATURATED VAPOUR (suffix g)	Vapour on the point of condensation or the vapour formed during evaporation of a liquid.
SATURATION TEMPERATURE (t_s)	The temperature at which evaporation or condensation take place *at a given pressure*; the boiling point; the temperature of a saturated liquid or saturated vapour at a given pressure.
SATURATION PRESSURE (p_s)	The pressure required for evaporation or condensation to take place *at a given temperature*; vapour pressure; the pressure of a saturated liquid or saturated vapour at a given temperature.
SUBCOOLED LIQUID	An UNSATURATED OR COMPRESSED LIQUID; liquid at a temperature below the saturation temperature or at a pressure above the saturation pressure.
SUPERHEATED VAPOUR	Vapour at a temperature greater than the saturation temperature corresponding to the pressure of the vapour.
DEGREE OF SUPERHEAT	*The difference* between the temperature of a superheated vapour and the corresponding saturation temperature.
SATURATED MIXTURE	A mixture of saturated liquid and saturated vapour; a saturated vapour containing small droplets of liquid.
DRYNESS FRACTION (x)	The proportion by mass of saturated vapour in a given quantity of wet vapour; also called the QUALITY of a wet vapour.
SPECIFIC LATENT HEAT OF VAPORIZATION (h_{fg})	Energy required as heat to completely evaporate unit mass of saturated liquid to saturated vapour; equal to ($h_g - h_f$)

Table 4.5 Terminology of phases of substances

Temperature–specific volume diagram

If the same experiment is repeated for different values of pressure, a series of constant pressure lines can be generated and regions of the diagram in which water or steam is in one or more phases can be identified. This is shown in Figure 4.31.

Joining the points at which evaporation begins for each pressure forms a line on the diagram called the **saturated liquid line**; repeating the task for the points at which evaporation is completed gives a second line called the **saturated vapour line**. As pressure is raised, the change in specific volume between the saturated liquid and saturated vapour states shrinks and the two lines curve towards a point where they meet, enclosing the **saturated mixture region** where saturated vapour and saturated liquid coexist. The point at which the saturated

Thermodynamics

liquid and saturated vapour lines meet is the **critical point** at which liquid and vapour states become indistinguishable; the pressure is equal to the critical pressure and the temperature is equal to the critical temperature. For water, the critical pressure P_{crit} is 221.1 bar and the critical temperature is 374.16 °C. If the temperature and pressure are raised above their critical values, the vapour is described as being **supercritical steam.** To the right of the saturated vapour line, or in the region where pressure is below the critical pressure but above the critical temperature, the vapour is described as being **superheated steam.** To the left of the saturated liquid line and at temperatures below the critical temperature, water is in the **subcooled liquid state.**

Any liquid, vapour or mixture in a saturated condition will be at a particular saturation pressure and saturation temperature. These are inseparable; fixing one fixes both values and we can refer to the saturation temperature at a given pressure or the saturation pressure at a given temperature. As the pressure is raised, the saturation temperature increases. At the same time, the volume change between the saturated liquid and saturated vapour states decreases and the heat transfer required to evaporate the liquid to vapour is reduced; that is, as pressure increases the latent heat of vaporization decreases.

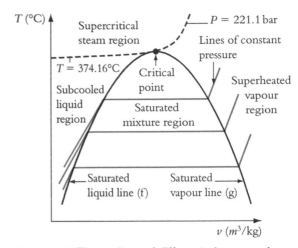

Figure 4.31 The regions of different phases and states of a temperature–specific volume diagram for water

Dryness fraction (x) has relevance only in defining the state of a saturated mixture at a given temperature or pressure. It provides a convenient measure of the vapour content in the mixture:

$$x = \frac{m_g}{m_g + m_f}$$

where m_g is the mass of saturated vapour and m_f is the mass of saturated liquid in the mixture. Dryness fraction is never less than zero, when the mixture is all saturated liquid, nor greater than 1, when the mixture is all saturated vapour and sometimes described as 'dry' saturated vapour, meaning the mixture contains no liquid. In the saturated condition, the pressure and temperature of the mixture are independent of the value of x, but properties which have an intensive form are not. This is because, per unit mass, these may have significantly different values for saturated liquid and saturated vapour. For example, the specific volume of saturated vapour can be 10–1000 times that of saturated liquid at the same temperature. The value of the specific volume of the mixture will be given by sum of a vapour contribution plus a liquid contribution. The saturated vapour will occupy a volume equal to its specific volume multiplied by its mass fraction, which will be x. The saturated liquid will occupy a volume equal to its specific volume multiplied by its mass fraction which will be $(1 - x)$. Thus if the specific volume of saturated vapour is v_g and the specific volume of saturated liquid is v_f, then the specific volume of the mixture will be

$$v = (1 - x)v_f + xv_g$$

The specific entropy s, specific internal energy u and specific enthalpy h of the mixture are also found as mass-weighted sums of saturated liquid and vapour values. In each case the suffix 'f' is used to denote a saturated liquid value and the suffix 'g' is used to denote a saturated vapour value:

$$s = (1 - x)s_f + xs_g$$
$$u = (1 - x)u_f + xu_g$$
$$h = (1 - x)h_f + xh_g$$

Expanding the last of these gives
$$h = h_f - xh_f + xh_g = h_f + x(h_g - h_f)$$

The difference $h_g - h_f$ is the **specific latent heat of vaporization** h_{fg} and, using this, the expression for specific enthalpy can be written in the alternative form

$$h = h_f + x h_{fg}$$

The corresponding increases in specific internal energy and specific entropy between the saturated liquid and saturated vapour states can also be used, although these are not referred to as latent values:

$$u = u_f + x u_{fg}$$
$$s = s_f + x s_{fg}$$

Evaluating steam properties

For steam, the mathematical functions used to describe the relationships between property values are far more complicated than they are for a perfect gas; there is no simple equivalent to the perfect gas equation. Because of this, the calculation of property values from the equation of state may require a computer, or the use of tables of values generated from the equation of state and other property relationships. The tables (Rogers and Mayhew, 1995) are widely used and here, data is taken from these to illustrate some of the calculations involved in establishing values of properties. When water is in the vapour phase, or is a mixture of vapour and liquid, it is usually referred to as 'steam'. The water and steam tables in the booklet are arranged to separately cover the regions of the temperature–specific volume diagram identified in Figure 4.31. In the simplified version of this, Figure 4.32, a series of state points are labelled by letter with sufficient information to determine the missing values.

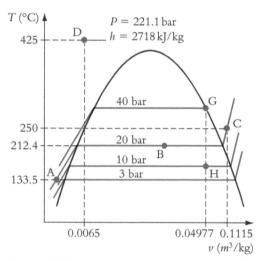

Figure 4.32 **Points on the temperature-specific volume diagram associated with property calculations described in the text**

Point A, determine the value of specific enthalpy.

This point is on the line of constant pressure for 20 bar but at a temperature below the saturation temperature. For 20 bar, this is 212.4 °C and point A is therefore in the subcooled liquid region. The specific enthalpy can be calculated using property values at 20 bar,

$$h = h_f - c_p(T_f - T_A)$$

where c_p is the value of specific heat at the average temperature between points f and A.

From the tables, h_f at 20 bar is 909 kJ kg^{-1} and c_p at 172.9 °C is 4.39 kJ kg^{-1}, and

$$h = h_f - c_p(T - T_A) = 909 - 4.39 \times (212.4 - 133.5) = 562 \text{ kJ kg}^{-1}$$

Alternatively, and to a good approximation, h is equal to the value for saturated liquid at the same temperature. In this case, the temperature of $T_A = 133.5$ °C is the saturation temperature at for a pressure of 3 bar. At this saturation condition h_f (and to a good approximation, h_A) is 561 kJ kg^{-1}.

Point B, determine the specific entropy and specific enthalpy.

This point also lies on the 20 bar constant pressure line. The dryness factor x is 0.6, meaning the mixture at point B contains 60 per cent saturated vapour and 40 per cent saturated liquid. From tables, at 20 bar, s_f is 2.447 kJ kg^{-1} K^{-1} and s_{fg} is 3.893 kJ kg^{-1} K^{-1}; the required specific enthalpy values are $h_f = 909$ kJ kg^{-1} and $h_{fg} = 1890$ kJ kg^{-1}. Substituting these values into the equations for specific entropy and specific enthalpy in the saturated mixture region:

$$s = s_f + xs_{fg} = 2.447 + 0.6 \times 3.893 = 4.783 \text{ kJ kg}^{-1}\text{K}^{-1}$$
$$h = h_f + xh_{fg} = 909 + 0.6 \times 1890 = 2043 \text{ kJ kg}^{-1}$$

Point C, determine the degree of superheat.

This is the difference between the superheated steam temperature at point C, and the saturation temperature at the same pressure, in this case 212.4 °C. The specific volume at point C is given as 0.1115 m^3 kg^{-1}, which together with the pressure (20 bar) defines the state. The required property value is the temperature at point C and from the superheat tables, this can be found to be 250 °C. The degee of superheat at point C is therefore $(250 - 212.4) = 37.6$ °C.

Point D, determine the specific internal energy.

In this example, the property values given in the tables do not include specific internal energy, but do give enough information for the u to be calculated from the definition of enthalpy, $h = u + pv$. Rearranging and substituting known values, taking care to use consistent units – in this case, working in units of kJ kg^{-1}:

$$u = h - pv = 2718 - (275 \times 10^5 \times 0.0065) \times 10^{-3} = 2539.3 \text{ kJ kg}^{-1}$$

Point H, determine the dryness fraction x.

On the diagram, point H lies at one end of the constant volume line connected to point G where it intersects the saturated vapour line at a pressure of 40 bar. At point G the specific volume is equal to the specific volume of saturated vapour at 40 bar or (from tables) 0.04977 m^3 kg^{-1}. Given that at H the pressure is 10 bar, and, from tables, v_g is 0.1944 m^3 kg^{-1} and v_f is 0.00113, the specific volume defines the second property required to fix the state and determine x from

$$v = (1 - x)v_f - xv_g$$

Rearranging this gives

$$x = \frac{(v - v_f)}{(v_g - v_f)} = \frac{(0.04977 - 0.00113)}{(0.1944 - 0.00113)} = 0.252$$

Use of interpolation

When the state of interest does not correspond to the entries given in the tables, property values can be determined from the nearest state conditions using **linear interpolation**. This assumes the variation in properties between listed conditions will be linear. A property value at an intermediate condition can be determined from the proportions of similar triangles illustrated in Figure 4.33.

Suppose we require the specific enthalpy h of superheated steam at a pressure of 60 bar and a temperature of 440 °C. At 60 bar, values of enthalpy are tabulated for temperatures $T_1 = 400$ °C ($h_1 = 3177 \text{ kJ kg}^{-1}$) and $T_2 = 450$ °C ($h_2 = 3301 \text{ kJ kg}^{-1}$), allowing points 1 and 2 to be plotted. The required enthalpy is given by point 3 on the line joining 1 and 2. From the tangent of the angle common to both triangles,

$$\tan \phi = \frac{a}{b} = \frac{A}{B}$$

and

$$h = h_1 + a = h_1 + \frac{Ab}{B}$$

$$h = h_1 + a = \frac{(h_2 - h_1) \cdot (440 - 400)}{(450 - 400)}$$

$$h = 3279 \text{ kJ kg}^{-1}$$

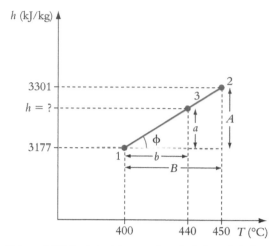

Figure 4.33 A plot of known local state points is useful in arriving at the correct interpolation relationship, in this case for the specific enthalpy at 60 bar and 440 °C

Here are two more examples, using data from Rogers and Mayhew (1995, p. 5, saturated water and steam table).

Find the specific internal energy of saturated liquid at 41.6 bar given the two known state points in the table.

State	P (bar)	u_f (kJ kg^{-1})
1	40	1082
2	42	1097

Table 4.6

This interpolation task is similar to the one given above, and the diagram is similar to Figure 4.33. Following the same procedure,

$$u_1 = u_{f1} + [u_{f2} - u_{f1}] \times \frac{41.6 - 40}{42 - 40} = 1082 + 15 \times \frac{1.6}{2} = 1094 \text{ kJ kg}^{-1}$$

Find the specific volume of saturated vapour at 123 bar.

The nearest state points tabulated are given below. Note that the specific volume of saturated vapour reduces as pressure increases. This is clear in the thumbnail figure, Figure 4.34, which helps avoid mistakes in the interpolation.

By inspection of the similar triangles in Figure 4.34,

$$v_g = v_{g2} + a = v_{g2} + \frac{Ab}{B}$$

$$v_g = v_{g2} + (v_{g1} - v_{g2}) \cdot \frac{(125 - 123)}{(125 - 120)}$$

$$v_g = 0.01380 \, \text{m}^3 \, \text{kg}^{-1}$$

State	P (bar)	v_n (m³ kg⁻¹)
1	120	0.01426
2	125	0.013449

Table 4.7

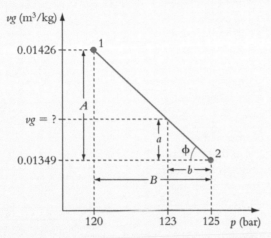

Figure 4.34 Specific volume–pressure diagram for the interpolation to find the specific volume of saturated vapour at 123 bar

Learning summary

✔ The working fluids commonly used in thermodynamic systems are gases and condensable vapours which may change phase at conditions of interest. The behaviour of working fluids must be understood as part of the analysis of system behaviour. This requires knowledge of the properties which distinguish one working fluid from another, and models of behaviour which define how the working fluid will respond to changes in state.

✔ The behaviour of air and many other gases used in engineering thermodynamic systems can be modelled as that of a perfect gas. A perfect gas obeys the perfect gas equation:

This is the equation of state of the gas. $pv = mRT$

✔ In addition, the specific heats of a perfect gas are constants which do not change as temperature or pressure changes.

✔ Water is the example of a condensable vapour covered in this section. The equation of state is more complex than the perfect gas equation and usually evaluated using a computer. Results are presented in tables (in Rogers and Mayhew, 1995).

4.6 Types of process and their analyses for work and heat transfer

Many of the processes undergone by closed and open systems fall into one of a small number of types. These are considered in this section together with some useful results for processes undergone by perfect gases. Understanding how a process will change the state of the working fluid and what work and heat transfer have taken place are tasks central to the analysis of many thermodynamic problems. Up to this point processes have been described as reversible or irreversible, and as being non-flow or steady flow processes. These terms alone do not completely define a process, however, or provide sufficient information for an analysis of changes in system states to be carried out.

> A **polytropic process** is one which obeys the polytropic law pv^n = constant, in which n is a constant called the **polytropic index**.

Many real processes undergone by gases or vapours are approximately polytropic with a polytropic index typically between 1.0 and 1.7, and several types of process obey the polytropic law with particular values of n. The variation of pressure with specific volume for positive values of polytropic index n greater than unity is illustrated in Figure 4.35. Note that increasing n increases the gradient of a polytrope, making it appear to rotate the process line clockwise; this is useful to remember when sketching a series of processes on a state diagram.

The polytropic law connects pressure and specific volume without reference to any restriction on the working fluid that can undergo a polytropic process: in general this will be a compressible gas or vapour; it can be a two-phase mixture of vapour and liquid, superheated steam or a gas which does not behave as a perfect gas. However, to determine the effect on other property values, the working fluid must be defined. If this is steam, steam tables can be used to find the values corresponding to a given pair of pressure and specific volume values. If the working fluid is a perfect gas, the perfect gas equation provides the connection between pressure, specific volume and temperature. For any polytropic process between states 1 and 2, the polytropic law can be written as:

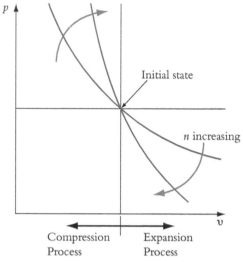

Figure 4.35 **Polytropic process lines (or polytropes) for positive values of polytropic index n equal to unity or greater**

$$p_1 v_1^n = p_2 v_2^n \qquad (4.32)$$

or

$$\frac{p_2}{p_1} = \left(\frac{v_2}{v_1}\right)^{-n} \qquad (4.33)$$

If the working fluid is a **perfect gas**, the perfect gas equation $pv = RT$ can be applied to states 1 and 2 to eliminate the specific gas constant R:

$$\frac{p_1 v_1}{T_1} = \frac{p_2 v_2}{T_2}$$

or

$$\frac{p_2}{p_1} = \left(\frac{T_2}{T_1}\right)\left(\frac{v_1}{v_2}\right) \qquad (4.34)$$

Equating (4.33) and (4.34) eliminates pressure and gives a relationship between **temperature and specific volume**:

$$\frac{T_2}{T_1} = \left(\frac{v_2}{v_1}\right)^{1-n} \qquad (4.35)$$

An illustration of the variation of temperature with specific volume is given in Figure 4.36.

Equation (4.35) can be rearranged to obtain

$$\frac{v_1}{v_2} = \left(\frac{T_2}{T_1}\right)^{\frac{1}{(n-1)}}$$

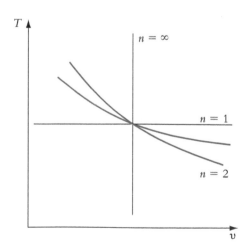

Figure 4.36 **Variation of temperature with specific volume during a polytropic process**

Thermodynamics

Substituting into (4.34) to eliminate volume gives the result connecting **pressure and temperature**:

$$\frac{p_2}{p_1} = \left(\frac{T_2}{T_1}\right)^{\frac{n}{(n-1)}} \qquad (4.36)$$

An illustration of the variation of pressure with temperature during a polytropic process is given in Figure 4.8, in Section 4.1. The value of the index n is 1.4 in this example.

Pressure, volume and temperature are not the only properties plotted on process diagrams (for example, refrigeration cycles are often plotted on enthalpy-entropy diagrams and open cycle gas turbine cycles are usually plotted on temperature–entropy diagrams) and being able to transform a process line from one to another diagram is a useful skill. An example is given in Figure 4.37; note that although the definition of a polytropic process relates pressure and specific volume, the working fluid must be defined to map this onto other process diagrams.

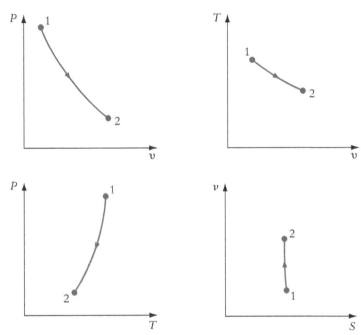

Figure 4.37 Example of plotting a polytropic process undergone by a perfect gas on different process diagrams

Several other types of process are special cases of the polytropic process. In the limit of n tending to zero, a polytropic becomes a **constant-pressure**, or **isobaric**, **process**. In the limit of n tending to infinity, this becomes a **constant-volume**, or **isochoric**, **process**. These bound the sectors in which all polytropic compression and expansion processes fall for positive values of n, as shown in Figure 4.38.

Other constant-property processes are the **isothermal process**, during which temperature is constant, the **isenthalpic process**, during which enthalpy is constant, and the **isentropic process**, during which entropy is constant. (The terms **isothermal** and **isentropic** are widely used and commonly confused, so it is worth making the effort to memorize which is which).

For a perfect gas undergoing one of these processes, the relationship between pressure and specific volume can be defined using the polytropic law with a different value of the polytropic index. For an isothermal process, since T is constant, the perfect gas equation gives pv = constant and the polytropic index $n = 1$. For a perfect gas, the

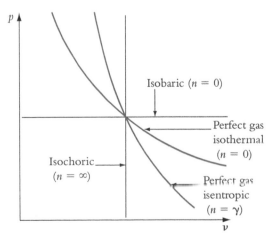

Figure 4.38 Constant-pressure (isobaric) and constant-volume (isochoric) processes are special cases of the polytropic process, with index values of zero and infinity respectively

isenthalpic process is also isothermal (since $dh = c_p \, dT$) so again the polytropic index $n = 1$. For an isentropic process, the change in entropy is zero, and substituting $s_2 - s_1 = 0$ (from Table 4.4) in

$$s_2 - s_1 = c_v \ln\left(\frac{p_2}{p_1}\right) + c_p\left(\frac{v_2}{v_1}\right)$$

yields, after rearranging,

$$\frac{p_2}{p_1} = \left(\frac{v_2}{v_1}\right)^{-\gamma} \tag{4.37}$$

or $pv^\gamma = $ constant, which is the polytropic law with the polytropic index $n = \gamma$. Since γ is always greater than unity, on the pressure–volume diagram, the isentropic polytrope will appear on the clockwise side of isothermal and isenthalpic polytropes.

The value of sketching processes on process diagrams

Descriptions of thermodynamic problems are commonly quite lengthy – it is in the nature of the subject – and to avoid the need to re-read the question, and minimize the likelihood of misinterpretation, it is useful to summarize the information: sketch details of the system in a conceptual diagram, list any property values given and keywords defining any process involved; identify what parameters need to be calculated. Importantly, sketching out the processes on a state or process diagram helps fix thoughts on how property values will change. Two examples of processes sketched on pressure–volume diagrams are given in Figures 4.39 and 4.40 below.

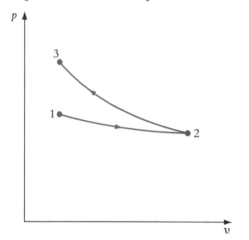

Figure 4.39 **An isothermal *expansion* from v_1 to v_2 followed by a polytropic compression with $n = 1.4$ back to v_1**

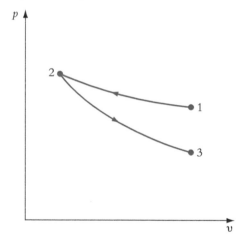

Figure 4.40 **An isothermal *compression* from v_1 to v_2 followed by a polytropic expansion with $n = 1.4$ to v_1**

The last type of process introduced here is the **adiabatic process**.

> An **adiabatic process** is one during which no heat transfer occurs. A reversible adiabatic process will be isentropic: no entropy is created if the process is reversible and none can be transferred to or from the system if no heat transfer occurs. For a perfect gas, reversible adiabatic and isentropic processes will obey the same polytropic law with $n = \gamma$.

Work done during reversible processes

Regardless of whether a process is reversible or irreversible, the first law can be applied to closed systems and, in the form of the SFEE, to open systems. This will usually be the only route to determining the work done when the process considered is irreversible. If the process is reversible, explicit expressions for the specific work done can be derived for various processes undergone by a perfect gas. The results are different for closed and open systems. The starting points for deriving the results are the equations for specific work done during a reversible process.

Thermodynamics

For a **closed system**:

$$w = -\int_1^2 p \, dv \tag{4.38}$$

Eliminating p using the polytropic law $p = \dfrac{c}{v^n}$, where c is a constant, yields

$$w = -\int_1^2 p \, dv = -c\int_1^2 \frac{dv}{v^n} = -c\left[\frac{v^{1-n}}{1-n}\right]_1^2 \tag{4.39}$$

Recalling that $c = p_2 v_2^n$ or $c = p_1 v_1^n$, and simplifying, we have

$$w = \frac{[p_2 v_2 - p_1 v_1]}{(n-1)} \quad \text{or} \quad \frac{R[T_2 - T_1]}{(n-1)} \tag{4.40}$$

The corresponding result for an **open system** is derived by integrating the result for specific work during a reversible flow process (4.23):

$$w = \int_1^2 v \, dp + \Delta KE + \Delta PE \tag{4.41}$$

and now, substituting to eliminate v using $v = \dfrac{c^\star}{p^{\frac{1}{n}}}$ where $c^\star$ is a second constant (in fact, $c^\star = c^{\frac{1}{n}}$), we obtain

$$w = c^\star \int_1^2 \frac{dp}{p^{\frac{1}{n}}} + \Delta KE + \Delta PE \tag{4.42}$$

or

$$w = \frac{n}{(n-1)}[p_2 v_2 - p_1 v_1] + \Delta KE + \Delta PE \tag{4.43}$$

Neglecting changes in kinetic and potential energy:

$$w = \frac{n}{(n-1)}[p_2 v_2 - p_1 v_1] \quad \text{or} \quad \frac{nR}{(n-1)}[T_2 - T_1] \tag{4.44}$$

If the changes in kinetic and potential energy can be neglected, then comparing (4.40) and (4.44) shows that the specific work done for a given change in the product pv is n times higher for an open system. This is because flow work is done at the boundaries of only the open system. By inspection it can be seen that both (4.40) and (4.44) will have a divide-by-zero if n is equal to 1 and the solution for specific work fails. This is because $n = 1$ is a special case, for which pv is constant, and the result of the integration has a different form.

For a perfect gas, when pv is constant, the perfect gas equation shows that temperature will also be constant:

$$RT = pv = c$$

Hence, from equation (4.38), for a **perfect gas undergoing an isothermal process in a closed system**:

$$w = -\int_1^2 p \, dv = -c\int_1^2 \frac{dv}{v} = -c \ln\left(\frac{v_2}{v_1}\right) = -RT \ln\left(\frac{v_2}{v_1}\right) \tag{4.45}$$

If kinetic and potential energy changes are neglected, then from (4.41), the corresponding result for **a perfect gas undergoing an isothermal process in an open system** is:

$$w = c^\star \int_1^2 \frac{dp}{p} = c^\star \ln\left(\frac{p_2}{p_1}\right) = RT \ln\left(\frac{p_2}{p_1}\right) \quad \text{or} \quad -RT \ln\left(\frac{v_2}{v_1}\right) \tag{4.46}$$

Interestingly, because the product of pressure and specific volume is constant during the process, the integrals of $\dfrac{dv}{v}$ and $\dfrac{dp}{p}$ are the same and so the specific work done on a perfect gas during an isothermal process is the same for both a closed system and an open system. This is an exception to the rule that specific work will be different for the closed and open system cases, as shown by the results given in Table 4.8.

Process	Closed systems	Open systems (neglecting changes in potential and kinetic energy)
Specific work done during a reversible process	$w = -\int_1^2 p\, dv$	$w = \int_1^2 v\, dp$
Isochoric (constant specific volume)	0	$w = v(p_2 - p_1)$
Isobaric (constant pressure)	$w = -p(v_2 - v_1)$	0
Isothermal (constant temperature)	$w = -RT \ln\left(\dfrac{v_2}{v_1}\right)$	$w = RT \ln\left(\dfrac{p_2}{p_1}\right)$
Isentropic (constant entropy)	$w = \dfrac{[p_2 v_2 - p_1 v_1]}{(\gamma - 1)}$	$w = \dfrac{\gamma}{(\gamma - 1)}[p_2 v_2 - p_1 v_1]$
Polytropic (pv^n = constant)	$w = \dfrac{[p_2 v_2 - p_1 v_1]}{(n - 1)}$	$w = \dfrac{n}{(n - 1)}[p_2 v_2 - p_1 v_1]$

Table 4.8 Specific work for reversible processes undergone by a perfect gas

Regardless of whether or not an explicit expression for work done exists, considerations of the first law and energy conservation are essential to the solution of most problems. (Be sure to distinguish between closed system and open system cases in these – confusing the two is a simple but fundamental mistake.) A comparison of results for closed system, non-flow processes and open system, flow processes with negligible changes in kinetic and potential energy are given in Table 4.9. For closed systems, work transfer is assumed to be by boundary work only and no work transfer occurs when the system has a constant volume. Shaft work is assumed to be zero, although (as illustrated in Section 4.2) there are practical applications, in types of braking system for example, when shaft work transfer to a closed system is used to dissipate mechanical forms of energy into heat. Generally, however, work transfer in closed systems is through boundary work. The reverse is true for open systems; in this case work transfer

Process	Closed systems	Open systems (neglecting changes in potential and kinetic energy)
Specific work done *if* the process is reversible	In the absence of shaft work $w = -\int_1^2 p\, dv$	$w = \int_1^2 v\, dp$
Isochoric (constant specific volume)	In the absence of shaft work $q = u_2 - u_1$	Enthalpy can change, and heat transfer can occur $q + w = h_2 - h_1$
Isobaric (constant pressure)	Internal energy can change, and heat transfer can occur $q + w = u_2 - u_1$	Enthalpy can change, and heat transfer can occur $q = h_2 - h_1$
Isothermal (constant temperature)	$u_2 - u_1 = 0$ $q = -w$	$h_2 - h_1 = 0$ $q = -w$
Isentropic (constant entropy)	If reversible, an isentropic process must be adiabatic $q = 0$ $w = u_2 - u_1$	If reversible, an isentropic process must be adiabatic $q = 0$ $w = h_2 - h_1$
Polytropic (pv^n = constant)	Internal energy can change, heat transfer can occur $q + w = u_2 - u_1$	Enthalpy can change, heat transfer can occur $q + w = h_2 - h_1$

Table 4.9 Summary of relationships which apply to non-flow and flow processes

through shaft work is most common and there are many designs of turbine and compressor which operate to extract or input work in a steady flow process. In many of these cases, the SFEE can be simplified. In flows through axial flow turbines and compressors, specific heat transfer is small enough for the process be considered adiabatic, and the change in enthalpy from upstream to downstream of the turbine or compressor can be large compared to the changes in kinetic and potential energy. In this case, to a good approximation, specific work transfer is given by:

$$w = h_2 - h_1$$

and the power input is

$$\dot{W} = \dot{m}(h_2 - h_1)$$

Air flows at a steady rate through an open system from 1 to 2. Gas velocity increases from $20\ \mathrm{m\ s^{-1}}$ at 1 to $40\ \mathrm{m\ s^{-1}}$ at 2. The temperature at 1 is 320 K and the pressure ratio $\dfrac{p_2}{p_1}$ is 1.2. The change in potential energy of the flow is negligible. Treat air as a perfect gas with $\gamma = 1.4$ and $cp = 1005\ \mathrm{J\ kg^{-1}\ K^{-1}}$. Calculate the specific work done on the gas and the specific heat transfer if the process is: (i) an adiabatic process and specific enthalpy increases by $500\ \mathrm{J\ kg^{-1}}$; (ii) a reversible adiabatic process; and (iii) a reversible polytropic process with a polytropic index value of 1.3.

(i) This requires the application of the SFEE. Heat transfer is zero because the process is adiabatic, and the potential energy change is negligible, so

$$w = (h_2 - h_1) + \frac{(C_2^2 - C_1^2)}{2} = 500 + 600 = 900\ \mathrm{J\ kg^{-1}}$$

(ii) Heat transfer is again zero, because the process is adiabatic. Because the process is also reversible, it will be isentropic and treating air as a perfect gas:

$$\frac{p_2}{p_1} = \left(\frac{T_2}{T_1}\right)^{\gamma/(\gamma - 1)} \quad \text{and} \quad T_2 = T_1\left(\frac{p_2}{p_1}\right)^{(\gamma - 1)/\gamma} = 337.1\ \mathrm{K}$$

Hence from the SFEE, the specific work done is

$$w = (h_2 - h_1) + \frac{(C_2^2 - C_1^2)}{2} = c_p(T_2 - T_1) + \frac{(C_2^2 - C_1^2)}{2}$$

$$= 1005(17.1) + 600 = 17\,797\ \mathrm{J\ kg^{-1}}$$

Alternatively, this could have been calculated from the formula given in Table 4.8:

$$w = \frac{\gamma}{(\gamma - 1)}[p_2 v_2 - p_1 v_1] + \Delta KE$$

or

$$\frac{\gamma R}{(\gamma - 1)}[T_2 - T_1] + \Delta KE$$

or

$$c_p(T_2 - T_1) + \frac{(C_2^2 - C_1^2)}{2} = 17\,797\ \mathrm{J\ kg^{-1}}$$

(iii) In this case, the process is a reversible polytropic one:

$$T_2 = T_1\left(\frac{p_2}{p_1}\right)^{\frac{(n - 1)}{n}} = 333.8\ \mathrm{K}$$

The SFEE can be applied as before, but now both work and heat transfer occur:

$$w + q = c_p(T_2 - T_1) + \frac{(C_2^2 - C_1^2)}{2} = 1005(13.8) + 600 = 14\,469\ \mathrm{J\ kg^{-1}}$$

The specific work transfer can still be calculated from the result given in Table 4.8 for a reversible polytropic process:

$$w = \frac{n}{(n-1)} (p_2 v_2 - p_1 v_1) + \Delta KE \quad \text{or} \quad \frac{nR}{(n-1)} (T_2 - T_1) + \Delta KE$$

After substituting $R = \frac{(\gamma - 1)c_p}{\gamma}$, and evaluating, we have

$$w = 17\,763\,\text{J}\,\text{kg}^{-1}$$

and the specific heat transfer is

$$q = 14\,469 - 17\,763 = -3294\,\text{J}\,\text{kg}^{-1}$$

Note that although work and heat transfer are not separated in the SFEE, increases in specific kinetic energy ΔKE and potential energy ΔPE increase the specific work input required.

Learning summary

✔ You should become familiar with the common types of process described in this section and the conditions which apply in each case. The names of the processes are independent of the working fluid: perfect gases and steam can undergo an isothermal process or an isentropic process, etc.

✔ For a polytropic process the relationship between pressure and volume changes is fixed by the definition as

$$p_1 v_1^n = p_2 v_2^n$$

✔ In general, however, changes in property values which occur when a working fluid undergoes a process depend on the type of process **and** the characteristics of the working fluid; these are different for a perfect gas and steam.

✔ The work done during reversible processes is different for closed and open systems. The results for a perfect gas are summarized in Table 4.8.

4.7 Modes of heat transfer and steady-state heat transfer rates

Heat transfer occurs as the result of a temperature difference, but how? And at what rate will it occur? These are the questions considered in this section. Heat transfer can occur through materials in any phase, whether solid, liquid or vapour, across interfaces between these, to or from surfaces, and across a vacuum. For heat transfer to occur, the source of heat must be at a higher temperature than the sink to which it is transferred. The mechanism acts always to redistribute energy from the higher temperature to the lower temperature, that is, in the direction of falling temperature. This is true for each of the three fundamental modes of heat transfer: **conduction**, **convection** and **radiation**. The *rate* of heat transfer by any mode increases if the temperature difference is raised. Usually, one mode will make the dominant contribution although it is possible for more than one mode to act simultaneously.

Conduction

Within solids, heat transfer occurs by **conduction**. Dip a cold metal spoon into hot water and the handle soon begins to feel warm because heat is being *conducted* up the handle from the dipped part. No energy is created in this process; it is simply transferred from the hot end by the conduction of thermal energy. The molecules within the handle are bound in position by the lattice structure of the metal. These are not free to migrate but they can vibrate, more energetically at higher temperatures, and through this transfer energy to surrounding cooler molecules. Free electrons contribute to the process through energy exchanges during collisions with molecules. On the macroscopic scale, the net effect is an energy flow through the solid from the hottest parts to the coolest parts at a rate which depends on the temperature difference.

The rate at which heat conduction occurs is given by **Fourier's law**, which can be expressed in mathematical form as:

$$\dot{Q} = -kA\frac{\mathrm{d}T}{\mathrm{d}x} \qquad (4.47)$$

where $\dot{Q}$ (W) is the rate of heat conduction through a surface of area A (m^2) owing to the temperature gradient $\frac{\mathrm{d}T}{\mathrm{d}x}$ (K m^{-1}) normal to the surface. The negative sign makes heat transfer positive in the direction of falling temperature. k is the **thermal conductivity** (W m^{-1} K^{-1}) of the material through which heat is being conducted. All solids, and matter in other phases, have some propensity to conduct heat, and thermal conductivity is a material property which varies widely with material type. A material with a low value of thermal conductivity, such as asbestos, will limit heat transfer to a relatively low rate and has been

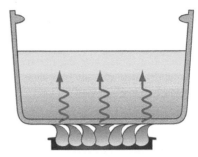

Figure 4.41 **Heat conduction is usually the dominant mode of heat transfer in solids. This illustrates how heat is transferred through the base of a cooking pot by heat conduction**

widely used in the past as a thermal insulating material. Conversely, materials with a high value of thermal conductivity, such as copper, might be used to achieve a high rate of heat transfer. For most engineering materials, thermal conductivity is a weak function of temperature. To a first approximation and over a moderate temperature range, the variation can often be neglected and an average value used. Typical values of thermal conductivity at room temperatures ($\sim$300 K) are given for a range of materials in Table 4.10. The thermal conductivity of gases is typically in the range 0.01–0.2 W m^{-1} K^{-1} and of the same order of magnitude as values for insulating solids; for non-metallic liquids, values in the range 0.07–0.6 W m^{-1} K^{-1} are representative and similar to values for non-metallic solids. In both gases and liquids, however, the second mode of heat transfer – convection – is usually more important.

Material	Thermal conductivity k (W m^{-1} K^{-1})
Asbestos (good insulator)	0.08-0.17
Wood shavings	0.06
Wool	0.05
Gypsum plaster	0.5
Building brick	0.4–1.3
Window glass	0.8
Iron	50–70
Alloy steels	10–80
Aluminium alloys	140–200
Copper (good conductor)	380

Table 4.10 Thermal conductivity values of a range of good and poor heat conducting materials

Convection

Although heat conduction occurs in liquids and gases as well as solids, if fluid motion occurs, heat transfer by **convection** is likely to be important and will usually be the dominant mode of heat transfer. Like conduction, heat transfer by convection increases with increasing temperature difference between the heat source and the heat sink. Unlike conduction, convection is an energy *transport* mechanism by which a net quantity of internal energy is *carried* from regions of high temperature to regions of low temperature by the movement of fluid between these. On a macroscopic scale, the motion might be driven by the use of a pump or fan, or it might be produced by buoyancy forces in the fluid. A distinction is made between these; heat transfer in the first case is by forced convection and in the second by natural convection.

The buoyancy forces producing conditions for **natural convection** are the result of density differences in the fluid. Fluid in regions of higher temperatures is less dense and experiences a body force acting upwards. As fluid moves in this direction it is replaced by cooler fluid in a continuous process. A central heating radiator in a room transfers heat to circulating air by natural convection, as depicted in Figure 4.42.

When fluid motion is produced by a pump or results from an imposed pressure gradient, heat transfer is by **forced convection**. Although the buoyancy forces associated with temperature gradients still exist, the effect of these on heat transfer rates can normally be neglected under the conditions of forced convection. A fan heater transfers heat to air passing over the heater element by forced convective heat transfer.

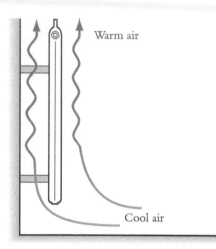

Figure 4.42 Heat transfer from central heating radiators depends on natural convection. Air circulation occurs as cool air is drawn in to replace the warm air which rises

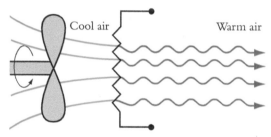

Figure 4.43 Electric fan heaters rely on forced convection to transfer heat from the heater element to the air flow through the heater

Convective heat transfer can be associated with large-scale, bulk motion of a gas or liquid as described above, and smaller-scale motion near to surfaces which are being cooled or heated. Close to a hot (or cold) surface the temperature gradient will be higher than in the bulk of the fluid. The temperature in fluid strata adjacent to the surface rises (or falls) to the value at the surface in a thin layer called the boundary layer. Packets of cool fluid which move close to the surface are heated and transport energy away from the surface into the bulk.

In many cases, details of fluid motion and local heat transport in the boundary layer are unknown and difficult to define. Instead, the heat transfer rate through the boundary layer is related to the overall temperature difference across it using a **surface** or **convective heat transfer coefficient**, h, which has units of $\text{W m}^{-2}\text{K}^{-1}$, and the rate of heat transfer across the layer to or from the surface is then determined using Newton's law of cooling:

$$\dot{Q} = hA\Delta T \qquad (4.48)$$

ΔT (K) is the temperature rise across the boundary layer, $\dot{Q}$ is transferred in the direction of falling temperature and A (m²) is the surface area. The value

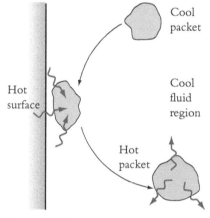

Figure 4.44 Convective heat transfer from a hot surface to a cool fluid through the movement of packets of fluid close to the surface

of the heat transfer coefficient h depends on flow regime, fluid properties, and surface geometry and roughness. When the fluid adjacent to the surface is a gas like air, h is typically in the range 5–40 $\text{W m}^{-2}\text{K}^{-1}$ for natural convection problems and between 20 and 2000 $\text{W m}^{-2}\text{K}^{-1}$ for forced convection problems. For liquids such as water, the heat transfer coefficients are typically more than an order of magnitude greater.

Radiation

The third mode of heat transfer is **radiation**, which is energy transfer by electromagnetic wave motion. Most thermal radiation takes place at wavelengths between 0.1 and 100 micrometres, which includes ultraviolet, visible light and infrared parts of the spectrum, and is the only mode of heat transfer which can occur across a vacuum. Radiative heat transfer is more sensitive to absolute temperatures than conduction and convective heat transfer, because thermal radiation emissions increase in proportion to T^4. All bodies *emit* radiant energy at a rate equal to or less than a maximum possible value given by the Stefan–Boltzmann law for a **black body**:

$$\dot{Q} = \sigma A T^4 \qquad (4.49)$$

where T is the *absolute* temperature (K) of the surface of the body, and σ is the Stefan–Boltzmann constant ($56.7 \times 10^{-9}\,\text{W m}^{-2}\,\text{K}^{-4}$).

(Making a small hole in a large and otherwise fully sealed box is a simple way to produce something which acts like a black body when viewed from the outside. Radiation entering the box through the hole has no escape and appears to be entirely 'absorbed' by the hole. This is a characteristic of a black surface. Radiation emerging through the hole also appears to be emitted by a black surface at the temperature of the interior of the box.)

Many bodies emit less than a black body at a given temperature. Emissivity ε is the fraction between 0 and 1 which is actually emitted. If emissivity does not depend on the wavelength, the body is described as a **grey body**. Typical emissivity values for grey bodies are given in Table 4.11.

At the same time as a body emits radiant energy, it can absorb all or part of the radiant energy incident upon it from other sources. If more radiant energy is absorbed than emitted, there will be a net rate of heat transfer to the body and this will be the rate of radiative heat transfer.

Material	Emissivity
Black paint	~1.0
White paint	0.9
Polished metals	0.02–0.17
Asphalt surfaces	0.9
Human skin	0.9

Table 4.11 Typical emissivity values for some common materials at ambient temperatures

A black body absorbs radiant energy at the rate of $5\,\text{kW m}^{-2}$. Calculate the temperature at which the rate of radiant heat transfer will be zero.

The net rate of heat transfer from the body will be

$$\dot{Q} = A(\sigma T^4 - 5 \times 10^3)$$

This will be zero when

$$T = \left(\frac{5 \times 10^3}{56.7 \times 10^{-9}}\right)^{0.25} = 544\,\text{K}$$

Whenever two bodies can be connected by a line of sight, part or all of the radiant energy leaving each will be incident on the other. How much of the radiant energy leaving one will be incident on the second depends on the geometry and relative positions of the two bodies. At the surface of the second body, the incident radiation can by reflected, absorbed, or transmitted. The fraction reflected is called the reflectivity, ρ, the fraction absorbed is called the absorptivity, α, and the fraction transmitted is called the transmissivity, τ. The sum of these parts must account for all the incident radiant energy, so the fractions must add up to 1:

$$\rho + \alpha + \tau = 1 \qquad (4.50)$$

The distribution between these depends on the properties of the material, conditions at the surface of the body such as surface roughness, and on radiation factors such as the angle of incidence and wavelength of the radiation. For most solids, transmissivity τ is taken to be zero, but window glass has a high transmissivity for radiation at wavelengths in the visible region.

A highly polished or silvered surface can have a high reflectivity ρ, absorbing little of the incident radiation, while a perfectly black surface has an absorptivity α of 1 and will absorb all the incident radiation. From the observation that there will not be any net heat exchange between two identical surfaces at the same temperature facing each other across a small gap, grey bodies must have equal absorptivity and emissivity values,

$$\alpha = \varepsilon$$

Hence for an opaque, grey body, the emissivity valve defines the absorptivity and also the reflectivity value, since τ is now zero and

$$\rho = 1 - \varepsilon$$

Although the analysis of many problems is complex, in general the solution for the net rate of radiative heat transfer $\dot{Q}$ from solid body 1 at the higher temperature T_1 to a second body 2 at the lower temperature T_2 can be expressed in the form

$$\dot{Q} = \sigma A_1 f_{12}(T_1^4 - T_2^4) \tag{4.51}$$

The **grey body view factor** f_{12} allows for the effects of surface properties, geometry and the view the surfaces have of each other. For diffuse surfaces, which reflect incident radiation equally in all directions regardless of the angle of incidence, if all the radiation leaving surface 1 must strike surface 2, then

$$f_{12} = \cfrac{1}{\cfrac{1}{\varepsilon_1} + \cfrac{A_1}{A_2}\left(\cfrac{1}{\varepsilon_2} - 1\right)} \tag{4.52}$$

There are a number of important problems this result can be applied to in simplified form, including the radiative heat transfer between plane parallel surfaces large enough for losses at the edge gaps to be neglected. In this case, the surface areas are equal and

$$f_{12} = \cfrac{1}{\cfrac{1}{\varepsilon_1} + \cfrac{1}{\varepsilon_2} - 1} \tag{4.53}$$

Two large parallel surfaces each have an emissivity of 0.8. Surface 1 is maintained at a temperature of 1000 K, surface 2 is maintained at 500 K. A thin sheet of material is placed midway between 1 and 2. The sheet (s) has an emissivity of 0.2 on both sides. Calculate the temperature of the sheet and the rate of radiant heat transfer per unit area from 1 to 2.

The rate of heat transfer from 1 to s will be

$$\dot{Q}_{1s} = \sigma A_1 f_{1s}(T_1^4 - T_s^4) \tag{4.54}$$

and from s to 2 it will be

$$\dot{Q}_{s2} = \sigma A_s f_{s2}(T_s^4 - T_2^4)$$

These rates of heat transfer must be equal if the temperature of the sheet is a constant, so per unit area of the surfaces

$$f_{1s}(T_1^4 - T_s^4) = f_{s2}(T_s^4 - T_2^4)$$

Because ε_1 and ε_2 are equal, f_{1s} and f_{s2} will be equal, and hence the sheet temperature will be

$$T_s = \left(\frac{T_1^4 + T_2^4}{2}\right)^{0.25} = 853.7 \text{ K}$$

From (4.53),

$$f_{1s} = \cfrac{1}{\cfrac{1}{\varepsilon_1} + \cfrac{1}{\varepsilon_s} - 1} = \cfrac{1}{1.25 + 5 - 1} = 0.190$$

and, hence, from equation (4.54) the heat transfer rate per unit area is

$$\frac{\dot{Q}}{A} = \frac{\dot{Q}_{1s}}{A_1} = \sigma f_{1s}(T_1^4 - T_s^4) = 56.7 \times 10^{-9} \times 0.190 \times (1000^4 - 853.7^4) = 5.74 \text{ kW m}^{-2}$$

Calculate the rate of heat transfer from a sphere to infinite surroundings at a temperature of 300 K. The sphere has a surface temperature of 700 K, a surface emissivity of 0.5 and a radius of 10 cm.

For a plane surface, cylinder or sphere in infinite surroundings, $A_2 \gg A_1$, where A_1 is the surface area of the body and equation (4.53) can be simplified to

$$f_{12} = \varepsilon_1$$

The surface area of the sphere, A_1, is $4\pi r^2$, and hence the heat transfer rate is

$$\dot{Q} = \sigma A_1 f_{12}(T_1^4 - T_2^4) = \sigma 4\pi r^2 \varepsilon_1 (T_1^4 - T_2^4)$$
$$= 56.7 \times 10^{-9} \times 2 \times \pi \times 0.1^2 \times 0.5 \times (700^4 - 300^4) = 413.3 \text{ W}$$

Using an effective radiant heat transfer coefficient

It is important to assess how significant radiant heat transfer might be when, for example, radiant heat transfer from a surface might be as or more significant than convective heat transfer occurring in parallel at the same time. One method of examining this is to examine the magnitude of an effective radiant heat transfer coefficient. This is defined using an expansion of the temperature terms inside the brackets of equation (4.51):

$$(T_1^4 - T_2^4) = (T_1^2 + T_2^2)(T_1^2 - T_2^2) = (T_1^2 + T_2^2)(T_1 + T_2)(T_1 - T_2)$$

and, comparing the modified equation (4.51) with the usual form of the convective heat transfer equation,

$$\dot{Q} = h^\star A_1 (T_1 - T_2)$$

From this, the effective radiant heat transfer coefficient is given by

$$h^\star = \sigma f_{12}(T_1^2 + T_2^2)(T_1 + T_2) \tag{4.55}$$

Clearly, the value of $h^\star$ has a strong dependence on temperatures T_1 and T_2, and this reflects the importance of radiative heat transfer at high temperatures.

Before leaving this section, consider the second question posed in the introduction:

What temperature will a disc brake rise to during braking?

This is a problem of practical importance. Disc brakes are used on vehicles of various types: cars, trucks, motorcycles. The discs are mounted on the wheel hubs and rotate with the wheels. When a disc brake is applied, pads are pressed against the faces of the disc causing frictional heating. The dissipated energy is transferred into the disc by heat conduction, raising its temperature. Energy is transferred from the disc by heat conduction into the wheel hub and by heat convection and radiation to the surroundings. If the combined rate of heat transfer is too low, the temperature of the disc and the pads will exceed working limits and brake fade or failure can occur.

The four disc brakes of, say, a medium-sized car weighing 1200 kg and travelling at 100 km h^{-1} must absorb about 463 kJ of kinetic energy in about 10 s under firm braking. The average rate of energy absorption will therefore be 46.3 kW or 11.6 kW per disc. The disc temperature during braking will continue to rise until the rate of heat transfer from the disc matches this. About 80 per cent of the heat transfer from the disc occurs by convection and radiation. If the effective surface area the disc is $0.4\ \text{m}^2$, the combined convective and radiative heat transfer coefficient is $80\ \text{W m}^{-2}\ \text{K}^{-1}$, and ambient temperature is $30\ ^\circ\text{C}$, then an estimate of the maximum disc temperature can be found from

$$\dot{Q} = (h + h^\star)A(T_{\text{disc}} - T_{\text{air}})$$

where $\dot{Q}$ is the heat transfer rate required (80 per cent of the rate of energy absorption in this case), $(h + h^\star)$ is the combined convective and radiative heat transfer coefficient and A is the effective area of the disc. Rearranging for the disc temperature gives

$$T_{\text{disc}} = T_{\text{air}} + \frac{\dot{Q}}{(h + h^\star)}A$$

Substituting the given values

$$T_{\text{disc}} = T_{\text{air}} + \frac{0.8 \times 11.6 \times 10^3}{80 \times 0.4} = 320\ ^\circ\text{C}$$

In the design of disc brakes, maximizing rates of heat transfer by conduction, convection and radiation is important. In other cases, engineers have long worked to find ways to minimize heat transfer rates by each mode of heat transfer. For example, a drinks vacuum flask for keeping drinks hot or cold has a combination of features designed to minimize heat transfer; this is illustrated in Figure 4.45. The inner flask containing the drink is surrounded by a second flask. The space between the flasks is evacuated to produce a partial vacuum, minimizing heat transfer across the gap by heat conduction and convection. The surfaces of the flasks are silvered to minimize radiative heat transfer across the gap. The flasks are usually made of glass, which has a low thermal conductivity, maximizing the resistance to heat transfer through the flask walls.

Steady-state heat transfer and thermal resistances

Under 'steady-state' heat transfer conditions, temperatures and rates of heat transfer do not change with time and this simplifies the analysis. Although there is an energy flow in the direction of falling temperature, the internal energy contained within a fixed volume remains constant and the heat transfer in must be balanced by heat transfer out. The assumption that steady-state conditions exist can be applied to many practical problems such as the calculation of heat losses from the steady flow of a fluid through a pipe, or heat transfer through the walls of a room maintained at a constant temperature. (Of course, there must be a heat source such as a central heating radiator in the room to balance the heat loss or the temperature would drop!)

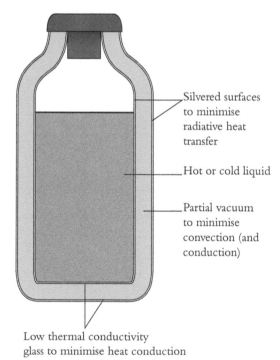

Silvered surfaces to minimise radiative heat transfer

Hot or cold liquid

Partial vacuum to minimise convection (and conduction)

Low thermal conductivity glass to minimise heat conduction

Figure 4.45 A drinks vacuum flask has features designed to minimize heat transfer to or from the liquid

The calculation of heat flows and temperature distributions in solids with surfaces exposed to fluids is common to many of these problems. Here, the analysis is restricted to steady-state conditions, and surface heat transfer coefficients (h) are assumed to account for any radiant ($h^\star$) heat transfer exchanges with fluids at a constant bulk or ambient temperature. Also, now is an

Thermodynamics

appropriate point to introduce another concept, that of **thermal resistance**, which is the resistance to heat transfer between two points at different temperatures. This is defined by

$$R_{th} = \frac{\Delta T}{\dot{Q}} \quad \text{or} \quad \dot{Q} = \frac{\Delta T}{R_{th}} \tag{4.56}$$

where is the rate of heat transfer and is the magnitude of the temperature difference. If the thermal resistance is known between points at known temperatures, the rate of heat transfer is easily calculated. Comparing equations (4.56) and (4.48) shows that the **thermal resistance of a surface** is

$$R_{th} = \frac{1}{hA} \tag{4.57}$$

The thermal resistances for other cases can derived in the same way. For a plane wall with a constant thermal conductivity, the temperature within the solid will vary linearly; the **thermal resistance of a plane wall** can be found from Fourier's law. If thermal conductivity is constant, and ΔT is the temperature change across the thickness Δx of the wall (there is no need for the minus sign in front of the ΔT, but $\dot{Q}$ *still* flows towards the lower temperature):

$$\dot{Q} = -kA\frac{dT}{dx} = kA\frac{\Delta T}{\Delta x} = \frac{\Delta T}{\left(\frac{\Delta x}{kA}\right)}$$

Hence, the **thermal resistance of a plane wall** is

$$R_{th} = \frac{\Delta x}{kA} \tag{4.58}$$

Steady heat conduction in the radial direction through the wall of a cylinder is a case when the temperature variation has a curved shape. This is because unlike a plane wall, in the cylinder wall heat is conducted through a surface area which increases with radius. Fourier's law still applies, now with the temperature gradient $\frac{dT}{dr}$, and for a circular cylinder of length L

$$\dot{Q} = -kA\frac{dT}{dr} = -k(2\pi rL)\frac{dT}{dr}$$

Rearranging to separate the variables, and integrating between limits of 1 for the inner surface and 2 for the outer surface, we have

$$\frac{\dot{Q}}{(2\pi kL)}\int_1^2 \frac{dr}{r} = -\int_1^2 dT$$

$$\frac{\dot{Q}}{(2\pi kL)}\ln\left(\frac{r_2}{r_1}\right) = -(T_2 - T_1)$$

$$\dot{Q} = \frac{-(T_2 - T_1)}{\dfrac{\ln\left(\dfrac{r_2}{r_1}\right)}{(2\pi kL)}} = \frac{\Delta T}{\dfrac{\ln\left(\dfrac{r_2}{r_1}\right)}{(2\pi kL)}}$$

and the **thermal resistance of a cylinder** is

$$R_{th} = \frac{\ln\left(\dfrac{r_2}{r_1}\right)}{(2\pi kL)} \tag{4.59}$$

Temperatures can be found for any location between the positions of known temperatures once the rate of heat transfer has been calculated. For the cylinder, the temperature T at radial position r will be

$$T = T_1 - \frac{\dot{Q}}{(2\pi kL)}\ln\left(\frac{r}{r_1}\right)$$

Calculate the temperature at the mid-point in: (i) a plane wall; and (ii) a cylinder with an inner radius of 0.01 m if the inner surface temperature is 1000 K, the outer surface temperature is 300 K, the material thermal conductivity $k = 200 \text{ W m}^{-1} \text{ K}^{-1}$ and the wall thickness is 0.02 m.

(i) Per unit area of wall surface, thermal resistance $\dfrac{\Delta x}{k}$ is $\dfrac{0.02}{200}$ or 10^{-4} K W^{-1}, and

$$\dot{Q} = \frac{\Delta T}{R_{th}} = \frac{(1000 - 300)}{0.0001} = 700 \times 10^4 \text{ W}$$

At the mid-point position,

$$T = T_1 - \dot{Q} \times \frac{\Delta x}{2k} = 1000 - \frac{700}{2} = 650 \text{ K}$$

(ii) Per unit length of cylinder, thermal resistance and heat transfer rate are

$$R_{th} = \frac{\ln\left(\dfrac{r_2}{r_1}\right)}{(2\pi k L)} = \ln\left(\frac{\left(\dfrac{0.2 + 0.1}{0.1}\right)}{(2\pi \times 200)}\right) = 8.74 \times 10^{-4} \text{ K W}^{-1}$$

$$\dot{Q} = \frac{\Delta T}{R_{th}} = \frac{(1000 - 300)}{8.74 \times 10^{-4}} = 80.07 \times 10^4 \text{ W}$$

At the mid-point position,

$$T = T_1 - \frac{\dot{Q}}{(2\pi k L)} \ln\left(\frac{r}{r_1}\right) = 1000 - \frac{80.07 \times 10^4}{2\pi \times 200} \ln\left(\frac{0.1 + 0.1}{0.1}\right) = 558.3 \text{ K}$$

Heat transfer through walls and surfaces in series

The thermal resistance approach is particularly useful in the analysis of steady-state heat transfer through walls and surfaces contributing resistances in series. By finding the overall resistance between points of known temperature, the rate of heat transfer can be determined and, subsequently, so can any values of the unknown temperatures along the heat transfer path.

Like electrical resistances in series, the overall thermal resistance is the sum of the individual contributions. Consider, for example, heat transfer through the unit area of a plane wall constructed from several layers, each with a different thermal conductivity as illustrated in Figure 4.46.

Supposing that the temperatures of the exposed surfaces 1 and 4 are known. The heat transfer rate through the layers will be

$$\dot{Q} = \frac{-(T_4 - T_1)}{R_{th}}$$

Under steady-state conditions, the rate of heat transfer through each layer must be the same (or the internal energy within the layers would not be constant and the temperature would change over time). Hence, applying Fourier's law to each layer gives an independent expression for the same heat transfer rate and:

$$\dot{Q} = \frac{-(T_2 - T_1)}{\left(\dfrac{\Delta x}{k}\right)_a} = \frac{-(T_3 - T_2)}{\left(\dfrac{\Delta x}{k}\right)_b} = \frac{-(T_4 - T_3)}{\left(\dfrac{\Delta x}{k}\right)_c}$$

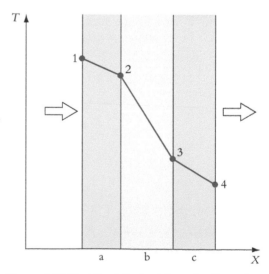

Figure 4.46 **Steady-state heat transfer through the layers of a plane composite wall from surface 1 to surface 4. If the layers have different thermal conductivities, the temperature gradient will be lowest in layers with highest thermal conductivity and the temperature gradient will change at the interfaces between layers**

Thermodynamics

If each expression is separated and rearranged, the following set is obtained:

$$\dot{Q}\left(\frac{\Delta x}{k}\right)_a = -(T_2 - T_1)$$

$$\dot{Q}\left(\frac{\Delta x}{k}\right)_b = -(T_3 - T_2)$$

$$\dot{Q}\left(\frac{\Delta x}{k}\right)_c = -(T_4 - T_3)$$

Adding these, and remembering that $\dot{Q}$ is the same for each, we have

$$\dot{Q}\left[\left(\frac{\Delta x}{k}\right)_a + \left(\frac{\Delta x}{k}\right)_b + \left(\frac{\Delta x}{k}\right)_c\right] = -(T_4 - T_1)$$

and

$$\dot{Q} = \frac{-(T_4 - T_1)}{\left[\left(\frac{\Delta x}{k}\right)_a + \left(\frac{\Delta x}{k}\right)_b + \left(\frac{\Delta x}{k}\right)_c\right]}$$

or

$$\dot{Q} = \frac{\Delta T}{\Sigma R_{\text{th}}}$$

and

$$\Sigma R_{\text{th}} = \left[\left(\frac{\Delta x}{k}\right)_a + \left(\frac{\Delta x}{k}\right)_b + \left(\frac{\Delta x}{k}\right)_c\right]$$

This approach can be simply extended to include the resistance of any surfaces exposed to fluid. The example illustrated in Figure 4.47 represents a cavity wall construction through which heat transfer occurs from a room at temperature T_R to the outside ambient at temperature T_A.

The heat transfer path passes though four surfaces and three layers of solid, each contributing to the total of seven heat transfer resistances. For unit area of the wall, the rate of heat transfer will be

$$\dot{Q} = \frac{-(T_A - T_R)}{\Sigma R_{\text{th}}} = \frac{\Delta T}{\Sigma \text{wall resistances} + \Sigma \text{surface resistances}} \qquad (4.60)$$

and

$$\Sigma R_{\text{th}} = \left[\frac{1}{h_1}\right] + \left[\frac{\Delta x}{k}\right]_a + \left[\frac{\Delta x}{k}\right]_b + \left[\frac{1}{h_2}\right] + \left[\frac{1}{h_3}\right] + \left[\frac{\Delta x}{k}\right]_c + \left[\frac{1}{h_4}\right]$$

If the wall area is A, the total rate of heat transfer would be $\dot{Q}_T = \dot{Q} \times A$. After $\dot{Q}$ is determined, the temperature at a point between the room and the outside ambient is calculated using the sum of resistances between a point of known temperature and the point of interest. If, for example, this is surface 2, the temperature could be calculated from

$$T_2 = T_R - \frac{\dot{Q}}{\left(\frac{\Delta x}{k}\right)_a + \frac{1}{h_1}}$$

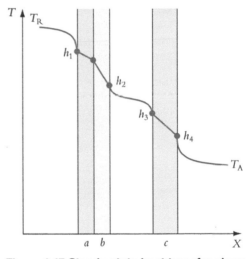

Figure 4.47 Steady-state heat transfer along a path with solid layers and surfaces exposed to fluids requires the surface heat transfer resistances of these to be taken into account

The glass in a single-glazed window of a house is to be replaced with a double glazed unit. Cross-sections through the old and new windows are shown in Figure 4.48. The heat transfer coefficient on the room-side surface of the glass (h_{room}) is 8 W m^{-2} K^{-1}; on the glass surface exposed to the outside ambient, the heat transfer coefficient (h_{out}) is 15 W m^{-2} K^{-1}. The original glass pane and the two glass panes in the double-glazed unit are each 7 mm thick and have a thermal conductivity of

$0.8\ \mathrm{W\,m^{-1}\,K^{-1}}$. Air fills the gap between the panes of the double-glazed unit and this offers negligible resistance to heat transfer across the gap by conduction compared with convective heat transfer resistances at interior glass surfaces. The convective heat transfer coefficient for these surfaces is $5\ \mathrm{W\,m^{-2}\,K^{-1}}$. If room temperature in the house is maintained at $20\,°\mathrm{C}$ and the outside air temperature is $0\,°\mathrm{C}$, compare the rate of heat loss per unit area through the original and new window and calculate the temperature air in the gap of the double-glazed unit.

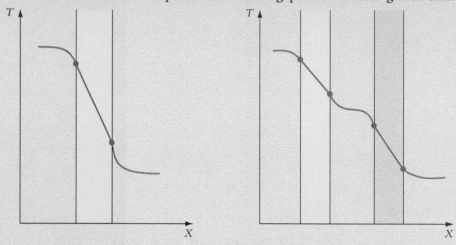

Figure 4.48 Double glazing a window achieves lower heat losses compared with single-glazed windows by increasing overall thermal resistance

For the original window, the thermal resistance between the room and the outside ambient is

$$\Sigma R_{\mathrm{th}} = \left[\frac{1}{h_{\mathrm{room}}}\right] + \left[\frac{\Delta x}{k_{\mathrm{glass}}}\right] + \left[\frac{1}{h_{\mathrm{out}}}\right] = \left[\frac{1}{8}\right] + \left[\frac{7 \times 10^{-3}}{0.8}\right] + \left[\frac{1}{15}\right]$$

Hence

$$\Sigma R_{\mathrm{th}} = (125 + 8.75 + 66.7) \times 10^{-3} = 0.200\ \mathrm{K\,m^2\,W^{-1}}$$

The rate of heat loss per unit area of window is

$$\dot{Q} = \frac{\Delta T}{\Sigma R_{\mathrm{th}}} = \frac{20}{0.200} = 100\ \mathrm{W\,m^{-2}}$$

For the replacement double-glazed window, the thermal resistance is

$$\Sigma R_{\mathrm{th}} = \left[\frac{1}{h_{\mathrm{room}}}\right] + \left[\frac{\Delta x}{k_{\mathrm{pane1}}}\right] + \left[\frac{1}{h_{\mathrm{int}}}\right] + \left[\frac{1}{h_{\mathrm{int}}}\right] + \left[\frac{\Delta x}{k_{\mathrm{pane2}}}\right] + \left[\frac{1}{h_{\mathrm{out}}}\right]$$

Hence

$$\Sigma R_{\mathrm{th}} = \left[\frac{1}{8}\right] + \left[\frac{7 \times 10^{-3}}{0.8}\right] + \left[\frac{1}{5}\right] + \left[\frac{1}{5}\right] + \left[\frac{7 \times 10^{-3}}{0.8}\right] + \left[\frac{1}{15}\right] = 0.609\ \mathrm{K\,m^2\,W^{-1}}$$

The rate of heat loss per unit area of the double-glazed window is

$$\dot{Q} = \frac{\Delta T}{\Sigma R_{\mathrm{th}}} = \frac{20}{0.609} = 32.8\ \mathrm{W\,m^{-1}}$$

For the temperature in the air gap between the two panes, the thermal resistance between a plane at which the temperature is known and a plane in the air gap is required. Working from the room, there are three resistance contributions:

$$\Sigma R_{\mathrm{th}} = \left[\frac{1}{h_{\mathrm{room}}}\right] + \left[\frac{\Delta x}{k_{\mathrm{pane1}}}\right] + \left[\frac{1}{h_{\mathrm{int}}}\right] = \left[\frac{1}{8}\right] + \left[\frac{7 \times 10^{-3}}{0.8}\right] + \left[\frac{1}{5}\right]$$

$$= 0.334\ \mathrm{K\,m^2\,W^{-1}}$$

The temperature in the air gap can now be found from

$$T_{gap} = T_{room} - \dot{Q} \times R_{th} = T_{room} - 32.8 \times 0.334 = 9.0\,°C$$

Alternatively, the temperature could be found by working from the outside ambient (0°C) to the air gap. In this case

$$R_{th} = \left[\frac{1}{h_{out}}\right] + \left[\frac{\Delta x}{k_{pane2}}\right] + \left[\frac{1}{h_{int}}\right] = \left[\frac{1}{15}\right] + \left[\frac{7 \times 10^{-3}}{0.8}\right] + \left[\frac{1}{5}\right] = 0.275\,K\,m^2\,W^{-1}$$

Giving the same air gap temperature as before, we have

$$T_{gap} = T_{out} + \dot{Q} \times R_{th} = T_{out} + 32.8 \times 0.275 = 9.0\,°C$$

The steady heat transfer rate through other combinations of surfaces and walls can be analysed using the thermal resistance approach. For radial heat transfer through concentric cylinders with inner and outer surfaces exposed to fluid, the thermal resistances must account for differences in the area through which heat is flowing; the outer surface of a pipe will have a larger surface area than the inner surface. This affects the resistance contribution of convective heat transfer at surfaces which, from equation (4.57) and for unit length of pipe, will be

$$R_{th} = \frac{1}{2\pi r h}$$

Heat transfer from a steam pipe lagged with a layer of thermal insulation is typical of problems involving radial heat flow. For the example shown in Figure 4.49, for a temperature difference ΔT between the steam and the surroundings, the rate of heat transfer per unit length of pipe will be

$$\dot{Q} = \frac{\Delta T}{\Sigma R_{th}} \tag{4.61}$$

and, working radially outwards from the inner surface, there are four contributions to the thermal resistance:

$$\Sigma R_{th} = \frac{1}{2\pi r_1 h_1} + \frac{\ln\left(\frac{r_2}{r_1}\right)}{2\pi k_a} + \frac{\ln\left(\frac{r_3}{r_2}\right)}{2\pi k_b} + \frac{1}{2\pi r_3 h_3}$$

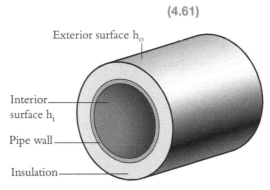

Figure 4.49 Cross-section through a steam pipe lagged with a layer of insulation with low thermal conductivity to reduce heat transfer losses

Calculate the reduction in heat transfer rate when a steel pipe conveying steam is lagged. The pipe has an inside diameter of 15 cm and an outside diameter of 20 cm. The lagging is 7 cm thick. The thermal conductivity of the pipe wall k_p is $60\,W\,m^{-1}\,K^{-1}$; the thermal conductivity of the lagging material k_l is $0.1\,W\,m^{-1}\,K^{-1}$. The heat transfer coefficient for the inner surface of the pipe is $80\,W\,m^{-2}\,K^{-1}$ and the steam temperature is 300 °C. The outer surface exposed to ambient (initially the outside surface of the pipe and, finally, the outside surface of the lagging) has a heat transfer coefficient of $18\,W\,m^{-2}\,K^{-1}$ and ambient temperature is 0 °C.

Before the lagging is applied, the thermal resistance per unit length of the unlagged pipe will be

$$\Sigma R_{th} = \frac{1}{2\pi r_1 h_1} + \frac{\ln\left(\frac{r_2}{r_1}\right)}{2\pi k_p} + \frac{1}{2\pi r_2 h_o}$$

where (per unit length of pipe)

$$\frac{1}{2\pi r_1 h_i} = \frac{1}{2 \times \pi \times 0.075 \times 80} = 26.52 \times 10^{-3}\,\text{K m W}^{-1}$$

$$\frac{\ln\left(\frac{r_2}{r_1}\right)}{2\pi k_p} + \frac{\ln\left(\frac{0.1}{0.075}\right)}{2\pi \times 60} = 3.435 \times 10^{-3}\,\text{K m W}^{-1}$$

$$\frac{1}{2\pi r_2 h_o} = \frac{1}{2 \times \pi \times 0.10 \times 18} = 88.42 \times 10^{-3}\,\text{K m W}^{-1}$$

The total resistance is

$$\Sigma R_{th} = 118.4 \times 10^{-3}\,\text{K m W}^{-1}$$

and the rate of heat transfer to the surroundings per unit length of pipe is:

$$\dot{Q} = \frac{\Delta T}{\Sigma R_{th}} = \frac{((300 + 273) - (0 + 273))}{118.4 \times 10^{-3}} = 2.53\,\text{kW m}^{-1}$$

After the lagging has been applied, the thermal resistance has an additional contribution and the resistance of the outside surface exposed to ambient is raised by the increase in radius by the thickness of the lagging. The thermal resistance per unit length of pipe is now

$$\Sigma R_{th} = \frac{1}{2\pi r_1 h_1} + \frac{\ln\left(\frac{r_2}{r_1}\right)}{2\pi k_p} + \frac{\ln\left(\frac{r_3}{r_2}\right)}{2\pi k_l} + \frac{1}{2\pi r_3 h_o}$$

where

$$\frac{1}{2\pi r_1 h_i} = \frac{1}{2 \times \pi \times 0.075 \times 80} = 26.52 \times 10^{-3}\,\text{K m W}^{-1}\,\text{(as before)}$$

$$\frac{\ln\left(\frac{r_2}{r_1}\right)}{2\pi k_p} + \frac{\ln\left(\frac{0.1}{0.075}\right)}{2\pi \times 60} = 3.435 \times 10^{-3}\,\text{K m W}^{-1}\,\text{(as before)}$$

$$\frac{\ln\left(\frac{r_3}{r_2}\right)}{2\pi k_l} + \frac{\ln\left(\frac{0.17}{0.1}\right)}{2\pi \times 0.1} = 88.45 \times 10^{-3}\,\text{K m W}^{-1}\,\text{(additional contribution)}$$

$$\frac{1}{2\pi r_3 h_o} = \frac{1}{2 \times \pi \times 0.17 \times 18} = 52.01 \times 10^{-3}\,\text{K m W}^{-1}\,\text{(modified contribution)}$$

The total resistance is now

$$\Sigma R_{th} = 166.4 \times 10^{-3}\,\text{K m W}^{-1}$$

and the new rate of heat transfer to the surroundings per unit length of pipe is

$$\dot{Q} = \frac{\Delta T}{\Sigma R_{th}} = \frac{((300 + 273) - (0 + 273))}{166.4 \times 10^{-3}} = 1.80\,\text{kW m}^{-1}$$

Comparing the values of heat transfer rate before and after the lagging has been applied shows that the percentage reduction in heat transfer rate was

$$\frac{\Delta \dot{Q}}{\dot{Q}_{unlagged}} = \left(\frac{2.53 - 1.80}{2.53}\right) \times 100 = 28.9\%$$

Thermodynamics

4.8 Cycles, power plant and engines

In this final section, some of the many cycles proposed for the production of mechanical power from a source of heat are described. The processes making up these cycles are treated here as being reversible and the working fluid is treated as a perfect gas or steam; all real cycles incur an efficiency penalty due to irreversibilities and the ideal case sets an upper limit on efficiency. In some cases the ideal cycle has a value of efficiency equal to the Carnot efficiency, others have a lower efficiency but are better suited to particular applications for practical and operational reasons. Recall that the Carnot efficiency is the maximum possible efficiency of a heat engine which will only be achieved if the heat engine is reversible and operates between only two heat reservoirs. This efficiency depends only on the temperatures of the two reservoirs (from equation (4.9)):

$$\eta_{cannot} = 1 - \frac{T_2}{T_1}$$

The efficiency of a cycle is reduced if any process in the cycle is irreversible or if some heat is supplied at a temperature below T_1 or rejected at a temperature above T_2.

Cycles which have the Carnot efficiency

Three ideal cycles which have an efficiency equal to the Carnot efficiency are the Carnot, Stirling and Ericsson cycles. The Carnot cycle is made up of four reversible processes: isentropic expansion (1–2), isothermal heat rejection (2–3), isentropic compression (3–4) and isothermal heat addition (4–1). The cycle is shown on a temperature–entropy (T–s) diagram commonly used for power cycles, and a pressure–specific volume diagram for a perfect gas in Figure 4.50.

That this cycle has the Carnot efficiency can be shown as follows. The heat supplied during the reversible isothermal process to will be $T_1(S_1 - S_4)$ and the magnitude of the heat rejected will be $T_2(S_2 - S_3)$. The first law applied to a closed system requires the network output of the cycle to be the difference between these, and noting that $(S_2 - S_3)$ is equal to $(S_1 - S_4)$:

$$\eta = \frac{W_{net}}{Q_{supplied}} = \frac{T_1(S_1 - S_4) - T_2(S_2 - S_3)}{T_1(S_1 - S_4)} = 1 - \frac{T_2}{T_1}$$

or, if the temperature ratio of upper to lower temperatures is $T_r = \frac{T_1}{T_2}$,

$$\eta = 1 - \frac{1}{T_r} \tag{4.62}$$

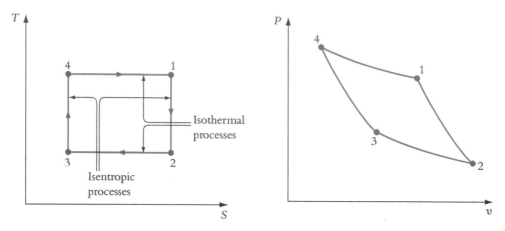

Figure 4.50 The Carnot cycle plotted on temperature–entropy and a pressure–specific volume diagram for a perfect gas

The Carnot cycle is an ideal that is difficult to utilize in a practical engine. There are, however, practical Stirling engines working on the **Stirling cycle** which in ideal form also has the Carnot efficiency. Like the Carnot cycle, the Stirling cycle is made up of four reversible processes. The heat supply and heat rejection processes at constant temperature are retained but the other two are constant volume processes where *internal* heat transfer occurs. Process diagrams for a perfect gas undergoing the Stirling cycle are shown in Figure 4.51. During the first constant volume process, heat is drawn from the working fluid causing this to cool to a smaller volume. The heat transferred is *stored within the system* and returned to the working fluid during the second constant volume process. The internal heat storage and return is described as **regeneration**. Because regeneration does not involve heat transfer across the system boundary, the Stirling engine operates only between two *external* heat reservoirs. If only two heat reservoirs are used and all the cycle processes are reversible, the heat engine cycle must have the Carnot efficiency.

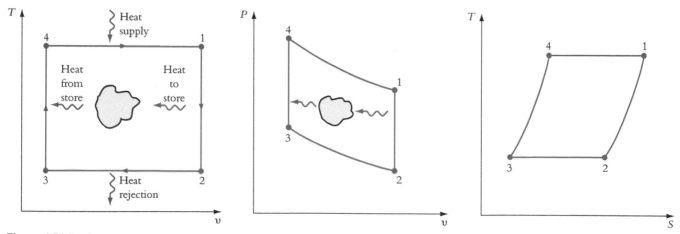

Figure 4.51 Perfect-gas process diagrams for a system undergoing the Stirling cycle

Before examining the third cycle which has the Carnot efficiency, it is worth considering how a cycle can be executed in more than one way. When a closed system formed by a piston–cylinder combination undergoes a cycle, the working fluid stays in the cylinder as it undergoes a sequence of processes. All the fluid undergoes the same process at the same time and one process is completed before the next begins. An amount of work will be output per cycle, but not continuously during the cycle. Reciprocating machines and engines operate in this way. The Stirling engine also produces work output only during part of the cycle. A power station, on the other hand, operates on a cycle in which the working fluid, usually steam, flows around a circuit through components with dedicated functions: heat exchangers, turbines and pumps. In this case, all the processes making up the cycle are steady flow processes and the work output is continuous. The **Ericsson cycle** is made up of four steady flow processes and is the third ideal cycle considered here which has the Carnot efficiency. Like the Stirling cycle, the Ericsson cycle exploits regeneration to achieve the Carnot efficiency. In the Ericsson cycle, regeneration is employed to cool and then heat the working fluid during two reversible constant pressure processes. The cycle is shown in Figure 4.52.

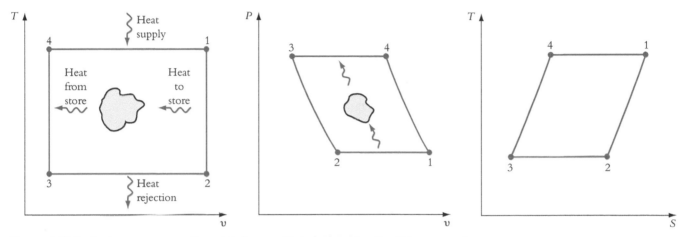

Figure 4.52 **Perfect-gas process diagrams for a system undergoing the Ericsson cycle**

Steam plant and the Rankine cycle with superheat

A **steam power plant** is shown schematically in Figure 4.53; this is a closed system in which steam circulates in a steady flow through the boiler, the turbine, the condenser and the pump in turn. The plant is designed to take thermodynamic advantage of steam being a condensable vapour to achieve high cycle efficiency. In principle, this can operate on the Carnot cycle and achieve the Carnot efficiency. The cycle is shown on a temperature–specific entropy (T–s) diagram, commonly used to present power cycle information, in Figure 4.54. For steam, the diagram is similar in appearance to a temperature–specific volume diagram. In the boiler, saturated vapour is generated from saturated liquid in a constant pressure heating process. The saturated vapour expands reversibly and adiabatically through the steam turbine to a low pressure; the steam drives the turbine in a process which converts some of the enthalpy, kinetic energy and potential energy of the steam into a power output from the turbine. The saturated mixture flows into the condenser where some of the vapour is condensed at constant pressure, lowering the dryness fraction. Finally, the saturated mixture flows from the condenser into a pump where its pressure is raised to the pressure of the boiler.

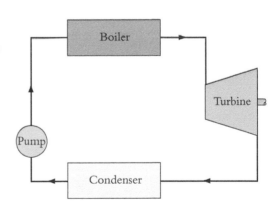

Figure 4.53 **The principal components of a steam power plant operating**

The Carnot cycle represents an ideal that poses practical difficulties: it is difficult to stop condensation in the condenser before all the steam is condensed to the saturated liquid state, or to pump a liquid and vapour mixture from condenser to boiler pressure without the phases separating. In the turbine expansion, if the steam dryness fraction falls too low, liquid droplets can cause severe erosion of the turbine blades. These problems are greatly eased by allowing steam to condense to the saturated liquid condition in the condenser and superheating the steam in the boiler. This modified cycle, called the **Rankine cycle with superheat**, is shown on a T–s diagram in Figure 4.55. In the ideal cycle, the turbine expansion and pump

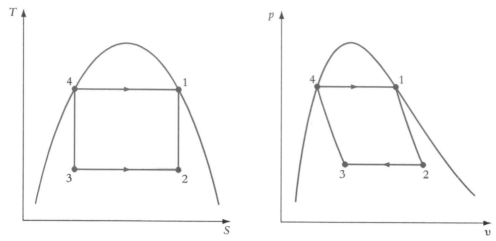

Figure 4.54 **The Carnot cycle plotted in the saturated mixture region of temperature–specific entropy and pressure–specific volume diagrams. Note the difference between this _p–v_ diagram and the diagram for a perfect gas given in Figure 4.50**

compression are still isentropic processes, and heat rejection still takes place at constant temperature. However, although heat addition in the boiler takes place at constant pressure as before, it is not an isothermal process in the Rankine cycle and efficiency is lower than for a Carnot cycle operating between the same upper and lower temperatures.

All the processes of the Rankine cycle are steady flow processes and the changes in potential and kinetic energy across each of the components in the circuit are small compared with the enthalpy changes, so the SFEE reduces to

$$\dot{Q} + \dot{W} = \dot{m}(\Delta h)$$

For the heat supply and rejection processes, no work transfers occur and $\dot{W} = 0$; the turbine expansion and pump compression processes are adiabatic, so for these $\dot{Q} = 0$.

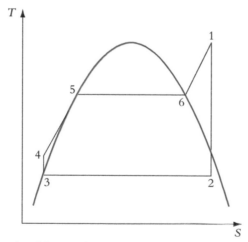

Figure 4.55 **The Rankine cycle with superheat plotted on a temperature–specific entropy diagram**

Calculate the efficiency of the Rankine cycle with superheat using the following information and tables given in (Rogers and Mayhew, 1995). Boiler pressure is 120 bar and condenser pressure is 0.3 bar. At inlet to the turbine (1) the steam temperature T_1 is 400 °C. The water is saturated liquid at inlet to the pump (3).

If the steam plant produces a net power output of 50 MW (50×10^6 W), what is the mass flowrate of steam leaving the boiler?

To calculate the efficiency of the cycle, the specific enthalpy values at state points 1 to 4 must be calculated. Using the superheat tables (in Rogers and Mayhew, 1995), the specific enthalpy at inlet to the turbine is

$$h_1 = 3052 \, \text{kJ kg}^{-1}$$

The specific entropy is

$$s_1 = 6.076 \, \text{kg}^{-1} \text{K}^{-1}$$

Because the turbine expansion is isentropic, the entropy value is the same at the turbine exist and at the given condenser pressure of 0.3 bar. This allows the dryness fraction of the steam at exit from the turbine to be calculated from

$$s_1 = s_2 = s_f + x.s_{fg}$$

Rearranging for x and substituting values for s_f and s_{fg} given in the tables,

$$x_2 = \frac{6.076 - 0.944}{7.767} = 0.661$$

and specific enthalpy can then be found using values for h_f and h_{fg} from the tables:

$$h_2 = h_f + x.h_{fg} = 289 + 0.661 \times 2336 = 1832 \, \text{kJ kg}^{-1}$$

If specific work *output* from the turbine is calculated as a positive value, then

$$w_t = (h_1 - h_2) = 3052 - 1832 = 1220 \, \text{kJ kg}^{-1}$$

At exit from the condenser and inlet to the pump, water is in the saturated state; $h_3 = h_f = 289 \, \text{kJ kg}^{-1}$. The specific work *input* by the pump to raise the pressure to boiler pressure is obtained from the SFEE:

$$w_p = (h_4 - h_3)$$

Because the compression is isentropic, state 4 is fixed by the known entropy ($s_4 = s_3$) and pressure values and h_4 can be found from the tables. More simply, to a good approximation, the specific enthalpy rise across the pump is given by $v_3\Delta p$, where v_3 is the specific volume of saturated liquid at the inlet pressure:

$$w_p = v_3(p_4 - p_3) = 0.00102 \times (120 - 0.3) \times 10^5 = 0.122 \times 10^5 \, \text{J kg}^{-1} = 12.2 \, \text{kJ kg}^{-1}$$

This work input is small compared with the turbine work output and sometimes neglected in calculations. The specific heat supplied in the boiler is

$$q_s = h_1 - h_4 = 3052 - 301.2 = 2750.8 \, \text{kJ kg}^{-1}$$

and finally, the cycle efficiency is

$$\eta = \frac{w_t - w_p}{q_s} = \frac{1220 - 12.2}{2750.8} = 43.9\%$$

The efficiency penalty connected with the supply of heat over a range of temperatures rather than the maximum temperature is significant, as can be seen by comparing this value and the Carnot efficiency for the same maximum and minimum values of temperature:

$$\eta_{cannot} = 1 - \frac{273 + 69.1}{273 + 400} = 49.2\%$$

The cycle produces a net work output of $(1220 - 12.2)$ kJ per unit mass of steam circulating in the plant. The mass flowrate is steady and the same at each point in the cycle, including the exit of the boiler. The mass flow of steam required to generate a net output of 50 MW is therefore

$$\dot{m} = \frac{50 \times 10^6}{(1220 - 12.2) \times 10^3} = 41.4 \, \text{kg s}^{-1}$$

Gas turbine engines and the Brayton cycle

In some markets for power-generating plant, gas turbine engines compete successfully with steam plant and steam turbines. Other applications include engines for ship propulsion and, most successfully, in aircraft propulsion. The main components of a gas turbine engine are shown schematically in Figure 4.56(a).

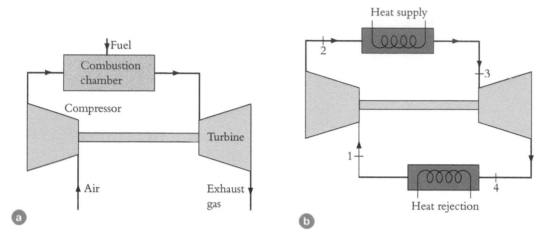

Figure 4.56 (a) A schematic of the components of a gas turbine engine and (b) an ideal gas turbine operating on a closed cycle

Unlike steam plant and Stirling engines which use external sources of heat which is transferred across a system boundary, the gas turbine engine is an **internal combustion engine** which utilizes heat released internally through the combustion of a fuel. In a gas turbine engine, the compressor feeds a steady flow of compressed air to the combustion chamber where fuel is burned to produce high-temperature gaseous products. These are expanded through the turbine produce shaft work output. The compressor and the turbine are axial-flow designs coupled together so that some of the shaft work produced by the turbine is used to drive the compressor; the remaining part is the net work output delivered by the engine. There is no recirculation of the working fluid in the gas turbine engine; downstream of the turbine the gas flow is directed into the atmosphere. This means the engine does not operate on a closed cycle, nor do the properties of the working fluid remain constant; upstream of the combustion chamber the working fluid is air, downstream it becomes the products produced by the combustion of fuel and air. Nevertheless, an ideal gas turbine engine operating on a closed cycle can be conjured as an analogue with similar performance. This is illustrated in Figure 4.56(b). The combustion chamber is replaced by a heat exchanger in which heat is transferred to the working fluid, and at exit from the turbine the working fluid is cooled as it passes through a second heat exchanger before re-entering the compressor. If the heat exchange processes take place at constant pressure and the turbine and compressor processes are isentropic, this is the **Brayton cycle.** The cycle is shown plotted on *T-s* and *p-v* diagrams in Figure 4.57.

When the working fluid is a perfect gas, the efficiency of the Brayton cycle depends only on the pressure ratio $\frac{p_2}{p_1}$ or $\frac{p_3}{p_4}$ and the ratio of specific heats γ. This is shown in the following analysis, starting from the application of the SFEE to the heat exchangers where heat is transferred to and from the working fluid. The potential and kinetic energy terms in the SFEE are small compared to the enthalpy terms so that the specific heat supplied is

$$q_s = (h_3 - h_2) = c_p(T_3 - T_2)$$

The heat rejected is

$$q_r = (h_4 - h_1) = c_p(T_4 - T_1)$$

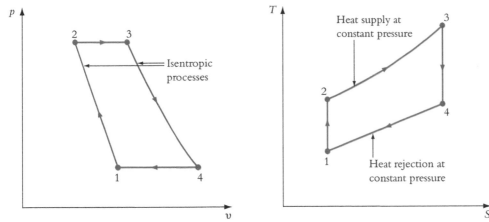

Figure 4.57 **The Brayton cycle plotted on T-s and p-v diagrams**

and hence the efficiency is

$$\eta = 1 - \frac{q_r}{q_s} = 1 - \frac{(T_4 - T_1)}{(T_3 - T_2)} = 1 - \frac{T_1\left(\dfrac{T_4}{T_1} - 1\right)}{T_2\left(\dfrac{T_3}{T_2} - 1\right)}$$

However, because the pressure ratios $\dfrac{p_2}{p_1}$ across the compressor and $\dfrac{p_3}{p_4}$ across the turbine are equal, the isentropic relationship between temperature and pressure ratios requires that

$$\frac{T_2}{T_1} = \frac{T_3}{T_4}$$

or, rearranging,

$$\frac{T_4}{T_1} = \frac{T_3}{T_2}$$

Therefore, the cycle efficiency becomes

$$\eta = 1 - \frac{T_1}{T_2}$$

By inspection, this is lower than the Carnot efficiency, $\left(1 - \dfrac{T_1}{T_3}\right)$, since $T_3 > T_2$, and, if $\left(\dfrac{p_2}{p_1}\right)$ is substituted for the temperature ratio $\dfrac{T_2}{T_1}$, then

$$\eta = 1 - \frac{1}{\left(\dfrac{p_2}{p_1}\right)^{(\gamma - 1)/\gamma}}$$

or, if P_r is the pressure ratio $\dfrac{p_2}{p_1}$, then

$$\eta = 1 - \frac{1}{P_r^{(\gamma - 1)/\gamma}} \tag{4.63}$$

The efficiency of the cycle increases with increasing pressure ratio. The pressure ratio is limited to around 20 by limits on the operating temperature of turbine blades. Air and the gas flowing from the combustion chamber into the turbine have similar values of γ, around 1.4. Substituting these values into (4.63) gives a value of efficiency of 57.5 per cent.

Reciprocating internal combustion engines and the Otto and diesel cycles

Although reciprocating internal combustion engines are used to power lawn mowers, power tools, generator sets and many other machines and devices, they are most closely identified with their use as car engines. They are particularly well suited to this application because they deliver power and torque over a broad range of engine speeds, providing flexibility, and have a high power-to-weight ratio and good fuel economy characteristics.

The two main types of engine are the spark ignition engine fuelled by gasoline (petrol) and the diesel engine fuelled by diesel. The designs used to power vehicles are almost invariably four-stroke engines, meaning they operate on an engine cycle that requires two engine revolutions, or four piston strokes, to complete. This is an *engine* cycle, not a thermodynamic cycle: the working fluid is modified by combustion and replaced by a fresh charge each time the cycle is completed. During the first part of the cycle, air (or premixed fuel and air) are induced into the engine cylinder during the induction stroke. This is followed by the compression stroke, during which the charge is compressed to a high pressure and temperature. Towards the end of the compression stroke, the charge is ignited and burns to release chemical energy from the fuel, raising the temperature and pressure still higher. The third stroke is the expansion (or power) stroke, during which some of the internal energy of the products of combustion is converted into work output. Finally, during the fourth and last stroke of the cycle, the exhaust stroke, cylinder gases are expelled into the engine exhaust. Only during the compression and expansion strokes of the cycle is the cylinder a closed system, and this part of the cycle is most readily represented by an ideal thermodynamic cycle.

The **Otto cycle** used to represent the closed part of the spark ignition engine cycle is illustrated in Figure 4.58. An isentropic compression process (1–2), is followed by heat addition at a constant volume (2–3). This is followed by an isentropic expansion process (3–4), followed finally by heat rejection at constant volume (4–1). Although the cycle is comprised of four processes, this represents only two strokes of the four-stroke cycle. At the end of the compression stroke, combustion is represented in the Otto cycle by the heat addition at constant volume (2–3), and the Otto cycle is closed by a heat rejection process which has no equivalent in the four-stroke cycle. Nevertheless, calculations of thermodynamic performance of the Otto cycle provide insights to the performance of the real engine.

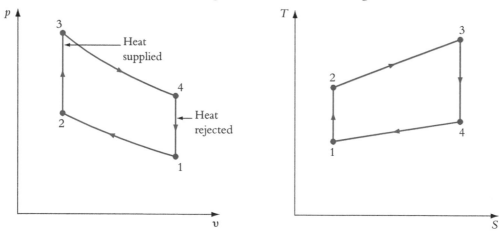

Figure 4.58 **The Otto cycle plotted on pressure-specific volume and temperature-specific entropy diagrams**

If the working fluid is a perfect gas, the efficiency of the Otto cycle depends only on the ratio of specific heats of the gas and the *compression ratio* $C_r = \dfrac{V_2}{V_1}$ or $\dfrac{v_2}{v_1}$ of the cycle. This can be shown using the first law and the definition of efficiency:

$$\eta = \frac{w_{net}}{q_s} = 1 - \frac{q_r}{q_s}$$

As no work is done during the heat rejection process 4–1, the specific heat rejected is equal to the reduction in specific internal energy $C_v(T_4 - T_1)$. Similarly, the specific heat supplied is $C_v(T_3 - T_2)$. Hence

$$\eta = 1 - \frac{(T_4 - T_1)}{(T_3 - T_2)} = 1 - \frac{T_1\left(\dfrac{T_4}{T_1} - 1\right)}{T_2\left(\dfrac{T_3}{T_2} - 1\right)} \tag{4.64}$$

Now, since process (1–2) is isentropic and the working fluid is a perfect gas, it follows that

$$\frac{T_2}{T_1} = \left(\frac{v_2}{v_1}\right)^{(\gamma - 1)} = C_r^{(\gamma - 1)} \tag{4.65}$$

Process (3–4) is also isentropic and $\dfrac{v_3}{v_4} = \dfrac{v_2}{v_1}$ so that

$$\frac{T_3}{T_4} = \frac{T_2}{T_1} = C_r^{(\gamma - 1)}$$

If the temperature ratios are rearranged to obtain

$$\frac{T_3}{T_2} = \frac{T_4}{T_1} \tag{4.66}$$

Substituting equations (4.65) and (4.66) in equation (4.64) gives

$$\eta = 1 - \frac{1}{C_r^{(\gamma - 1)}} \tag{4.67}$$

A single-cylinder, reciprocating piston engine operates on a machine cycle that takes two engine revolutions to complete. The working fluid is a perfect gas with $\gamma = 1.4$. During the first revolution, the gas in the cylinder is expelled and replaced by a fresh charge during which no net work is done. During the second revolution, the engine operates on the ideal Otto cycle. The maximum safe engine speed is 2400 rev min^{-1}. The compression ratio is 9. Calculate the maximum power output of the engine if the heat supplied during the Otto cycle is 800 J cycle^{-1}.

This describes an imaginary engine operating on a machine cycle similar to the four-stroke cycle of reciprocating internal combustion engines. The imaginary engine completes is machine cycle in two engine revolutions, and the working fluid undergoes the Otto cycle on the second of these.

If 800 J cycle^{-1} of heat is supplied during each Otto cycle, the net work output from this will be

$$W = \eta \times Q_s$$

where

$$\eta = 1 - \frac{1}{C_r^{(\gamma - 1)}} = 1 - \frac{1}{9^{(1.4 - 1)}} = 58.5\%$$

Hence

$$W = 0.585 \times 800 = 467.8 \, \text{J cycle}^{-1}$$

The engine undergoes the Otto cycle, producing the work output only every second engine revolution, so the power output of the engine will be

$$\dot{W} = W \times \frac{N}{2}$$

where N is the number of engine revolutions per second. Maximum power output will be achieved at the maximum safe engine speed, so

$$\dot{W} = W \times \frac{N}{2} = 467.8 \times \frac{2400}{60 \times 2} = 9356 \, \text{W} = 9.36 \, \text{kW}$$

The **Diesel cycle**, illustrated in Figure 4.59, is sometimes used to represent the closed part of the diesel *engine* cycle. The Diesel and Otto cycles differ in the way combustion is represented as a process of heat addition at constant *pressure* in the Diesel cycle rather than at constant volume as in the Otto cycle. In modern designs, the duration of combustion is similar for both types of engine, and neither volume nor pressure is truly constant, so the association with engine type is weak.

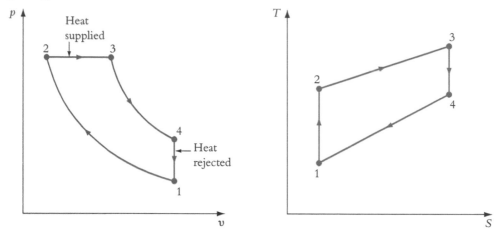

Figure 4.59 The Diesel cycle plotted on pressure-specific volume- and temperature-specific entropy diagrams

The efficiency of the Diesel cycle depends on the compression ratio of the cycle and a new, second parameter called the *cut-off ratio* $C_o = \dfrac{v_3}{v_2}$ which defines the duration of the constant pressure heat addition process. The derivation of the result for efficiency is similar to that for the Otto cycle, but the specific heat addition is now $c_p(T_3 = T_2)$, and equation (4.64) becomes

$$\eta = 1 - \frac{T_1\left(\dfrac{T_4}{T_1} - 1\right)}{\gamma T_2\left(\dfrac{T_3}{T_2} - 1\right)} \tag{4.68}$$

Making use of the definition of cut-off ratio, and remembering that processes (1–2) and (3–4) are isentropic processes, we have

$$\eta = 1 - \frac{1}{C_r^{(\gamma-1)}}\left[\frac{C_o^{\gamma} - 1}{\gamma(C_o - 1)}\right] \tag{4.69}$$

Comparing this with equation (4.67) for the Otto cycle shows that, for a given compression ratio, the Diesel cycle has a lower efficiency than the Otto cycle. That is, because temperature T_3 is higher, heat addition at constant volume in the Otto cycle produces higher cycle efficiency than the heat addition process at constant pressure in the Diesel cycle.

Learning summary

- ✔ The cycles analysed in this section are ideal, thermodynamic cycles. These provide insights to the types of cycle used to generate mechanical power.

- ✔ No cycle can have a higher efficiency than the Carnot efficiency, but the ideal Stirling and Ericsson cycles achieve an efficiency value equal to this. In these cycles, heat transfer at temperatures between the maximum and minimum available takes place internally, through regeneration. External heat transfer across system boundaries occurs isothermally and only at these maximum and minimum temperatures, meeting conditions for the Carnot efficiency to be achieved.

Thermodynamics

✔ The Rankine cycle with superheat is the basis for a practical cycle for the generation of power output using steam as the working fluid. The cycle is less efficient than the Carnot cycle because heat supply takes place over a range of temperatures rather than the maximum possible.

✔ The efficiencies of the Brayton, Otto and Diesel cycles depend on the ratio of specific heats and the pressure ratio (Brayton cycle), compression ratio (Otto cycle) or compression ratio and cut-off ratio (Diesel cycle). There is no need to remember the particular results for these cycles, but students should understand how these are derived.

✔ There are similarities between the ideal thermodynamic cycles and real engine and power plant cycles. Plant and engines operating on the Stirling, Ericsson and Rankine cycles have external heat supply and heat rejection and the working fluids do undergo continuous thermodynamic cycles. Differences between the ideal and the real cycles are more marked for internal combustion engines. In these, fuel is burned within the working fluid, changing its composition as well as releasing chemical energy. The working fluid is replaced during successive cycles of the machine, so internal combustion engines such as gas turbines and reciprocating internal combustion engines do not operate on true thermodynamic cycles, but on a machine cycles.

Notation

A	m^2	area
C	$m\ s^{-1}$	velocity
c_p	$J\ kg^{-1}\ K^{-1}$	specific heat capacity at constant pressure
c_v	$J\ kg^{-1}\ K^{-1}$	specific heat capacity at constant volume
g	$m\ s^{-2}$	gravitational constant ($=9.81$)
$h\ (H)$	$J\ kg^{-1}$	specific enthalpy (enthalpy, J)
h	$W\ m^{-2}\ k^{-1}$	heat transfer coefficient
k	$W\ m^{-1}\ K^{-1}$	thermal conductivity
$m\ (\dot{m})$	kg	mass (mass flowrate, kg s^{-1})
$\tilde{m}$	$kg\ kmol^{-1}$	molar mass
n	$kmol$	number of kilomoles of substance
n		polytropic index
p	$N\ m^{-2}$	absolute pressure
q	$J\ kg^{-1}$	heat transfer per unit mass
Q	J	heat transfer
$\dot{Q}$	W	rate of heat transfer
R	$J\ kg^{-1}\ K^{-1}$	specific gas constant
$\tilde{R}$	$J\ kmol^{-1}\ K^{-1}$	universal gas constant ($=8.3145 \times 10^3\ J\ kmol^{-1}\ K^{-1}$)
$s\ (S)$	$J\ kg^{-1}\ K^{-1}$	specific entropy (J K^{-1}, entropy)
T	K	absolute temperature ($°$C, temperature in degrees Celsius $= T\ (K) - 273$)
$u\ (U)$	$J\ kg^{-1}$	specific internal energy (internal energy, J)
$v\ (V)$	$m^3\ kg^{-1}$	specific volume (volume, m^3)
w	$J\ kg^{-1}$	specific work
W	J	work
$\dot{W}$	Ts^{-1} or watts	rate of work or power
x		dryness fraction
z	m	height above datum
γ		ratio of specific heat capacities $\left(\dfrac{c_p}{c_v}\right)$
μ	$kg\ s^{-1}\ (=N\ s\ m^{-2})$	dynamic viscosity
ρ	$kg\ m^{-3}$	density $\left(\dfrac{1}{v}\right)$

Reference

Rogers, G.F.C. and Y. R. Mayhew (1995) *Thermodynamic and Transport Properties of Fluids, SI Units,* 5th edition, Blackwell.

Electrical and electronic systems

Alan Howe

UNIT OVERVIEW

- Introduction
- Direct current circuits
- Electromagnetic systems
- Capacitance
- Alternating current circuits
- Three-phase circuits
- Semiconductor rectifiers
- Amplifiers
- Digital electronics
- Transformers
- AC induction motors

5.1 Introduction

Why does an Introduction to Mechanical Engineering include a chapter on Electrical and Electronic Systems? The reason is that most modern mechanical systems incorporate some electrical or electronic equipment. One only has to think of cars with on-board computers, high-speed electric trains, processing plants, printing presses and even the domestic multiprogramme washing machine. The list is endless. It is therefore essential for the mechanical engineer to understand how electrical and electronic devices work. This chapter begins by looking at some the key principles of electrical and electronic systems before moving on to examine some practical examples that most mechanical engineers will encounter during their careers.

Direct current circuits studies the relationship between voltage, current and resistance and examines ways to analyse simple circuits. We also look at how bridge circuits can be used to measure resistance and strain.

Electromagnetic systems investigates the relationship between electric currents and magnetic fields. It examines how a force is exerted on a current-carrying conductor contained within a magnetic field, the principle behind the design of electric motors. It also investigates how a varying current flowing through an inductor will induce an electromagnetic force in the inductor.

Capacitance examines how capacitors work and the effect of connecting capacitors in series and in parallel and connecting a capacitor in series with a resistor.

Alternating current circuits are widespread. This section demonstrates how we analyse circuits comprising any combination of resistors, inductors and capacitors supplied by an alternating voltage. It investigates how the different components affect the current and power dissipation.

Three-phase circuits are used to supply electricity to industrial users. As most industrial machines are connected to such systems it is in the interests of safety that mechanical engineers should understand how such systems work. This part examines the two most common forms of three-phase connection: star and delta.

Semiconductor rectifiers are used to convert alternating current to direct current. Electricity supplies are generally alternating but most electronic devices require direct voltage supplies. Simple rectifiers produce a series of unidirectional current pulses. Where the magnitudes of the pulses are too large for an application, current smoothing or voltage stabilization is used.

Amplifiers are used to reproduce voltage signals several times larger than the original. Simple amplifiers may be constructed using discrete transistors, more advanced designs incorporate a form of integrated circuit called an operational amplifier. With a few components it is possible to construct an amplifier that will add a number of independent signals or one that will integrate a signal, for example to convert the output from an accelerometer to a velocity.

Digital electronics begins with an investigation of combinational logic and how logic gates may be used to construct control circuits. Sequential logic introduces the bistable, the simplest form of memory cell and the rudimentary component in any computer or microprocessor. Analogue-to-digital converters and digital-to-analogue converters transfer data between analogue and digital systems and vice versa. They are essential for the control of complex machines.

Transformers are used to convert alternating voltage and current levels. They have no moving parts, consume little power and are used in a wide variety of applications from chargers and power supplies for small electronic equipment to the transmission and distribution of electrical power. They may also be used for electrical isolation where there is a risk of personal injury.

AC induction motors provide the drive for many mechanical systems, such as industrial pumps, fans, compressors and mills. This section explains how these motors work and discusses variations in the torque versus speed characteristics, to help users choose the most suitable motor for each application.

5.2 Direct current circuits

The direct current circuit, or dc circuit for short, is one of the simplest circuits to analyse. The basic circuit comprises an energy source, load, connecting wires and a switch. The energy source may be a battery or generator, the load could be lamp or heater. A simple example of a dc circuit is the torch light.

Torch light

A torch or flashlight comprises a battery, light bulb and switch connected by conducting wires made of aluminium or copper.

A battery is a chemical system in which positive and negative electric charges are separated, creating an electromotive force or emf, E. The emf is measured in volts (V) and is identified on the circuit diagram by an arrow pointing towards the positive terminal of the battery.

Electrical and electronic systems

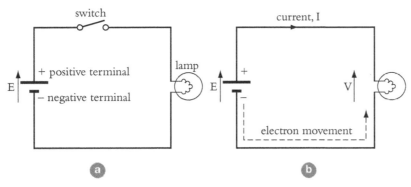

Figure 5.1

While the switch is open there is no charge movement, but once the switch is closed the battery acts as a source of energy and the current, I begins to flow. By convention it is assumed that the current flows in the direction of the emf from the positive terminal of the battery through the circuit back to the negative terminal. In fact, the current is the movement of electrons, which have negative charges and travel in the opposite direction. Current is measured in amperes (A) or amps for short.

The movement of electric charges through the circuit transfers energy from the battery to the load where it is dissipated making the bulb glow. This creates a potential difference, V, across the load which equals the battery emf.

$$V = E$$

To show that energy is dissipated in the load the potential difference is indicated by an arrow pointing in the opposite direction to the conventional current flow.

Ohm's law

The potential difference, V, across the load equals the current, I, multiplied by a constant, called the resistance, R, where resistance has the unit ohms (Ω):

$$V = IR \tag{5.1}$$

This relationship is named after its discoverer and is called Ohm's law.

Power

The rate at which energy is dissipated is called power, P, and is found from the equation

$$Power = voltage \times current$$

$$P = VI \tag{5.2}$$

Power is measured in watts (W).

Substituting for V from (5.1), power can also be expressed as

$$P = I^2R \tag{5.3}$$

Rewriting Ohm's law

$$I = \frac{V}{R}$$

and substituting in (5.2), enables us to write the equation for power

$$P = \frac{V^2}{R}$$

Energy

The energy transferred from the source to the load may be calculated from

$$Energy = power \times time$$
$$W = Pt$$

Time is measured in seconds (s) and energy has the units joules (J).

An electric heater with a resistance of 15 Ω is connected to a 120 V dc supply. Calculate the current, power dissipation and energy transferred from the supply to the heater in 1 h.

$$E = supply\ voltage = 120\,V$$
$$V = E$$
$$V = IR$$
$$I = \frac{V}{R} = \frac{120}{15} = 8\,A$$
$$P = VI$$
$$= 120 \times 8 = 960\,W$$

Alternatively

$$P = I^2R = 8^2 \times 15 = 960\,W$$
$$W = Pt = 960 \times 60 \times 60 = 3.456 \times 10^6\,J = 3.456 \times 10^6\,MJ$$

Resistivity

The resistance of a conductor depends on its length, l, cross-sectional area, A, and a constant for the material from which it is made, called the resistivity, ρ, such that

$$R = \frac{\rho l}{A} \qquad (5.4)$$

where ρ has the units Ω m, l is measured in m and A is in m^2.

The resistivities of some common conducting materials are listed in Table 5.1

Conducting material	Resistivity at 0 °C (Ω m)	Temperature coefficient (°C)$^{-1}$
Aluminium	2.7×10^{-8}	0.0038
Copper	1.59×10^{-8}	0.0043
Platinum	1.17×10^{-7}	0.0039
Silver	1.58×10^{-8}	0.0040
Carbon	6.5×10^{-5}	-0.0005

Table 5.1

Calculate the resistance at 0 °C of a copper wire 15 m long and 1 mm in diameter.

From Table 5.1, $\rho_{Cu} = 1.59 \times 10^{-8}\,\Omega$ m

$$R = \frac{\rho l}{A}$$
$$= \frac{1.59 \times 10^{-8} \times 15}{\frac{\pi}{4} \times (1 \times 10^{-3})^2} = 0.30\,\Omega$$

Temperature coefficient of resistance

For most conducting materials the resistivity is dependent on temperature. Generally, the resistivity of metals increases with temperature although the resistivity of some alloys are independent of temperature. The resistivity of carbon falls as the temperature rises.

Values for the temperature coefficients of some common conducting materials are shown in Table 5.1

The resistance of a conductor at temperature θ, measured in degrees Celsius (°C), can be calculated from

$$R_\theta = R_0(1 + \alpha\theta) \tag{5.5}$$

where R_0 is the resistance of the conductor at $0\,°C$ and α is the temperature coefficient of temperature.

Calculate the resistance of the copper wire in the previous worked example at 50 °C.

From the previous example, $R_0 = 0.30\ \Omega$.

From Table 5.2, $\alpha = 0.0043\ (°C)^{-1}$

$$R_{50} = R_0(1 + \alpha 50) = 0.30(1 + 0.0043 \times 50) = 0.36\ \Omega$$

A carbon resistor has a resistance of 47 kΩ at 15 °C. Calculate its resistance at 60 °C.

$$R_{15} = R_0(1 + \alpha 15)$$

$$47 \times 10^3 = R_0(1 - 0.0005 \times 15)$$

$$R_0 = 47.355 \times 10^3\ \Omega$$

$$R_{60} = R_0(1 + \alpha 60)$$

$$R_{60} = 47.355 \times 10^3 \times (1 - 0.0005 \times 60)$$

$$R_{60} = 45.935\ k\Omega$$

Series circuits

Electric circuits normally have several components which may be connected in a wide variety of ways. Figure 5.2 shows an example with three resistors connected in series so that the current, I, from the supply source flows through them all.

From Ohm's law, the potential differences across the resistors R_1, R_2 and R_3 are respectively

$$V_1 = IR_1, \quad V_2 = IR_2, \quad V_3 = IR_3$$

From the circuit diagram it can be seen that

$$E = V_1 + V_2 + V_3$$

Substituting for V_1, V_2 and V_3

$$E = I(R_1 + R_2 + R_3)$$

The total resistance

$$R_{eq} = \frac{E}{I} = R_1 + R_2 + R_3 = \Sigma R \tag{5.6}$$

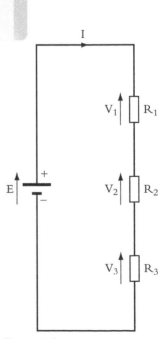

Figure 5.2

Parallel circuits

Figure 5.3 shows the three resistors connected in parallel with the voltage supply, E.

The voltage across each resistor is E. Applying Ohm's law, we have

$$E = I_1 R_1, \quad E = I_2 R_2, \quad E = I_3 R_3$$

$$I_1 = \frac{E}{R_1} \quad I_2 = \frac{E}{R_2} \quad I_3 = \frac{E}{R_3}$$

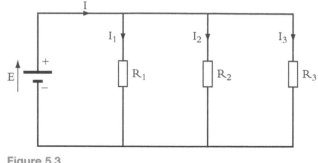

Figure 5.3

from the circuit diagram it is can be seen that the supply current, I equals

$$I = I_1 + I_2 + I_3$$

substituting for I_1, I_2 and I_3

$$I = E\left(\frac{1}{R_1} + \frac{1}{R_2} + \frac{1}{R_3}\right)$$

The total resistance of the circuit

$$R_{eq} = \frac{E}{I}$$

$$\frac{1}{R_{eq}} = \frac{1}{R_1} + \frac{1}{R_2} + \frac{1}{R_3} = \sum \frac{1}{R}$$

(5.7)

Kirchhoff's Laws

To analyse more complex circuits, it is necessary to use more sophisticated methods of analysis. These are based on two fundamental laws attributed to Kirchhoff.

Kirchhoff's Current Law

The parallel circuit shown in Figure 5.3 is an example of the application of Kirchhoff's Current Law.

> Kirchhoff postulated that in any circuit the sum of currents entering a node or junction equals the sum of the currents leaving that node.

This is demonstrated by the example shown in Figure 5.4(a).

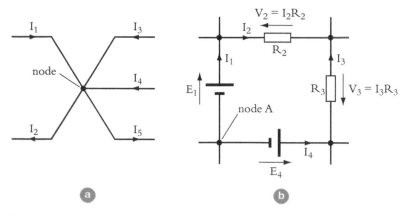

Figure 5.4

$$I_1 + I_3 + I_4 = I_2 + I_5$$

$$\text{Rewriting } I_1 - I_2 + I_3 + I_4 - I_5 = 0$$

This may expressed mathematically

$$\Sigma I = 0 \qquad\qquad (5.8)$$

and the current law summarized:

> The algebraic sum of currents entering and leaving a node is zero

Kirchhoff's Voltage Law

The series circuit shown in Figure 5.2 is an example of Kirchhoff's Voltage Law.

> Kirchhoff postulated that the algebraic sum of emfs in a closed loop equals the algebraic sum of potential differences (or volt drops) across components in that loop.

Another example is shown in Figure 5.4(b). In this case

$$E_1 - E_4 = I_2 R_2 - I_3 R_3 \qquad\qquad (5.9)$$

Voltage E_1 is driving current around the loop in a clockwise direction, while voltage E_4 is driving current in the opposite direction. The emfs are opposing each other reducing their overall effect, hence the need to subtract E_4 from E_1. It has been assumed that I_2 and I_3 are in opposite directions, producing potential differences $(I_2 R_2)$ and $(I_3 R_3)$ in opposite directions. The directions chosen for the currents shown are arbitrary and are only for the purposes of analysis but once chosen must be strictly adhered to. If any current is actually flowing in the opposite direction, then it will be calculated as a negative value.

Another way of writing equation (5.9) is:

$$E_1 - I_2 R_2 + I_3 R_3 - E_4 = 0$$

This may be derived directly from the circuit, starting at node A and working round in a clockwise direction, assuming that all voltages acting in that direction are positive and any in the opposite direction are negative. Kirchhoff summarized this for any loop in his Voltage Law thus:

$$\Sigma V = 0 \qquad\qquad (5.10)$$

> The algebraic sum of voltages around a closed loop is zero.

A 10 V battery supplies the circuit shown in Figure 5.5. Calculate the voltages across the resistors R_2, R_3 and R_4 and the currents in each branch of the network I_1, I_2 and I_3.

Right loop:

$$E - V_2 = 0$$

where

$$V_2 = I_2 R_2$$
$$E = V_2 = I_2 R_2$$
$$V_2 = 10\,\text{V}$$
$$10 = I_2 \times 5$$
$$I_2 = 2\,\text{A}$$

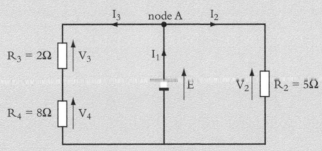

Figure 5.5

Left loop:

$$V_4 + V_3 - E = 0$$
$$E = V_3 + V_4$$
$$= I_3 R_3 + I_3 R_4 = I_3 (R_3 + R_4)$$
$$10 = I_3 (2 + 8)$$
$$I_3 = 1 \text{ A}$$
$$V_3 = I_3 R_3 = 1 \times 2 = 2 \text{ V}$$
$$V_4 = I_3 R_4 = 1 \times 8 - 8 \text{ V}$$

At node A:

$$I_1 = I_2 + I_3 = 2 + 1 = 3 \text{ A}$$

Mesh analysis

A more complex circuit is shown in Figure 5.6(a). This network has two nodes, A and B and three loops, or meshes as they are usually called, 1, 2 and 3. Using Kirchhoff's Laws we need to write a current equation for each node and a voltage equation for each mesh; in this case a total of five equations.

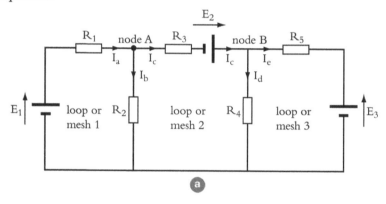

a

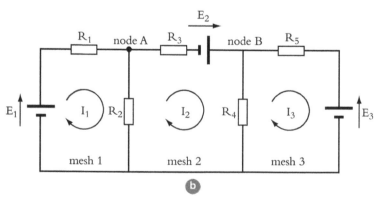

b

Figure 5.6

Node A	$I_a = I_b + I_c$
Node B	$I_c = I_d + I_e$
Left loop (mesh 1)	$E_1 = I_a R_1 + I_b R_2$
Centre loop (mesh 2)	$E_2 = -I_b R_2 + I_c R_3 + I_d R_4$
Right loop (mesh 3)	$E_3 = I_d R_4 - I_e R_5$

Electrical and electronic systems

These equations must be solved simultaneously to find the current distribution, quite a difficult task. The problem can be eased by using mesh analysis. The method assumes that currents circulate clockwise around each loop with I_1 flowing in mesh 1, I_2 in mesh 2, etc. as shown in Figure 5.6(b). I_1 flows through resistor R_1 and I_2 passes through R_3. But both I_1 and I_2 travel through R_2 flowing in opposite directions making the net current in the resistor $(I_1 - I_2)$. Similarly I_3 flows in R_5 and $(I_2 - I_3)$ passes through R_4. Kirchhoff's Voltage Law is then used to derive equations for the voltage drops around each mesh.

Mesh 1
$$0 = E_1 - (I_1 R_1) - (I_1 - I_2) R_2$$
$$E_1 = I_1 R_1 + (I_1 - I_2) R_2$$
$$E_1 = (R_1 + R_2) I_1 - R_2 I_2 \tag{5.11}$$

Mesh 2
$$0 = -(I_2 - I_1) R_2 - I_2 R_3 + E_2 - (I_2 - I_3) R_4$$
$$-E_2 = I_1 R_2 - I_2(R_2 + R_3 + R_4) + I_3 R_4$$
$$E_2 = -R_2 I_1 + (R_2 + R_3 + R_4) I_2 - R_4 I_3 \tag{5.12}$$

Mesh 3
$$0 = -(I_3 - I_2) R_4 - I_3 R_5 - E_3$$
$$E_3 = I_2 R_4 - I_3 (R_4 + R_5)$$
$$-E_3 = -R_4 I_2 + (R_4 + R_5) I_3 \tag{5.13}$$

Equations (5.11) to (5.13) may be rearranged
$$E_1 = (R_1 + R_2) I_1 \qquad - R_2 I_2$$
$$E_2 = \qquad -R_2 I_1 + (R_2 + R_3 + R_4) I_2 \qquad - R_4 I_3$$
$$-E_3 = \qquad -R_4 \qquad\qquad I_2 + (R_4 + R_5) I_3$$

and solved by substitution as demonstrated in Example 5.6

In the circuit in Figure 5.6(b) E_1 is 48 V, E_2 16 V and E_3 24 V. R_1 is 4 Ω, R_2 1 Ω, R_3 5 Ω, R_1 2 Ω and R_5 2 Ω. Calculate the three mesh currents.

Mesh 1 $48 = (4 + 1) I_1 \qquad\qquad - (1) I_2$

Mesh 2 $16 = -(1) I_1 + (1 + 5 + 2) I_2 \qquad - (2) I_3$

Mesh 3 $-24 = -(2) \qquad\qquad I_2 + (2 + 2) I_3$

$$48 = 5 I_1 \qquad\qquad - I_2$$
$$16 = -I_1 \qquad + 8 I_2 \qquad - 2 I_3$$
$$-24 = \qquad\qquad - 2 I_2 \qquad + 4 I_3$$

$$I_1 = \frac{48 + I_2}{5} = 9.6 + 0.2 I_2$$

$$I_3 = \frac{-24 + 2 I_2}{4} = -6 + 0.5 I_2$$

substituting for I_1 and I_3

$$16 = -9.6 - 0.2 I_2 + 8 I_2 + 12 - I_2$$
$$13.6 = 6.8 I_2$$
$$I_2 = 2 \text{ A}$$
$$I_1 = 9.6 + 0.4 = 10 \text{ A}$$
$$I_3 = -6 + 1 = -5 \text{ A}$$

I_3 is negative, indicating that the current circulates around mesh 3 in an anticlockwise direction.

There is another way to solve the equations. This method is more popular for solving large networks where users can employ computer software to do the hard work. The equations are first rewritten as a matrix equation.

$$
\begin{vmatrix} E_1 \\ E_2 \\ -E_3 \end{vmatrix} = \begin{vmatrix} (R_1 + R_2) & -R_2 & \\ -R_2 & (R_2 + R_3 + R_4) & -R_4 \\ & -R_4 & (R_4 + R_5) \end{vmatrix} \begin{vmatrix} I_1 \\ I_2 \\ I_3 \end{vmatrix}
$$

On the left of the equation are the driving voltages in each mesh $\begin{vmatrix} E_1 \\ E_2 \\ -E_3 \end{vmatrix}$. The voltages are shown as positive when they drive current clockwise around the mesh and negative when they drive current in the opposite direction, like E_3 in this example. On the extreme right of the equation we have the mesh currents to be calculated. Also on the right side of the equation is the resistance matrix. It is conventional with matrices like this to identify individual elements by their row and column, counting rows from the top and columns from the left. The element in the top left corner is called a_{11}. The element to its immediate right is a_{12} and the one immediately below is a_{21}. The full list of elements in three rows by three-column matrix is:

$$
\begin{vmatrix} a_{11} & a_{12} & a_{13} \\ a_{21} & a_{22} & a_{23} \\ a_{31} & a_{32} & a_{33} \end{vmatrix}
$$

The value or determinant, Δ of this matrix is given by:

$$\Delta = a_{11}(a_{22}\,a_{33} - a_{32}\,a_{23}) - a_{12}(a_{21}\,a_{33} - a_{31}\,a_{23}) + a_{31}(a_{21}\,a_{32} - a_{31}\,a_{22}) \tag{5.14}$$

The diagonal terms, a_{11} to a_{33}, in the resistance matrix equal the sum of the resistances in each mesh, $(R_1 + R_2)$, $(R_2 + R_3 + R_4)$ and $(R_4 + R_5)$ respectively. The off-diagonal terms are all negative and have a magnitude equal to the resistance linking two meshes. a_{12} and a_{21} are both equal to $-R_2$, the resistance linking meshes 1 and 2, whilst a_{23} and a_{32} are both equal to $-R_4$ the resistance between meshes 2 and 3. There is no direct link between meshes 1 and 3, so a_{13} and a_{31} are both zero.

The simplest way to explain this method is through a practical example.

Using the circuit in Figure 5.6(b) again but with E_1 is 2 V, E_2 11 V and E_3 10 V. R_1 is 2Ω, R_2 1Ω, R_3 1Ω, R_1 1Ω and R_5 5Ω. Calculate the three mesh currents.

Mesh 1	2		(2 + 1)	−1	0		I_1
Mesh 2	11	=	−1	(1 + 1 + 1)	−1		I_2
Mesh 3	−10		0	−1	(1 + 5)		I_3

$$
\begin{vmatrix} 2 \\ 11 \\ -10 \end{vmatrix} = \begin{vmatrix} 3 & -1 & 0 \\ -1 & 3 & -1 \\ 0 & -1 & 6 \end{vmatrix} \begin{vmatrix} I_1 \\ I_2 \\ I_3 \end{vmatrix}
$$

By substituting the elements of the resistance matrix in equation (5.14), we can find its determinant.

$$\Delta = 3[(3 \times 6) - (-1 \times -1)] - (-1)[(-1 \times 6) - (0 \times -1)]$$
$$+ 0[(-1 \times -1) - (0 \times 3)]$$
$$= 3[17] + 1[-6]$$
$$= 45$$

Using a technique called Cramer's Rule the next step is to replace the first column of the resistance matrix by the three voltages and to calculate the determinant of the new matrix, ΔI_1

$$\Delta I_1 = \begin{vmatrix} 2 & -1 & 0 \\ 11 & 3 & -1 \\ -10 & -1 & 6 \end{vmatrix}$$

$$\Delta I_1 = 2[(3 \times 6) - (-1 \times -1)] - (-1)[(11 \times 6) - (-10 \times -1)] \\ + 0[(11 \times -1) - (-10 \times 3)]$$

$$= 2[17] + 1[56]$$

$$= 90$$

ΔI_1 = equals Δ multiplied by I_1 so

$$I_1 = \frac{\Delta I_1}{\Delta} = \frac{90}{45} = 2 \text{ A}$$

This process is repeated to find I_2 by replacing the second column of the resistance matrix by the three voltages.

$$\Delta I_2 = \begin{vmatrix} 3 & 2 & 0 \\ -1 & 11 & -1 \\ 0 & -10 & 6 \end{vmatrix}$$

$$\Delta I_2 = 3[(11 \times 6) - (-10 \times -1)] - 2[(-1 \times 6) - (0 \times -1)] \\ + 0[(-1 \times -10) - (0 \times 11)]$$

$$= 3[56] - 2[-6]$$

$$= 180$$

$$I_2 = \frac{\Delta I_2}{\Delta} = \frac{180}{45} = 4 \text{ A}$$

Finally to find I_3

$$\Delta I_3 = \begin{vmatrix} 3 & -1 & 2 \\ -1 & 3 & 11 \\ 0 & -1 & -10 \end{vmatrix}$$

$$\Delta I_3 = 3[(3 \times -10) - (-1 \times 11)] - (-1)[(-1 \times -10) - (0 \times 11)] \\ + 2[(-1 \times -1) - (0 \times 3)]$$

$$= 3[-19] + 1[10] + 2[1]$$

$$= -45$$

$$I_3 = \frac{\Delta I_3}{\Delta} = \frac{-45}{45} = -1 \text{ A}$$

To check that you have fully understood mesh analysis, you should try to solve this example using the substitution method and then solve the previous worked example using the matrix method. You should get the same answers.

Bridge measurements

There is a special type of dc circuit, known as the Wheatstone bridge, which is used to measure accurately the resistance of an unknown component. R_A and R_B in Figure 5.7 are fixed resistors of known value called standards resistances. R_C is a variable resistor the value of which may be adjusted in known steps and R_D is the unknown resistance. The value of R_D is found by adjusting R_C until the voltmeter reading is zero, at which point no current flows through the voltmeter, the voltages of nodes 1 and 2 are equal and the bridge is balanced.

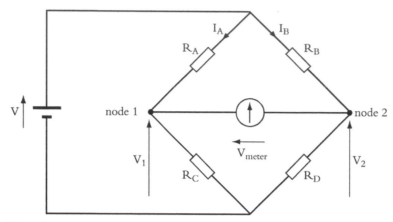

Figure 5.7

With $V_1 = V_2$

Voltage across $R_C = V_1 = I_A R_C$ and voltage across $R_D = V_1 = I_B R_D$

Voltage across $R_A = (V - V_1) = I_A R_A$ and voltage across $R_B = (V - V_1) = I_B R_B$

Dividing the equations

$$\frac{(V - V_1)}{V_1} = \frac{I_A R_A}{I_A R_C} = \frac{I_B R_B}{I_B R_D}$$

$$\frac{R_A}{R_C} = \frac{R_B}{R_D}$$

As R_A, R_B and R_C are all known, R_D can be calculated.

$$R_D = \frac{R_B R_C}{R_A}$$

(5.15)

Strain measurement

The Wheatstone bridge may be used with a strain gauge to measure the strain in a mechanical member, such as a steel girder. The most common form of strain gauge is the bonded foil type. This comprises a very thin layer of metal bent into a zigzag shape and bonded to the surface of a non-conducting tape. The strain gauge is fixed rigidly to the member but electrically isolated from it so that no current passes through the member. As force is applied to the member the strain increases and causes the resistance of the gauge to rise. The changes are small and are recorded using a modified Wheatstone bridge. The strain gauge replaces R_B in the circuit shown in Figure 5.7 and R_A, R_C and R_D are all set equal to the initial gauge resistance to ensure that the bridge is balanced before the member is mechanically loaded. When the load is applied and the resistance of the strain gauge changes, the voltmeter deflects. The strain, ε is calculated from the voltmeter reading using the equation derived below.

If R is the resistance of the unstrained gauge and the three other resistors and ΔR is the increase in the resistance of the strain gauge when the mechanical member is loaded, then

$$I_A = \frac{V}{2R} \quad \text{and} \quad I_B = \frac{V}{(2R + \Delta R)}$$

$$V_1 = I_A R = \frac{V}{2} \quad \text{and} \quad V_2 = I_B R = \frac{VR}{(2R + \Delta R)}$$

voltmeter reading,

$$v = V_1 - V_2$$

$$= \frac{V}{2} - \frac{VR}{(2R + \Delta R)}$$

$$= V\frac{(2R + \Delta R - 2R)}{2(2R + \Delta R)}$$

$$= V\frac{\Delta R}{2(2R + \Delta R)}$$

with $R \gg \Delta R$

voltmeter reading,

$$v = \frac{V}{4}\frac{\Delta R}{R}$$

$$\frac{\Delta R}{R} = \varepsilon G$$

where ε is the measured strain and G is the strain gauge factor provided the manufacturer.

Therefore, the voltmeter reading, $v = \frac{V}{4}\varepsilon G$

and the measured strain,

$$\varepsilon = \frac{4v}{VG}$$

(5.16)

Extreme care is needed when taking measurements. Change in strain gauge resistance is very small; for example, for the bonded foil type it is around 0.25 mΩ per microstrain. The resistance of gauges are also very sensitive to temperature change. To eliminate any errors due to changes in ambient conditions and to ensure that voltmeter deflections are due entirely to changes in strain in the mechanical member the resistor R_D may be replaced by a compensating strain gauge. This has identical characteristics to the first gauge and is fixed near the former but not on the loaded member, so that the second gauge is only subjected to changes in temperature. Temperature changes will cause the resistance of both gauges to change by the same amount, ensuring that the balance of the bridge is unaffected by changes in temperature.

Measurement accuracy can be further improved when one side of the mechanical member is in tension and the other in compression. By fixing one strain gauge on the side in tension and a matched device directly opposite on the side in compression, the resistance of the first gauge increases as the member is loaded while the resistance of the second gauge falls by an equal amount, thereby doubling the voltmeter reading. With both gauges subjected to the same changes in temperature, the balance of the bridge will remain independent of temperature.

5.3 Electromagnetic systems

The drive in many mechanical systems is provided by an electric motor. Motors are electromagnetic devices. To understand how they work it is first necessary to study the basic principles of electromagnetism.

Magnetic field around a current carrying conductor

A magnetic field is established around any current carrying conductor. This is demonstrated by pushing a conductor through the middle of a piece of paper covered in iron filings and turning on a direct current. The filings immediately form concentric circles around the wire, indicating the existence of a magnetic field, as illustrated in Figure 5.8(a).

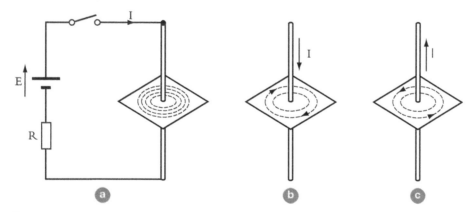

Figure 5.8

By convention it is assumed that a magnetic field circulates clockwise around a conductor carrying current into the page and anticlockwise around a conductor where the current flows out of the page. This is known as the right hand screw rule

Solenoid

If a current carrying conductor is wound into a cylindrical solenoid then the magnetic fields around each conductor merge to produce a field pattern similar to that associated with a bar magnet, as demonstrated in Figure 5.9.

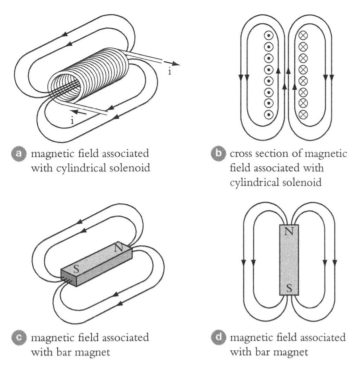

a magnetic field associated with cylindrical solenoid

b cross section of magnetic field associated with cylindrical solenoid

c magnetic field associated with bar magnet

d magnetic field associated with bar magnet

Figure 5.9

In the solenoid cross-section, (Figure 5.9(b)) the crosses indicate current flowing into the page and dots show current flowing out. The right hand screw rule is used to find the direction of the field through the centre of the solenoid, which in this case is upwards.

Toroid

The magnetic fields associated with both the single conductor and the solenoid are non-uniform and difficult to calculate. It is much easier to analyse a closed system like a toroid, a ring shaped core of magnetic or non-magnetic material wrapped with a conducting coil insulated from the core as in Figure 5.10. In this case when a current flows in the coil a uniform magnetic field is established that circulates around the ring core.

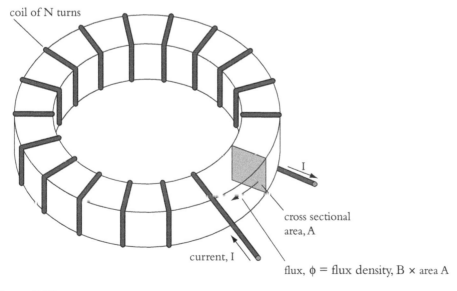

coil of N turns

current, I

I

cross sectional area, A

flux, φ = flux density, B × area A

Figure 5.10

Flux and flux density

The magnitude of the magnetic field is called the flux. This has the symbol Φ and the units webers (Wb).

In a uniform field, the flux is distributed evenly across the cross-section of the core and the flux density, B can be calculated from

$$B = \frac{\Phi}{A}$$

(5.17)

where B has the units T (tesla); (where 1T is defined as equal to $1\ \mathrm{Wb\ m^{-2}}$)

and A is the cross-sectional area of the core in $\mathrm{m^2}$.

Magnetomotive force (mmf)

The magnitude of the flux is proportional to the current, I multiplied by the number of turns in the coil, N. This multiple is called the magnetomotive force or mmf for short. It has the symbol F_{mmf} and has the units amperes (A). However, to recognize that its magnitude depends on both the current and the number of turns the units are often referred to as ampere-turns.

$$F_{mmf} = IN$$

(5.18)

The mmf is the 'driving force' in the magnetic circuit, and is analogous to the emf in the electric circuit.

Reluctance

It was shown in section 5.1 that in an electric circuit the current, I equals the emf, E divided by the circuit resistance, R. By analogy in the magnetic circuit the flux, Φ equals the mmf, F_{mmf} divided by the reluctance, S.

$$\Phi = \frac{F_{mmf}}{S}$$

(5.19)

$$S = \frac{F_{mmf}}{\Phi}$$

where Φ is in Wb

and S is in $\mathrm{H^{-1}}$ (Henries)$^{-1}$ (where 1 H is defined as equal to $1\ \mathrm{Wb\ A^{-1}}$).

Permeability

Continuing the analogy between electric circuits and magnetic systems it was shown in equation 5.4 that the resistance of an electric circuit, R is given by

$$R = \frac{\rho l}{A}$$
$$= \frac{l}{\sigma A}$$

where σ is the electrical conductivity. This is the reciprocal of the resistivity, ρ

The corresponding equation for the reluctance of a magnetic circuit is

$$S = \frac{l}{\mu A}$$

(5.20)

where l is the length of the magnetic path in m, (for the toroid, this is the mean circumference of the core), A is the cross-sectional area of the magnetic path in m^2, (i.e. cross-sectional area of the toroid core), and μ is the permeability of the core material in Wb $m^{-1} A^{-1}$ or Hm^{-1}.

The permeability of a vacuum or free space, μ_0 equals $4\pi \times 10^{-7}$ H m^{-1}. This value can also be used for all other non-magnetic materials. The permeabilities of magnetic materials, μ are found by multiplying μ_0 by the relative permeability, μ_r, which has no units.

$$\mu = \mu_0 \, \mu_r$$

μ_r is non-linear; its value depends on the flux density. Some examples are shown in Figure 5.11.

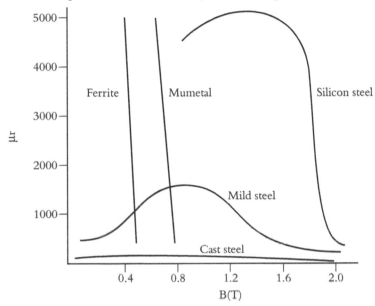

Figure 5.11

Magnetic field strength

If the reluctance of the core is uniform, then the effect of the magnetomotive force (mmf) will be distributed evenly throughout the length of the magnetic path and the magnetic field strength, H can be calculated from

$$H = \frac{F_{mmf}}{l} \tag{5.21}$$

H has the units A m^{-1}.

B versus H characteristic

From equations (5.17) and (5.19)

$$\Phi = BA = \frac{F_{mmf}}{S}$$

where

$$F_{mmf} = H \, l$$

$$S = \frac{l}{\mu A}$$

thus

$$BA = \frac{\mu A}{l} H l$$

$$B = \mu H \tag{5.22}$$

This relationship is extremely important and is used extensively in magnetic circuit analysis.

In circuits with non-magnetic materials B versus H is a straight line with the slope μ_0 but when the core is made of a magnetic material the relationship is non-linear. Figure 5.12 shows the B versus H characteristics for the materials shown in the previous graph. It is usual for manufacturers of magnetic materials to provide B versus H graphs for their products, leaving the user to calculate μ from the slopes of curves.

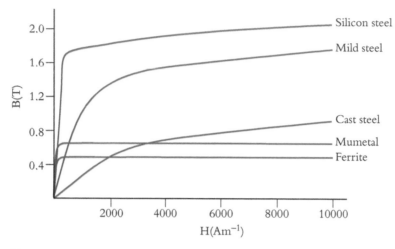

Figure 5.12

The following worked examples demonstrate how to analyse magnetic circuits.

A toroid with a wooden core, mean circumference 20 cm and cross-sectional area 2 cm^2 is wound with a coil of 1000 turns in which a current of 0.5 A flows. Calculate the mmf, magnetic field strength, reluctance, flux and flux density.

mmf,
$$F_{mmf} = IN = 0.5 \times 1000$$
$$= 500 \text{ A}$$

magnetic field strength,
$$H = \frac{F_{mmf}}{l}$$
$$= \frac{500}{20} \times 10^{-2}$$
$$= 2500 \text{ Am}^{-1}$$

reluctance,
$$S = \frac{l}{\mu_0 A}$$
$$= \frac{20 \times 10^{-2}}{4\pi \times 10^{-7}} \times 2 \times 10^{-4}$$
$$= 7.96 \times 10^8 \text{ H}^{-1}$$

flux,
$$\Phi = \frac{F_{mmf}}{S}$$
$$= \frac{500}{7.96} \times 10^8$$
$$= 62.8 \times 10^{-8} \text{ Wb}$$
$$= 0.628 \text{ } \mu\text{Wb}$$

flux density, $\qquad\qquad\qquad B = \dfrac{\Phi}{A}$

$$= \dfrac{62.8 \times 10^{-8}}{2 \times 10^{-4}}$$

$$= 3.14 \times 10^{-3}\,\text{T}$$

$$= 3.14\,\text{mT}$$

A toroid with a silicon steel core has identical dimensions to that in the previous worked example. The number of turns and current are also the same. Calculate the mmf, magnetic field strength, flux density, flux and reluctance.

mmf, $F_{mmf} = IN = 0.5 \times 1000 = 500\,\text{A}$ (as before)

magnetic field strength, $H = \dfrac{F_{mmf}}{l} = \dfrac{500}{20 \times 10^{-2}} = 2500\,\text{Am}^{-1}$ (as before)

flux density, B is found from the B versus H characteristic for silicon steel (see Figure 5.12)

$$B = 1.85\,\text{T}$$

flux, $\Phi = BA = 1.85 \times 2 \times 10^{-4} = 3.7 \times 10^{-4}\,\text{Wb} = 0.37\,\text{mWb}$

reluctance, $S = \dfrac{F_{mmf}}{\Phi} = \dfrac{500}{3.7 \times 10^{-4}} = 1.35 \times 10^{6}\,\text{H}^{-1}$

When the results for these two examples are compared, it is clear that for the same mmf and magnetic field strength, the reluctance of a toroid with a magnetic core is far lower but the flux and flux density are far higher.

Electromagnetic induction

Electromagnetic induction was discovered by Michael Faraday. He wound two coils on a toroid, connected one through a switch to a battery and the other he wired to a galvanometer, which measures small voltages.

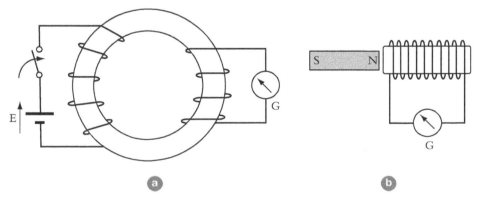

Figure 5.13

When he closed the switch, he observed that the galvanometer needle deflected immediately and then returned to zero, and when he opened the switch, the needle moved in the opposite direction before dropping back to zero.

In a second experiment Faraday moved a bar magnet towards a solenoid and noticed that the galvanometer needle deflected. When he moved the magnet away, the needle deflected in the opposite direction. Faraday also observed that the magnitude of the needle deflection depended on the speed at which he moved the magnet.

> From his experiments Faraday was able to deduce that the *magnitude of the emf, e induced in the coil* (as measured by the galvanometer) *was proportional to the rate of change of magnetic flux linkages* with the coil.

This phenomenon can be expressed mathematically

$$e = \frac{d\psi}{dt} \text{ volts}$$

where ψ is the magnetic flux linkages with the coil, and t is time.

Note that lower case symbols are used in this equation to indicate that values are instantaneous and continually changing.

Flux linkage, ψ has the units Wb.

Flux linkage is equal to the flux multiplied by the number of turns in the coil, N. Therefore

$$\psi = \phi N$$

where ϕ is the instantaneous flux.

Thus:

$$e = N \frac{d\phi}{dt}$$

(5.23)

> This equation is known as Faraday's Law.

Inductance

It was shown in equations (5.17), (5.21) and (5.22) that

$$\phi = BA$$

$$B = \mu H$$

$$H = \frac{F_{mmf}}{l} = \frac{IN}{l}$$

Using instantaneous values and substituting in equation (5.23)

$$e = N \frac{d}{dt}\left(\mu \frac{iN}{l} A\right)$$

$$= \left(N^2 \frac{\mu A}{l}\right)\frac{di}{dt}$$

but

$$S = \frac{l}{\mu A}$$

$$e = \frac{N^2}{S}\frac{di}{dt}$$

Let

$$L = \frac{N^2}{S}$$

(5.24)

where L is called the inductance and has the units henries (H).

$$e = L \frac{di}{dt}$$

(5.25)

Putting this equation into words:

> The emf induced in a coil is equal to its inductance multiplied by the rate of change of current flowing through it.

A wooden toroid with a mean circumference of 30 cm and cross-sectional area 4 cm² is wound with a coil of 2000 turns. A current flows in the coil. It starts at zero and increases steadily at a rate of 0.1 As⁻¹. Calculate the reluctance, inductance and induced emf.

$$\text{Reluctance, } S = \frac{l}{\mu_0 A} = \frac{30 \times 10^{-2}}{4\pi \times 10^{-7}} \times 4 \times 10^{-4} = 5.97 \times 10^8 \, \text{H}^{-1}$$

$$\text{Inductance, } L = \frac{N^2}{S} = \frac{2000^2}{5.97 \times 10^8} = 6.7 \times 10^{-3} \, \text{H} = 6.7 \, \text{mH}$$

$$\text{Induced emf, } e = L\frac{di}{dt} = 6.7 \times 10^{-3} \times 0.1 = 6.7 \times 10^{-4} \, \text{V} = 0.67 \, \text{mV}$$

For the toroid in the previous worked example, find the magnitudes of the mmf, flux and flux linkages after 5 s.

$$\text{Current after 5 s, } i_5 = 5 \times \frac{di}{dt} = 5 \times 0.1 = 0.5 \, \text{A}$$

$$\text{mmf after 5 s, } F_{mmf5} = i_5 N = 0.5 \times 2000 = 1000 \, \text{A}$$

$$\text{flux after 5 s, } \phi_5 = \frac{F_{mmf5}}{S} = \frac{1000}{5.97 \times 10^8} = 1.675 \times 10^{-6} \, \text{Wb} = 1.675\mu \, \text{Wb}$$

$$\text{Flux linkages after 5, } \psi_5 = \phi_5 N = 1.675 \times 10^{-6} \times 2000 = 3.35 \times 10^{-3} \, \text{Wb}$$

$$= 3.35 \, \text{mWb}$$

Coil with inductance and resistance

A real coil has both inductance and resistance. To examine how the resistance affects the response when an emf is applied, the two components have to be separated as shown in Figure 5.14(a).

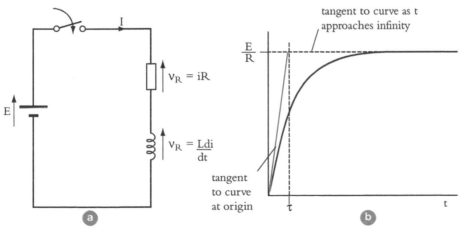

Figure 5.14

The moment the switch is closed a current starts to flow, producing a voltage across the inductance

$$v_L = L\frac{di}{dt}$$

and a voltage across the resistor

$$v_R = iR$$

Instantaneous values are used in these equations as the current and voltages are constantly changing.

By Kirchhoff's Voltage Law

$$E = v_R + v_L$$

$$= iR + L\frac{di}{dt}$$

$$E - iR = L\frac{di}{dt}$$

$$\frac{E}{R} - i = \frac{L}{R}\frac{di}{dt} \tag{5.26}$$

Eventually everything will settle down and the current will reach a constant steady state value. When this is achieved the rate of change of current $\frac{di}{dt}$ will be zero and the current will equal $\frac{E}{R}$. If we call this the steady state current, I then:

$$\frac{E}{R} = I$$

Substituting for I in equation (5.26)

$$I - i = \frac{L}{R}\frac{di}{dt}$$

$$\frac{R}{L}dt = \frac{di}{I - i}$$

Integrating both sides

$$\frac{R}{L}t = -\ln(I - i) + K \tag{5.27}$$

where K is the constant of integration.

When the switch is closed at $t = 0$ the current, $i = 0$.

$$0 = -\ln I + K$$

$$K = \ln I$$

Substituting for K in equation (5.27)

$$\frac{Rt}{L} = -\ln(I - i) + \ln I$$

$$= -\ln\frac{(I - i)}{I}$$

$$e^{\frac{-Rt}{L}} = \frac{I - i}{I}$$

$$Ie^{\frac{-Rt}{L}} = I - i$$

$$i = I\left(1 - e^{\frac{-Rt}{L}}\right)$$

Substituting for $I = \dfrac{E}{R}$

$$i = \frac{E}{R}\left(1 - e^{\frac{-Rt}{L}}\right)$$ (5.28)

Figure 5.14(b) shows the graph of current versus time. The current rises exponentially towards the asymptote $\dfrac{E}{R}$.

To find what determines the speed at which the current rises draw the tangent to the current at the origin. This is the dashed line in Figure 5.14(b). The time at which this line intersects the horizontal line $i = \dfrac{E}{R}$ is called the time constant, τ.

From the graph the slope of tangent $= \dfrac{E}{R} \cdot \dfrac{1}{\tau}$

At $t = 0$, $i = 0$ and there is no volt drop across the resistance, R, so $E = L\dfrac{di}{dt}$

$$\frac{di}{dt} = \frac{E}{L}$$

but $\dfrac{di}{dt}$ is the slope of the graph, therefore

$$\frac{E}{L} = \frac{E}{R\tau}$$

and the time constant,

$$\tau = \frac{L}{R} \text{ s}$$ (5.29)

It is this ratio $\dfrac{L}{R}$ that defines the speed at which the current rises. The larger the ratio of inductance to resistance the slower the current will rise.

In the circuit in Figure 5.14(a), $E = 10$ V, $L = 1$ H and $R = 5\ \Omega$. Calculate the time constant of the circuit, τ and hence the find the current when the time equals τ, 2τ, 3τ, 5τ and 8τ.

Time constant, $\tau = \dfrac{L}{R}$

$$= \frac{1}{5} = 0.2 \text{ s}$$

$$i = \frac{E}{R}\left(1 - e^{\frac{-Rt}{L}}\right) = \frac{E}{R}\left(1 - e^{\frac{-t}{\tau}}\right)$$

when $\quad t = \tau = 0.2$ s

$$i = \frac{10}{5}(1 - e^{-1}) = 2(1 - 0.368)$$

$$= 1.264 \text{ A}$$

Repeating for $t = 2\tau, 3\tau, 5\tau$ and 8τ.

	t (s)	i (A)
2τ	0.4	1.729
3τ	0.6	1.900
5τ	1.0	1.987
8τ	1.6	1.999

Table 5.2

Energy stored in an inductor

When current flows through an inductor, the instantaneous power supplied to the inductor is given by:

instantaneous power = instantaneous voltage across the inductor × instantaneous current

The energy supplied to the inductor in time $dt = p\, dt = L\dfrac{di}{dt}i\, dt$

$$= Li\, di$$

This means that the total energy supplied to the inductor is independent of time and depends only on the current flowing through it.

If the inductor current is I, then the

$$\text{total energy supplied} = \int_0^I Li\, di$$

$$= \tfrac{1}{2}LI^2 \text{ joules}$$

(5.30)

This energy is stored in the inductor. As the stored energy is a function of current through the inductor, energy is only stored while current flows. When the current starts to fall, the rate of change of current, $\dfrac{di}{dt}$ and hence the induced emf, e become negative returning energy to the supply. All the energy is returned when the current reaches zero. No energy is dissipated by an inductor, unlike resistors which dissipate energy.

Force on a current-carrying conductor

You may have tried to push together two bar magnets with like poles opposing but are prevented by the force created by the distortion of the magnetic fields.

The interaction of magnetic fields always produces force. Another example occurs when a current-carrying conductor is brought in the vicinity of another magnetic field. The field around the conductor interacts with the other field exerting a force on the conductor.

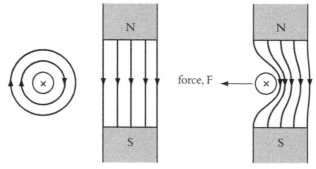

Figure 5.15

To find the magnitude of this force it is necessary to consider the work done to move the conductor.

Figure 5.16 shows a conductor carrying a current, I in a magnetic field with a flux density, B. The depth of the field is l and the distance moved by the conductor is dx.

Mechanical work done to move the conductor	=	force on conductor	×	distance moved by the conductor

$$= F\, dx$$

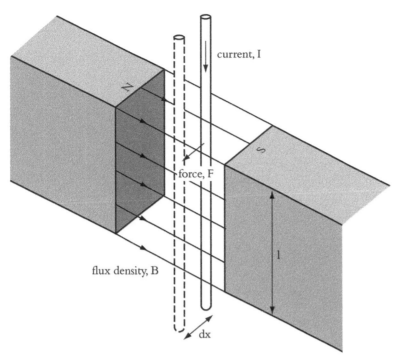

Figure 5.16

The movement of the conductor in a magnetic field induces an emf in the conductor in accordance with Faraday's Law which opposes the current flow.

$$e = N\frac{d\phi}{dt} = \frac{d\psi}{dt}$$

with only one conductor, $N = 1$ and $d\phi = d\psi$

The flux linkages produced in moving the conductor a distance dx through a magnetic field with a flux density, B and depth l is

$$d\psi = Bldx = d\phi$$

This induces an emf in the conductor, $e = Bl\dfrac{dx}{dt}$

To maintain the current an external voltage source equal to but opposing e must be applied to the ends of the conductor.

In the time, dt, that it takes to move the conductor dx, the electrical energy supplied by the external source is $e\,I\,dt$

This energy is equal to the mechanical work done in moving the conductor, thus:

Mechanical work done = electrical energy supplied

$$F\,dx = e\,I\,dt$$

$$F\,dx = Bl\frac{dx}{dt}It$$

$$F = Bl\,I$$

(5.31)

where F is measured in newtons (N).

The electrical energy supplied is converted to mechanical energy to make the conductor move. This form of energy conversion is exploited in the design of electrical motors, described in section 5.10.

Fleming's Left-hand Rule

To determine the direction of the force on the conductor, we use Fleming's Left-hand Rule. Point the thumb, first finger and second finger of the left hand at right angles to each other as shown in Figure 5.17. Align the **first** finger with the **field** and the **second** finger with the **current**. The thumb shows the direction of the force and hence the direction of **motion**.

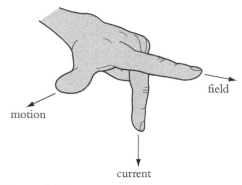

field

motion

current

Figure 5.17

Magnetic circuits

Earlier in this section we calculated the flux in a homogeneous toroid. Unfortunately, for most practical applications the magnetic path is not uniform. The cross-sectional area may vary along the magnetic path or the magnetic flux may travel through media of different permeability. Figure 5.18(a) shows the core of a two-pole dc motor.

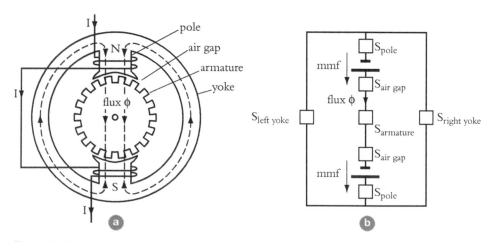

Figure 5.18

The core has two parts, a stationary outer section and a rotating armature. The outer section comprises a closed ring called the yoke and two poles which are wrapped with current-carrying coils to form north (N) and south (S) poles. When current flows in the coils, magnetic flux circulates around the magnetic path. It leaves the N-pole, crosses an air gap, passes through the armature, crosses a second air gap and enters the S-pole. When the flux reaches the yoke, it divides; half travels up the left side of the yoke; the remainder returns via the right side to the N-pole, where the two halves of the flux recombine.

Electrical and electronic systems

The analysis of a system such as this has to take into account both the geometry and the permeabilities of the elements. Using the analogy between electric and magnetic circuits engineers are able to represent magnetic circuits with equivalent electric networks and to modify formulae relating to electric circuits to deal with magnetic systems. Figure 5.18(b) shows the electrical equivalent circuit for the two-pole dc machine. The voltage sources represent the magnetomotive forces (mmfs), the currents substitute for the fluxes, the resistors stand for the reluctances and Kirchhoff's Voltage and Current Laws are modified to analyse the magnetic circuit. To demonstrate this, two examples will be used. These are simpler than the dc machine. In the first example we will consider the case a toroid with a radial air gap, where the flux has to travel through two different media in series. In this case we will use a modified form of Kirchhoff's Voltage Law.

In the second example we will examine the case where the flux is divided into two and travels along parallel paths. Analysing this situation requires modifying Kirchhoff's Current Law.

A toroid with a rectangular cross-section is manufactured of silicon steel with a mean diameter of 10 cm, radial thickness 3 cm, depth 2 cm and a radial cut 1 mm wide as illustrated in Figure 5.19. The toroid is wrapped with a coil of 1000 turns. Calculate the current required to maintain a steady state flux of 1.2 mWb in the core.

In Section 5.2 on dc circuits Kirchhoff's Voltage Law was given as: 'the algebraic sum of voltages around a closed loop is zero' or as written mathematically in equation (5.10)

$$\Sigma V = 0$$

Using the analogy between electrical and magnetic systems this law may be rewritten for magnetic circuits to say that: 'the algebraic sum of mmfs around a closed loop is zero' or

$$\Sigma F_{mmf} = 0 \qquad (5.32)$$

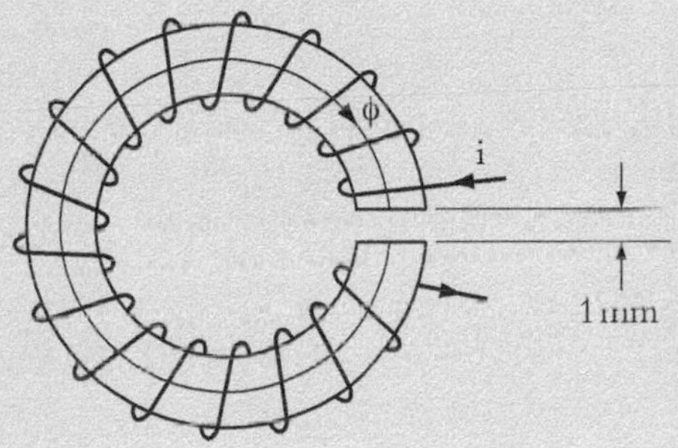

Figure 5.19

> The 'driving' mmf in a circuit, IN must equal the sum of the mmf drops, ΣHl around the network

where H is the magnetic field strength in each component in the system, and l is the length of that component.

In this case: 'driving' mmf, $IN = H_{air} l_{air} + H_{steel} l_{steel}$
considering first the air gap: $l_{air} = 1 \times 10^{-3}$ metres
assuming that the flux crosses the air gap uniformly, the flux density in the air gap will be the same as that in the steel, i.e.

$$B = \frac{\Phi}{A} = \frac{1.2 \times 10^{-3}}{3 \times 10^{-2} \times 2 \times 10^{-2}} = 2.0\text{T}$$

for air, $B = \mu_0 H_{air}$

$$H_{air} = \frac{2}{4\pi \times 10^{-7}} = 1.592 \times 10^6 \, \text{Am}^{-1}$$

$$H_{air} l_{air} = 1.592 \times 10^6 \times 1 \times 10^{-3}$$

$$= 1592 \, \text{A}$$

Considering now the steel:

$$l_{steel} = \pi D - l_{air}$$

$$= 0.1\pi - (1 \times 10^{-3}) = 0.3132\,\text{m}$$

The flux density in the steel is the same as that in the air gap, 2.0 T and H_{steel} is found from Figure 5.12 to equal 7200 Am^{-1}

$$H_{steel}\,l_{steel} = 7200 \times 0.3132$$

$$= 2255\,\text{A}$$

$$IN = H_{air}\,l_{air} + H_{steel}\,l_{steel}$$

$$IN = 1592 + 2255$$

$$= 3847$$

$$I = \frac{3847}{N} = \frac{3847}{1000}$$

$$I = 3.85\,\text{A}$$

A 1000-turn coil is wound around the centre limb of a symmetrical eight-shaped silicon steel core shown in Figure 5.20. The B–H characteristic for the steel is in Figure 5.12. The length of the centre limb is 4 cm and its cross-sectional area is 4 cm^2. The outer limbs both have a length of 10 cm and the cross-sectional area of 2.1 cm^2. Calculate the current needed to establish a flux of 0.8 mWb in the centre limb.

Consider first the centre limb:

Flux, $\phi_c = 0.8\,\text{mWb} = 8 \times 10^{-4}\,\text{Wb}$

Flux density, $B_c = \dfrac{\phi_c}{A_c} = \dfrac{8 \times 10^{-4}}{4 \times 10^{-4}} = 2.0\text{T}$

from B–H characteristic (Figure 5.12) magnetic field strength, $H_c = 7200\,\text{Am}^{-1}$

mmf drop in centre limb = $H_c\,l_c = 7200 \times 4 \times 10^{-2} = 288\,\text{A}$

At the end of the centre limb the flux divides. At this point we have to apply the modified form of Kirchhoff's Current Law for magnetic circuits: 'the sum of fluxes entering a node or junction equals the sum of the fluxes leaving that node',

$$\Sigma\phi = 0 \tag{5.33}$$

$$\phi_c = \phi_{ol} + \phi_{or}$$

by symmetry $\phi_{ol} = \phi_{or} = \phi_o$

$$\phi_c = 2\,\phi_o$$

$$\phi_o = 0.4\,\text{mWb}$$

flux density in outer limbs, $B_o = \dfrac{\phi_o}{A_o} = \dfrac{4 \times 10^{-4}}{2.1 \times 10^{-4}} = 1.9\text{T}$

from B-H characteristic (Figure 5.12) magnetic field strength, $H_o = 3660\,\text{Am}^{-1}$

mmf drop in the outer limbs = $H_o\,l_o = 3660 \times 10 \times 10^{-2} = 366\,\text{A}$

Applying Kirchhoff's modified voltage law, equation (5.32)

'driving' mmf, $IN = \Sigma$ mmf drops around a closed loop

'driving' mmf, $iN = H_c\, l_c + H_o\, l_o$

$$IN = 288 + 366$$

$$= 654\,\text{A}$$

$$I = \frac{654}{N} = \frac{654}{1000}$$

$$I = 0.654\,\text{A}$$

Learning summary

You have completed the section on electromagnetic systems and now should:

✔ be familiar with the constituents of a magnetic circuit;

✔ understand the concepts of flux, flux density, magnetomotive force (mmf), reluctance, magnetic field strength and permeability;

✔ be able to analyse magnetic field in toroids with both magnetic and non-magnetic cores;

✔ understand the principles of electromagnetic induction;

✔ understand Faraday's Law;

✔ calculate the current in a direct current circuit with inductance and resistance;

✔ compute the energy stored in an inductor;

✔ analyse the force on a current carrying conductor in a magnetic field;

✔ analyse systems with magnetic and non-magnetic elements in series;

✔ calculate magnetic fields in systems with parallel paths.

5.4 Capacitance

There are only three types of electrical component that are passive; that is, devices whose operation is independent of the polarity of the applied voltage. Resistors and inductors are both passive devices. The third group are capacitors. In its simplest form the capacitor comprises two metal plates separated by a thin layer of insulation which is used to store electric charge for short periods.

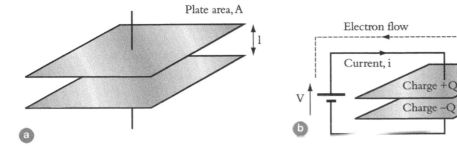

Plate area, A

Electron flow

Current, i

Charge +Q

Charge −Q

ⓐ ⓑ

Figure 5.21

When a battery is connected across the capacitor negatively charged electrons flow from the plate connected to the positive terminal of the battery, through the battery to the other capacitor plate leaving the first plate positively charged and making the second negatively charged. This creates a potential difference across the capacitor equal to the applied voltage.

> The movement of electrons produces a current, which by convention flows in the opposite direction to the electrons (as explained in Section 5.2). This current, i equals the rate at which charge, q moves through the circuit.

Put mathematically, at any instant:

$$i = \frac{dq}{dt}$$

(5.34)

The total charge stored on the plates, Q depends on the magnitude of the applied voltage V, the cross sectional area of the capacitor plates A, the distance between the plates l, and a characteristic of the insulating medium, called the permittivity, ε such that:

$$Q = \frac{A\varepsilon}{l} V$$

Charge is measured in coulombs (C).

A, ε and l are all related to the construction of the capacitor and are combined into a constant called the capacitance, C, where

$$C = \frac{A\varepsilon}{l}$$

(5.35)

Capacitance is measured in farads (F).

The permittivity of a vacuum or free space, ε_0 equals $8.85 \times 10^{-12}\,\text{Fm}^{-1}$. For all other insulating materials, or dielectrics, as they are often known, the permittivity is a multiple of ε_0. This multiple is called the relative permittivity ε_r.

$$\varepsilon = \varepsilon_0\,\varepsilon_r$$

(5.36)

Normally ε_r is quoted in data sheets and ε has to be calculated from the formula above. Values of ε_r for some common insulating materials are given in Table 5.3:

Substituting for ε in equation (5.35)

$$C = \frac{A\varepsilon_0\,\varepsilon_r}{l}$$

(5.37)

and

$$Q = CV$$

(5.38)

	ε_r
Air	1.0
Paper	2.0 to 3.7
Glass	3.8 to 10.0
Mica	3.0 to 7.0
Polythene	2.3
Insulating oil	3.5
PVC	3.0 to 4.0
Ceramics	6 to 1000

Table 5.3

The capacitor shown in Figure 5.21(a) has a relatively low capacitance. Capacitance is increased by enlarging the area of the plates, reducing the separation between the plates, using a dielectric with a high relative permittivity or any combination of these. Figure 5.22 shows examples of how the area of the plates can be increased. The capacitor in Figure 5.22(a) has several plates connected in parallel. In Figure 5.22(b) an increase in area is achieved by rolling two thin strips of metal foil sandwiched between layers of insulating material into a spiral.

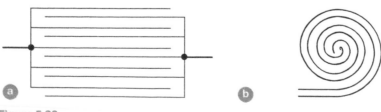

Figure 5.22

Capacitors connected in parallel

From equation 5.38 it can be seen that the charge, Q stored in a capacitor equals the capacitance, C multiplied by the voltage across the capacitor, V so when two capacitors C_1 and C_2 are connected in parallel with the supply voltage, V_s the charge stored on each will be respectively:

$$Q_1 = C_1 V_s$$

and

$$Q_2 = C_2 V_s$$

The total charge stored in the capacitors,

$$Q = Q_1 + Q_2$$
$$= C_1 V_s + C_2 V_s$$
$$= (C_1 + C_2) V_s$$

If C is the equivalent capacitance of the circuit, then

$$Q = C V_s$$
$$C V_s = (C_1 + C_2) V_s$$
$$C = C_1 + C_2$$

The total capacitance of the circuit equals the sum of the individual capacitances. Generally in a circuit with n capacitors in parallel the total capacitance, C is given by:

$$C = C_1 + C_2 + C_3 + ,..., + C_n = \Sigma C_n \qquad (5.39)$$

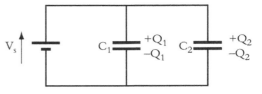

Figure 5.23

2.5 nF, 1 nF and 4 nF capacitors are connected in parallel with a 100 V battery. Calculate the equivalent capacitance of the circuit and the charge on the capacitors.

$$C_1 = 2.5 \, nF \quad C_2 = 1 \, nF \quad C_3 = 4 \, nF$$
$$C = (2.5 + 1 + 4) \times 10^{-9} = 7.5 \times 10^{-9} = 7.5 \, nF$$

Charge on C_1 $\quad Q_1 = C_1 V_s = 2.5 \times 10^{-9} \times 100 = 0.25 \, \mu C$

Charge on C_2 $\quad Q_2 = C_2 V_s = 1 \times 10^{-9} \times 100 = 0.1 \, \mu C$

Charge on C_3 $\quad Q_3 = C_3 V_s = 4 \times 10^{-9} \times 100 = 0.4 \, \mu C$

Adding the charges on the individual capacitors the total charge equals 0.75 μC.

Alternatively, the total charge on the capacitors can be found from $Q = C V_s$

$$Q = 7.5 \times 10^{-9} \times 100 = 0.75 \, \mu C$$

Capacitors connected in series

When two capacitors C_1 and C_2 are connected in series with the supply voltage, V_s as shown in Figure 5.24, the voltages across the capacitors will be V_1 and V_2 respectively. To satisfy Kirchhoff's Voltage Law (equation (5.10))

$$V_s = V_1 + V_2$$

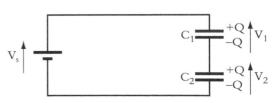

Figure 5.24

Assuming that the charge on the upper plate of C_1 is $+Q$, then the charge on the lower plate will be $-Q$. The lower plate is connected directly to the upper plate of C_2. As these plates are electrically isolated from the rest of the circuit electronics will flow from the upper plate of C_2 leaving a charge of $+Q$ on the upper plate of C_2. This is equal and opposite to the charge on the lower plate of C_1. The $+Q$ charge on the upper plate of C_2 will in turn induce a charge of $-Q$ on the lower plate of C_2.

$$Q = C_1 V_1 \quad \text{and} \quad Q = C_2 V_2$$

$$V_1 = \frac{Q}{C_1} \quad \text{and} \quad V_2 = \frac{Q}{C_2}$$

If C is the equivalent capacitance of the circuit, then

$$Q = CV_s$$

$$V_s = \frac{Q}{C}$$

But

$$V_s = V_1 + V_2$$

$$\frac{Q}{C} = \frac{Q}{C_1} + \frac{Q}{C_2}$$

$$\frac{1}{C} = \frac{1}{C_1} + \frac{1}{C_2}$$

Generally for a circuit with n capacitors in series the equivalent capacitance can be found from:

$$\frac{1}{C} = \frac{1}{C_1} + \frac{1}{C_2} + \frac{1}{C_3} + \ldots\ldots + \frac{1}{C_n} = \sum \frac{1}{C_n} \tag{5.40}$$

2.5 nF, 1 nF and 4 nF capacitors are connected in series with a 100 V battery. Calculate the equivalent capacitance of the circuit and the charge on and the voltage across each capacitor.

$$\frac{1}{C} = \frac{1}{C_1} + \frac{1}{C_2} + \frac{1}{C_3}$$

$$= \frac{10^9}{2.5} + \frac{10^9}{1} + \frac{10^9}{4}$$

$$= (0.4 + 1 + 0.25) \times 10^9 = 1.65 \times 10^9$$

$$C = 0.606 \times 10^9 \, \text{F} = 0.606 \, \text{nF}$$

Charge on each capacitor, $Q = CV_s = C_1 V_1 = C_2 V_2 = C_3 V_3$

Using $Q = CV_s$

$$Q = 0.606 \times 10^9 \times 100 = 0.0606 \, \mu\text{C}$$

$$V_1 = \frac{Q}{C_1} = \frac{0.0606 \times 10^{-6}}{2.5 \times 10^{-9}} = 24.2 \, \text{V}$$

$$V_2 = \frac{Q}{C_2} = \frac{0.0606 \times 10^{-6}}{1 \times 10^{-9}} = 60.6 \, \text{V}$$

$$V_3 = \frac{Q}{C_3} = \frac{0.0606 \times 10^{-6}}{4 \times 10^{-9}} = 15.2 \, \text{V}$$

Series/parallel combination of capacitors

Capacitors may be connected in configurations which involve both parallel and series combinations. In these cases it is necessary to break the circuit into sections and to analyse each part separately as shown in the following example.

Capacitors C_2 and C_3 are connected in parallel and the combination connected in series with a capacitor C_1, as shown in Figure 5.25. Given that C_1 equals 2.5 nF, C_2 is 1 nF and C_3 is 4 nF, calculate the equivalent circuit capacitance.

A 100 V dc supply is connected across the network. Calculate the charge on each capacitor.

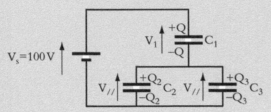

Figure 5.25

Let $C_{//}$ be the equivalent of C_2 and C_3 capacitors in parallel

$$C_{//} = C_2 + C_3 = (1 + 4) \times 10^{-9} = 5 \times 10^{-9} = 5\,\text{nF}$$

Let C_{eq} = circuit capacitance (i.e. the equivalent of $C_{//}$ in series with C_1)

$$\frac{1}{C_{eq}} = \frac{1}{C_{//}} + \frac{1}{C_1} = \frac{1}{5 \times 10^{-9}} + \frac{1}{2.5 \times 10^{-9}}$$

$$= (0.2 + 0.4) \times 10^9 = 0.6 \times 10^9$$

$$C_{eq} = 1.667 \times 10^{-9} = 1.667\,\text{nF}$$

$$Q = C_{eq}\,V_s = 1.667 \times 10^{-9} \times 100 = 0.1667\,\mu\text{C}$$

This is the charge on C_1, therefore:

$$Q = C_1\,V_1$$

$$V_1 = \frac{0.1667 \times 10^{-6}}{2.5 \times 10^{-9}} = 66.67\,\text{V}$$

By Kirchhoff's Voltage Law $V_s = V_1 + V_{//}$

$$V_{//} = 100 - 66.67 = 33.33\,\text{V}$$

Charge on C_2 $\qquad Q_2 = C_2\,V_{//} = 1 \times 10^{-9} \times 33.33 = 0.0333\,\mu\text{C}$

Charge on C_3 $\qquad Q_3 = C_3\,V_{//} = 4 \times 10^{-9} \times 33.33 = 0.1333\,\mu\text{C}$

Check $\qquad Q = Q_2 + Q_3$

$$0.1667 \times 10^{-6} = 0.0333 \times 10^{-6} + 0.1333 \times 10^{-6}$$

Capacitor in series with a resistance

To see what happens when a resistor is connected in series with a capacitor consider Figure 5.26(a). When the switch is closed, current starts to flow, producing voltages across both capacitor C and resistor R.

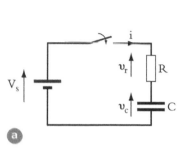

Figure 5.26

From equation (5.34) instantaneous current,

$$i = \frac{dq}{dt}$$

and from equation (5.38)

$$q = Cv_C$$

$$i = C\frac{dv_c}{dt} \qquad (5.41)$$

Lower-case symbols are used to indicate parameters which are functions of time, the magnitudes of which are continually varying.

Voltage across resistor,

$$v_R = iR$$

$$= RC\frac{dv_c}{dt}$$

By Kirchhoff's Voltage Law

$$V_s = v_R + v_C$$

$$V_s = RC\frac{dv_c}{dt} + v_C$$

$$V_s - v_C = RC\frac{dv_c}{dt}$$

$$\frac{dt}{RC} = \frac{dv_c}{(V_s - v_C)}$$

Integrating both sides

$$\frac{t}{RC} = -\ln(V_s - v_C) + K$$

where K is the constant of integration.

Assuming that the capacitor is initially uncharged, i.e. when $t = 0$, $v_C = 0$

$$0 = -\ln V_s + K$$

$$K = \ln V_s$$

Thus $\dfrac{t}{RC} = -ln\dfrac{(V_s - v_C)}{V_s}$

$$e^{\frac{-t}{RC}} = \frac{V_s - v_C}{V_s}$$

$$V_s e^{\frac{-t}{RC}} = V_s - v_C$$

$$v_C = V_s\left(1 - e^{\frac{-t}{RC}}\right)$$

(5.42)

v_C rises exponentially towards the asymptote V_s.

By differentiating equation (5.42) we can find the slope of the graph at any point in time

$$\frac{dv_c}{dt} = \frac{V_s}{RC} e^{\frac{-t}{RC}}$$

To find the slope at the origin, we put $t = 0$

$$\frac{dv_c}{dt}\bigg|_{t=0} = \frac{V_s}{RC}$$

but from Figure 5.26(b) the slope of the graph at the origin $= \dfrac{V_s}{\tau}$

so

$$\frac{V_s}{\tau} = \frac{V_s}{RC}$$

$$\tau = RC \text{ seconds}$$

(5.43)

where τ is called the time constant of the circuit.

The time constant determines the speed at which v_C rises. As the charge on the capacitor, q equals Cv_C, then the time constant also determines the rate at which the capacitor is charged. The larger the value of RC, the slower the rate of rise of v_C and charge.

Energy stored in a capacitor

We saw above that when the capacitor in Figure 5.26 is being charged, the instantaneous current i is given by equation (5.41):

$$i = C\frac{dv_c}{dt}$$

where v_c is the instantaneous voltage across the capacitor.

Let p be the instantaneous power supplied to the capacitor, thus
p = instantaneous current $\times$ instantaneous capacitor voltage

$$p = iv_c$$

$$p = C\frac{dv_c}{dt}v_c$$

Let energy supplied to the capacitor in time, $dt = dW$

$$dW = pdt$$

$$= v_c C\frac{dv_c}{dt}dt$$

$$= v_c C\, dv_c$$

The total energy supplied in charging the capacitor to $Vs = W$

$$W_s = \int_0^{V_s} Cv_c \,.\, dv_c$$

is

$$W = \tfrac{1}{2}CV_s^2 \text{ joules (J)}$$

(5.44)

This energy is stored in the capacitor. The total energy depends entirely on the capacitance and the supply voltage and is independent of the time taken to charge the capacitor. If the supply voltage is reduced, then energy is returned to the supply. No energy is dissipated in the capacitor.

A capacitor comprises two strips of aluminium 30×15 cm separated by a strip of mica 2 mm thick. The relative permittivity of mica is 6.0. Calculate the capacitance.

A voltage of 250 V is applied across the capacitor. Calculate the stored energy.

$$C = \frac{\varepsilon_0 \varepsilon_r A}{l} = \frac{8.85 \times 10^{-12} \times 6 \times 30 \times 10^{-2} \times 15 \times 10^{-2}}{2 \times 10^{-3}}$$

$$= 1195 \times 10^{-12} \, F = 1195 \, pF$$

Energy stored $= \frac{1}{2}CV_s^2 = \frac{1}{2} \times 1195 \times 10^{-12} \times (250)^2$

$$= 37.3 \, \mu J$$

Learning summary

You have completed the section on capacitance and you should now:

✔ understand the concepts of charge, permittivity and capacitance

✔ be able to calculate capacitance

✔ be able to compute the equivalent capacitance of capacitors in parallel, series and combinations thereof

✔ calculate the voltage across a capacitor in a direct current circuit with capacitance and resistance

✔ understand the concept of the time constant of a circuit

✔ compute the stored energy in a capacitor.

5.5 Alternating current circuits

In the previous sections we examined circuits with direct voltage supplies but most electricity companies supply their customers with alternating voltage. It is therefore essential for engineers to understand how such systems work and to be able to analyse them.

Single-phase alternating current generators

It was explained in Section 5.3 that when a conductor is moved through a magnetic field, an emf is induced in the conductor. Mechanical energy is used to produce electrical energy. The magnitude of the emf depends on the rate of change of flux linkages. This is Faraday's Law which we saw in equation (5.23) could be written mathematically as:

$$e = N\frac{d\phi}{dt} = \frac{d\psi}{dt}$$

To determine the direction of the induced emf we use Fleming's Right-hand Rule. The thumb represents the direction of **motion**, the first finger aligns with the **field** and the second finger shows the direction of the **emf**.

Electrical and electronic systems

It is important to remember when to use Fleming's Right-hand Rule and when to use his Left-hand Rule. The Right-hand Rule is used in cases like this when mechanical energy is being converted to electrical energy. The Left-hand Rule is employed when electrical energy is being turned into mechanical energy, as we had in Section 5.3 when we calculated the force on a current carrying conductor.

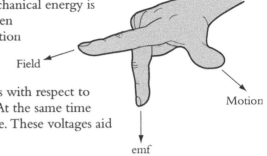

Figure 5.27

Figure 5.28 shows a coil rotating within a magnetic field. Rotating the coil anticlockwise, conductor 1, the left arm of the coil, moves downwards with respect to the field. This induces an emf, e in the conductor which is into the page. At the same time conductor 2, on the right, travels upwards inducing an emf out of the page. These voltages aid (i.e. add) to make a coil voltage,

$$e_{coil} = 2e$$

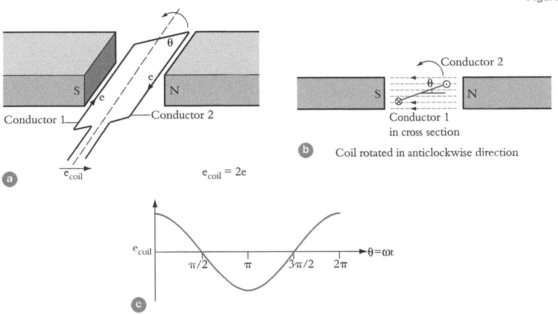

Figure 5.28

As the coil rotates, the magnitude of the flux linkage varies sinusoidally. The maximum flux linkage ψ_m occurs when the coil is perpendicular to the field ($\theta = 90°$) and there is zero flux linkage when the coil is parallel to the field ($\theta = 0°$). Generally

Flux linkages, $\psi = \psi_m \sin \theta$

where θ is the angle of rotation

If the coil rotates at a speed of n revolutions per second it will have an angular velocity, $\omega = 2\pi n$ rad s^{-1} and will move through (ωt) radians in t seconds. As $\omega t = \theta$ the flux linkages at time t will be given by:

$$\psi = \psi_m \sin \omega t$$

The emf induced in the coil can then be found by substituting for ψ in Faraday's Law (equation (5.23))

$$e = \frac{d\psi}{dt} = N\frac{d\phi}{dt}$$

As the coil has two conductors, $N = 2$, the induced emf will be given by:

$$e_{coil} = 2\frac{d}{dt}(\phi_m \sin \omega t)$$

$$e_{coil} = 2\omega\phi_m \cos \omega t \tag{5.45}$$

and the maximum emf, $e_{coil\,m} = 2\omega\phi_m$

The maximum emf is induced when $\omega t = 0°$; and the plane of the coil is parallel to the magnetic field.

It is difficult to construct a coil like that shown in Figure 5.28(a) which is sufficiently robust to rotate at high speeds. A far simpler system for producing sinusoidal voltages is to rotate a magnet within a stationary coil as shown in Figure 5.29(a). The maximum voltage occurs when the magnet is parallel to the coil plane.

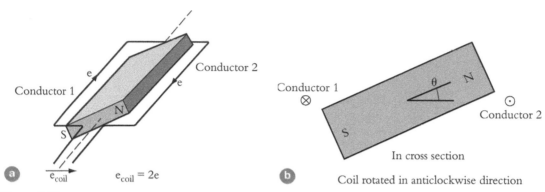

Figure 5.29

This arrangement is a rudimentary form of single-phase alternating current (ac) generator. The instantaneous voltage generated, v is given by:

$$v = V_m \sin (\omega t)$$

where V_m is the maximum (or peak) voltage, measured in volts (V), ω is the angular velocity in rads s^{-1} and t is the time in seconds (s).

But the angular velocity, $\omega = 2\pi n$

where n = revolutions per second, giving:

$$v = V_m \sin (2\pi n t)$$

As every revolution of the magnet induces one complete sine wave of voltage, known as a cycle and the number of cycles generated per second equals the frequency, f

$$f = n$$

the instantaneous voltage, v can be expressed by:

instantaneous voltage, $v = V_m \sin (2\pi f t)$

where V_m is the maximum (or peak) voltage, measured in volts (V), f is the supply frequency in hertz (Hz) and t is the time in seconds (s).

Resistance connected to an alternating voltage supply

When a resistor is connected to the alternating voltage supply a current flows.

Using lower case letters to indicate instantaneous values the instantaneous current, i can be calculated from Ohm's Law, equation (5.1).

$$i = \frac{v}{R}$$

but

$$v = V_m \sin (2\pi f t)$$

therefore

$$i = \frac{V_m}{R} \sin (2\pi f t)$$

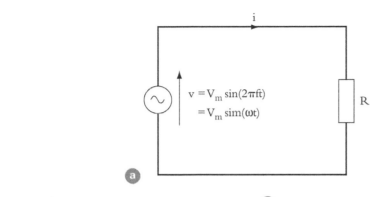

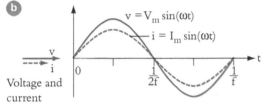

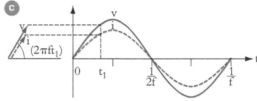

Voltage and
current

Figure 5.30

letting

$$I_m = \frac{V_m}{R}$$

$$i = I_m \sin (2\pi ft)$$

The current and voltage waveforms are shown in Figure 5.30(b).

To avoid drawing voltage and current waveforms every time we want to analyse a circuit, we use phasors. The lengths of the phasors are equal to the amplitudes of the sine waves. They rotate anticlockwise one revolution every cycle. The phasors in Figure 5.30(b) are shown in their starting position at time, $t = 0$. To find the instantaneous voltage and current at any other time, t_1 the phasors are rotated through an angle $(2\pi ft_1)$ radians. The instantaneous values are equal to the vertical displacements of the phasors. This is illustrated in Figure.5.30(c).

Root mean square (rms) voltage and current

As current flows through a resistor power is dissipated in the resistor in the form of heat. We saw in Section 5.2 that the power dissipated is given by equation (5.3).

$$p = i^2 R$$

As alternating currents and voltages are continually changing the power also varies over the cycle as in Figure 5.31.

Instantaneous current,

$$i = I_m \sin 2\pi ft$$

but

$$\omega = 2\pi ft$$

Instantaneous power,

$$p = R \, I_m{}^2 \sin^2 \omega t$$

Using the trignometrical relationship

$$\sin^2\theta = \tfrac{1}{2}(1 - \cos 2\theta)$$

$$p = \frac{I_m{}^2 R}{2} (1 - \cos 2\omega t)$$

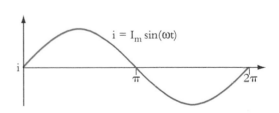

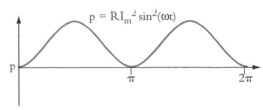

Figure 5.31

the average power dissipated over one cycle, P is given by:

$$P = \frac{1}{2\pi} \int_0^{2\pi} \frac{I_m^2 R}{2}(1 - \cos 2\omega t)\, d(\omega t)$$

$$= \frac{1}{2\pi} \frac{I_m^2 R}{2}\left[\omega t - \frac{\sin 2\omega t}{2}\right]_0^{2\pi}$$

$$= \frac{1}{2\pi} \frac{I_m^2 R}{2} 2\pi$$

$$P = \frac{I_m^2 R}{2}$$

Let I equal the direct current that would dissipate the same power, P and hence produce the same heating effect

$$I^2 R = \frac{I_m^2 R}{2}$$

$$I = \frac{I_m}{\sqrt{2}} = 0.707 I_m \qquad\qquad (5.46)$$

I is the 'effective' value of the alternating current. It is called the root mean square or rms current and is the value normally measured and used for calculations.

The rms voltage is found using Ohm's Law

$$v = iR$$

$$V_m \sin \omega t = I_m \sin \omega t\, R$$

$$V_m = I_m R$$

$$\frac{V_m}{\sqrt{2}} = \frac{I_m}{\sqrt{2}}R$$

$$V = IR \qquad\qquad (5.47)$$

where

$$V = \text{rms voltage} = \frac{V_m}{\sqrt{2}} \qquad\qquad (5.48)$$

Inductance connected to an alternating voltage supply

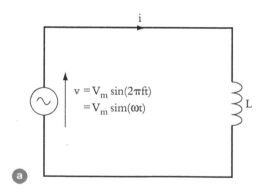

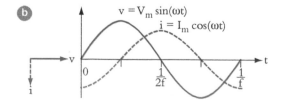

Figure 5.32

Electrical and electronic systems

From equation (5.25) we know that the instantaneous current, i, through an inductor is related to the voltage by

$$v = L\frac{di}{dt}$$

with sinusoidal voltage, $v = V_m \sin(2\pi ft)$

The current under steady state conditions will equal

$$i = \frac{-V_m}{2\pi fL}\cos(2\pi ft)$$

letting

$$\omega = 2\pi f$$

$$i = \frac{-V_m}{\omega L}\cos(\omega t) \tag{5.49}$$

Figure 5.32(b) shows the voltage and current waveforms. It can be seen that **at all times the current waveform is 90° behind the voltage.** At time, $t = 0$, when the voltage is zero, the current is negative with maximum amplitude. To represent this in the phasor diagram, we draw the current phasor pointing downwards at right angles to the voltage, so that as they rotate anticlockwise the **current phasor is always lagging the voltage phasor by 90°.**

From equation (5.49)

$$I_m = \frac{V_m}{\omega L}$$

rms current, $I = 0.707\,I_m$ and rms voltage, $V = 0.707\,V_m$

therefore

$$I = \frac{V}{\omega L} = \frac{V}{2\pi fL}$$

Inductive reactance, X_L is defined as

$$X_L = 2\pi fL = \omega L \tag{5.50}$$

So

$$I = \frac{V}{X_L} \tag{5.51}$$

This only gives the amplitude of the current. It does not give any indication that the current always lags the voltage by 90°.

j operator

To show the phase difference between voltage and current phasors we use complex algebra and the j operator (90° operator). The j operator indicates an anticlockwise rotation of 90°. Mathematicians call this operator i but as engineers use i for current they call the operator j. To describe a phasor that is two units long in a direction 90° anticlockwise to the positive real axis, we write j2. This direction is called the positive imaginary axis; see Figure 5.33(a).

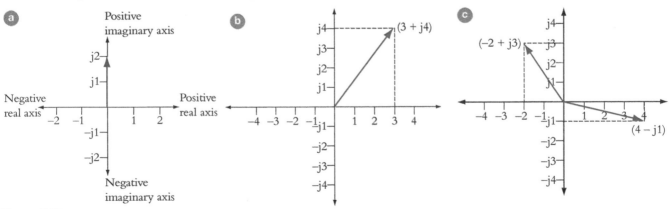

Figure 5.33

If the phasor is turned from $j2$ another $90\,°$ anticlockwise it would face along the negative real axis and would reach the point -2, therefore

$$j \times (j2) = -2$$
$$j \times j = j^2$$

so

$$j^2 2 = -2$$
$$j^2 = -1$$
$$j = \sqrt{(-1)} \tag{5.52}$$

Rotating a further $90°$ from -2 we reach the negative imaginary axis, therefore

$$j \times j^2 2 = j^3 2 = -j2$$

Finally, moving another $90°$ brings us back to 2 on the positive real axis and

$$j \times j^3 2 = j^4 2 = 2$$

When a phasor points in a direction which is not along either axis its dimension and position are described by its coordinates to the two axes. For example, the phasor shown in Figure 5.33(b) is described as $(3 + j4)$ and the phasors in Figure 5.33(c) as $(-2 + j3)$ and $(4 - j1)$.

Figure 5.32(b) shows the voltage and current phasors at time $t = 0$. The voltage phasor is in the direction of the positive real axis and the current is pointing downwards in the direction of the negative imaginary or $(-j)$ axis. To represent this phase difference mathematically, the current is multiplied by $(-j)$ and equation (5.51) becomes

$$I = \frac{-jV}{X_L} \tag{5.53}$$

multiplying numerator and denominator by j

$$I = \frac{(-j)jV}{jX_L}$$

but

$$j^2 = -1$$

$$I = \frac{V}{jX_L} \tag{5.54}$$

> The j operator is linked to the inductive reactance, X_L to ensure that both the magnitude and the phase angle of the current are calculated simultaneously.

This is demonstrated in the following example.

A 100 V rms 1000 Hz alternating voltage supply is connected to a 159 mH inductor. Calculate the current.

$$jX_L = j\,2\pi fL$$
$$= j\,2\pi \times 1000 \times 159 \times 10^{-3}$$
$$= j\,1000\text{ V}$$

$$I = \frac{V}{jX_L}$$
$$= \frac{100}{j\,1000}$$
$$= -j\,0.1\text{A}$$

Capacitance connected to an alternating voltage supply

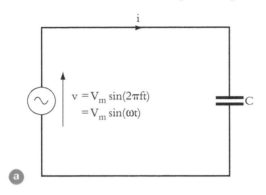

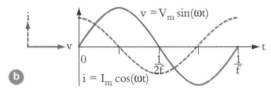

Figure 5.34

We saw in Section 5.4 that the instantaneous current through a capacitor, i, is related to the voltage across it, v by equation (5.41)

$$i = C\frac{dv}{dt}$$

here

$$v = V_m \sin(2\pi ft)$$

and so the steady state current will be given by:

$$i = 2\pi fCV_m \cos(2\pi ft)$$

letting

$$\omega = 2\pi f$$
$$i = \omega CV_m \cos(\omega t)$$
$$I_m = \omega CV_m$$

rms current,

$$I = \omega CV = 2\pi fCV$$

The waveforms and phasor diagram are shown in Figure 5.34(b).

> This time the current **leads** the voltage by 90°. To incorporate the phase difference in the current equation we have to insert $+j$.

$$I = +j\omega CV \qquad (5.55)$$

capacitive reactance,

$$X_C = \frac{1}{2\pi fC} = \frac{1}{\omega C} \qquad (5.56)$$

therefore

$$I = +j\frac{V}{X_C} = +j\frac{(-jV)}{(-jX_c)}$$

By linking the $-j$ to the capacitive reactance X_C as shown in equation (5.57) we are able to calculate simultaneously the amplitude and phase of the current.

$$I = \frac{V}{(-jX_C)}$$

(5.57)

A 100 V rms 1000 Hz alternating voltage supply is connected to a 159 nF capacitor. Calculate the current.

$$-jX_C = \frac{-j}{2\pi fC}$$

$$= \frac{-j}{2\pi \times 1000 \times 159 \times 10^{-9}}$$

$$= -j\,1000\,\Omega$$

$$I = \frac{V}{-jX_C}$$

$$= \frac{100}{-j\,1000}$$

$$= +j\,0.1\,\text{A}$$

Resistance and inductance in series with an alternating voltage supply

Most circuits incorporate two or more components. To demonstrate how we can analyse such circuits we will start with a resistor and inductor in series with an alternating voltage supply.

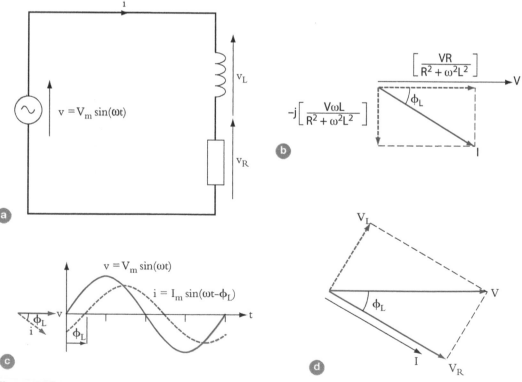

Figure 5.35

Electrical and electronic systems

Using rms quantities

From equation (5.47)

$$V_R = IR$$

From equation (5.54)

$$V_L = I \cdot jX_L$$

By Kirchhoff's Voltage Law

$$V = V_R + V_L \tag{5.58}$$

$$V = I(R + jX_L)$$

$$I = \frac{V}{R + jX_L}$$

The current magnitude and phase angle are affected by both the resistance, R and inductive reactance, X_L

As

$$X_L = \omega L = 2\pi f L$$

$$I = \frac{V}{R + j\omega L}$$

This equation may be rationalized by multiplying both the numerator and denominator by the complex conjugate of $(R + j\omega L)$ which is $(R - j\omega L)$.

$$I = \frac{V}{R + j\omega L} \cdot \frac{R - j\omega L}{R - j\omega L}$$

$$(R + j\omega L) \cdot (R - j\omega L) = R^2 + \omega^2 L^2$$

and the current equals:

$$I = \left[\frac{VR}{R^2 + \omega^2 L^2} \right] - j \left[\frac{V\omega L}{R^2 + \omega^2 L^2} \right] \tag{5.59}$$

> Describing the current in terms of these components is known as the Cartesian form. The first term in this equation is in phase with the voltage and is called the real component of current. The second term is, because of the $-j$, at $90\,°$ to the voltage, facing in the direction of the negative imaginary axis and is called the imaginary component of current.

Figure 5.35b shows the two components and the current and voltage phasors.

> The $-j$ before the second bracket indicates that the current lags the voltage.

As Figure 5.35(b) shows, the angle, ϕ is somewhere between $0\,°$ and $90\,°$. Using trigonometry we find that

$$\tan \phi_L = \frac{\left[\dfrac{V\omega L}{R^2 + \omega^2 L^2} \right]}{\left[\dfrac{VR}{R^2 + \omega^2 L^2} \right]}$$

$$\tan \phi_L = \frac{\omega L}{R}$$

$$\phi_L = \tan^{-1}\left(\frac{\omega L}{R} \right)$$

The magnitude of the current is calculated from the two components, using Pythagoras' Theorem, adding the squares of the components and taking the square root such that

$$|I| = \sqrt{\left[\frac{VR}{R^2 + \omega^2 L^2}\right]^2 + \left[\frac{V\omega L}{R^2 + \omega^2 L^2}\right]^2}$$

$$|I| = \frac{V}{R^2 + \omega^2 L^2}\sqrt{R^2 + \omega^2 L^2}$$

$$|I| = \frac{V}{\sqrt{R^2 + \omega^2 L^2}}$$

> The equations for current magnitude and phase angle may then be put together to express the current in what is called the **polar form**.

$$I = \frac{V}{\sqrt{R^2 + \omega^2 L^2}} \angle -\phi_L$$

(5.60)

where

$$\phi_L = \tan^{-1}\left(\frac{\omega L}{R}\right)$$

(5.61)

> The minus sign before the ϕ_L indicates that the current **lags** the voltage.

As the voltage and current waveforms sweep out the phasors, shown in Figure 5.35(c), rotate anticlockwise, maintaining the angle between them at all times.

To complete the phasor diagram we need to add the voltages across the resistor and the inductor. From equations (5.47), (5.54) and (5.58)

voltage across the resistor,

$$V_R = IR$$

voltage across the inductor,

$$V_L = IjX_L$$

and the supply voltage,

$$V = V_R + V_L$$

The voltage across the resistor, V_R is in phase with the current and the voltage across the inductor, V_L is, because of the j operator, at right angles to the current as shown in Figure 5.35(d). The vectorial sum of V_R and V_L equals the supply voltage, V.

To simplify circuit analysis, the circuit resistance and inductance are often combined into one parameter called the impedance or complex impedance, Z such that:

complex impedance

$$Z = R + j\omega L$$

(5.62)

where the magnitude of Z, is given by

$$|Z| = \sqrt{R^2 + \omega^2 L^2}$$

(5.63)

and the phase angle, ϕ_L by

$$\phi_L = \tan^{-1}\left(\frac{\omega L}{R}\right)$$

(5.64)

Z can be written in polar form

$$Z = \sqrt{R^2 + \omega^2 L^2} \angle + \phi_L$$

(5.65)

It is unusual to quote the phase angle in degrees or radians. It is more common to express it in terms of its cosine and call it the power factor or p.f. for short.

$$\text{power factor, p.f.} = \cos(\phi_L) = \cos\left(\tan^{-1}\frac{\omega L}{R}\right)$$

A 200 V rms ac supply feeds a 100 mH inductor in series with a 800 Ω resistor. Calculate the circuit impedance, current, phase angle and power factor when the frequency is 955 Hz.

This problem may be solved using either Cartesian or polar forms. Both are demonstrated here.

$$jX_L = j2\pi f L = j2\pi \times 955 \times 100 \times 10^{-3} = j600 \, \Omega$$

Using the Cartesian form

circuit impedance, $\quad Z = (R + jX_L) = (800 + j600) \, \Omega$

$$I = \frac{V}{Z} = \frac{V}{(R + jX_L)}$$

$$I = \frac{200}{(800 + j600)}$$

$$= \frac{200 \times (800 - j600)}{(800^2 + 600^2)}$$

$$= (0.16 - j0.12) \, \text{A} \qquad \text{the } -j \text{ indicates that the current lags the voltage}$$

$$\phi_L = \tan^{-1}\frac{(-0.12)}{(0.16)}$$

$$= 36.9° \text{ lagging}$$

power factor, $p.f. = \cos \phi_L = \cos (36.9°)$

$$= 0.8$$

Alternatively, solving using polar form

$$Z = \sqrt{R^2 + X_L^2} \angle + \phi_L$$

where $\quad \phi_L = \tan^{-1}\left(\frac{X_L}{R}\right)$

$$Z = \sqrt{800^2 + 600^2} \angle \tan^{-1}\left(\frac{600}{800}\right)$$

$$= 1000 \, \Omega \angle +36.9°$$

$$I = \frac{V}{Z} = \frac{V}{\sqrt{R^2 + X_L^2} \angle + \phi_L}$$

$$= \frac{V \angle - \phi_L}{\sqrt{R^2 + X_L^2}}$$

$$= \frac{200}{\sqrt{800^2 + 600^2}} \angle \tan^{-1}\left(\frac{600}{800}\right)$$

$$= \frac{200}{1000} \angle -36.9°$$

$$= 0.2A \angle -36.9° \qquad \text{current lags the voltage}$$

$$\phi_L = 36.9° \text{ lagging}$$

$$\text{power factor} = \cos(\phi_L) = \cos(36.9°)$$

$$= 0.8$$

Resistance and capacitance in series with an alternating voltage supply

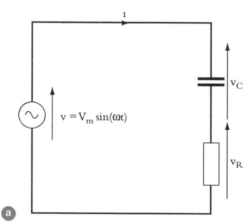

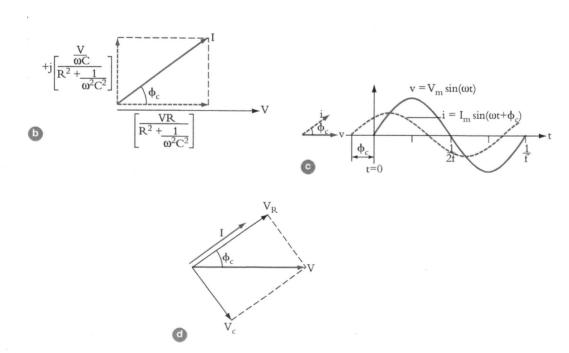

Figure 5.36

Electrical and electronic systems

When the inductor is replaced by a capacitor

$$V_R = IR$$

and

$$V_C = I(-jX_C)$$

using rms quantities.

By Kirchhoff's Voltage Law

$$V = V_R + V_C$$

$$V = I(R - jX_C)$$

$$I = \frac{V}{R - jX_C}$$

but

$$X_C = \frac{1}{\omega C} = \frac{1}{2\pi f C}$$

and so

$$I = \frac{V}{R - \left(\dfrac{j}{\omega C}\right)}$$

Rationalizing the current equation, I is given in Cartesian form as

$$I = \left[\frac{VR}{R^2 + \dfrac{1}{\omega^2 C^2}} \right] + j \left[\frac{\dfrac{V}{\omega C}}{R^2 + \dfrac{1}{\omega^2 C^2}} \right]$$

(5.66)

> The imaginary component has $(+j)$ which signifies that the current leads the voltage, as shown in Figure 5.36(b)

Using trigonometry and Pythagoras' Theorem (as we did in the case of inductor and resistor) we find that the current in polar form is:

$$I = \frac{V}{\sqrt{R^2 + \dfrac{1}{\omega^2 C^2}}} \angle + \phi_C$$

(5.67)

where

$$\phi_C = \tan^{-1}\left(\frac{1}{\omega CR}\right)$$

(5.68)

The current and voltage waveforms are shown in Figure 5.36(c).

> The current leads the voltage by an angle of $+\phi_C$.

The power factor is equal to the cosine of ϕ_C

$$\text{p.f.} = \cos\left(\tan^{-1}\left(\frac{1}{\omega CR}\right)\right)$$

(5.69)

To complete the phasor diagram we need to add the voltages across the resistor and capacitor, V_R and V_C.

$$V_R = IR$$

$$V_C = I(-jX_C)$$

and

$$V = V_C + V_R$$

The current is in phase with V_R and leads V_C by 90°. When the two voltages are added vectorially they equal the supply voltage, V as shown in Figure 5.36(d).

The resistance, R and capacitive reactance, $-jX_C = \dfrac{-j}{\omega C}$ are in series and so the total impedance of the circuit, Z is equal to their sum such that:

in Cartesian form,

$$Z = \left(R - \frac{j}{\omega C} \right)$$

and in polar form, where

$$Z = |Z| \angle -\phi_c$$

$$|Z| = \sqrt{R^2 + \frac{1}{\omega^2 C^2}}$$

(5.70)

and

$$\phi_C = \tan^{-1}\left(\frac{1}{\omega C R} \right)$$

A 200 V rms ac supply feeds a 800 Ω resistor in series with a 159 nF capacitor. Calculate the circuit impedance, current, phase angle and power factor when the frequency is 1668 Hz.

$$-jX_C = \frac{-j}{2\pi f C}$$

$$= \frac{-j}{2\pi \times 1668 \times 159 \times 10^{-9}}$$

$$= -j600 \ \Omega$$

Using the Cartesian form $Z = (R - jX_C) = (800 - j600) \ \Omega$

$$I = \frac{V}{Z} = \frac{V}{(R - jX_C)}$$

$$= \frac{200}{(800 - j600)}$$

$$= \frac{200 \times (800 + j600)}{(800^2 + 600^2)}$$

$$= (0.16 + j0.12)\text{A} \qquad \text{current leads voltage}$$

In polar form

$$Z = \sqrt{R^2 + X_C^2} \angle \tan^{-1}\left(\frac{-X_C}{R} \right)$$

$$= \sqrt{800^2 + 600^2} \angle \tan^{-1}\left(\frac{-600}{800} \right)$$

$$= 1000 \ \Omega \angle -36.9°$$

$$I = \frac{V}{Z} = \frac{V}{\sqrt{R^2 + X_C^2} \angle \tan^{-1}\left(\dfrac{-X_C}{R} \right)}$$

$$= \frac{V \angle -\tan^{-1}\left(\dfrac{-X_C}{R} \right)}{\sqrt{R^2 + X_C^2}}$$

$$= \frac{200}{\sqrt{800^2 + 600^2}} \angle -\tan^{-1}\left(\frac{-600}{800}\right)$$

$$= \frac{200}{1000} \angle + 36.9°$$

$$= 0.2\text{A} \angle + 36.9° \qquad \text{current leads voltage}$$

$$\phi_C = 36.9° \text{ leading}$$

$$\text{power factor} = \cos(\phi_C) = \cos(36.9°)$$

$$= 0.8$$

Resistance, inductance and capacitance in series with an alternating voltage supply

We are now in a position to explore the situation when we have a resistor, an inductor and a capacitor all connected in series.

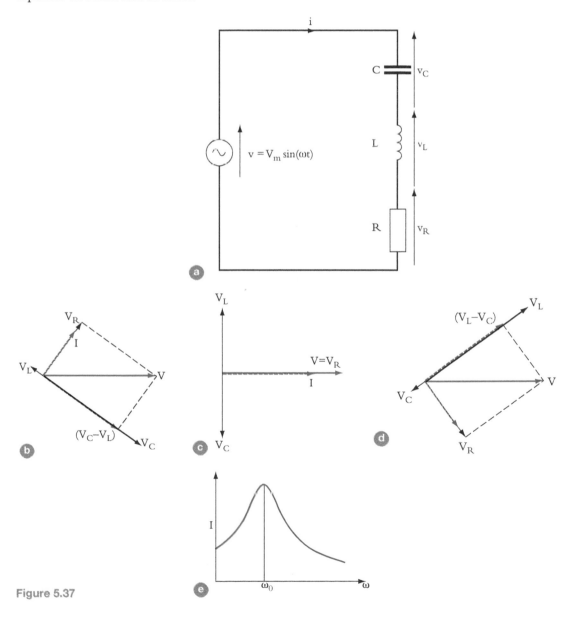

Figure 5.37

333

The voltage drops across the three components are

$$V_R = IR \tag{5.71}$$

$$V_L = I(jX_L) \tag{5.72}$$

$$V_C = I(-jX_C) \tag{5.73}$$

and the total voltage across all three components is equal to

$$V = V_R + V_L + V_C \tag{5.74}$$

Combining these equations

$$V = I[R + jX_L - jX_C] = I[R + j(X_L - X_C)]$$

But

$$X_L = \omega L \quad \text{and} \quad X_C = \frac{1}{\omega C}$$

Therefore

$$I = \frac{V}{R + j\left(\omega L - \dfrac{1}{\omega C}\right)}$$

The denominator is the complex impedance, $Z = R + j\left(\omega L - \dfrac{1}{\omega C}\right)$ which has a magnitude

$$|Z| = \sqrt{R^2 + \left(\omega L - \frac{1}{\omega C}\right)^2}$$

Rationalizing the equation for the current

$$I = \frac{VR}{R^2 + \left(\omega L - \dfrac{1}{\omega C}\right)^2} - j\frac{V\left(\omega L - \dfrac{1}{\omega C}\right)}{R^2 + \left(\omega L - \dfrac{1}{\omega C}\right)^2}$$

which can be simplified in Cartesian form to

$$I = \frac{V}{R^2 + \left(\omega L - \dfrac{1}{\omega C}\right)^2}\left[R - j\left(\omega L - \frac{1}{\omega C}\right)\right] \tag{5.75}$$

and re-written in polar form

$$I = \frac{V}{\sqrt{R^2 + \left(\omega L - \dfrac{1}{\omega C}\right)^2}} \angle -\phi_s \tag{5.76}$$

where

$$\phi_s = \tan^{-1}\left(\frac{\omega L - \dfrac{1}{\omega C}}{R}\right) \tag{5.77}$$

The phasor diagram (and hence the waveforms) depend on the relative values of $\omega L \ (= 2\pi f L)$ and $\dfrac{1}{\omega C} \left(= \dfrac{1}{2\pi f C}\right)$. Three cases must be considered:

1. at low frequencies when $2\pi f L < \dfrac{1}{2\pi f C}$;
2. when $2\pi f L = \dfrac{1}{2\pi f C}$
3. at high frequencies when $2\pi f L > \dfrac{1}{2\pi f C}$.

In all cases the magnitudes of V_R, V_L and V_C and their individual directions with respect to the current are defined by equations (5.71) to (5.73). The voltage across the resistor, V_R is in phase with the current. The current lags the voltage in an inductor, so the voltage across the inductor V_L is 90° ahead of the current. Likewise the current in a capacitor leads the voltage across it so V_C is 90° behind the current. To satisfy equation (5.74) the vector sum of the three voltages is equal to the supply voltage.

In case 1 at low frequencies when ωL is much less than $\left(\dfrac{1}{\omega C}\right)$ the voltage across the capacitor is far larger than that across the inductor and the current leads the supply voltage as shown in Figure 5.37(b).

In case 3 at high frequencies when ωL is much greater than $\left(\dfrac{1}{\omega C}\right)$ the voltage across the inductor is dominant and the current lags the supply voltage. See Figure 5.37(d).

In case 2 shown in Figure 5.37(c) when $\omega L = \left(\dfrac{1}{\omega C}\right)$ the voltages across the inductor and capacitor are equal cancelling each other, leaving the current in phase with the supply voltage and the voltage across the resistor V_R equal to the supply voltage. Thus:

$$I = \frac{V}{R} \angle 0°$$

If the resistance is low, the current will be large and produce very high voltages across both the inductor and the capacitor which may be far larger than the supply voltage. This phenomenon is called series resonance and is utilized in analogue radio tuners to centre on the frequency of the chosen station.

At resonance $2\pi fL = \dfrac{1}{2\pi fC}$

If f_0 is the series resonant frequency

$$f_0 = \frac{1}{2\pi\sqrt{LC}} \tag{5.78}$$

The variation in current magnitude with frequency is illustrated in Figure 5.57(e).

A 250 V rms ac supply feeds a 250 Ω resistor in series with a 100 mH inductor and a 159 nF capacitor. Calculate the current and phase angle when the frequency is (i) 1078.6 Hz, (ii) 1262.1 Hz, and (iii) 1476.6 Hz.

At 1078.6 Hz

$$X_L = j2\pi fL$$
$$= j2\pi \times 1078.6 \times 100 \times 10^{-3}$$
$$= j678 \ \Omega$$

$$-jX_C = \frac{-j}{2\pi fC}$$
$$= \frac{-j}{(2\pi 1078.6 \times 159 \times 10^{-9})}$$
$$= -j928 \ \Omega$$

$$I = \frac{V}{R + jX_L - jX_C} = \frac{250}{250 + j678 - j928}$$
$$= \frac{250}{250 - j250}$$

$$= \frac{250}{250^2 + 250^2} (250 + j250)$$

$$= (0.5 + j0.5)\text{A}$$

$$= 0.707\text{A} \angle + 45°$$

The current leads the voltage

At 1262.1 Hz

$$jX_L = j2\pi \times 1262.1 \times 100 \times 10^{-3} = j793$$

$$-jX_C = \frac{-j}{(2\pi \times 1262.1 \times 159 \times 10^{-9})} = -j793$$

$$I = \frac{V}{R + jX_L - jX_C} = \frac{250}{250 + j793 - j793}$$

$$= \frac{250}{250}$$

$$= (1.0 + j0)\text{A}$$

$$= 1.0\,\text{A}\ \angle 0°$$

The current is in phase with the voltage. This is resonance.

At 1476.6 Hz

$$jX_L = j2\pi \times 1476.6 \times 100 \times 10^{-3} = j927.8$$

$$-jX_C = \frac{-j}{(2\pi \times 1476.6 \times 159 \times 10^{-9})} = j677.8$$

$$I = \frac{V}{R + jX_L - jX_C} = \frac{250}{250 + j927.8 - j677.8}$$

$$= \frac{250}{250 - j250}$$

$$= \frac{250}{250^2 + 250^2} (250 - j250)$$

$$= (0.5 - j0.5)\text{A}$$

$$= 0.707\text{A}\ \angle - 45°$$

The current lags the voltage.

Power dissipation

The instantaneous power in any circuit is given by:

> *instantaneous power dissipation = instantaneous voltage × instantaneous current*

$$p = v \times i$$

In alternating current circuits the instantaneous voltage is

$$v = V_m \sin(\omega t),$$

When the current is leading, the instantaneous current is $i = I_m \sin(\omega t + \phi)$ and when the current is lagging, $i = I_m \sin(\omega t - \phi)$, so generally

$$i = I_m \sin(\omega t \pm \phi)$$

Electrical and electronic systems

instantaneous power,

$$p = vi = V_m \sin(\omega t) I_m \sin(\omega t \pm \phi)$$

using the trigonometrical relationship

$$\cos(A - B) - \cos(A + B) = 2 \sin A \sin B$$

and setting

$$A = \omega t \text{ and } B = (\omega t \pm \phi)$$

$$p = \frac{V_m I_m}{2} [\cos(\pm\phi) - \cos(2\omega t \pm \phi)]$$

This equation is plotted in Figure.5.38 where it is assumed that the current is lagging.

The first term in the bracket is a constant. The second term represents an oscillation of power to and from the load throughout each cycle. As a result, over a complete cycle there is a net power transfer given by

$$P = \frac{V_m I_m}{2} \cos\phi$$

but rms voltage,

$$V = \frac{V_m}{\sqrt{2}}$$

and rms current,

$$I = \frac{I_m}{\sqrt{2}}$$

hence net power transfer,

$$\boxed{P = VI \cos \phi}$$

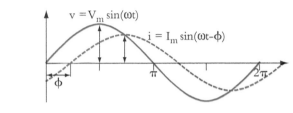

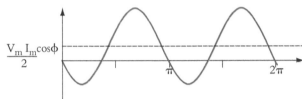

Figure 5.38

(5.79)

A 200 V rms ac supply feeds a 600 V resistor in series with a 100 mH inductor. Calculate the power dissipated when the frequency is 1273 Hz.

$$jX_L = j2\pi f L$$
$$= j2\pi \times 1273 \times 100 \times 10^{-3}$$
$$= j800 \ \Omega$$

$$Z = \sqrt{R^2 + X_L^2} \ \angle \tan^{-1}\frac{X_L}{R}$$
$$= \sqrt{600^2 + 800^2} \ \angle \tan^{-1}\left(\frac{800}{600}\right)$$
$$= 1000 \ \Omega \ \angle 53.1°$$

$$I = \frac{V}{Z} = \frac{200}{1000 \angle 53.1°}$$
$$= 0.2\text{A} \ \angle -53.1°$$

$$\phi_L = 53.1° \text{ lagging}$$
$$\cos \phi_L = 0.6$$

$$\text{Power dissipated} = VI \cos \phi$$
$$= 200 \times 0.2 \times 0.6$$
$$= 24 \text{ W}$$

Alternating current bridges

A special form of alternating current circuit is the ac bridge. It is used to measure capacitance very accurately.

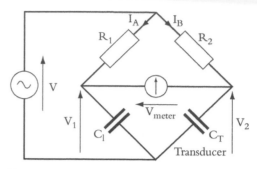

Figure 5.39

The resistors R_1 and R_2 are fixed and of known value. C_1 is a variable standard capacitor where its value can be adjusted in small known steps. The value of C_T is unknown. C_1 is adjusted until there is no current through the voltmeter making the voltages, V_1 and V_2 equal. At this point

$$I_A R_1 = I_B R_2$$

$$I_A \frac{1}{j\omega C_1} = I_B \frac{1}{j\omega C_T}$$

dividing the first equation by the second

$$j\omega C_1 R_1 = j\omega C_T R_2$$

$$C_T = \frac{C_1 R_1}{R_2}$$

(5.80)

Hence the value of C_T is found.

Displacement measurement

An ac bridge can be used in conjunction with a capacitance or displacement transducer to measure accurately very small movements. The transducers have a part which moves the distance to be measured. Moving this part changes the transducer capacitance. By measuring the capacitance both before and after the part is moved, the displacement can be calculated. It was shown in equation (5.37) that the capacitance C of a parallel plate capacitor is given by:

capacitance,
$$C = \frac{\varepsilon_0 \varepsilon_r A}{l}$$

where ε_0 is the permittivity of free space; ε_r is the relative permittivity of dielectric; A is the area of overlap of the plates; and l is the distance between the plates.

The capacitance can be changed by altering any of the last three parameters. Figure 5.40 shows three types of transducer. The capacitance of the first transducer is changed by adjusting the separation of the plates. This type is the most sensitive. In the second device the capacitance is changed by altering the area of overlap of the plates. This is the most linear device. The third type is the least common; here the capacitance is altered by varying the permittivity of the dielectric in the space between the plates.

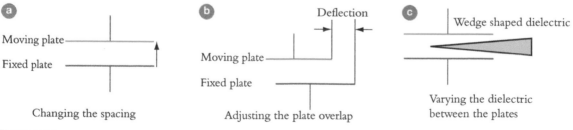

Figure 5.40

Before tests begin the transducer is connected in the circuit shown in Figure 5.39. The bridge is balanced and the transducer capacitance found using equation (5.80). The moving part is displaced by the distance to be measured, the bridge is rebalanced and the new transducer capacitance calculated. The displacement is found from the change in capacitance, as illustrated in the following worked example.

The plates in the capacitance transducer shown in Figure 5.40(a) are initially 2 mm apart. The transducer is connected in a bridge circuit as shown in Figure 5.39. The supply voltage is 5 V, frequency is 10 kHz, R_1 is 100 kΩ and R_2 is 1 MΩ. To balance, the bridge C_1 is set at 44.25 pF. Calculate the transducer capacitance.

The movable plate of the transducer is then displaced from its original position and C_1 readjusted to 40 pF to rebalance the bridge. Calculate the new value of transducer capacitance and displacement of the movable plate.

from equation 5.80

$$C_T = C_1 \times \frac{R_1}{R_2}$$

with the transducer plates 2 mm apart

$$C_T = 44.25 \times 10^{-12} \times \frac{100 \times 10^3}{1 \times 10^6}$$

$$C_T = 4.425 \text{ pF}$$

When moveable plate is displaced C_1 reset to 40 pF and the new value of transducer capacitance, C'_T is given by

$$C'_T = 40 \times 10^{-12} \times \frac{100 \times 10^3}{1 \times 10^6}$$

$$C'_T = 4 \text{ pF}$$

from equation 5.37 capacitance of a parallel plate

$$\text{capacitor} = \frac{\varepsilon_0 \varepsilon_r A}{l}$$

in this case l is variable and $\varepsilon_0, \varepsilon_r, A$ are all constant

let $\varepsilon_0 \varepsilon_r A = K$

∴
$$C_T = 4.425 \times 10^{-12} = \frac{K}{l}$$

and
$$C'_T = 4 \times 10^{-12} = \frac{K}{l'}$$

Dividing these equations

$$1.10625 = \frac{l'}{l}$$

$$l = 2 \text{ mm}$$

∴
$$l' = 2.2125 \text{ mm}$$

displacement $= l' - l = 0.2125 \text{ mm}$

For maximum bridge sensitivity, the two capacitors should be equal and the resistances equal to the capacitive reactance at the measuring frequency. A high-quality transducer should be used with a pure capacitor and a voltage supply that produces a true sine wave to ensure that the point at which the bridge is balanced can be clearly determined. Errors may be incurred if it is difficult to identify a good balance. Precautions must also be taken to prevent/minimize the effects of stray capacitance between the leads and the ground. This can be checked by reversing connections to the transducer, detector and power source systematically. Negligible differences in measurement will be detected if the effects are small.

5.6 Three-phase circuits

While domestic electricity users have their electricity delivered via a single phase supply, it is more economic to provide industrial users with a three-phase supply. As a result most machines are driven by three-phase electric motors. During their careers, most mechanical engineers will use three-phase systems and need to appreciate some important differences to single-phase networks.

Three-phase generator

In Section 5.5 it was shown that an elementary single-phase alternating current (ac) generator comprises a magnet rotating within a coil inducing a sinusoidal voltage in the coil. A three-phase generator is similar, but has two additional coils, each displaced from the original and from each other by 120° as illustrated in cross section in Figure 5.41(a). It is conventional to label the coil conductors red–red′ (or R–$R′$ for short), yellow–yellow′ (Y–$Y′$) and blue–blue′ (B–$B′$). As the magnet rotates, sinusoidal voltages are induced in each of the coils. The magnitudes of the three voltages are the same, but because of the physical displacement of the coils, there is a phase shift of 120° $\left(\dfrac{2\pi}{3}\ \text{radians}\right)$ between the voltages as shown in Figure 5.41(b). At time $t = 0$ no voltage is being induced in the coil R–$R′$, while $V_m \sin(-120°)$ is being induced in Y–$Y′$ and $V_m \sin(-240°)$ in B–$B′$. This phase shift is depicted in the phasor diagram, Figure 5.41(c) by showing the phasors 120° apart.

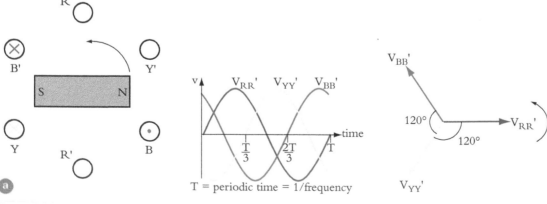

Figure 5.41

Three-phase connections

Loads may be connected to the three generator coils in a number of ways. The simplest method is to connect each coil to a separate load as shown in Figure 5.42. This requires six conductors.

Alternative methods reduce the number of conductors needed and are thus more economical. The two most common connections are star and delta. These names are not universal and in some countries they are called wye and mesh respectively

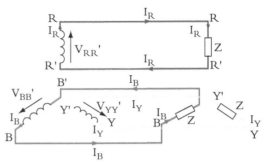

Figure 5.42

Star connection

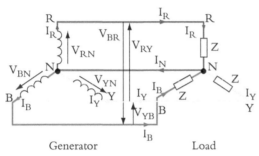

Generator   Load

Figure 5.43

In star connection, the ends of the return conductors R', Y' and B' are combined to form a common return called the neutral, N, as illustrated in Figure 5.43. All the voltages have an arrow to show the assumed voltage direction and are labelled with two suffixes to identify the points between which the voltage is being measured. For example, V_{RY} is the voltage of the red line R (the first suffix) with respect to yellow line Y (the second suffix). Similarly V_{RN} is the voltage of R with respect to (wrt) the neutral, N and V_{YN} is voltage of Y wrt N. All these voltages are marked in Figure 5.43. The initial selection of voltage directions is arbitrary and is only for the purposes of analysis; however, once chosen must be strictly adhered to.

Applying Kirchhoff's Voltage Law to this circuit we can see that

$$V_{RY} = V_{RN} - V_{YN}$$

These voltages are not in phase and so the equation must be solved using phasors as shown in Figure 5.44.

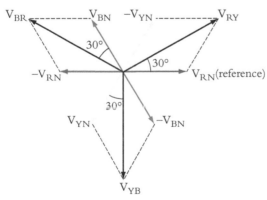

Figure 5.44

Using the cosine rule

$$V_{RY}^2 = V_{RN}^2 + V_{YN}^2 - 2\,V_{RN}\,V_{YN}\cos 120°$$

with

$$|\,V_{YN}\,| = |\,V_{RN}\,|$$

$$V_{RY}^2 = 3\,V_{RN}^2$$

$$|\,V_{RY}\,| = \sqrt{3}\,|\,V_{RN}\,| = \sqrt{3}\,|\,V_{YN}\,|$$

V_{RY} is called a line voltage as it is measured between two lines (red and yellow). V_{RN} is a phase voltage as it is measured across one of the phases, between one line (in this case the red line) and the neutral, N.

Phasors always rotate anticlockwise, so V_{RY} leads V_{RN} by 30°.

Similarly, V_{YB} can be found from

$$V_{YB} = V_{YN} - V_{BN}$$

$$|\,V_{YB}\,| = \sqrt{3}\,|\,V_{YN}\,| = \sqrt{3}\,|\,V_{BN}\,|$$

where V_{YB} leads V_{YN} by 30°.

V_{BR} is found from

$$V_{BR} = V_{BN} - V_{RN}$$

$$|\,V_{BR}\,| = \sqrt{3}\,|\,V_{BN}\,| = \sqrt{3}\,|\,V_{RN}\,|$$

where V_{BR} leads V_{BN} by 30°.

Combining these equations, it is found that

$$|\,V_{RY}\,| = |\,V_{YB}\,| = |\,V_{BR}\,| = \sqrt{3}\,|\,V_{RN}\,| = \sqrt{3}\,|\,V_{YN}\,| = \sqrt{3}\,|\,V_{BN}\,|$$

The line voltages V_{RY}, V_{YB} and V_{BR} are all equal in magnitude and equal to $\sqrt{3}$ times the magnitude of the phase voltages V_{RN}, V_{YN} and V_{BN}. This may be generalized as:

$$|\,\text{line voltage}\,| = \sqrt{3}\,|\,\text{phase voltage}\,|$$

which is usually written as

$$V_l = \sqrt{3}\,V_p \tag{5.81}$$

A load impedance, Z, is connected across each phase of the generator, so currents will flow from the generator through the lines to the load such that:

$$I_R = \frac{V_{RN}}{Z} \qquad I_Y = \frac{V_{YN}}{Z} \qquad I_B = \frac{V_{BN}}{Z}$$

These currents all leave a generator phase, flow along a line and enter a phase of the load. It is the same current flowing in the line as in the phase from which we can deduce:

$$\text{line current} = \text{phase current}$$

$$I_l = I_p \tag{5.82}$$

When the supply is balanced (i.e. the three-phase voltages are of equal magnitude and displaced from each other by 120°) and the load is balanced (i.e. each phase of the load has the same impedance) the magnitudes of three currents are the same and we only have to calculate the current in one line. The other currents can be deduced from the result as illustrated in the following example.

A balanced three-phase 415 V (line) 50 Hz supply feeds three coils each with an inductance of 20 mH and resistance 4 Ω connected in star. Calculate the line current and phase angle.

The analysis begins by converting all line quantities to phase values. The main calculations are done using only phase quantities. Only at the end do we convert back to line quantities.

line voltage, $V_l = 415\,\text{V}$

From equation (5.81)

phase voltage,

$$V_p = \frac{V_l}{\sqrt{3}} = \frac{415}{\sqrt{3}} = 239.6\,\text{V}$$

inductive reactance,

$$jX_L = j2\pi fL$$
$$= j2\pi 50 \times 20 \times 10^{-3}$$
$$= j6.28\,\Omega$$

complex impedance of one phase of the load,

$$Z = R + jX_L$$
$$= (4 + j6.28)\,\Omega$$

Assume that the red phase voltage, V_{RN} is the reference

phase current,
$$I_p = \frac{V_p}{R + jX_L} = \frac{239.6}{(4 + j6.28)}$$

$$= \frac{239.6}{(4 + j6.28)} \times \frac{(4 - j6.28)}{(4 - j6.28)}$$

$$= \frac{958.4 - j1504.7}{55.44}$$

$$= (17.29 - j27.14)\,\text{A in Cartesian coordinates or}$$

$$= 32.18\text{A} \angle -57.5° \text{ in polar coordinates}$$

From equation (5.82)

line current I_l = phase current, I_p

$$= 32.18\text{A} \angle -57.5°$$

With V_{RN} as the reference this is the current in the red phase, I_R.

$$I_Y = \frac{V_{YN}}{Z} \quad \text{and} \quad I_B = \frac{V_{BN}}{Z}$$

V_{YN} lags V_{RN} by 120° and V_{BN} lags V_{RN} by 240°

so $\qquad V_{YN} = V_{RN} \angle -120°$ and $V_{BN} = V_{RN} \angle -240°$

Therefore, to find I_Y and I_B, all we have to do is subtract 120° and 240° respectively from the angle of I_R such that:

$$I_Y = 32.18\text{A} \angle -177.5°$$

$$I_B = 32.18\text{A} \angle -297.5° = 32.18\text{A} \angle +62.5°$$

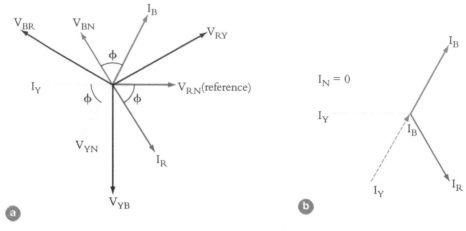

Figure 5.45

These currents can then be added to the phasor diagram as shown in Figure 5.45(a). The currents converge at the neutral, N, to form the neutral current, I_N. By Kirchhoff's Current Law

$$I_N = I_R + I_Y + I_B$$

This is a vectorial sum, which is solved graphically in Figure 5.45(b). I_N is zero. This is always the result when we have a balanced supply and load. As a result we can dispense with the neutral conductor and use only three conductors to supply the load, saving the cost of a conductor, which on a long run can be quite significant.

Power dissipation in a star-connected load

It was shown in Section 5.5 that the net power transferred to and dissipated in a single-phase load is given by

$$P = VI \cos \phi$$

Hence the power dissipated in one phase of a balanced three-phase load will be

$$P = V_p \, I_p \cos \phi$$

where V_p is the phase voltage, I_p is the phase current, and ϕ is the phase angle between the phase voltage and phase current, and the total power dissipated will be three times this.

Total power dissipated $= 3 V_p I_p \cos \phi$

In star connection where

$$V_l = \sqrt{3} V_p$$

and

$$I_l = I_p$$

the total power dissipated is

$$\frac{3 V_l}{\sqrt{3}} I_l \cos \phi$$

$$= \sqrt{3} V_l I_l \cos \phi$$

(5.83)

Calculate the power dissipated in the star connected load in the previous worked example.

line voltage, $V_l = 415\,\text{V}$

From the same worked example the line current $I_l = 32.18\,\text{A}$ and phase angle, $\phi = -57.5°$

Total power $= \sqrt{3}\,V_l I_l \cos \phi = \sqrt{3} \times 415 \times 32.18 \times \cos(57.5°)$

$\qquad = 12.43\,\text{kW}$

Delta connection

Delta or mesh connection is another way to connect the three generator coils. This is shown in Figure 5.46.

This time, $V_{RY} = V_{RR'}$; $V_{YB} = V_{YY'}$; $V_{BR} = V_{BB'}$

> line voltage = phase voltage

> $$V_l = V_P$$

(5.84)

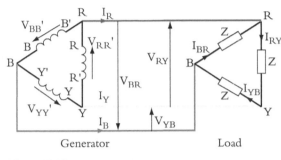

Figure 5.46

The line currents divide at the load, such that

$$I_R = I_{RY} - I_{BR}$$
$$I_Y = I_{YB} - I_{RY}$$
$$I_B = I_{BR} - I_{YB}$$

With a balanced voltage supply and balanced load

$$|I_{RY}| = |I_{YB}| = |I_{BR}|$$

These currents are mutually displaced by 120° as shown in the phasor diagram Figure 5.47.

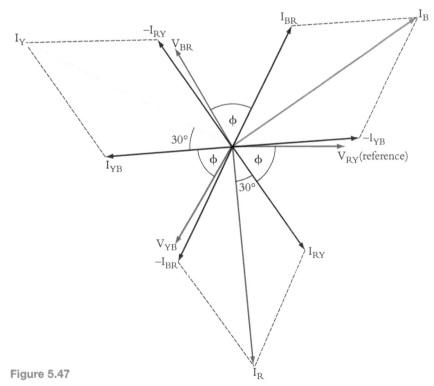

Figure 5.47

Using the cosine rule, we can find the magnitude of the line current, I_R relative to the phase currents.

$$I_R{}^2 = I_{RY}{}^2 + I_{BR}{}^2 - 2\,I_{RY}\,I_{BR}\cos 120°$$

$$I_R{}^2 = 3\,I_{RY}{}^2$$

$$|\,I_R\,| = \sqrt{3}\,|\,I_{RY}\,|$$

similarly

$$|\,I_Y\,| = \sqrt{3}\,|\,I_{YB}\,|$$

$$|\,I_B\,| = \sqrt{3}\,|\,I_{BR}\,|$$

which may be generalized

$$\boxed{|\,\text{line current}\,| = \sqrt{3}\,|\,\text{phase current}\,|}$$

$$\boxed{|\,I_l\,| = \sqrt{3}\,|\,I_p\,|}$$

(5.85)

Each of the line current lags its respective phase current by 30°.

As with star connection the magnitudes and phase angles of the phase currents are calculated using the phase voltage and complex impedance of the load. The line currents are found at the end. This is demonstrated in the following worked example.

A balanced three–phase 415 V (line) 50 Hz supply feeds three coils each with an inductance of 20 mH and resistance 4 Ω connected in delta. Calculate the line current and phase angle.

Taking V_{RY} as reference

line voltage = phase voltage = 415 V

inductive reactance,

$$jX_L = j2\pi fL$$

$$= j2\pi 50 \times 20 \times 10^{-3}$$

$$= j6.28\,\Omega$$

complex impedance,

$$Z = R + jX_L$$

$$= (4 + j6.28)\ \Omega$$

$$I_p = \frac{V_p}{R + jX_L} = \frac{415}{(4 + j6.28)}$$

$$= \frac{415(4 - j6.28)}{55.44}$$

$$= (29.94 - j47.01)\ \text{A in Cartesian coordinates or}$$

$$= 55.73\text{A}\ \angle{-57.5°}\ \text{in polar coordinates}$$

where the phase angle, $\phi = -57.5°$

line current, $|\,I_l\,| = \sqrt{3} \times$ phase current $|\,I_p\,|$

$$|\,I_l\,| = \sqrt{3} \times 55.73 = 96.54\text{A}$$

The line current lags the phase current by 30°, so

$$I_l = 96.54 \text{ A} \angle -87.5°$$

This is the current in the red line, I_R. The currents in the yellow and blue lines lag by 120° and 240° respectively.

$$I_Y = 96.54\text{A} \angle -207.5° = I_l = 96.54\text{A} \angle +152.5°$$

$$I_B = 96.54\text{A} \angle -327.5° = 96.54\text{A} \angle +32.5°$$

Comparing this worked example with that after equation (5.82) with the same line voltage and same load impedance, the line currents are three times larger in delta connection than in star.

Power dissipation in a delta connected load

Like star connection, the power dissipated in one phase of a balanced three-phase load is:

$$P = V_p I_p \cos \phi$$

and the total power dissipated is $3V_p I_p \cos \phi$

In delta connection

$$V_l = V_p$$

and

$$I_l = \sqrt{3} I_p$$

therefore, the total power dissipated $= \dfrac{3V_l I_l}{\sqrt{3}} \cos \phi$

$$= \sqrt{3} V_l I_l \cos \phi \qquad \text{(5.86)}$$

This is the same formula as for star connection.

Calculate the power dissipated in the load in the previous worked example.

line voltage $= 415\,\text{V}$

From the previous worked example,

$I_l = 96.54$ A and phase angle, $\phi = -57.5°$

Total power $= \sqrt{3} V_l I_l \cos\phi$

$\qquad\qquad = \sqrt{3} \times 415 \times 96.54 \times \cos(-57.5°)$

$\qquad\qquad = 37.29\,\text{kW}$

We saw earlier that with the same line supply voltage and same load impedance, the line currents were three times larger in delta than in star. As a result, the power dissipated in a delta-connected load is also three times larger than that dissipated in a star-connected load. One important consequence of this for mechanical engineers is that some large three-phase induction motors (see Section 5.11) are started with motor windings connected in star, to avoid tripping the circuit and then, when the motor is running at a steady speed, reconnected into delta, using a special switch, to drive full power to the load.

You have completed the section on three phase circuits and should now:

✔ know how three phase voltages are generated;

✔ remember that in a balanced supply the voltages are of equal magnitude and are mutually 120° apart;

✔ be able to distinguish between star and delta connections;

✔ be able to differentiate between phase and line currents and voltages;

✔ remember that in star connection line voltage equals $\sqrt{3}$ phase voltage and line current equals phase current;

✔ remember that in delta connection line voltage equals phase voltage and line current equals $\sqrt{3}$ phase current;

✔ be able to analyse star and delta connected systems;

✔ be able to draw a phasor diagram for a three phase network;

✔ calculate the power dissipated in a three phase load;

✔ remember that with the same supply voltage and same load impedance, the power dissipated in a delta connected load is three times larger than in a star load.

5.7 Semiconductor rectifiers

Mechanical engineers occasionally need to design electronic circuits to control equipment. Generally, such circuits require direct voltage supplies but most electricity supplies are alternating. Ac can be converted to dc using a rectifier. Manufactured from silicon, rectifiers are non-linear semiconductor devices which produce an output that is a series of unidirectional current pulses To minimize the effects of these pulses on electronic devices it is common to use either current smoothing or voltage stabilization. This section begins by looking at how semiconductor rectifiers operate, before examining a range of practical rectifier circuits.

Semiconductors

The most commonly used material for the manufacture of semiconductors is silicon (chemical symbol, Si) which has a valency of 4. Each atom has a nucleus with an effective positive charge of $+4e$ and four valence electrons, each with a negative charge $-e$, where e equals 1.6×10^{-19} C. The four electrons are shared with the atom's four nearest neighbours to form a crystal lattice in which the nuclei are held rigidly, as shown in Figure 5.48(a). This is pure or intrinsic silicon, as it is usually called. At room temperature it has a resistivity of 2000 Ωm. This is higher than the resistivities of common conductors such as copper, 1.8×10^{-8} Ωm and aluminium, 2.4×10^{-8} Ωm. Hence, silicon is called a semiconductor.

The silicon may be doped by adding a small quantity, between one part per million and one part per 100 million, of a valency 5 element, phosphorus, arsenic or antimony, to form n-type silicon. Valency 5 atoms each have a nucleus with an effective charge of $+5e$ and five valence electrons. Only four electrons are needed to link with the adjacent silicon atoms to complete the rigid lattice, leaving one electron to move freely through the crystal, as shown in Figure5.48(b) thereby increasing conduction.

Normally, the movement of free electrons is random, but when an electric voltage is applied across the silicon, the negatively charged electrons drift towards the positive terminal as illustrated in Figure 5.48(c).

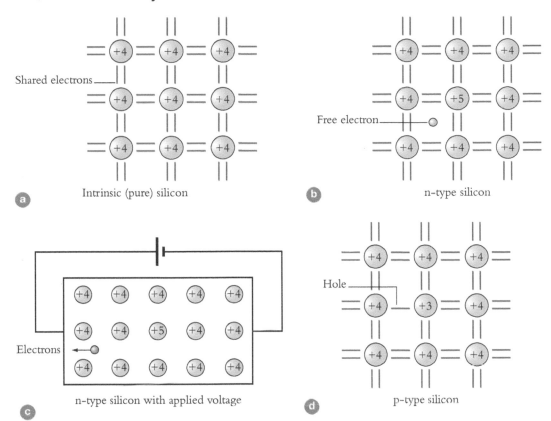

a Intrinsic (pure) silicon

Shared electrons

b n-type silicon

Free electron

c n-type silicon with applied voltage

Electrons

d p-type silicon

Hole

Figure 5.48

Silicon may also be doped with a valency 3 element, for example boron, aluminium, gallium or indium, to form p-type silicon. The impurity atoms only have three positive charges in the nuclei and three valence electrons to link with the neighbouring silicon atoms. This leaves holes in the bonding as shown in Figure 5.48(d). The bonding forces between atoms are very strong and incomplete bonds attract electrons from other bonds, thereby creating new holes. In the absence of an electric field, the holes drift through the silicon randomly, but when an electric voltage is applied the holes act like positive charges and move towards the negative terminal.

Junction diode

A junction diode is formed by making a crystal which is one half n-type silicon and the other half p-type. In the absence of an electric field, free electrons will move randomly through the n-type region. Some cross the boundary and enter the p-type where they fill the gaps left in the atomic bonding, creating a negatively charged zone close to the junction as shown in Figure 5.49. Similarly, holes cross the boundary from p-type to n-type, where they combine with free electrons close to the junction, producing a positively charged zone in the n-type. The zones each side of the boundary are both left without any free charge carriers. As a result, this region is called the depletion layer. The build-up of charge close to the boundary establishes a potential difference between the regions and inhibits further charge migration across the junction.

When an external voltage is applied across the crystal, the current that flows depends on the polarity of the supply. Connecting the positive terminal of the supply to the p-type, which is called the

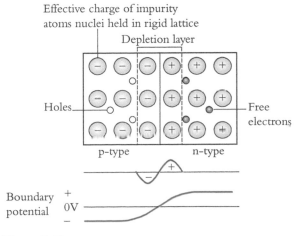

Effective charge of impurity atoms nuclei held in rigid lattice

Depletion layer

Holes

Free electrons

p-type   n-type

Boundary potential

Figure 5.49

anode and the negative terminal to the *n*-type, named the cathode, the applied voltage opposes the potential difference across the boundary, thereby closing the depletion layer. The depletion layer disappears when the applied voltage exceeds 0.7 V, allowing both holes and electrons to move freely, establishing a current in the diode and giving it a low resistance. This is illustrated in Figure 5.50(a). Figure 5.50(b) shows the circuit redrawn using conventional circuit symbols. This connection is called forward biased.

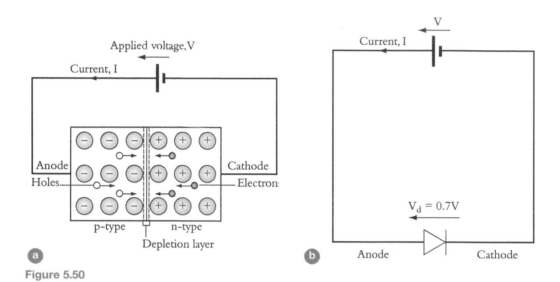

Figure 5.50

When the supply terminals are reversed as shown in Figure 5.51(a), the depletion layer is widened as holes and electrons are both attracted away from the junction, leaving no holes and no electrons near the boundary. With no free charge carriers in the depletion layer, current cannot flow, the diode has a high resistance and the supply voltage appears across the diode. This connection is called reverse biased.

> Devices that conduct in only one direction are called rectifiers.

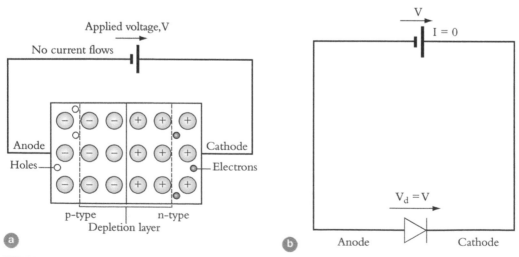

Figure 5.51

Electrical and electronic systems

Figure 5.52(b) shows the idealized current *versus* voltage characteristic for the junction diode.

To avoid destroying the diode when it is forward biased, the current has to be limited by other components in the circuit, such as a resistor, as shown in Figure 5.52(a).

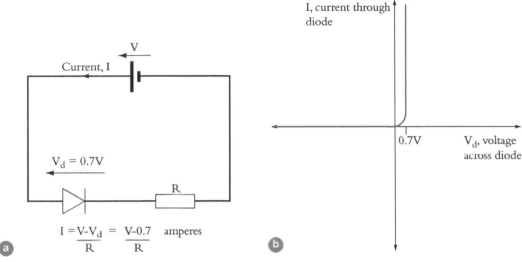

$$I = \frac{V-V_d}{R} = \frac{V-0.7}{R} \quad \text{amperes}$$

a
b

Figure 5.52

Rectifying alternating currents

When the battery in Figure 5.52(a) is replaced by an alternating (ac) voltage supply as shown in Figure 5.53(a), the diode conducts only during the positive half cycles (Figure 5.53(b)),when the potential of the anode is positive with respect to that of the cathode. As a result, the current waveform comprises a series of positive half sine waves, as illustrated in Figure 5.53(c).

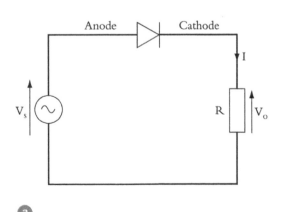

a

Figure 5.53

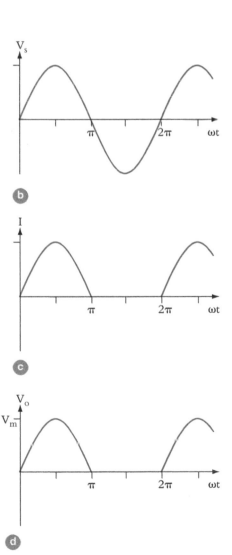

b

c

d

The voltage across the resistor is given by:

$$V_0 = iR$$

This voltage is continually changing. The average value is calculated by dividing the area under the graph in Figure 5.53(d) by the periodic time (i.e. the time elapsed before the signal repeats). Hence

$$\text{average dc output voltage, } V_o = \frac{1}{2\pi}\left[\int_0^\pi V_m \sin(\omega t)\,\mathrm{d}(\omega t) + \int_\pi^{2\pi} 0\,\mathrm{d}(\omega t)\right]$$

$$= \frac{V_m}{2\pi}[-\cos\omega t]_0^\pi$$

$$= \frac{V_m}{\pi} \tag{5.87}$$

> This circuit is called an *uncontrolled half wave rectifier.*

A 20 V (rms) 50 Hz supply feeds the circuit in Figure 5.53(a). Calculate the average dc output voltage. The load resistance is 900 Ω. Calculate the average load current.

Rms supply voltage, $V = 20\,\text{V}$

Maximum supply voltage, $V_m = \sqrt{2} \times 20 = 28.3\,\text{V}$

Average dc output voltage, $V_{dc} = \dfrac{V_m}{\pi}$

$$= 9.0\,\text{V}$$

average dc load current, $I = \dfrac{V_{dc}}{R}$

$$= \frac{9}{900}$$

$$= 10 \times 10^{-3}\,\text{A} = 10\,\text{mA}$$

Although the average dc output voltage is 9 V, it is clear from Figure 5.53(d) that the instantaneous voltage varies from a peak of 28.3 V to zero where it remains for half of the time.

Thyristor

The thyristor is similar to the diode but has a third terminal called a gate, as shown in Figure 5.54(a).

The thyristor acts like a diode when it is reverse biased, i.e. no current flows from cathode to anode, but it operates differently when forward biased. With the anode potential positive with respect to the cathode (Figure 5.54(b), no current will flow between the terminals until the gate is triggered by a small pulse from another, usually low, voltage source independent of the main supply, Figure 5.54(c). Once the current in the main circuit has been triggered, it quickly establishes itself flowing from anode to cathode, Figure 5.54(d). Current continues flowing until it falls to zero, when it will be extinguished. Current cannot flow in the main circuit when the thyristor is reversed biased, so triggering the gate during this period has no effect. Current will only be re-established when the thyristor is forward biased and the gate has been triggered again. The waveforms in Figure 5.54(d) and (e) show the effect of triggering the thyristor at an angle α after every positive supply voltage zero.

Electrical and electronic systems

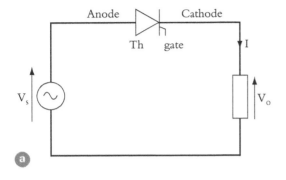

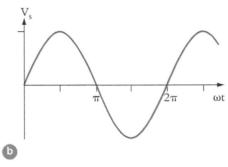

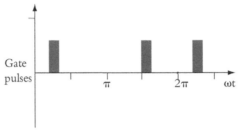

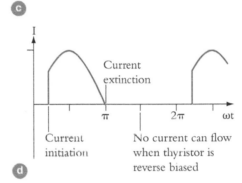

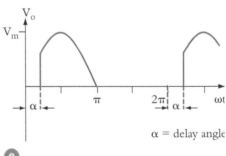

α = delay angle

Figure 5.54

The average dc voltage across the resistive load is calculated in a similar way to that for the diode, dividing the area under the graph in Figure 5.54(e) by the periodic time 2π.

Average dc output voltage, $V_o = \dfrac{1}{2\pi}\left[\int_0^\alpha 0\,d(\omega t) + \int_0^\pi V_m \sin(\omega t)\,d(\omega t) + \int_\pi^{2\pi} 0\,d(\omega t)\right]$

$\qquad = \dfrac{V_m}{2\pi}[-\cos \omega t]_\alpha^\pi$

$\qquad = \dfrac{V_m}{\pi}[1 + \cos \alpha]$ (5.88)

By varying α from 0 to π the average output voltage can be adjusted from $\dfrac{V_m}{\pi}$ (the same as for the diode rectifier) to 0.

> The thyristor circuit is called a *controlled half-wave rectifier*.

Full-wave rectifiers

It was seen above that a single diode blocks the negative half cycles of current and allows only the positive half cycles to flow through the load. Half the input signal is lost in the rectification process. The bridge rectifier, shown in Figure 5.55(a), overcomes this problem, doubling the number of half sine waves in the output waveform and with it the average dc voltage. When the potential of point A is positive with respect to B the current flows through diode D_1 down through the load and diode D_3 back to the supply. When B is positive with respect to A the current flow is through diode D_2, the resistive load and diode D_4. The direction of the current through the load is the same.

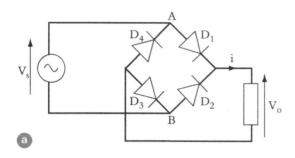

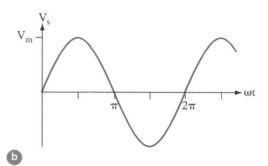

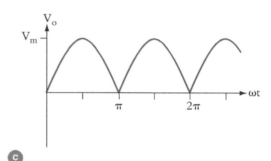

Figure 5.55

$$\text{Average dc output voltage} = \frac{1}{\pi}\left[\int_0^\pi V_m \sin(\omega t)\,\mathrm{d}(\omega t)\right]$$

$$= \frac{2V_m}{\pi} \tag{5.89}$$

> As both the positive and negative halves of the input signal are rectified and the output voltage cannot be adjusted, this circuit is called an *uncontrolled full wave rectifier*.

Controlled full-wave rectification

To achieve controlled full wave rectification, the diodes in the bridge must be replaced by thyristors as shown in Figure 5.56. Thyristors Th_1 and Th_3 are triggered at an angle α after every positive going supply voltage zero and Th_2 and Th_4 are triggered at the same angle after the negative going voltage zeros.

$$\text{Average dc output voltage} = \frac{1}{\pi}\left[\int_\alpha^\pi V_m \sin(\omega t)\,\mathrm{d}(\omega t)\right]$$

$$= \frac{V_m}{\pi}[1 + \cos \alpha] \tag{5.90}$$

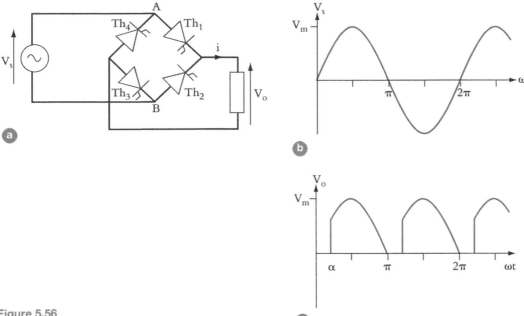

Figure 5.56

Smoothed supplies

For many applications where direct voltages are needed, the variation in output voltage from rectifiers such as these is too large. In these situations it is necessary to smooth the output voltage by connecting a capacitor in parallel with the resistive load.

Consider the example of the smoothed uncontrolled half-wave rectifier in Figure 5.57. When the supply is first switched on, the voltage is positive and rising, so the diode will conduct

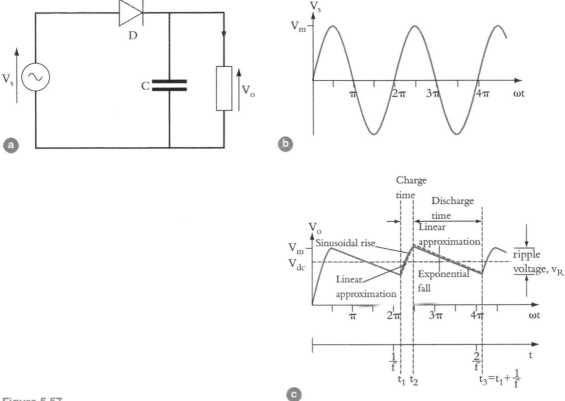

Figure 5.57

current charging the capacitor to the maximum voltage, V_m. Immediately the supply voltage starts to fall, the diode becomes reversed biased and the capacitor starts to discharge through the load resistor. This continues until the positive supply voltage again exceeds the capacitor voltage when the diode starts to conduct again enabling the capacitor to be recharged to V_m volts and the process repeats.

The variation in capacitor voltage (and hence output voltage) is called the ripple voltage. Manufacturers of equipment which require dc supplies will specify both the average dc voltage and the maximum ripple voltage for their products, sometimes expressing the latter as a percentage of the average dc voltage.

To calculate the precise value of smoothing capacitance required for a particular application can be difficult as the rise in capacitor voltage is sinusoidal and the fall is exponential. However, a reasonably accurate approximation can be found by assuming that both the rise and fall are linear. Making this assumption, it can be seen from Figure 5.57(c) that the average dc output voltage, V_{dc} is equal to

$$V_{dc} = V_m - \frac{V_R}{2} \tag{5.91}$$

where V_m = peak ac supply voltage

V_R = ripple voltage

The capacitor charges between t_1 and t_2 and discharges between t_2 and t_3

giving a charging time $= (t_2 - t_1)$

and discharge time $= (t_3 - t_2)$

but $t_3 = \dfrac{1}{f} + t_1$

where f = supply frequency

making the discharge time $= \dfrac{1}{f} + (t_1 - t_2) = \dfrac{1}{f} -$ charging time $\tag{5.92}$

When the capacitor discharges current only flows through the capacitor and load resistor. Using equations (5.34) and (5.41)

$$i = \frac{dq}{dt} = C\frac{dv}{dt}$$

By assuming that the fall in capacitor voltage, v is linear, then both $\dfrac{dv}{dt}$ and the capacitor current, i are constant. As a result this equation may be simplified to:

$$i = \frac{q}{t} = C\frac{v}{t} \tag{5.93}$$

where v is the total fall in capacitor voltage, and t is the discharge time.

However, the total fall in capacitor voltage = ripple voltage, V_r

During discharge the capacitor current flows through the load resistor hence:

Capacitor current, i = load resistor current $= \dfrac{V_{dc}}{R}$

Substituting in equation (5.93)

$$\frac{V_{dc}}{R} = \frac{CV_r}{t}$$

$$C = \frac{V_{dc}}{V_r}\frac{t}{R} \tag{5.94}$$

From equation (5.92)

discharge time $= \dfrac{1}{f} -$ charging time

If the ripple voltage is small compared to the average dc output voltage, then the charging time will be small in comparison to the discharge time. It may therefore be assumed that the former can be ignored and the discharge time is approximately equal to $\frac{1}{f}$ giving:

$$C = \frac{V_{dc}}{V_r}\frac{1}{fR}\text{Farads}$$

(5.95)

The average dc output from the smoothed uncontrolled half wave rectifier shown in Figure 5.57 is 9 V, the ripple voltage is 0.7 V and the load resistance is 900 Ω. The supply frequency is 50 Hz. Calculate the supply rms voltage, the load current and capacitance.

Using equation (5.91)

$$V_{dc} = V_m - \frac{V_R}{2}$$

$$9 = V_m - \frac{0.7}{2}$$

$$V_m = 9.35\,\text{V}$$

Supply rms voltage $= \dfrac{V_m}{\sqrt{2}} = \dfrac{9.35}{\sqrt{2}} = 6.61\,\text{V}$

load current $= \dfrac{V_{dc}}{R} = \dfrac{9}{900}$

$= 10 \times 10^{-3}\,\text{A} = 10\,\text{mA}$

from equation (5.95)

$$C = \frac{V_{dc}}{V_r}\frac{1}{fR}$$

$$C = \frac{9}{0.7} \times \frac{1}{50 \times 900} = 286\,\mu\text{F}$$

Even a 0.7 V ripple can be too large for many electronic applications. In the next section ways to reduce the ripple voltage to less than 0.1 V will be examined.

Zener diode

The characteristic of an ideal diode, illustrated in Figure 5.52(b) shows that the diode only conducts when it is forward biased. When the diode is reversed biased no current flows. In practice, real diodes will conduct in the reverse direction if the reverse voltage is sufficiently large for the diode to breakdown, as shown in Figure 5.58(b).

The voltage at which the reverse current suddenly rises is called the breakdown or zener voltage, V_z. This voltage depends on the levels of doping in the silicon semiconductor. By regulating the doping levels, manufacturers are able to produce devices, known as zener diodes, that break down at prescribed voltages between 3 V and 20 V. Beyond the 'knee point' the voltage remains constant over a wide variation in current. If a device is used outside this region, small changes in current will produce large changes in voltage. As the main purpose for a zener diode is to provide a stable voltage, manufacturers specify a minimum reverse operating current below which a device should not be used.

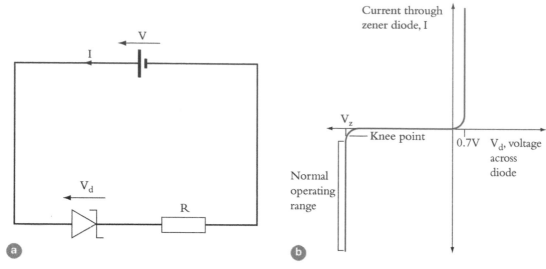

Figure 5.58

Voltage stabilizer

By connecting a zener diode in series with a resistor, R across the output of a smoothed rectifier and connecting the load resistor directly across the terminals of the zener diode, as shown in Figure 5.59, it is possible to produce dc output voltages with a very small ripple. V_{dc} is the average dc output voltage of the smoothed rectifier, I_s is the rectifier output current, V_z is the zener voltage, I_z is the zener diode current, R_z is the effective resistance of the diode when it is conducting, I_o is the load current, V_o is the voltage across the load, R is the series resistance and V_R is the voltage across it.

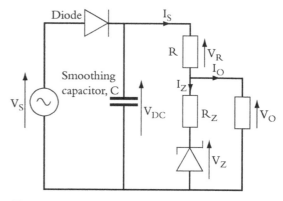

Figure 5.59

From Kirchhoff's Laws

$$I_s = I_o + I_z \qquad (5.96)$$

$$V_{dc} = I_s R + I_z R_z + V_z$$

$$V_{dc} = I_s R + (I_s - I_o) R_z + V_z$$

$$I_s R = V_{dc} - (I_s - I_o) R_z - V_z$$

R is chosen to be significantly larger than R_z so that the second term on the right side of the equation may be ignored and I_s is approximately equal to

$$I_s = \frac{V_{dc} - V_z}{R}$$

As V_{dc}, V_z and R are all constant, I_s will also be constant. As a result as the load current, I_o is increased the current through the zener diode, $I_z = (I_s - I_o)$ falls. If the drop is too large, then the diode current falls below the minimum specified by the manufacturer, the voltage across the diode starts to fall and voltage stabilization is lost. The following example demonstrates how a circuit is designed to ensure that this does not happen.

The voltage stabilizing circuit in Figure 5.59 is supplied by a smoothed rectifier with an average dc output voltage of 13.8 V. The zener diode has a zener voltage, V_z of 9 V, minimum current of 2mA and an effective resistance when conducting of 10 Ω. The maximum stabilizer output current, I_o is 10 mA. Calculate the series resistance, R, the ripple voltage and the average dc voltage across the load.

Operating the voltage stabilizer at maximum output current, $I_0 = 10$ mA:

minimum current, $I_{z\,min} = 2$ mA flows through the zener diode

$$V_o = V_z + I_z R_z$$
$$= 9 + 0.002 \times 10$$
$$= 9.02\,\text{V}$$

from equation (5.96)

$$I_s = I_0 + I_z$$
$$= 0.01 + 0.002 = 0.012\,\text{A} = 12\,\text{mA}$$
$$V_{dc} = I_s R + V_0$$
$$R = \frac{V_{dc} - V_0}{I_s}$$
$$= \frac{13.8 - 9.02}{0.012} = 398\,\Omega$$

Operating the voltage stabilizer on no load $I_0 = 0$:

$$I_s = I_z$$
$$V_0 = I_s R_z + V_z$$

But

$$I_s = \frac{V_{dc} - V_z}{R + R_z}$$
$$= \frac{(13.8 - 9.0)}{(398 + 10)}$$
$$= 0.0118\text{A} = 11.8\,\text{mA}$$
$$V_0 = 0.0118 R_z + V_z$$
$$= 0.0118 \times 10 + 9.0$$
$$= 0.118 + 9.0$$
$$= 9.118\,\text{V}$$

The variation in V_0 between no load ($I_o = 0$) and full load ($I_o = 10$ mA)

$$= 9.118 - 9.02 = 0.098\,\text{V}$$

This variation in V_0 is the ripple voltage,

therefore, the ripple voltage $= 0.098$ V

average load voltage, $V_0 = \dfrac{9.118 + 9.02}{2} = 9.069\,\text{V}$

The ripple voltage is significantly lower than the 0.7 V ripple from the smoothed half wave rectifier and the 28.28 V variation in the output of the unsmoothed half wave rectifier. 0.098 V ripple is satisfactory for many applications, but may still be too large for some electronic systems. Where this is the case a voltage regulator should be used.

Learning summary

You have now completed the section on semiconductor rectifiers and should:

✔ know how a junction diode works;

✔ be able to distinguish between half and full wave rectifiers and controlled and uncontrolled rectifiers;

✔ recognize the output waveforms for each of these rectifiers;

✔ be able to calculate the average dc output voltage for these rectifiers;

✔ understand the principles of voltage smoothing;

✔ be able to calculate the average dc output voltage and ripple voltage from a smoothed supply;

✔ know how a zener diode works;

✔ understand the principles of operation of a voltage stabilizer;

✔ be able to calculate the average dc output voltage from a stabilized supply.

5.8 Amplifiers

The control of large machines needs continual monitoring. The signals from measuring devices are often small and require amplification. Sometimes several signals have to be added together, other times a signal may have to be integrated, for example to convert the output from an accelerometer into a velocity. Amplifiers are used for this. Simple amplifiers may be constructed using discrete transistors, resistors and capacitors but most are manufactured using 'micro-chips' called operational amplifiers, which comprise several transistors, resistors and capacitors on a single chip. This section begins by examining how one of the most common forms of transistor, the bipolar transistor, works.

Bipolar transistor

The *npn* bipolar transistor has a very thin slice of *p*-type silicon about 25 μm thick, called the base, sandwiched between two larger slices of *n*-type silicon known as the emitter and the collector.

When direct voltage supplies are connected to the transistor as in Figure 5.60(a) the junction between the base and the emitter is forward biased encouraging electrons to move from the emitter to the base. Around 97–8 per cent of the electrons pass straight through the thin base and reach the collector, where they are attracted to the positive terminal of the battery. The 2–3 per cent of electrons that do not cross the base flow out of the base terminal to form the base current. As the electrons are negatively charged, it is conventional to show the currents flowing in the opposite direction, as explained in Section 5.2 and illustrated in Figure 5.60(b). Figure 5.60(c) shows this circuit redrawn using the conventional symbol for a transistor.

It can be seen from these diagrams that:

$$I_E = I_B + I_C$$

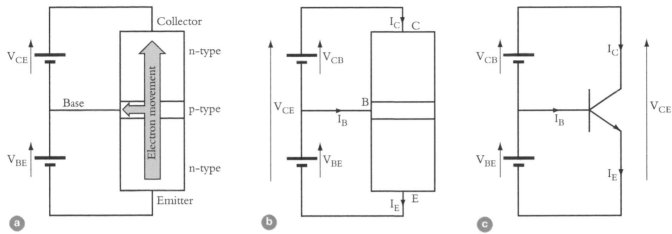

Figure 5.60

With between 2 and 3 per cent of the electrons from the emitter flowing out through the base to form the base current and the remainder going on to the collector to form the collector current

$$0.02I_E < I_B < 0.03I_E$$

$$0.97I_E < I_C < 0.98I_E$$

thus

$$33I_B < I_E < 50I_B$$

and

$$32I_B < I_C < 49I_B$$

By regulating the base current it is possible to control the far larger current flowing between the collector and emitter. Figure 5.61 shows a typical characteristic for a small transistor. The graph shows the collector current, I_C *versus* collector-emitter voltage, V_{CE} for various values of base current, I_B. For collector-emitter voltages greater than 0.7 V the collector current is virtually constant for any given value of base current and in this case is approximately 40 times larger than the base current

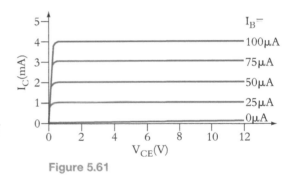

Figure 5.61

Transistor amplifier

These characteristics are used to design simple amplifiers such as that shown in Figure 5.62(a). Here the voltage across R_C, $V_{RC} = I_C R_C$.

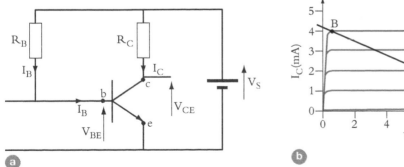

a

Figure 5.62

By Kirchhoff's Voltage Law, the voltage across the collector resistor,

$$V_{RC} = V_S - V_{CE}$$

$$I_C = \frac{V_S - V_{CE}}{R_C}$$

rearranging

$$I_C = \frac{-1}{R_C} V_{CE} + \frac{V_S}{R_C} \qquad (5.97)$$

This equation is of the form $y = mx + C$ which defines a straight line. In this case, the slope of the graph equals $\frac{-1}{R_C}$. The graph intersects the collector current, I_C axis at $\frac{V_S}{R_C}$ and crosses the collector–emitter voltage, V_{CE} axis at V_S.

To find the ideal value for the collector resistance, R_C, a straight line is drawn on the transistor characteristic that passes through the supply voltage V_S on the V_{CE} axis and through the 'knee' of the curve for the largest base current, in this case 100 μA, as illustrated in Figure 5.62(b). The required value for R_C is then calculated from where the line, known as the load line, crosses the vertical I_C axis. In this example, this is 4.2 mA.

From equation (5.97), when $V_{CE} = 0$

$$I_C = \frac{V_S}{R_C}.$$

$$R_C = \frac{V_S}{I_C} = \frac{12}{4.2 \times 10^{-3}} = 2.9 \ k\Omega$$

The transistor will operate at a point along the load line defined by the base current. To amplify signals which normally have positive and negative fluctuations, this point should be in the middle of the transistor characteristic. Ideally, the base current should equal $\frac{1}{2}I_{Bmax}$, making Q the best operating point.

By Kirchhoff's Voltage Law

$$V_S = I_B R_B + V_{BE}$$

$$I_B = \frac{V_S - V_{BE}}{R_B}$$

The base current is dependent on V_{BE} and R_B.

$$R_B = \frac{V_S - V_{BE}}{I_B}$$

When the transistor is conducting the base emitter junction is forward biased with 0.7 V across the junction i.e. $V_{BE} = 0.7$ V.

$$R_B = \frac{V_S - 0.7}{I_B}$$

For I_B to equal $\frac{1}{2}I_{Bmax}$, which in this example is 50 μA, R_B must be set at

$$R_B = \frac{12 - 0.7}{50 \times 10^{-6}} = 226 \ k\Omega$$

Figure 5.62(b) shows that when $I_B = 50$ μA the collector current, I_C equals 2.0 mA and the collector emitter voltage V_{CE} is 6.2 V.

When an input signal is applied, additional currents flow as shown in Figure 5.63(a). The signal current i_B combines with I_B flowing through R_B making the base current equal to $(I_B + i_B)$. It is the total base current that determines the operating point at any instant. As the signal current varies, the operating point moves along the load line. With a positive signal current, the base

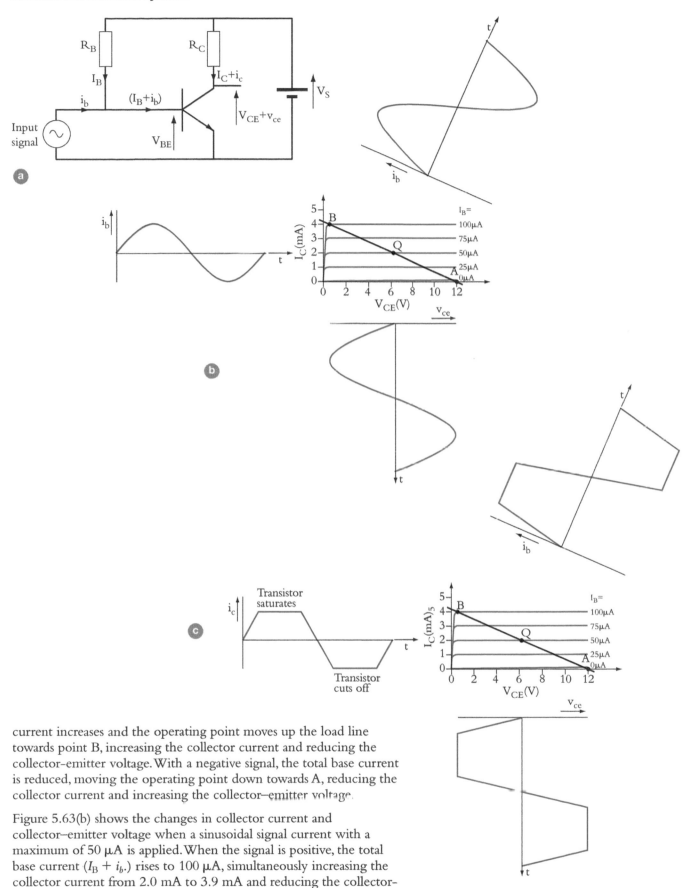

current increases and the operating point moves up the load line towards point B, increasing the collector current and reducing the collector–emitter voltage. With a negative signal, the total base current is reduced, moving the operating point down towards A, reducing the collector current and increasing the collector–emitter voltage.

Figure 5.63(b) shows the changes in collector current and collector–emitter voltage when a sinusoidal signal current with a maximum of 50 μA is applied. When the signal is positive, the total base current ($I_B + i_b$.) rises to 100 μA, simultaneously increasing the collector current from 2.0 mA to 3.9 mA and reducing the collector–emitter voltage from 6.2 V to 0.7 V.

Figure 5.63

Change in collector current, i_c = collector current at point B – collector current at point Q

$\Delta i_c = 3.9$ mA – 2.0 mA = 1.9 mA

Change in collector–emitter voltage, v_{ce}
= collector-emitter voltage at B – collector–emitter voltage at Q

$\Delta v_{ce} = 0.7$ V – 6.2 V = –5.5 V

The minus sign indicates that a rise in input signal produces a fall in output voltage, i.e. the output voltage is the inverse of the input signal and the transistor is operating as an inverting amplifier.

As the signal current falls back to zero, the operating point moves back to Q, reducing the collector current and increasing the collector-emitter voltage. When the signal current reaches –50 µA, the operating point falls to A, where the collector current is 0.1 mA and collector-emitter voltage is 11.7 V.

Change in collector current, i_c = collector current at point Q – collector current at point A

$$= 2.0 \text{ mA} - 0.1 \text{ mA} = 1.9 \text{ mA}$$

Change in collector-emitter voltage, v_{ce}
= collector-emitter voltage at Q – collector–emitter voltage at A

$$= 6.2 \text{ V} - 11.7 \text{ V} = -5.5 \text{ V}$$

These changes are the same for both halves. Amplification is linear and there is no distortion in the output signal.

> The ratio of $\dfrac{\text{change in collector current}}{\text{change in base current}}$ is called the current gain, which is usually given the symbol h_{fe}

in this case current gain, $h_{fe} = \dfrac{1.9 \times 10^{-3}}{50 \times 10^{-6}} = 38$

If the peak signal current exceeds $\frac{1}{2}I_{\text{Bmax.}}$ the output voltage signal becomes distorted. The transistor will saturate on the positive peaks when the total base current exceeds the maximum and will cut off on the negative peaks when the total base current falls below zero, as demonstrated in Figure 5.63(c).

To prevent interference between the small signal currents and the dc currents, known as transistor-biasing currents, capacitors are included in a practical circuit, as shown in Figure 5.64. These capacitors are chosen to have low reactance at the signal frequency to allow the alternating signal to pass from the input source to the base and from the collector to the output while blocking the direct bias currents from flowing to the signal source and to the load where they may cause signal distortion.

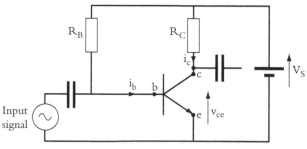

Figure 5.64

Analysing a transistor amplifier

To analyse a transistor amplifier, it is convenient to use a small signal equivalent circuit. The most simple equivalent circuit for a bipolar transistor is shown in Figure 5.65.

i_b is the small signal component of the base current and v_{ce} the corresponding change in collector-emitter voltage. h_{ie} is the transistor input resistance, that is the resistance of the base region and of the junction between the base and emitter. It is generally in the range 800 Ω to 2000 Ω. (h_{fe} i_b) is a current generator that produces a current equal to the base current i_b multiplied by the current gain h_{fe}. This equivalent circuit replaces the transistor in the practical circuit (Figure 5.64) as shown in Figure 5.66(a).

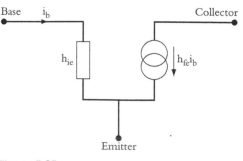

Figure 5.65

Electrical and electronic systems

The capacitors have a low reactance at the signal frequency and the dc sources offer negligible impedance to alternating signals so both may be replaced by short circuits allowing the circuit to be simplified to Figure 5.66(b).

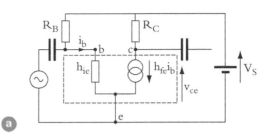

Using a typical value for h_{ie} of 1200 Ω and peak input current of 50 μA the peak base current, i_b can be calculated.

Peak base current $i_b = 50 \times 10^{-6} \times \dfrac{226 \times 10^3}{(226 \times 10^3 + 1200)} = 49.7\ \mu A$

The current gain, h_{fe}, calculated previously, is 38; hence,

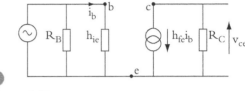

peak current produced by current generator, $(h_{fe}\ i_b) = 38 \times 49.7$ $\mu A = 1.9\ mA$

and peak collector–emitter voltage,

$v_{ce} = (h_{fe}\ i_b)\ R_c = -1.9 \times 10^{-3} \times 2.9 \times 10^3 = -5.5\,V$

Figure 5.66

These values are the same as those found when the signal waveform was superimposed on the transistor characteristic and load line, Figure 5.63(b).

To convert peak currents and voltages to rms quantities, they must be divided by $\sqrt{2}$.

In practice, to achieve the desired amplification it may be necessary to connect a number of transistors in cascade. These circuits can be analysed using the principles described above.

Operational amplifiers

Today, amplifiers usually comprise a large number of transistors, resistors and capacitors all on a single chip or integrated circuit. One of the most common is the 741 which contains 20 transistors, 11 resistors and one capacitor in one small package known as an operational amplifier or op-amp for short. Although the actual circuit is quite complex, the 741 can be represented by the equivalent circuit shown in Figure 5.67(a). It has an inverting input identified by a − sign, a non-inverting input shown with a + and an output terminal.

> The ratio of output voltage, $\dfrac{V_0}{v}$
> where v = potential of the inverting input − potential of the non-inverting input
> is called voltage gain of the amplifier, or the open loop gain, A.

Figure 5.67(b) shows the graph of voltage gain versus frequency. It is customary to draw this graph with logarithmic scales on both axes. It shows that the amplification is high at low frequencies but falls away as the frequency is increased.

The amplifier has a high resistance, R_i connected between the input terminals. This is often assumed to be infinite. The controlled voltage generator has an output equal to A multiplied by v and is connected to the output terminal via a low resistance, R_0, which is usually taken to be zero.

The large variation of gain with frequency is unacceptable for most applications. Most signals are made up of a wide range of frequencies, for example music and speech. To prevent distortion of the signal, it is necessary to amplify all frequencies uniformly. This is achieved by using negative feedback, connecting a resistor between the output terminal and inverting input.

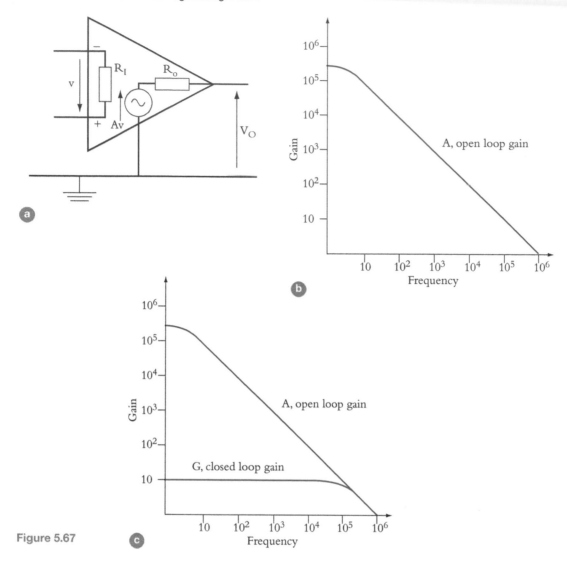

Figure 5.67

Negative feedback extends the range of frequencies over which the gain remains constant at the cost of reducing the gain at low frequencies as illustrated in Figure 5.67(c). The overall voltage gain of the amplifier with feedback is called the closed loop gain, G and is defined as

$$G = \frac{\text{output voltage, } V_0}{\text{input voltage, } V_i}$$

The magnitude of G depends on the feedback and must be calculated for each case but is always equal to or less than that of A. Here are some common examples of practical circuits.

Inverting amplifier

The inverting amplifier, shown in Figure 5.68(a) has two external resistors, the feedback resistor R_f and input resistor R_1.

Assuming that there is an infinite resistance between the inverting and non-inverting inputs to the op-amp

$$i = 0 \text{ and } i_1 = -i_f$$

using Kirchhoff's Voltage Law

$$V_i = i_1 R_1 - v$$
$$V_o = i_f R_f - v$$

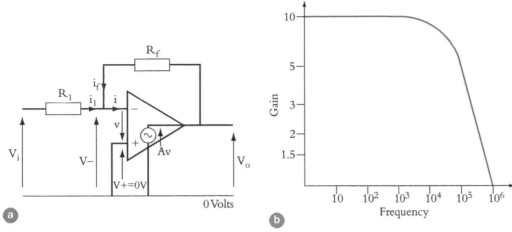

Figure 5.68

The controlled voltage generator output voltage $= Av$

with zero op-amp output resistance $V_0 = Av$

$$v = \frac{V_0}{A}$$

$$V_o = i_f R_f - \frac{V_0}{A}$$

$$i_f = \frac{V_0}{R_f}\left[1 + \frac{1}{A}\right] = -i_1$$

$$V_i = i_1 R_1 - \frac{V_0}{A}$$

$$V_i = -\frac{V_0 R_1}{R_f}\left[1 + \frac{1}{A}\right] - \frac{V_0}{A}$$

$$V_i = -\frac{V_0[R_1(A+1) + R_f]}{R_f A}$$

Closed loop gain, G, (i.e. gain of the amplifier with feedback)

$$= \frac{V_0}{V_i}$$

$$G = \frac{V_0}{V_i} = \frac{-R_f A}{[R_1(A+1) + R_f]}$$

(5.98)

The negative sign indicates that the output voltage is inverted. Using the graph for the open loop gain, A in Figure 5.67(b) G may be calculated for any combination of R_1 and R_f. Figure 5.68(b) shows the closed loop gain when $R_1 = 100$ kΩ and $R_f = 1$ MΩ.

In this case, the closed loop gain is reasonably constant up to 1 kHz but falls beyond this frequency. From this it may be concluded that this amplifier is only suitable for the reproduction of low frequency signals when A is much greater than 1 in which case equation (5.98) may be simplified to:

overall gain,
$$G = \frac{V_0}{V_i} = -\frac{R_f}{R_1}$$

To operate at frequencies higher than 1 kHz it would be necessary to replace the 741 by an op-amp with a wider frequency range to achieve this overall gain.

Summing amplifier

There are many mechanical engineering applications, such as in control systems, where it is necessary to add two or more signals. In these situations a multiple input version of the inverting amplifier, called the summing amplifier is used. The example in Figure 5.69 has three inputs.

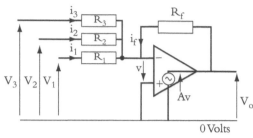

Figure 5.69

Using Kirchhoff's Voltage Law

$$V_1 = i_1 R_1 - v$$

$$V_2 = i_2 R_2 - v$$

$$V_3 = i_3 R_3 - v$$

$$V_o = i_f R_f - v$$

If the op-amp output resistance is zero $Av = V_0$ and $v = \dfrac{V_o}{A}$

As A is very large v tends to zero and the four voltage equations may be simplified

$$V_1 = i_1 R_1$$

$$V_2 = i_2 R_2$$

$$V_3 = i_3 R_3$$

$$V_o = i_f R_f$$

As with the inverting amplifier, if the op-amp input resistance is infinite the current into the inverting input

$$i = 0$$

and

$$i_1 + i_2 + i_3 = -i_f$$

hence

$$\frac{V_1}{R_1} + \frac{V_2}{R_2} + \frac{V_3}{R_3} = -\frac{V_0}{R_0}$$

$$V_0 = -R_f\left[\frac{V_1}{R_1} + \frac{V_2}{R_2} + \frac{V_3}{R_3}\right] \tag{5.99}$$

If $R_f = R_1 = R_2 = R_3$ then

$$V_0 = -[V_1 + V_2 + V_3]$$

i.e. output voltage $= -\Sigma$(input voltages)

The output signal is the inverse of all the input signals added together.

Non-inverting amplifier

There are many situations where the output signal must not be inverted. In such cases a non-inverting amplifier like that shown in Figure 5.70(a) is used. With this type of amplifier the input is connected to the non-inverting input of the op-amp and a fraction of the output voltage is fed back to the inverting input of the op-amp.

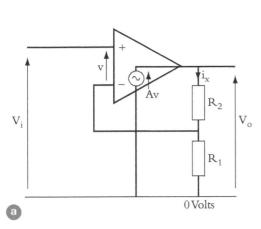

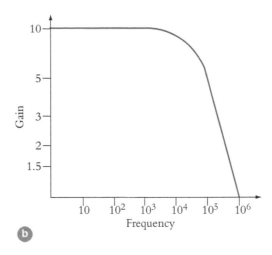

Figure 5.70

Output voltage, $V_0 = i_x (R_2 + R_1)$

With infinite op-amp input resistance the current entering the inverting input is zero and the potential of the inverting input, V_- is given by:

$$V_- = i_x R_1 = \frac{V_0 R_1}{(R_2 + R_1)}$$

By Kirchhoff's Voltage Law

Input voltage, $V_i = V_- + v$

But

$$V_0 = Av$$

$$v = \frac{V_0}{A}$$

$$V_i = \frac{V_0 R_1}{(R_2 + R_1)} + \frac{V_0}{A}$$

$$= V_0 \left[\frac{R_1}{R_2 + R_1} + \frac{1}{A} \right]$$

$$G = \frac{V_o}{V_i} = \frac{(R_2 + R_1)A}{R_1(A+1) + R_2} \tag{5.100}$$

G is positive and the output voltage is not inverted.

Using the open loop gain, A versus frequency characteristic for the 741, Figure 5.67(b), the closed loop gain, G may be calculated for any combination of R_1 and R_2. Figure 5.70(b) shows the closed loop gain when $R_1 = 100\ k\Omega$ and $R_2 = 900\ k\Omega$. The overall gain of the amplifier falls at frequencies above 1 kHz, making it unsuitable for operation at high frequencies. Like the inverting amplifier, the fall is due to the limited frequency range of the 741. To operate at higher frequencies the 741 would have to be replaced by an op-amp with a wider frequency range.

Integrating amplifier

There are situations where it is necessary to integrate a signal, for example to derive the velocity from the output of an accelerometer. In such cases an integrating amplifier, such as that shown in Figure 5.71, has to be used. This is similar to the inverting amplifier but with the feedback resistor replaced by a capacitor. Like the previous examples, this amplifier must be operated at low frequencies where the open loop gain, A is high.

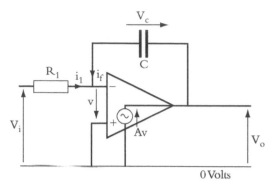

Figure 5.71

Using equation (5.41)

$$i_f = C\frac{dV_C}{dt}$$

$$V_C = \frac{1}{C}\int i_f\,dt \tag{5.101}$$

output from the voltage controlled generator, $Av = V_0$

$$v = \frac{V_0}{A}$$

$$V_0 = V_C - v = V_C - \frac{V_0}{A}$$

$$V_i = i_1 R_1 - v = i_1 R_1 - \frac{V_0}{A}$$

when

$$A \gg 1$$

$\dfrac{V_0}{A}$ tends to zero, so that

$$V_0 = V_C \tag{5.102}$$

and

$$V_i = i_1 R_1$$

$$i_1 = \frac{V_i}{R_1} \tag{5.103}$$

the op-amp has infinite resistance between its input terminals, so

$$i_f = -i_1 \tag{5.104}$$

substituting (5.102), (5.103) and (5.104) in (5.101) .

Hence,

$$V_0 = -\frac{1}{C}\int_0^t \frac{V_i}{R_1}\,dt$$

$$\boxed{V_0 = -\frac{1}{CR_1}\int_0^t V_i\,dt} \tag{5.105}$$

The amplifier output voltage is minus the integral of the input voltage multiplied by $\dfrac{1}{CR_1}$

5.9 Digital electronics

Digital electronics are used in computers, microprocessors and control systems. These differ from analogue systems in one important aspect. In analogue systems currents and voltages continually vary but in digital electronics only two states exist, on and off. The two conditions are mutually exclusive. This is known as two–state binary logic. Logic devices are divided into two categories – combinational logic and sequential logic. Combinational logic gates are used to construct control circuits. The basic element in sequential logic is the bistable. This is a memory cell and the rudimentary component in any computer and microprocessor.

Combinational logic

The easiest way to understand how combinational logic systems work is to look at a number of simple lighting circuits. Figure 5.72a shows a battery supplying a lamp via a switch. The lamp is only illuminated when the switch is closed, the light is off when the switch is open.

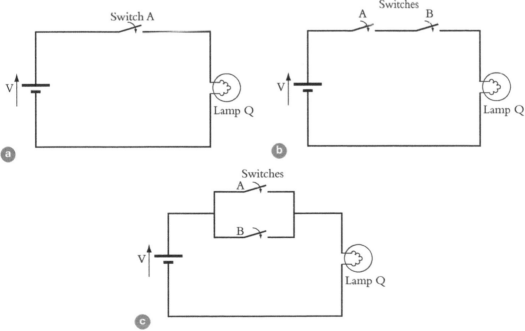

Figure 5.72

This may be summarized in the form of a truth table:

Switch A	Lamp Q
open	off
closed	on

If 0 is used to represent open or off and 1 used to represent closed or on, then the operation of the circuit may be described thus:

Switch A	Lamp Q
0	0
1	1

Any system in which the state of all points must be either 0 or 1 is known as a binary system.

A special type of algebra is used to analyse such systems. It is called Boolean algebra.

This case, where the state of Q is always equal to the state of A, is expressed in Boolean algebra as:

$$Q = A$$

Consider now what happens when two switches are connected in series as in Figure 5.72(b). The lamp is only lit when both switches are closed. This may be summarized:

Switch A	Switch B	Lamp Q
open	open	Off
open	closed	Off
closed	open	Off
closed	closed	On

and rewritten in binary form using only 1s and 0s thus:

A	B	Q
0	0	0
0	1	0
1	0	0
1	1	1

As the lamp is only lit when switch A *and* switch B are both closed, this is an example of AND logic, which may be expressed in Boolean algebra as:

$$Q = A \cdot B$$

Boolean algebra has its own collection of symbols. AND is represented by a dot (not as one might expect by a plus sign). The logic symbol for the AND gate is shown in Table 5.4

When the switches are connected in parallel as in Figure 5.72(c) the lamp is illuminated when either one or both switches are closed. Writing this in a truth table:

A	B	Q
0	0	0
0	1	1
1	0	1
1	1	1

The output is 1 when either A or B or both are 1. This is an example of OR logic for which the Boolean expression is:

$$Q = A + B$$

In Boolean algebra the $+$ sign is used to represent OR. The logic symbol for the OR gate is shown in Table 5.4.

Type of gate	Truth table	Boolean expression	Logic symbol
NOT	A B 0 1 1 0	$Q = \bar{A}$	
AND	A B Q 0 0 0 0 1 0 1 0 0 1 1 1	$Q = A.B$	
NAND (not AND)	A B Q 0 0 1 0 1 1 1 0 1 1 1 0	$Q = \overline{A.B}$	
OR	A B Q 0 0 0 0 1 1 1 0 1 1 1 1	$Q = A + B$	
NOR (not OR)	A B Q 0 0 1 0 1 0 1 0 0 1 1 0	$Q = \overline{A + B}$	
XOR (exclusive OR)	A B Q 0 0 0 0 1 1 1 0 1 1 1 0	$Q = A \oplus B$	

Table 5.4

Other common types of logic gate are:

NOT (the output is the inverse of the input)
NOT AND (or NAND as it is more usually called)
NOT OR (or NOR) and
Exclusive OR (XOR for short).

These are all listed in Table 5.4 with their logic symbol, Boolean expression and truth table.

> The bar over the A in the Boolean expression for the NOT gate indicates inversion, i.e. equals NOT A.

For the same inputs as the AND gate the NAND produces the opposite outputs i.e. where an AND would produce a 1 a NAND will produce a 0 and vice versa. Similarly the NOR produces the opposite outputs to the OR for the same inputs. The XOR differs from the OR in that it only produces a 1 at the output when one input is 1. When both inputs are 1, the output is 0.

Derive the truth table for a logic circuit

To derive the truth table for a logic circuit comprising a number of devices the first step is to identify the inputs and outputs for each device with a unique label and to write these on the circuit diagram as shown in the example in Figure 5.73.

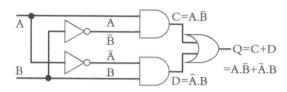

Figure 5.73

The truth table is constructed with a column for every device input and output and a row for every combination of circuit input states, as shown in Table 5.5. With two inputs A and B there will be four combinations of input states – both equal to 0, $A = 0$ and $B = 1$, $A = 1$ and $B = 0$ and both equal to 1.

A	B	$\bar{A}$	$\bar{B}$	$C = A.\bar{B}$	$D = \bar{A}.B$	$Q = A.\bar{B} + \bar{A}.B$
		not A	not B	A and not B	not A and B	C or D
0	0	1	1	0	0	0
0	1	1	0	0	1	1
1	0	0	1	1	0	1
1	1	0	0	0	0	0

Table 5.5

At any instant every point in the circuit will be in one of two states, 0 or 1. To determine the outputs of the AND gates C and D the truth table for an AND gate, shown in Table 5.4, is used. To find the output of the OR gate, Q the truth table for an OR gate, also shown in Table 5.4, is used

The first row in Table 5.5 shows the situation when inputs A and B are both 0. When B equals 0 NOT B (which is identified by $\bar{B}$) will be 1. The inputs to the AND gate C are A and $\bar{B}$. Using the truth table for an AND gate in Table 5.4 when one input is 0 and the other is 1, the output is 0 so C will equal 0. Similarly for AND gate D $\bar{A} = 1$ and $B = 0$ so the output D will also be 0. C and D form the inputs to the OR gate. From the OR gate truth table in Table 5.4 when both inputs are 0 the output is 0, so with $C = 0$ and $D = 0$ the OR gate output Q will be 0.

Construction of a combinational logic circuit

To demonstrate how combinational logic is used in the construction of electronic circuits, consider the example of a simple burglar alarm, which has two detection circuits A and B.

A: An infrared detector: this is used to identify when an intruder is present. It normally produces a low output (i.e. =0) but goes high (=1) when an intruder is detected.

B: Switches fitted to all the doors and windows and connected in series with a battery. This circuit produces a high output when all the doors and windows are closed but goes low when any door or window is open.

The outputs from the detection systems are fed to a control box, Q which goes high to sound an alarm when either security network is breached.

To solve this problem, the first step is to summarize this information (Table 5.6) in a format from which the truth table can be extracted.

Infra red detector A	Doors and windows detector B	Control box Q
No intruders detected $A = 0$ intact $Q = 0$	Door or window open $B = 0$	Both detection circuits are intact $Q = 0$
Intruder detected $A = 1$ is breached $Q = 1$	All doors and windows are closed $B = 1$	Either detection circuit network is breached $Q = 1$

Table 5.6

The results are then tabulated in the form of a truth table with a row for every combination of states for A and B, Table 5.7.

A	B	Q	
0	0	1	(door or window open)
0	1	0	
1	0	1	(intruder detected, door or window open)
1	1	1	(intruder detected)

Table 5.7

The next stage is to determine the Boolean expression for Q. This may be done by trial and error but a more systematic way is to draw a Karnaugh map.

Karnaugh Map

The Karnaugh Map is a matrix with a cell for every combination of the input states and populated with the output states (Q) identified in the truth table. In the case of the burglar alarm the Karnaugh map has 0 in the cell corresponding to $A = 0$ $B = 1$ and 1s in the other three cells, as shown in Table 5.8(a):

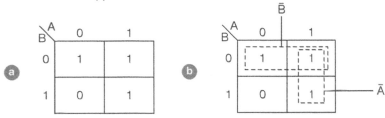

Table 5.8

The Boolean expression for Q is found by grouping together in rows and columns (but not diagonals) all the cells containing 1s, as illustrated in Table 5.8(b).

The two 1s in the top row indicate that $Q = 1$ when $B = 0$ irrespective of the state of A. The two 1s in the right column depend only on A being equal to $\underline{1}$. Putting these conditions together Q will equal 1 when $A = 1$ or $B = 0$ (and therefore $\bar{B} = 1$). This is expressed in Boolean algebra as

$$Q = A + \bar{B}$$

Using this expression the control circuit may be designed as shown in Figure 5.74(a). First the input B must be inverted to produce $\bar{B}$. This is then combined with A in an OR gate to form Q.

Figure 5.74

Alternatively, the Boolean expression for Q may be found by selecting the cells in the Karnaugh Map containing 0s, which correspond to $\bar{Q}$, and inverting the result. In this case there is a 0 in the cell $A = 0$ and $B = 1$ so

$$\bar{Q} = \bar{A}B$$

hence

$$Q = \overline{\bar{A}B}$$

$$\overline{\bar{A}B} = A + \bar{B}$$

It is therefore possible to construct an alternative circuit, Figure 5.74(b), that will perform the same function as Figure 5.74(a).

Combinational logic is used is in ticketing and vending machines where the customer has options. The following example demonstrates how the logic circuit is devised.

Motorists using a car park must purchase a £4 ticket from a machine that accepts only £1 and £2 coins and does not give change. Coins are inserted into the machine through the coin slot and roll down the chute to the coin-sorting platform where the £1 and £2 coins are separated into different columns, as illustrated in Figure 5.75(a).

Sensors detect when coins are in the columns. The sensors normally produce a low output but go high when they are in contact with a coin. Sensor B indicates that the first £2 coin has been entered. Sensor A responds when a second £2 coin has been inserted. Sensor D shows that at least two £1 coins have been put in and sensor C indicates that four £1 coins have been entered.

The output from each sensor is fed to a controller which normally produces a low output, Q_1 but goes high to issue a ticket when a minimum of £4 has been paid. Should a fault occur in the machine a second output, Q_2 is derived. This is normally low but goes high when a fault is detected, to illuminate a light to show that the machine is out of order.

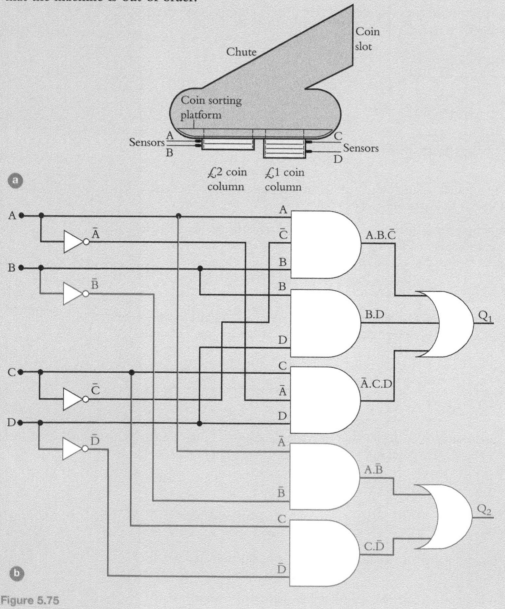

Figure 5.75

The first step is to establish the truth table for the controller output, Q_1 for all combinations of the four sensor outputs, Table 5.9.

A Two x £2	B One x £2	C Four x £1	D Two x £1	Q1 Issue ticket	Comments	Q2 Machine out of order
0	0	0	0	0	Nothing paid	0
0	0	0	1	0	Only £2 paid	0
0	0	1	0	0	Impossible	1
0	0	1	1	1	OK £4 paid	0
0	1	0	0	0	Only £2 paid	0
0	1	0	1	1	OK £4 paid	0
0	1	1	0	0	Impossible	1
0	1	1	1	1	£6 paid	0
1	0	0	0	0	Impossible	1
1	0	0	1	0	Impossible	1
1	0	1	0	0	Impossible	1
1	0	1	1	0	Impossible	1
1	1	0	0	1	OK £4 paid	0
1	1	0	1	1	£6 paid	0
1	1	1	0	0	Impossible	1
1	1	1	1	1	£8 paid	0

Table 5.9

It can be seen from Table 5.9 that certain combinations are impossible, as A cannot equal 1 (two £2 coins inserted) when B, which should go high when the first £2 coin is entered, is 0. Similarly C cannot equal 1 (four £1 coins inserted) when D, which goes high to identify that the first two £1 coins have been entered, is 0. Should these situations occur, the machine has a fault and the 'out of order' light should be illuminated, but this will be dealt with later. The next step is to draw the Karnaugh map for the controller output, Q_1, Table 5.10a.

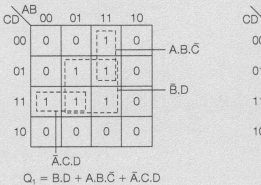

$$Q_1 = B.D + A.B.\bar{C} + \bar{A}.C.D$$

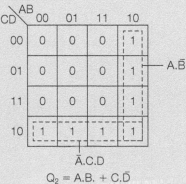

$$Q_2 = A.B. + C.\bar{D}$$

Table 5.10

The 1s in the Karnaugh Map are combined into the largest groups possible. The four 1s which form a square in the centre indicate that $Q_1 = 1$ when $B = 1$ and $D = 1$ irrespective of the values of A and C. This corresponds to $B.D$ in Boolean algebra.

Similarly there is a column of two 1s corresponding to the Boolean expression $A.B.\overline{C}$ and a row of two 1s corresponding to $\overline{A}.C.D$.

These three expressions are then combined to give the Boolean expression for Q_1 which covers all the cases when Q_1 equals 1:

$$Q_1 = B.D + A.B.\overline{C} + \overline{A}.C.D$$

From this equation the most basic logic circuit for output Q_1 to issue a ticket can be devised as shown in black in Figure 5.75(b).

Should an 'impossible' state arise the second output Q_2 should go high to illuminate the 'out of order' to warn other motorists not to use the machine. The required value for Q_2 for each input combination is shown in the far right column of the truth table, Table 5.9. Using this column the Karnaugh map for Q_2 is drawn, as shown in Table 5.10(b), the largest groups of 1s identified, the corresponding Boolean expression found and the additional logic circuitry devised. The extra wiring required is shown in red in Figure 5.75(b).

Binary notation

It is important to have a thorough understanding of binary notation. Everyone is familiar with decimal notation where numbers are represented by a series of digits from 0 to 9 with each digit in the series having a different weighting, with the highest on the left and the lowest on the right. For example, in the number 749 the first digit represents seven hundreds, the second four tens and the last, nine units. Expressing this mathematically, we have

$$7 \times 100 + 4 \times 10 + 9 \times 1 = 749$$

which may be rewritten:

$$7 \times 10^2 + 4 \times 10^1 + 9 \times 10^0 = 749$$

Binary is very similar but uses only 0s and 1s. Numbers larger than 1 are represented by a series of 0s and 1s, where each digit, or bit as they are usually called, in the chain has a different weighting. If there are x bits in the series then the first bit has a weighting of 2^{x-1}, the second 2^{x-2}, down to the last which has a weighting of 2^0. For example, to find the value of the 3-bit chain 101, the first bit is multiplied by 2^2, the second by 2^1 and the last by 2^0.

$$1 \times 2^2 + 0 \times 2^1 + 1 \times 2^0 = 5$$
$$4 + 0 + 1 = 5$$

To distinguish binary numbers from decimal the former are usually followed by the suffix $_2$ and decimal numbers by the suffix $_{10}$. For example, $1100_2 = 12_{10}$.

Calculate the decimal equivalent of of the binary number 101110_2

$$1 \times 2^5 + 0 \times 2^4 + 1 \times 2^3 + 1 \times 2^2 + 1 \times 2^1 + 0 \times 2^0 = 46_{10}$$

Sequential logic

The basic element in a sequential logic circuit is the bistable or flip-flop. This is the simplest form of memory cell. The bistable has two stable output states 0 and 1 and once set, the output will remain unchanged until the device is reset. One of the most common forms of bistable is the JK. It has two data inputs J and K, a clock input (marked >), a clear (CLR) input and two complementary outputs Q and $\overline{Q}$. See Figure 5.76.

Electrical and electronic systems

The outputs will only change when either

(a) a reset <u>pulse</u> is received at the clear input when the output Q is set to 1 and output Q goes to 0, or

(b) a clock pulse is received.

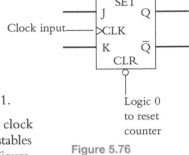

Figure 5.76

In the second case the output states depend on the states of the inputs J and K immediately before the clock pulse is received as described in the state table, Table 5.11.

Q_n is the state of output Q before the clock pulse arrives and Q_{n+1} is its state after the clock pulse. Unlike gates where outputs respond instantaneously to changes at the inputs, bistables only change **after** a clock pulse. To demonstrate this examine the timing diagram in Figure 5.77. It has been assumed that initially $J = 1$, $K = 0$ and $Q = 0$. The output remains low (i.e. equal to 0) until the end of the first clock pulse when the output, Q rises to 1, as defined in the state table, Table 5.10. Before the second clock pulse J falls to 0 but the output remains high (=1). With both outputs low, there is no change at the output after the second clock pulse. Before the third clock pulse K rises to 1 but Q stays high until after the clock pulse is complete, when Q resets to 0. The reader is left to check the outcomes after the next five clock pulses.

J	K	Q_{n+1}
0	0	Q_n
0	1	0
1	0	1
1	1	Q_n

Table 5.11

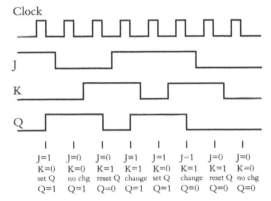

Figure 5.77

Asynchronous binary counter

JK bistables may be cascaded as shown in Figure 5.78(a) to form an asynchronous binary counter. This particular example has two bistables and is known as a 2-bit counter.

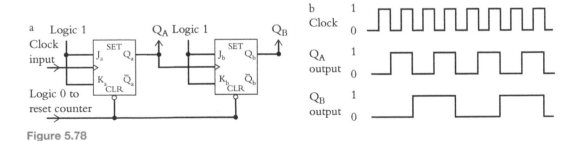

Figure 5.78

With the J and K inputs of both bistables permanently connected to logic 1 (i.e. kept high) the outputs will change after a pulse at their clock input. As a result the second bistable will only change when the output of the first bistable Q_a falls to 0. Assuming that the outputs have been cleared at the start Q_a will rise to + after the first clock pulse and Q_b will remain unchanged. After the second clock pulse to bistable A Q_a will drop to 0 triggering Q_b to rise to 1. The third clock pulse will see Q_a rise again and Q_b remain unchanged. The fourth clock pulse will cause Q_a and hence Q_b to fall to 0. The timing diagram is shown in Figure 5.78(b).

The output from each bistable represents one digit in a binary number and has the effective weighting of 2^{x-1} where x is its position in the chain. The first bistable (A) has a weighting of $2^0 = 1$, the second (B) $2^1 = 2$, etc. The count after each clock pulse can then be found by multiplying the state of Q for each bistable by their respective weighting and adding the results, as shown in Table 5.12. After the fourth clock pulse, both outputs fall to 0 resetting the counter to 0.

The output from the bistable with the lowest weighting is called the least significant bit (LSB) and the output from the bistable with highest weighting is called the most significant bit (MSB). In this case Q_a is the LSB and Q_b the MSB.

Number of clock pulses received	Q_a LSB	Q_a multiplied by weighting of $A = 2^0 = 1$	Q_b MSB	Q_b multiplied by weighting of $B = 2^1 = 2$	Total count (= column 2 + column 4)
0	0	0	0	0	0
1	1	1	0	0	1
2	0	0	1	2	2
3	1	1	1	2	3
4	0	0	0	0	0

Table 5.12

The count can be displayed directly in binary by rearranging the table to display the output columns in the order of the bistable weightings with the highest weighted (MSB) column on the left descending to the lowest weighted (LSB) column on the right, as demonstrated in Table 5.13. For example, when the MSB is 1 and LSB is 0, as in the third row of the table, the count in binary is 10, which corresponds to 2 in decimal.

MSB	LSB	Count in binary	Count in decimal
0	0	00	0
0	1	01	1
1	0	10	2
1	1	11	3
0	0	00	0

Table 5.13

Unfortunately, it takes a finite time for the output of the first bistable to change and to clock the second bistable. As a result a ripple effect is created as each bistable operates in turn. This can cause problems when there is a large number of bistables in cascade and is a major disadvantage with this type of counter.

Synchronous binary counter

The problem is overcome in the synchronous binary counter, shown in Figure 5.79 where the bistables are clocked simultaneously. The state table is shown in Table 5.14.

First a CLR (clear) pulse resets both the bistable Q outputs to 0. The inputs to bistable A, J_a and K_a are permanently connected to logic1 but the inputs to bistable B, J_b and K_b are derived from the output of the AND gate for which the Boolean expression equals $(1 . Q_a)$. Initially $Q_a = 0$ and the gate output will be 0, see the AND gate in Table 5.4. As a result both J_b and K_b will be 0. This data is recorded in the first row of the state table.

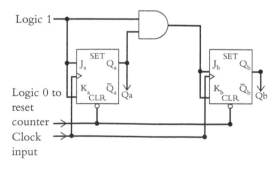

Figure 5.79

Electrical and electronic systems

	2-bit synchronous counter					
	J_a state $=1$	K_a $=1$	Q_a LSB	J_b $=1.Q_a$	K_b $=1.Q_a$	Q_b MSB
Both bistable Q outputs reset Calculate bistable inputs	1	1	0	0	0	0
Q outputs after 1st clock pulse Recalculate bistable inputs	1	1	1	1	1	0
Q outputs after 2nd clock pulse 0 Recalculate bistable inputs	1	1	0	0	0	1
Q outputs after 3rd clock pulse Recalculate bistable inputs	1	1	1	1	1	1
Q outputs after 4th clock pulse **Cycle repeats modulo = 4** Recalculate bistable inputs	1	1	0	0	0	0
Q outputs after 5th clock pulse Recalculate bistable inputs	1	1	1	1	1	0
Q outputs after 6th clock pulse Recalculate bistable inputs	1	1	0	0	0	1

Table 5.14

The input data is then used to calculate the bistable outputs after the first clock pulse. With $J_a = K_a = 1$ Q_a will change after the clock pulse rising to 1. However, $J_b = K_b = 0$ so Q_b will remain unchanged at 0. These data are then used to recalculate the bistable inputs. J_a and K_a stay at 1, but with Q_a now at 1, both inputs to the AND gate are 1 making its output equal 1 and causing J_b and K_b to both rise to 1, as recorded in the second row of the state table.

With the inputs to both bistables set at 1 the outputs will change after the second clock pulse, Q_a will fall to 0 and Q_b will rise to 1. The process continues as shown in the completed state table, Table 5.14. The sequence repeats after clock 4 pulses. The number of clock pulses required for the sequence to repeat is called the counter modulo.

Listing only the bistable outputs, re-ordered with the most significant bit on the left, the state table may be simplified as in Table 5.15.

Weighting → Clock pulse ↓	Q_b $2^1 = 2$ MSB	Q_a $2^0 = 1$ LSB	Count in decimal	Comments
0	0	0	0	
1	0	1	1	
2	1	0	2	
3	1	1	3	
4	0	0	0	Sequence repeats after 4 clock pulses
5	0	1	1	

Table 5.15

Figure 5.80 shows a 3-bit synchronous counter. The Q outputs for all the bistables are initially reset to 0. Derive the state table with the columns arranged to display the binary count directly. Hence find the counter modulo.

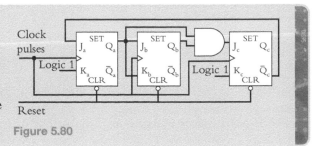

Figure 5.80

Using the method described above, first determine the bistable inputs before every clock pulse and then, after the clock pulses, calculate the bistable outputs, using the JK bistable state table.

Clock pulse		$J_a=$	$K_a=1$	Q_a	$J_b=Q_a$	$K_b=Q_a$	Q_b	$J_c=Q_a.Q_b$	$K_c=1$	Q_c	$\bar{Q}_c$	
0	outputs			0			0			0	1	
	inputs	1	1		0	0		0	1			
1	outputs			1			0			0	1	
	inputs	1	1		1	1		0	1			
2	outputs			0			1			0	1	
	inputs	1	1		0	0		0	1			
3	outputs			1			1			0	1	
	inputs	1	1		1	1		1	1			
4	outputs			0			0			1	0	
	inputs	0	1		0	0		0	1			
5	outputs			0			0			0	1	Cycle repeats
	inputs	1	1		0	0		0	1			
6	outputs			1			0			0	1	
	inputs	1	1		1	1		0	0			

Table 5.16

Clock pulse	MSB Q_c	Q_b	LSB Q_a	Count in decimal
0	0	0	0	0
1	0	0	1	1
2	0	1	0	2
3	0	1	1	3
4	1	0	0	4
5	0	0	0	0
6	0	0	1	1

Table 5.17

Counter modulo = number of clock pulses needed to make the count repeat = 5

To transmit the binary output from the counter correctly to the next device in the network the output from each stage of the counter must be connected to the appropriate bit input of the following device, with the MSB output being joined to the MSB input etc. The digital-to-analogue converter, described in the next section, is an example of a device with a binary input. It has a separate input lead for each bit.

Digital-to-analogue converters

There are many applications where an engineer wants to use a computer or microprocessor to control a piece of equipment and it is necessary to convert a binary signal into an analogue voltage. The simplest form of digital to analogue converter is the weighted resistor D/A converter, shown in Figure 5.81.

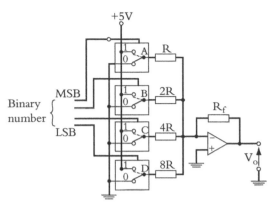

Figure 5.81

This is basically a summing amplifier with a set of electronic switches that connect the input resistors to either logic 1 or logic 0, depending on the binary number to be converted. The 4-bit device shown has four input leads. A is the most significant bit (MSB), corresponding to 1000_2 in binary (that is, equivalent to 8_{10}) and D is the least significant bit (LSB) corresponding to 0001_2 (i.e. 1_{10}). The following worked example shows how a binary number is converted to an analogue voltage.

A 4-bit digital to analogue (D/A) converter, shown in Figure 5.81, has logic 1 set at 5 V and a resistor ratio $\dfrac{R_f}{R} = \dfrac{8}{5}$. The binary input is 1101_2. Calculate the analogue output voltage.

To convert 1101_2 (i.e. 13_{10}) on the input leads switches A, B and D must be connected to logic 1 and switch C to logic 0 (ground).

Using equation (5.99) the output voltage of the summing amplifier, V_o is given by:

$$V_o = -R_f\left[\frac{V_A}{R} + \frac{V_B}{2R} + \frac{0}{4R} + \frac{V_D}{8R}\right]$$

$$= -\frac{8}{5}\left[5 + \frac{5}{2} + 0 + \frac{5}{8}\right]$$

$$V_o = -13\,\text{V}$$

The binary input of 1101_2 (equivalent to 13 in decimal) has been converted to $-13\,\text{V}$.

The major problem with this design is that for the D/A converter to function accurately, the resistor values must be exact. This is difficult to achieve in converters with a large number of bits and errors are inherent.

R–2R ladder digital-to-analogue converter

The problem is avoided in the R–$2R$ ladder, shown in Figure 5.82, which uses only two resistance values R and $2R$. The current entering each node is divided equally with half going through the shunt resistor and the remainder passing to the next stage. The effective resistance of each stage of the ladder is R. Consider the circuit to the right of node D_2 where there are two resistors, both $2R$, connected in parallel with an effective resistance of R. This resistance is in series with a resistor R, making the total resistance to the right of node C_2 equal to $2R$. This resistance is in parallel with the ground resistor $2R$, giving an effective resistance at C_2 of R. The reader is left to check that the effective resistances at nodes B_2 and A_2 are also equal R.

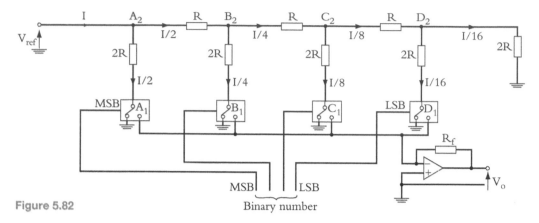

Figure 5.82

Binary number

The shunt resistors are connected to electronic switches, which depending on the input binary number being converted, join the resistors to either the inverting input of the amplifier or to ground to produce an amplifier output voltage that is the analogue equivalent to the binary signal. This is demonstrated in the following worked example.

A 4-bit R-$2R$ ladder D/A converter is shown in Figure 5.82. It has a reference voltage of $+5\,V$. The effective resistance at each stage of the ladder, R is $10\,k\Omega$ and the amplifier feedback resistance, R_f is $32\,k\Omega$. The binary input is 1010_2. Calculate the analogue output voltage.

To convert the binary 1010_2 $(= 10_{10})$ only the first (corresponding to the most significant bit) and third switches are connected to the op-amp input, producing an amplifier input current, I_{in} which is given by:

$$I_{in} = \frac{I}{2} + \frac{0}{4} + \frac{I}{8} + \frac{0}{16}$$

$$= \frac{5I}{8}$$

$$I = \frac{V_{ref}}{R} = \frac{5}{10 \times 10^{-3}} = 0.5\,mA$$

$$I_{in} = \frac{5}{8}I = 0.3125\,mA$$

For an inverting amplifier

$$V_{out} = I_f R_f \quad \text{and} \quad I_f = -I_{in}$$

hence,

$$V_{out} = -I_{in} R_f$$

$$= -0.3125 \times 10^{-3} \times 32 \times 10^3$$

$$V_{out} = -10\,V$$

The binary data word 1010_2 $(= 10_{10})$ has been converted into an analogue voltage of $10\,V$.

Analogue-to-digital converter

Analogue to digital (A/D) converters are used to convert analogue voltages to a digital format. Figure 5.83(a) shows a simple system called a counter-ramp A/D converter.

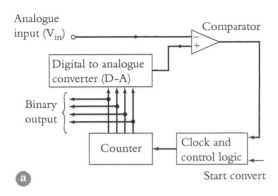
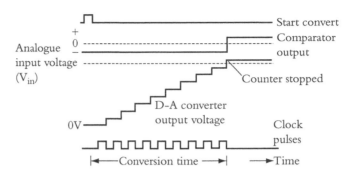

Figure 5.83

The analogue voltage to be converted is connected to the inverting input of the comparator, which is an operational amplifier with a very high gain. Its output is zero when the two input voltages are **exactly** equal. The output voltage is negative when the voltage applied to the inverting input is greater than that applied to the non-inverting input and positive when the non-inverting input voltage is larger. When the process starts, the count is set to zero, making the output voltage from the D/A converter equal to 0 V and the comparator output negative. The first clock pulse increments the binary counter output to 0001_2. This is converted to an analogue voltage by the D/A converter which rises one voltage step, where:

$$\text{voltage step} = \frac{\text{maximum analogue voltage that may be applied to the inverting input}}{\text{maximum binary count}}$$

If this is smaller than the analogue input voltage, the comparator output remains negative and the binary counter increments by one, making the D/A converter output rise another step. Eventually the count becomes sufficiently large that the output from the D/A converter exceeds the analogue voltage and the comparator output becomes positive, at which point the count stops and the binary equivalent of the analogue voltage is held at the digital output. The whole process is illustrated in Figure 5.83(b).

A 4-bit analogue to digital counter has a maximum analogue input voltage of 15 V and a clock frequency of 100 kHz. Calculate the time taken to convert an analogue voltage of 9 V.

A binary counter can count from 0 to $(2^x - 1)$ where x is the number of bits. With four bits the maximum count is $(2^4 - 1) = 15$.

$$\text{voltage step} = \frac{\text{maximum analogue input voltage}}{\text{maximum binary count}}$$

$$= \frac{15}{15}$$

$$= 1 \text{ V}$$

To convert an analogue voltage of 9 V, nine steps will be needed.

$$\text{time taken for one step} = \frac{1}{\text{clock frequency}} = \frac{1}{100 \times 10^3} = 10 \text{ μs}$$

$$\text{time for nine steps} = 9 \times 10 \text{ μs} = 90 \text{ μs}$$

5.10 Transformers

Transformers convert alternating voltage and current levels. They have no moving parts and consume little power. They are used in a wide variety of applications including chargers and power supplies for small electronic equipment, televisions, medical apparatus and electrical power transmission and distribution. Transformers may also be used for electrical isolation where there is a risk of personal injury, for example in the use of portable electric tools.

Turns ratio

A single-phase transformer has a core of insulated steel laminations around which two separate coils of conducting wire are wound, each covered with insulation, as illustrated in Figure 5.84. One coil is connected to an alternating current (ac) supply. This is called the primary winding. The other normally has a different number of turns and is known as the secondary. In this case the primary has N_1 turns and the secondary N_2 turns. N_1 and N_2 may be equal in isolation transformers.

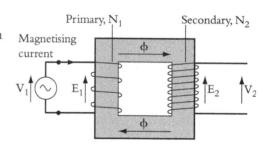

Figure 5.84

When an alternating voltage V_1 is applied across the primary winding, a small magnetizing current flows in that winding establishing an alternating flux ϕ in the core, such that:

$$\phi = \phi_m \sin(2\pi ft)$$

where ϕ_m is the maximum flux, f is the frequency and t is time.

In accordance with Faraday's Law (equation (5.23)) this flux induces an emf e_1 in the primary winding:

$$e_1 = N_1\frac{d\phi}{dt} \qquad\qquad (5.106)$$

substituting for ϕ

$$e_1 = N_1 \frac{d}{dt}\{\phi_m \sin(2\pi ft)\}$$

$$= N_1\, 2\pi f\, \phi_m \cos(2\pi ft)$$

maximum induced emf, $e_{1m} = N_1\, 2\pi f\, \phi_m$

from equation (5.48)

$$\text{rms voltage} = \frac{\text{maximum voltage}}{\sqrt{2}}$$

thus rms induced emf,

$$E_1 = \frac{2\pi f}{\sqrt{2}} N_1 \phi_m$$

$$E_1 = 4.44 \times f N_1 \phi_m \tag{5.107}$$

If A is the cross-sectional area of the core

$$E_1 = 4.44 \times f N_1 A B_m \tag{5.108}$$

where B_m is the maximum flux density in the core.

As the flux is common to both coils, it also induces an emf e_2 in the secondary winding such that

$$e_2 = N_2 \frac{d\phi}{dt} \tag{5.109}$$

and

$$E_2 = 4.44 \times f N_2 A B_m \tag{5.110}$$

Dividing equation (5.108) by (5.110)

$$\frac{E_1}{E_2} = \frac{N_1}{N_2} \tag{5.111}$$

The ratio of $\left(\dfrac{\text{primary turns, } N_1}{\text{secondary turns, } N_2} \right)$ is called the 'turns ratio' or 'transformation ratio'.

In practice the windings will have a small finite resistance and there will be some leakage of flux (that is, a very small fraction of flux will fail to link with both coils) but these effects are generally very small, so it may be assumed with reasonable accuracy that:

E_1 is the V_1, applied voltage

E_2 is the V_2, secondary terminal voltage

Hence,

$$\frac{V_1}{V_2} = \frac{N_1}{N_2} \tag{5.112}$$

Impedance connected to secondary winding

When a load impedance is connected to the secondary winding, as in Figure 5.85, a current I_2 will flow in the secondary circuit. This current will set up its own flux, in a direction which opposes the main flux ϕ, tending to reduce the latter. Any reduction in the main flux produces a corresponding fall in E_1 (see equation (5.107)). This reduction in E_1 increases the potential difference ($V_1 - E_1$) in the primary circuit, raising the primary current I_1 and with it the main flux, ϕ.

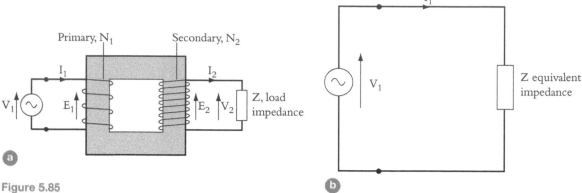

Primary, N_1 Secondary, N_2

Z equivalent impedance

Z, load impedance

Figure 5.85

Equilibrium is achieved when the primary current has increased sufficiently to restore the flux to its original value, at which point the magnetomotive force (mmf) produced in the primary winding, $I_1 N_1$ equals the mmf created in the secondary winding, $I_2 N_2$.

$$I_1 N_1 = I_2 N_2 \qquad (5.113)$$

$$\frac{I_1}{I_2} = \frac{N_2}{N_1} \qquad (5.114)$$

A 200 V: 50 V, 60 Hz transformer supplies a load comprising a 10 Ω resistor in series with a 100 mH inductor. Calculate the primary current.

Load reactance, $jX_l = j2\pi \times 60 \times 100 \times 10^{-3} = j37.7 \ \Omega$

Load impedance, $Z = R + jX_l = (10 + j37.7) \ \Omega$

$$I_2 = \frac{V_2}{Z}$$

$$I_2 = \frac{50}{10 + j37.7} = \frac{50}{10 + j37.7}\left(\frac{10 - j37.7}{10 - j37.7}\right)$$

$$= \frac{50}{1521}(10 - j37.7)$$

$$= (0.329 - j1.239) \ \text{A}$$

From equation (5.114)

$$\frac{I_1}{I_2} = \frac{N_2}{N_1}$$

Comparing this with equation (5.112)

$$\frac{I_1}{I_2} = \frac{V_2}{V_1} \qquad (5.115)$$

$$V_1 = 200 \, \text{V and } V_2 = 50 \, \text{V}$$

$$I_1 = I_2 \frac{V_2}{V_1} = \frac{50}{200}(0.329 - j1.239)$$

$$= (0.082 - j0.309)\text{A}$$

$$= 0.319 \text{A} \ \angle -75.10°$$

Equivalent impedance

Engineers use equivalent circuits to simplify circuit analysis. Equivalent circuits replicate the effect of the components they replace. The voltage across the terminals of an equivalent circuit and the current entering it must be the same as those for the original components. Currents and voltages in the rest of the network must be unaffected by the use of an equivalent circuit.

The transformer with a load impedance connected to its secondary winding, shown in Figure 5.85(a), will have an equivalent circuit which has a voltage V_1 across its terminals and a current, I_1 entering it, see Figure 5.85(b).

Effective impedance of the equivalent circuit, $Z_e = \dfrac{V_1}{V_1}$

From equations (5.112) and (5.114) for a transformer with a load impedance connected to its secondary winding

$$V_1 = \frac{V_2 N_1}{N_2}$$

and

$$I_1 = \frac{I_2 N_2}{N_1}$$

$$Z_e = \frac{V_1}{I_1} = \frac{V_2 \dfrac{N_1}{N_2}}{I_2 \dfrac{N_2}{N_1}}$$

$$= \frac{V_2}{I_2} \times \left(\frac{N_1}{N_2}\right)^2$$

by Ohm's Law (equation (5.1)) the secondary impedance,

$$Z = \frac{V_2}{I_2}$$

Hence, the equivalent impedance,

$$Z_e = Z\left(\frac{N_1}{N_2}\right)^2 \tag{5.116}$$

The transformer and load impedance may be represented by impedance Z_e, equal to the load impedance Z, multiplied by the turns ratio squared. This impedance is sometimes called the referred impedance. The following example demonstrates how it may used in calculations.

A 200 V: 50 V, 60 Hz transformer supplies a load comprising a 10 Ω resistor in series with a 100 mH inductor. Calculate the equivalent impedance for the transformer and load and hence find the primary current.

From the previous worked example

Load impedance, $Z = R + jX_l = (10 + j37.7)\,\Omega$

equivalent impedance, $Z_e = Z\left(\dfrac{N_1}{N_2}\right)^2$

$$= (10 + j37.7) \times \left(\frac{200}{50}\right)^2$$

$$= 160 + j603.2$$

$$I_1 = \frac{V_1}{Z_e}$$

$$= \frac{200}{(160 + j603.2)}$$

$$= \frac{200}{(160 + j603.2)} \times \frac{(160 - j603.2)}{(160 - j603.2)}$$

$$= \frac{200 \times (160 - j603.2)}{389450}$$

$$= (0.082 - j0.309)A$$

$$= 0.319A \angle -75.10°$$

Voltage and voltampere ratings

The voltage rating of an electrical device is the maximum rms voltage at which it may be safely operated. This is determined by the manufacturer and printed on the rating plate fixed to the device. Also on the rating plate is the voltampere or VA rating. This is used to calculate the maximum current that the equipment can safely conduct, which is called the full load current. The VA rating is defined as:

VA rating, $S = V I_{fl}$

where V is the voltage rating and I_{fl} is the full load current.

For a transformer:

VA rating of the primary, $S_1 = V_1 I_{1fl}$

and

VA rating of the secondary, , $S_2 = V_2 I_{2fl}$

From equation (5.112)

$$\frac{I_1}{I_2} = \frac{V_2}{V_1}$$

$$V_1 I_1 = V_2 I_2 \tag{5.117}$$

VA rating of primary, $S_1 =$ VA rating of secondary, S_2

A 3.3 kV: 240 V single phase transformer has a voltampere rating of 200 kVA. Calculate the full load primary and secondary currents.

Primary current, $I_1 = \dfrac{S}{V_1} = \dfrac{200 \times 10^3}{3.3 \times 10^3}$

$$= 60.6 \, A$$

Secondary current, $I_2 = \dfrac{S}{V_2} = \dfrac{200 \times 10^3}{240}$

$$= 833.3 \, A$$

Transformer losses

There are two forms of power loss in a transformer; both produce internal heating. The currents flowing through the primary and secondary windings, which have finite resistance, generate I^2R losses in the coils. These losses are often called the 'copper losses'.

Power loss is also produced by the alternating flux in the steel core. This loss is minimized by constructing the core of laminations of special low loss steel, each coated with an insulating varnish. Core loss is independent of load current and may be considered constant.

Efficiency

Efficiency is defined as the ratio of output power to input power, that is:

$$\text{Efficiency, } \eta = \frac{\text{output power, } P_2}{\text{input power, } P_1} \tag{5.118}$$

$$\text{input power, } P_1 = \text{output power, } P_2 + \text{losses} \tag{5.119}$$

for a transformer,

$$\text{input power} = \text{output power} + (I^2R \text{ losses}) + \text{core losses} \tag{5.120}$$

from equation (5.79)

$$\text{output power, } P_2 = V_2 I_2 \cos\phi_2 \tag{5.121}$$

where V_2 is the secondary voltage, I_2 is the secondary full load current, and $\cos\phi_2$ is the power factor of the load.

If the primary and secondary winding resistances are R_1 and R_2 respectively

I^2R loss in primary, $P_{11} = I_1^2 R_1$
I^2R loss in secondary, $P_{12} = I_2^2 R_2$
Total I^2R losses $= P_{11} + P_{12} = I_1^2 R_1 + I_2^2 R_2$

Let the core loss $= P_c = $ constant

substituting in (5.120)

input power, $P_1 = P_2 + (P_{11} + P_{12}) + P_c$

$$\text{efficiency, } \eta = \frac{P_2}{P_1} = \frac{P_2}{P_2 + (P_{11} + P_{12}) + P_c}$$

$$\eta = \frac{V_2 I_2 \cos\phi_2}{V_2 I_2 \cos\phi_2 + (I_1^2 R_1 + I_2^2 R_2) + P_c} \tag{5.122}$$

A 60 kVA 1200 V : 240 V single-phase transformer dissipates 1.6 kW when the secondary is open-circuited. The resistance of the primary (high-voltage) winding is 0.8 Ω and the resistance of secondary is 0.04 Ω. Calculate the efficiency of the transformer when it supplies full load at 0.8 power factor lagging.

Full load primary current, $I_{1fl} = \dfrac{S}{V_1}$

$$= \frac{60 \times 10^3}{1200}$$

$$= 50 \text{ A}$$

Full load secondary current, $I_{2fl} = \dfrac{S}{V_2}$

$$= \dfrac{60 \times 10^3}{240}$$

$$= 250 \text{ A}$$

At full load:

Primary I^2R loss $= 50^2 \times 0.8 = 2000\,\text{W}$

Secondary I^2R loss $= 250^2 \times 0.04 = 2500\,\text{W}$

Power dissipation on no load $=$ (constant) core loss, $P_c = 1600\,\text{W}$

Output power, $P_2 = V_2 I_2 \cos\phi_2$

$$= 240 \times 250 \times 0.8$$

$$= 48{,}000\,\text{W}$$

Input power, $P_1 =$ output power, $P_2 + I^2R$ losses $+$ core loss, P_c

$$= 48{,}000 + (2000 + 2500) + 1600$$

$$= 54{,}100\,\text{W}$$

From equation (5.118)

Efficiency, $\eta = \dfrac{\text{output power, } P_2}{\text{input power, } P_1}$

$$= \dfrac{48{,}000}{54{,}100}$$

$$= 0.887 \text{ per unit}$$

It is usual to multiply this figure by 100 to quote efficiency as a percentage, that is:

Efficiency, $\eta = 88.7$ per cent

Learning summary

You have completed the section on transformers and should now:

✔ understand how transformers work;

✔ be able to use the turns ratio to calculate the primary or secondary voltage;

✔ be familiar with the concept of magnetomotive force (mmf) balance;

✔ be able to use the mmf balance to find the primary or secondary current;

✔ know how to compute the equivalent impedance of a transformer with a load;

✔ be able to use the voltampere (VA) rating to find the full load primary and secondary currents;

✔ know how to calculate the I^2R losses;

✔ remember that the core losses are constant;

✔ be able to compute the efficiency.

5.11 AC induction motors

The drives for many mechanical systems are provided by electric motors. The most common and most rugged motor is the ac induction motor. There are both single-phase and three-phase versions of this motor. Single-phase motors are used in many household appliances such as lawn mowers, food processors, washing machines and refrigerators. Three-phase motors are used to drive industrial pumps, fans, compressors and mills. For clarity, this section will concentrate on the operation of the three-phase induction motor.

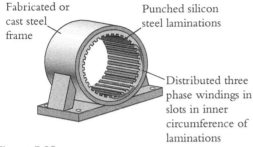

Fabricated or cast steel frame

Punched silicon steel laminations

Distributed three phase windings in slots in inner circumference of laminations

Aluminium or copper bars in slots

Castellations on end rings act as a fan to cool rotor

Bars welded to aluminium end rings

Figure 5.86

Construction

All induction motors have two parts; the outer stationary frame called the stator and inner rotating section known as the rotor. The stator has a large number of circular silicon steel laminations with slots cut in the inner circumference mounted in a fabricated or cast steel frame as shown in Figure 5.86(a). In the slots of three-phase motors are wound three separate coils. The rotor has silicon steel laminations keyed to a central shaft, see Figure 5.86(b). Slots are cut in the laminations at or close to the outer circumference. In these slots aluminium or copper conductors are fitted. The simplest rotor is called the squirrel cage rotor. It has the ends of the conductors welded to aluminium end rings. These are sometimes castellated to facilitate cooling during running. There are no electrical connections to the rotor.

Operating principles

Balanced three-phase currents are supplied to the stator windings with the object of producing a uniform magnetic field that rotates in the air gap between stator and rotor. To understand how this is achieved, consider Figure 5.87. This shows the cross-section of a motor with a rudimentary stator that has three coils distributed around its circumference 120° apart, carrying the three phase currents. Conductors R and R' represent the coil in the red phase, Y–Y' represents the yellow phase coil and B–B' the blue coil.

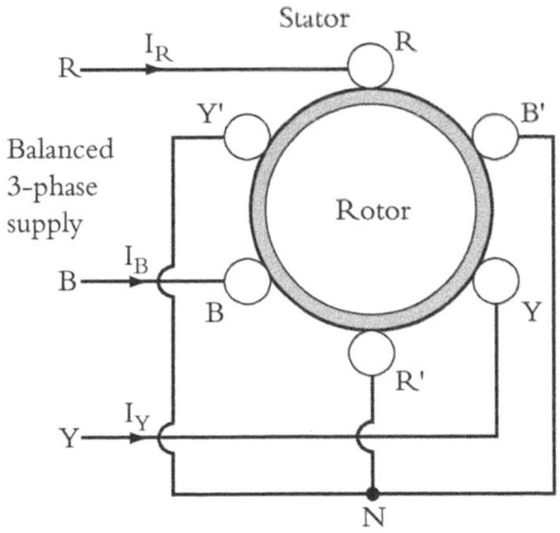

The three current waveforms are shown in Figure 5.88(a). At time, $t = 0$ maximum current flows in the red phase. The polarity of the current is positive, so it will be assumed that current enters the winding through R and returns via R'. At the same time the magnitude of the current in the yellow phase is half maximum but its polarity is negative, so current enters the winding through Y' and leaves via Y. Similarly, for the blue phase, the magnitude of the current is half maximum current and the polarity is negative, so it enters via B' and leaves by B. Each of these currents produces a magnetic field. These interact to form a magnetic field that crosses the air gap and passes through the rotor core, as shown in Figure 5.88b. The resultant magnetic field resembles that associated with a two-pole bar magnet. As a consequence the machine is called a two-pole motor.

With time, the current distribution changes. The red current falls; the yellow current becomes less negative eventually becoming positive and the blue current approaches a maximum with negative polarity. As these changes take place, the magnetic field rotates clockwise, maintaining a constant magnitude. Figure 5.88(c) shows the position after a one-third cycle (120 electrical degrees) when the yellow current has reached maximum magnitude with positive polarity and the currents in the red and blue phases are both at half maximum with negative polarity. At this time, the field has rotated 120° from its original position.

Figure 5.87

Stator

R

I_R

R

Y'

B'

Balanced 3-phase supply

Rotor

I_B

B

B

Y

R'

I_Y

Y

N

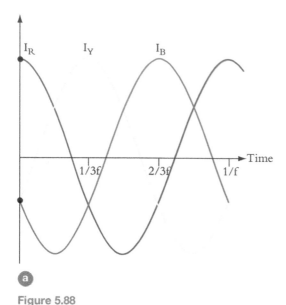

Figure 5.88

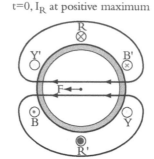

t=0, I_R at positive maximum

b

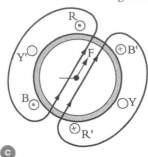

t=1/3f, (1/3rd of a cycle later)
I_Y at positive maximum
field has moved 120 degrees clock·

c

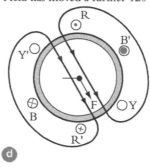

t=2/3f, I_B at positive maximum
Field has moved a further 120°

d

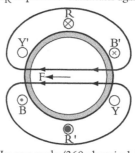

t=1/f, (one cycle after start)
I_R at positive maximum again

In one cycle (360 electrical degrees)
the field has moved through
360 mechanical degrees

e

After $\frac{2}{3}$rd cycle, see Figure 5.88(d), the blue
phase current is at maximum magnitude and
positive polarity, the red and yellow phase
currents are at half maximum with negative
polarity and the field has moved a total of 240°.
After one complete cycle (Figure 5.88(e)) the
currents have returned to their initial values
and the field is back in its original position. The field always moves in synchronism
with the stator currents, whatever angle the currents have moved through the field
will have rotated through the same angle.

The speed at which the field rotates is called the synchronous speed, n_s. Measured in revolutions
per second it is equal to the frequency of the stator currents, f, in hertz (Hz).

$$n_s = f$$

(5.123)

Rotor slip

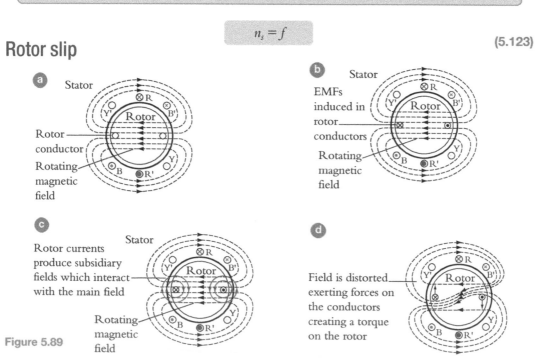

Figure 5.89

The rotating magnetic field interacts with the rotor. The rotor shown in Figure 5.89a has two conductors, the ends of which are connected to aluminium end rings. With the magnetic field rotating at constant speed and the rotor initially stationary, emfs will be induced in the rotor conductors in accordance with Faraday's Law, equation 5.23:

emf induced in each conductor = rate of change of flux linkages

Figure 5.89b shows the induced emfs at time, $t = 0$. With the rotor conductors short circuited by the end rings, the emfs produce currents in the rotor conductors. These currents create their own subsidiary magnetic fields (Figure 5.89(c)) which interact with the rotating stator field distorting the latter to produce forces on the rotor conductors. These forces create a torque on the rotor making it turn (Figure 5.89(d)). This torque is called the electrical or driving torque.

The rotor starts to accelerate, lowering the relative speed between the rotating field and rotor conductors. This reduces the magnitudes of the induced emfs, conductor currents and subsidiary magnetic fields, decreasing the forces on the conductors and the electrical torque on the rotor.

> The rotor will continue to accelerate until the electrical torque produced exactly equals the mechanical load torque on the shaft. At this point the rotor will be running at a speed, n, slightly slower than the synchronous speed, n_s; i.e. at a speed slightly slower than that of the rotating magnetic field. This difference in speed is expressed as a ratio and is known as the slip or per unit slip, s

$$s = \frac{n_s - n}{n_s} = 1 - \frac{n}{n_s}$$

(5.124)

> The small difference between the speed of the rotating field and that of the rotor is fundamental to the operation of the induction motor.

For a torque to be developed, there has to be some distortion of the resultant magnetic field and this will only happen when currents flow in the rotor conductors. These depend on emfs being induced in the conductors, which in turn depend on there being a difference between the speed at which the conductors rotate and the speed of the rotating magnetic field. For most motors, the value of the slip varies between around 0.01 on no load (when the only torque required is to overcome friction at the bearings) and 0.10 when the motor is driving full load.

Rotor frequency

While the rotor is stationary, the rotating flux links with each rotor conductor once every cycle and induces in the conductor an emf of the same frequency as the stator currents. However, when the rotor starts to move the relative speed between the field and rotor, $(n_s - n)$ falls and the stator flux has to rotate more than one full revolution to complete a linkage cycle. This lengthens the periodic time, T, of both the flux linkage, the emf induced in a rotor conductor and hence the rotor current. As frequency equals $\frac{1}{T}$ the frequency of the rotor currents, f_r is reduced to:

$$f_r = (n_s - n) \text{ Hz}$$

(5.125)

combining equations (5.124) and (5.125)

$$(n_s - n) = f_r = sn_s = sf$$

$$f_r = sf$$

(5.126)

Rotor emf

The induction motor may be considered to be a special type of transformer, where the stator replaces the primary. The rotor forms the secondary winding, mounted on a rotating drum, with its terminals short circuited. The emfs induced in both windings can be derived from equations 5.108 and 5.110, letting suffixes 1 and 2 represent stator and rotor respectively.

emf induced in stator winding, $E_1 = 4.44 f N_1 \phi_m$ (5.127)

emf induced in rotor winding $= 4.44 f_r N_2 \phi_m$ (5.128)

substituting for f_r using equation (5.126)

emf induced in rotor winding $= 4.44\, s f N_2 \phi_m$ (5.129)

Defining E_2 as:

$E_2 = 4.44 f N_2 \phi_m$ (5.130)

emf induced in rotor winding $= s E_2$ (5.131)

E_2 equals the emf induced in the rotor winding when $s = 1$, i.e. when the rotor is stationary.

Equivalent circuit

The equivalent circuit for one phase of a three-phase induction motor is shown in Figure 5.90(a).

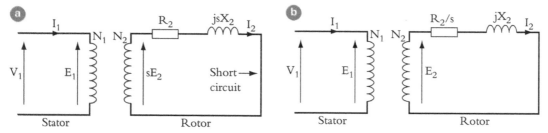

Figure 5.90

V_1 is the applied stator phase voltage

E_1 is the emf induced in stator winding

If the stator winding resistance and reactance are negligible, $E_1 = V_1$

N_1 is the stator turns

N_2 is the effective rotor turns

sE_2 is the emf induced in rotor winding

R_2 is the rotor resistance per phase

Rotor reactance per phase $= j2\pi f_r L$

where L_2 is the rotor winding inductance per phase.

From equation (5.126)

$$f_r = sf$$

Rotor reactance per phase $= j2\pi sf L = jsX_2$

where $\qquad X_2 = 2\pi f L$

From Figure 5.90(a) it can be deduced that

$$I_2 = \frac{sE_2}{R_2 + jsX_2}$$ (5.132)

Dividing the numerator and denominator by s

$$I_2 = \frac{E_2}{\dfrac{R_2}{s} + jX_2}$$ (5.133)

Making this change enables a simpler equivalent circuit (Figure. 5.90(b)) to be drawn. This has a fixed transformation ratio $E_1:E_2$, fixed rotor reactance, X_2 and variable rotor resistance, $\dfrac{R_2}{s}$.

Mechanical output power

The power supplied via the stator to each phase of the rotor, P_1 is given by

$$P_1 = |I_2|^2 \frac{R_2}{s}$$ (5.134)

Some of this power is dissipated as heat in the rotor winding, P_2. The remainder is converted to mechanical output power, P_m.

$$P_1 = P_2 + P_m$$ (5.135)

Power dissipated in the rotor winding as heat, $P_2 = |I_2|^2 R_2$ (5.136)

substituting equations (5.134) and (5.136) in (5.135)

$$|I_2|^2 \frac{R_2}{s} = |I_2|^2 R_2 + P_m$$

$$P_m = |I_2|^2 R_2 \left(\frac{1}{s} - 1 \right)$$

$$P_m = |I_2|^2 R_2 \frac{(1-s)}{s}$$ (5.137)

From equation (5.133) the modulus of I_2 is found to equal:

$$|I_2| = \frac{E_2}{\sqrt{\left(\left[\dfrac{R_2}{s} \right]^2 + X_2^2 \right)}}$$

and

$$|I_2|^2 = \frac{E_2^2}{\left[\dfrac{R_2}{s} \right]^2 + X_2^2}$$ (5.138)

substituting in equation (5.137)

$$P_m = \frac{E_2^2}{\left[\dfrac{R_2}{s} \right]^2 + X_2^2} \frac{R_2 (1-s)}{s}$$

Writing

$$a = \frac{R_2}{X_2}$$

i.e.

$$R_2 = aX_2$$

$$P_m = \frac{E_2^2}{X_2^2 \left(\left[\dfrac{a}{s} \right]^2 + 1^2 \right)} \frac{aX_2 (1-s)}{s}$$

$$P_m = \frac{E_2^2 as}{X_2(a^2 + s^2)} (1-s)$$ (5.139)

This is the mechanical output power per phase. It is measured in watts (W).

The same mechanical power is produced in each phase so the total mechanical output power developed by the motor, P_{mech} is three times larger.

$$P_{mech} = 3P_m = \frac{3E_2^2 as}{X_2(a^2 + s^2)}(1 - s)$$

(5.140)

Torque produced by a two-pole motor

The torque developed per phase, T_p is given by:

$$T_p = \frac{P_m}{2\pi n}$$

(5.141)

from equation (5.124)

$$n = n_s(1 - s)$$

$$T_p = \frac{P_m}{2\pi n_s(1 - s)}$$

using equation (5.123)

$$n_s = f = \frac{\omega}{2\pi}$$

$$T_p = \frac{P_m}{\omega(1 - s)}$$

substituting in equation (5.139)

$$T_p = \frac{1}{\omega} \times \frac{E_2^2 as}{X_2(a^2 + s^2)}$$

The total torque developed in three phases is equal to three times that produced in one phase: i.e. total torque, T is given by:

$$T = 3T_p$$

(5.142)

$$T_p = \frac{3}{\omega} \times \frac{E_2^2 as}{X_2(a^2 + s^2)}$$

(5.143)

This is the torque produced by a two pole motor. It is measured in Newton metres (Nm).

A three-phase 415 V two-pole 50 Hz induction motor has an effective stator:rotor turns ratio of 2:1, rotor resistance 0.15 Ω per phase and rotor standstill reactance 0.75 Ω per phase.

The motor runs at 2900 rpm^{-1}. Calculate

(a) per unit slip
(b) total torque
(c) total mechanical output power.

$$n_s = f$$

$$s = 1 - \frac{n}{n_s}$$

$$s = 1 - \frac{\left(\frac{2900}{60}\right)}{50}$$

$$= 0.0333$$

emf induced in each phase of stator winding $= E_1 = \dfrac{415}{\sqrt{3}}$

Dividing equation (5.130) by equation (5.127)

$$\frac{E_2}{E_1} = \frac{N_2}{N_1},$$

emf induced in each phase of rotor winding at standstill, $E_2 = \left(\dfrac{415}{\sqrt{3}}\right) \times \dfrac{1}{2} = 119.8\ V$

using equation (5.143)

total torque, $T = \dfrac{3}{\omega} \times \dfrac{E_2^2 as}{X_2(a^2 + s^2)}$

where

$$a = \frac{R_2}{X_2} = \frac{0.15}{0.75} = 0.2$$

$$T = \frac{3}{2\pi 50} \times \frac{119.8^2 \times 0.2 \times 0.0333}{0.75 \times (0.2^2 + 0.0333^2)}$$

$$= 29.6\ \text{Nm}$$

using equation (5.140)

total mechanical output power,

$$P_{mech} = \frac{3E_2^2 as}{X_2(a^2 + s^2)}(1 - s)$$

$$= \frac{3 \times 119.8^2 \times 0.2 \times 0.0333}{0.75 \times (0.2^2 + 0.0333^2)}(1 - 0.0333)$$

$$= 8991\ \text{W}$$

Starting torque produced by a two-pole motor

When choosing a motor for a particular application, the user must ensure that the motor will start with the mechanical load connected. To achieve this, the starting torque must exceed the load torque. On starting the rotor speed:

$$n = 0 \quad \text{and} \quad \text{slip}, s = \left(1 - \frac{n}{n_s}\right) = 1$$

Substituting for s in equation (5.143)

Starting torque, $\qquad\qquad T_{start} = \dfrac{3}{\omega} \times \dfrac{E_2^2 a}{X_2(a^2 + 1)}$ $\qquad\qquad$ (5.144)

Maximum torque produced by a two-pole motor

The condition for maximum torque is found by differentiating equation (5.143) by s and setting it to zero, i.e.

$$\frac{dT}{ds} = \frac{3}{\omega} \times \frac{E_2^2}{X_2} \times \frac{a(a^2 - s^2)}{(a^2 + s^2)^2} = 0$$

$$a = s$$

Maximum torque occurs when $a = s$

substituting this in equation (5.143), the maximum torque is found to be:

$$T_m = \frac{3}{2\omega} \times \frac{E_2^2}{X_2}\ \text{Nm} \qquad\qquad (5.145)$$

For the motor described in the previous worked example, calculate the starting torque and maximum torque.

$$T_{start} = \frac{3}{2\pi 50} \times \frac{119.8^2 \times 0.2}{0.75 \times (0.2^2 + 1)}$$

$$T_{start} = 35.1 \text{ Nm}$$

This is greater than the load torque of 29.6 Nm, so the motor will start.

$$T_m = \frac{1}{2} \times \frac{3}{2\pi 50} \times \frac{119.8^2}{0.75}$$

$$T_m = 91.4 \text{ Nm}$$

Torque–slip characteristics

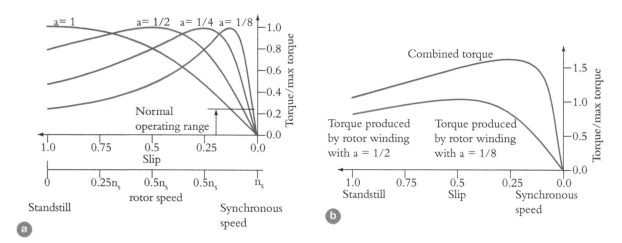

Figure 5.91

Figure 5.91(a) shows the torque versus slip curves for motors with different values of $a = \left(\dfrac{R_2}{X_2}\right)$. It is customary to draw the slip along the horizontal axis rather than the rotor speed, although both are shown here. To highlight the shapes of the characteristics, the torque axis has been normalized (graphs showing $\left(\dfrac{T}{T_m}\right)$ rather than absolute values of T). The graphs show that when a is less than 1 the driving torque increases with speed, reaching a maximum when $a = s$. Further increases in speed produce a fall in driving torque. A steady speed is achieved when the driving torque matches the load torque on the rotor shaft. This should be in the normal operating range shown on the graph. If the motor is operated outside this region, excessive currents will be drawn from the supply and power wasted as heat.

The graphs show that motors with a low ratio of rotor resistance to rotor standstill reactance (i.e. small values of a) have low starting torques but will run at virtually constant speed over the normal operating range, at close to the synchronous speed. Motors with a high ratio (i.e. large values of a) have good starting torques but suffer the disadvantage that their normal operating speed is dependent on the load torque. It is therefore necessary to select a motor that has a characteristic that falls between the two extremes with sufficient starting torque for the motor to start.

Double cage rotors

Manufacturers have overcome these problems by fitting the rotors in some machines with two separate windings – one with a large value of a to create a high starting torque and the other with a low value of a to achieve the best running conditions. The total driving torque is the sum of the torques produced by the individual windings giving a torque–slip characteristic similar to that shown in Figure 5.91(b).

Multi-pole motors

It was shown in equation (5.123) that for a two-pole induction motor:

$$\text{synchronous speed, } n_s \text{ (in rev s}^{-1}) = \text{frequency, } f \text{ (in Hz)}$$

As all induction motors should be run at speeds in the normal operating range, slightly below the synchronous speed, this means that a two-pole motor connected to a 50 Hz supply should run at a steady state speed somewhere between 45 rev s^{-1} (when $s = 0.1$) and 49.5 rev s^{-1} ($s = 0.01$). There are, however many applications for which these speeds are too high. To reduce the operating speed of a motor significantly, it is necessary to reduce the synchronous speed. This is achieved by manufacturers winding the stator coils in a way that increases the number of poles in the rotating magnetic field. Figure 5.92 shows the cross-section of a four-pole motor. In each phase two coils are connected in series. For example, in the red phase, conductor R on the right of the stator and conductor R' at the bottom form the first coil. Conductor R on the left and R' at the top make the second coil.

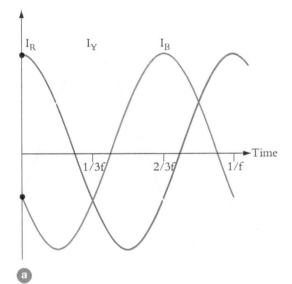

Figure 5.92

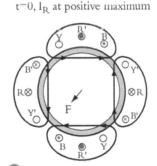

t=0, I$_R$ at positive maximum

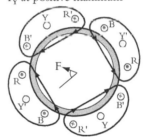

t=1/3f, (1/3rd of cycle later)
I$_Y$ at positive maximum

Field has moved clockwise through 60 mechanical degrees

Figures 5.92 (b)–(e) show how the field rotates as the three phase currents (Figure 5.92(a)) progress through one cycle. Figure 5.92(b) shows the magnetic field at the start of the current cycle, Figure 5.92(c) shows the field after a one-third cycle, Figure 5.92(d) demonstrates the situation after a two-thirds cycle and Figure 5.92(e) illustrates the position at the end of the cycle. Although the currents have moved through 360 electrical degrees, the magnetic field has only rotated through 180 mechanical degrees, halving the synchronous speed and producing a commensurate reduction in rotor speed.

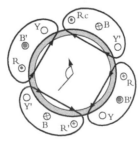

t=2/3f, I$_R$ at positive maximum

Field has moved through a further 60 degrees

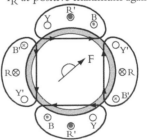

t=1/f, (one cycle after start)
I$_R$ at positive maximum again

In one cycle (360 degrees electrical) field has moved through 180 mechanical degrees

Further increases in the number of stator coils in each phase increases the number of poles and reduces the synchronous speed even more, as demonstrated in Table 5.18.

Number of poles	Pairs of poles	Synchronous speed, n_s as a fraction of frequency, f	Synchronous speed in rev s^{-1} when $f = 50$ Hz	Synchronous speed in rev s^{-1} when $f = 60$ Hz
2	1	$n_s = f$	50	60
4	2	$n_s = \dfrac{f}{2}$	25	30
6	3	$n_s = \dfrac{f}{3}$	16.67	20
8	4	$n_s = \dfrac{f}{4}$	12.5	15
2p	p	$n_s = \dfrac{f}{p}$	$\dfrac{50}{p}$	$\dfrac{60}{p}$

Table 5.18

Torque produced in multi-pole motor

For the same mechanical output power this reduction in synchronous speed, and hence rotor speed, will produce an increase in torque delivered to the shaft.

From equations (5.124) , (5.141) and (5.142)

Total torque, $T = \dfrac{P_{\text{mech}}}{2\pi n} = \dfrac{3P_m}{2\pi n} = \dfrac{3P_m}{2\pi n_s(1-s)}$

for a motor with p pairs of poles, $n_s = \left(\dfrac{f}{p}\right)$

$$T = \frac{3}{2\pi\left(\dfrac{f}{p}\right)} \times \frac{P_m}{(1-s)} = \frac{3p}{\omega} \times \frac{P_m}{(1-s)} \tag{5.146}$$

Substituting for P_m using equation (5.139)

Torque,
$$T = \frac{3p}{\omega} \times \frac{E_2^2 as}{X_2(a^2 + s^2)} \tag{5.147}$$

Similarly from equation (5.144)

Starting torque,
$$T_{\text{start}} = \frac{3p}{\omega} \times \frac{E_2^2 a}{X_2(a^2 + 1)} \tag{5.148}$$

and from equation (5.145)

Maximum torque,
$$T_m = \frac{3p}{2\omega} \times \frac{E_2^2}{X_2} \tag{5.149}$$

A three-phase 415 V six-pole 50 Hz inductor motor has effective stator:rotor turns ratio of 3:1, rotor resistance 0.1 Ω per phase and rotor standstill reactance 0.4 Ω per pahse. When the motor delivers full load, it runs with a slip of 0.04. Calculate the torque produced and the total mechanical output power.

$$T = \frac{3p}{\omega} \times \frac{E_2^2 as}{X_2(a^2 + s^2)}$$

$$p = 3$$

$$\omega = 2\pi50$$

$$E_2 = \frac{E_1 N_1}{N_2}$$

$$= \frac{415}{\sqrt{3}} \times \frac{1}{3} = 79.86\,\text{V}$$

$$a = \frac{R_2}{X_2} = \frac{0.1}{0.4} = 0.25$$

at slip, $s = 0.04$

$$T = \frac{3 \times 3}{2\pi50} \times \frac{79.86^2 \times 0.25 \times 0.04}{0.4 \times (0.25^2 \times 0.04^2)} = 71.3\,\text{Nm}$$

$$P_{\text{mech}} = T2\pi n$$

$$= T2\pi n_s(1 - s)$$

$$= T2\pi\left(\frac{f}{p}\right)(1 - s)$$

$$= 71.3 \times 2\pi \times \left(\frac{50}{3}\right) \times 0.96$$

$$P_{\text{mech}} = 7168\,\text{W}$$

Learning summary

You have now completed the section on ac induction motors and should:

✔ understand how three-phase induction motors work;

✔ comprehend the concept of slip and how it relates to rotor speed;

✔ understand how the ratio of rotor resistance to rotor standstill reactance affects the torque-slip characteristics;

✔ comprehend how the torque-slip characteristic may be modified by fitting two rotor windings;

✔ understand how the number of magnetic poles affects the synchronous speed and rotor speed;

✔ calculate the driving torque;

✔ compute the starting torque;

✔ calculate the mechanical output power.

Unit 6

Machine dynamics

Stewart McWilliam

UNIT OVERVIEW

- Introduction
- Basic mechanics
- Kinematics of a particle in a plane
- Kinetics of a particle in a plane
- Kinematics of rigid bodies in a plane
- Kinematics of linkage mechanisms in a plane
- Mass properties of rigid bodies
- Kinetics of a rigid body in a plane
- Balancing of rotating masses
- Geared systems
- Work and energy
- Impulse, impact and momentum

6.1 Introduction

Mechanical engineers apply their scientific knowledge to solve problems and design machines that help us to enjoy a better lifestyle. Acquiring the skills necessary to design and analyse machines is a fundamental part of the education required to become a mechanical engineer. A key element of this education is the need to develop a sound understanding of the physical principles that govern the behaviour of machines.

Machines consist of rigid bodies connected together in such a way that some components are fixed while others are moving. Dynamics provides the Mechanical Engineer with the basic tools necessary to analyse machines, and involves studying the motion of rigid bodies. The subject of dynamics involves two distinct studies: i) the geometric properties of motion, known as kinematics, and (ii) the relationship between the applied forces and the resulting motion, known as kinetics.

This chapter is intended for first and second-year students studying for a mechanical engineering degree at university, who already have a sound understanding of mathematics and physics to Advanced (A) level standard or equivalent. Throughout the chapter attention is given to developing an understanding of the basic principles of dynamics and their application to mechanical engineering machines. Section 6.2 provides a review of the basic concepts of dynamics. Sections 6.3 to 6.8 build-up the tools required to analyse rigid bodies in a systematic way, and should be studied in order. Sections 6.3 and 6.4 consider the kinematics and kinetics of particles moving in a plane. Sections 6.5 and 6.6 consider the kinematics of individual and connected rigid bodies (including linkage systems) moving in a plane. Section 6.7 defines the mass properties of rigid bodies – centre of mass and mass moment of inertia. Section 6.8 analyses the kinetics of rigid bodies moving in a plane. Sections 6.9 and 6.10 use some of the concepts developed in earlier sections to analyse practical engineering applications, and can be studied independently of each other. Section 6.9 considers the process of balancing rotating machinery, while Section 6.10 considers the basic analysis tools needed to perform design calculations for geared systems. Sections 6.11 and 6.12 use some of the ideas discussed in Sections 6.3 to 6.8 to develop related concepts, and can be studied in any order. Section 6.11 introduces the topics of work and energy, and the results developed are used to perform flywheel design calculations. Section 6.12 introduces the topics of impulse and momentum.

6.2 Basic mechanics

Introduction

This section introduces the basic concepts that are used throughout this chapter. It is expected that the reader has some understanding of these concepts prior to studying later sections. The topics introduced include: vectors; Newton's laws of motion, friction, motion in a straight line, and rotational motion.

Vector quantities

Vector quantities possess both magnitude and direction. Examples include force, displacement, velocity, acceleration and impulse. A vector is represented by an arrow in a particular direction. The head of the arrow indicates the sense of the vector, and the length represents the magnitude of the vector.

When adding vectors together, it is often convenient to construct the resultant vector diagrammatically, as shown in Figure 6.1.

The order in which the vectors are added does not change the resultant vector.

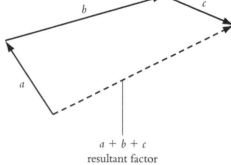

$a + b + c$
resultant factor

Figure 6.1 Vector addition

Newton's laws

Newton's laws form the basis for much of engineering dynamics analysis. As applied to a particle of constant mass they can be expressed as follows:

Newton's first law
A particle will maintain its state of rest or of uniform motion (at constant speed) along a straight line unless acted upon by some force to change that state. In other words, a particle has no acceleration if there is no resultant force acting on it.

Newton's second law
The acceleration of a particle is proportional to the resultant applied force acting on it, and is in the same direction as the resultant force.

This is often written as

$$\mathbf{F} = m\mathbf{a} \qquad \text{(vector equation)}$$

(units)   (N) (kg) (ms^{-2})

where **F** is the applied force (vector), m is the mass of the particle, and a is the acceleration (vector).

More generally, Newton's second law can be expressed in terms of the rate of change of linear momentum as $\mathbf{F} = \mathrm{d}(m\mathbf{V})/\mathrm{d}t$, where **V** is the velocity)

When applying Newton's second law, it is convenient to express it in components, for example

$$\xrightarrow{\;+\;} \quad F_X = ma_X \tag{6.1}$$

$$\uparrow^{+} \quad F_Y = ma_Y \tag{6.2}$$

where F_X is the X component of the resultant applied force, F_Y is the Y component of the resultant applied force, a_X is the X component of the resulting acceleration, and a_Y is the Y component of the resulting acceleration.

In equations (6.1) and (6.2) the arrows indicate that a particular sign convention must be adhered to when applying Newton's second law. Equation (6.1) states that the resultant applied force (F_X) on the left-hand side is assumed to be positive in the left to right direction, while the resulting acceleration a_X on the right-hand side is also assumed to be positive in the left to right direction. This sign convention ensures that the resultant acceleration is in the same direction as the resultant force.

Section 6.4 reconsiders equations (6.1) and (6.2) in more detail.

Newton's third law
Forces acting between two particles in contact are equal and opposite. In other words, if a particle exerts a force on a second particle, then the second particle exerts a numerically equal and oppositely directed force on the first particle.

Static equilibrium of a body in a plane

Consider the planar rigid body shown in Figure 6.2, where P is an arbitrary point and $\mathbf{F}_1, \mathbf{F}_2, \mathbf{F}_3, \ldots$ are externally applied forces.

The resultant externally applied force **F** is the vector sum of the applied forces (see 'Vector quantities'), i.e.

$$\mathbf{F} = \sum_i \mathbf{F}_i$$

Each externally applied force produces a moment about P. The moment $\mathbf{M}_{Pi}$ of the force $\mathbf{F}_i$ about P is given by

$$\mathbf{M}_{Pi} = r_i \mathbf{F}_i$$

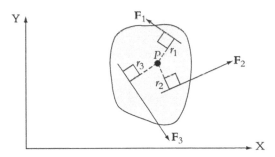

Figure 6.2 **Static equilibrium of a planar rigid body**

where r_i is the perpendicular distance from P to the line of action of $\mathbf{F}_i$. Moments are vector quantities and have both magnitude and direction – direction here refers to the sense of the rotation (clockwise or anticlockwise) about an axis through P. The resultant externally applied moment $\mathbf{M}_P$ is the vector sum of the applied moments about P, i.e.

$$\mathbf{M}_P = \sum_i \mathbf{M}_{Pi}$$

The conditions for static equilibrium of the body are that for an arbitrary point P in the body:

(i) the resultant applied force acting on P in the X-direction is zero

$$F_X = 0$$

(ii) the resultant applied force acting on P in the Y-direction is zero

$$F_Y = 0$$

(iii) the resultant applied moment about P is zero

$$\mathbf{M}_P = 0$$

Conditions (i) and (ii) are a special case of equations (6.1) and (6.2), where the accelerations are zero, while condition (iii) ensures that the body does not rotate.

Section 6.8 considers the dynamics of rigid bodies and derives equations that generalize equations (i) to (iii) to rigid bodies that are both translating and rotating.

Coulomb (dry) friction between two surfaces

When two surfaces are in contact, the contact force between the surfaces will have normal and tangential components. The normal force corresponds to the reaction force, while the tangential force corresponds to the friction force and acts in a direction that always opposes the motion that would occur when there is no contact.

Consider a body of mass M lying on a horizontal surface that is subjected to a horizontal applied force P, see Figure 6.3(a).

When the body is in contact with the horizontal surface, a normal reaction force acts upwards on the body to counteract the weight Mg. Using Newton's third law, an 'equal and opposite' normal reaction force N acts downwards on the surface.

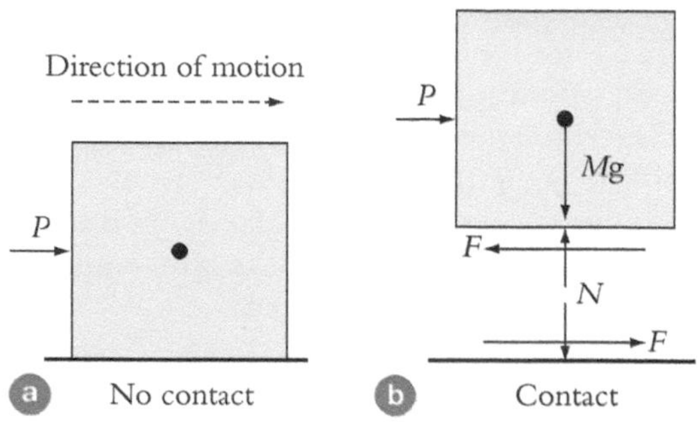

When there is no contact with the surface, the applied force P causes the body to move in the left to right direction, see Figure 6.3(a). When the body is in contact with the horizontal surface, friction forces act parallel to the contact surface, provided that the contact is not frictionless. The friction force F acting on the body opposes any motion arising from the applied force P (when there is no contact) and acts in the right to left direction, see Figure 6.3b). Using Newton's third law, the surface experiences an 'equal and opposite' friction force F acting in the left to right direction.

Figure 6.3 **Schematic and free body diagram of a body lying on a horizontal surface**

The Coulomb friction force F is the friction force resisting motion, and always satisfies

$$F \leqslant \mu N$$

where μ is the coefficient of friction and N is the normal reaction force. The coefficient of friction is independent of the normal force and the area of contact.

Sliding motion will occur if the applied force P is greater than the maximum possible friction force, i.e. if

$$P > \mu N$$

When sliding occurs the friction force takes its maximum value (i.e. limiting friction), such that

$$F = \mu N$$

The friction force between two bodies after sliding occurs is referred to as kinetic friction.

Two practical cases involving friction when a wheel makes contact with a horizontal surface are considered below. In these cases it is important to understand the direction of the friction forces.

Driven wheel in contact with a horizontal surface

This situation can be used to analyse the driven wheels of a car. To determine the direction of the friction force it is convenient to consider the no contact situation first, when the wheel is acted on by an applied torque T, see Figure 6.4(a).

When there is no contact with the surface, the centre of the wheel is stationary and the applied torque causes the wheel to rotate in the clockwise direction, see Figure 6.4(a). When the wheel is in contact with the horizontal surface, the friction force F acting on the wheel opposes the rotation of the wheel (when there was no contact), resulting in a friction force that acts in the anticlockwise sense (direction). In Figure 6.4(b) this friction force F is shown to act in the left

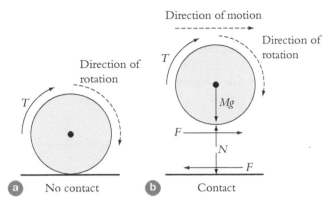

Figure 6.4 Schematic and free body diagram of a driven wheel making contact with a horizontal surface

to right direction at the point of contact, and causes the wheel to move to the right. Using Newton's third law, the surface experiences an 'equal and opposite' friction force F acting in the right to left direction.

Non-driven wheel subjected to a horizontal applied force in contact with a horizontal surface

This situation can be used to analyse the freely rotating (non-driven) wheels of a car. To determine the direction of the friction force it is convenient to consider the no contact situation first, when the wheel is acted on by a horizontal applied force P, see Figure 6.5(a).

When there is no contact, the (non-rotating) wheel moves in the left to right direction, see Figure 6.5(a). When the wheel makes contact with the horizontal surface, the friction force F acting on the wheel opposes the motion of the wheel (when there was no contact), resulting in a friction force that acts in the right to left direction at the point of contact. This friction force F is indicated in Figure 6.5(b), and causes the wheel to rotate in a clockwise direction. Using Newton's third law, the surface experiences an 'equal and opposite' friction force F acting in the left to right direction.

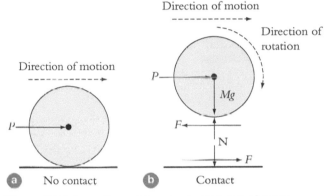

Figure 6.5 Schematic and free body diagram of a non-driven wheel making contact with a horizontal surface under the action of a horizontal force

Variable (non-constant) acceleration

Consider the situation when a body is moving in a straight line with variable (non-constant) acceleration a. The position of the particle at any time is expressed in terms of the distance s from a fixed origin (starting point) along the straight line, as shown in Figure 6.6.

The distance s is positive if the body lies to the right of O, and negative if it lies to the left of O.

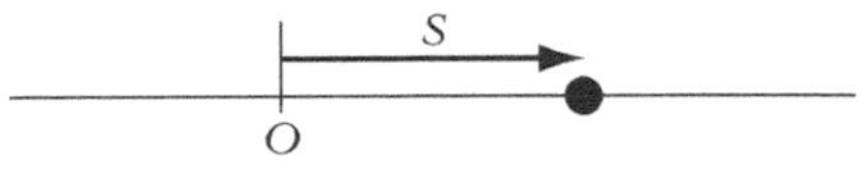

Consider the situation when the body changes its position from s to $s + \Delta s$ in the time interval t and $t + \Delta t$. The average velocity v_{av} of the body during this time interval is $\dfrac{\Delta s}{\Delta t}$. Mathematically this is written as

Figure 6.6 Position of a particle on a line

$$v_{av} = \frac{\Delta s}{\Delta t}$$

The instantaneous velocity v at time t is the limit of the average velocity as the increment of time Δt approaches zero. Mathematically this is written as

$$v = \lim_{\Delta t \to 0} \frac{\Delta s}{\Delta t} = \frac{ds}{dt}$$

The velocity is positive if the body is moving from left to right, and negative if it moves from right to left. The velocity is often written as $v = \dot{s}$, where the overdot represents differentiation with respect to time.

Similarly the instantaneous acceleration a at time t is the limit of the average acceleration as the increment of time Δt approaches zero. Mathematically this is written as

$$a = \lim_{\Delta t \to 0} \frac{\Delta v}{\Delta t} = \frac{dv}{dt}$$

The acceleration is positive if the velocity is increasing, and negative if the velocity is decreasing.

It can also be shown that

$$a = \frac{d^2 s}{dt^2} = \frac{dv}{ds}\frac{ds}{dt} = v\frac{dv}{ds}$$

The acceleration is often written as $a = \ddot{s}$, where the double overdot represents double differentiation with respect to time.

Constant acceleration

Consider the special case when a body is moving in a straight line with constant acceleration a, (e.g. vertical motion under gravity with negligible air resistance). In this case the acceleration a, velocity v and distance travelled s can be plotted against time t as shown in Figure 6.7.

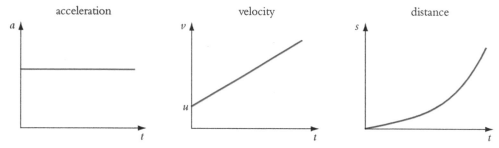

Figure 6.7 **Acceleration, velocity and distance travelled under constant acceleration**

At time $t = 0$, the body is assumed to have an initial velocity u. The distance s travelled from the starting point, and the velocity v at any time t can be written as follows

$$s = ut + \tfrac{1}{2}at^2, \; v = u + at, \; v^2 = u^2 + 2as$$

Angular (rotational) motion

The angular position of a straight line can be defined by the angle made between the line and some reference direction (e.g. the OX axis), as shown in Figure 6.8.

For the case shown, the angle θ in radians (rad) is used to define the angular position at time t, where θ is defined to be positive in the anticlockwise direction (from OX). At time $t + \delta t$ the angle will have increased to $\theta + \delta\theta$, and the instantaneous angular velocity ω of the line is defined as

$$\omega = \lim_{\delta \to 0} \frac{\delta\theta}{\delta t} = \frac{d\theta}{dt} = \dot{\theta}$$

The units of angular velocity are rad s^{-1}.

A similar argument can be used to determine the instantaneous angular acceleration α of the line, such that

$$\alpha = \lim_{\delta t \to 0} \frac{\delta\omega}{\delta t} = \frac{d\omega}{dt} = \dot{\omega} = \ddot{\theta}$$

The units of angular acceleration are rad s^{-2}.

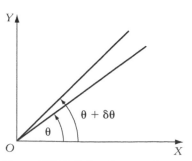

Figure 6.8 **Defining the angular position of a straight line**

Machine dynamics

The equations presented here are analogous to those for variable acceleration discussed in 'Variable (non-constant) acceleration'. The only difference is that that section refers to motion in a straight line, while here rotational (angular) quantities are considered. As in that section, the sign of the angular quantities is sufficient to indicate its direction. In accordance with rotations being considered to be positive in the anticlockwise direction, the angular velocity and angular acceleration are positive in the anticlockwise direction.

Relating rotation about a fixed axis to translational motion

In some mechanical systems it is necessary to relate the rotational motion of a circular drum, pulley or gear about a fixed axis to the motion of a cable or chain. Figure 6.9 shows an example in which a cable that has one end wrapped around a drum is used to raise a load mass.

Anti-clockwise rotation of the drum through an angle θ about an axis through O raises the load mass through a distance x. Provided that the cable does not slip, the distance x is equal to the arc length indicated in Figure 6.9, and the relationship between x and θ is given by

$$x = r\theta$$

where r is the (constant) radius of the drum.

The velocity of the load mass is obtained by differentiating the above equation with respect to time to give

$$\dot{x} = r\dot{\theta} = r\omega$$

where $\dot{\theta}$ is the angular velocity of the drum (see the previous section).

The acceleration of the load mass $\ddot{x}$ is obtained by differentiating the above velocity equation with respect to time to give

$$\ddot{x} = r\ddot{\theta} = r\alpha$$

where $\ddot{\theta}$ is the angular acceleration of the drum (see previous section).

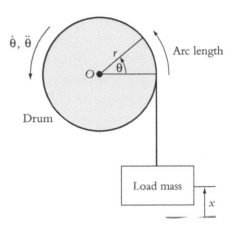

Figure 6.9 **Drum-cable system used to raise a load mass**

Acceleration of a particle down an inclined surface

Figure 6.10 shows a particle of mass m sliding down a frictionless inclined surface.

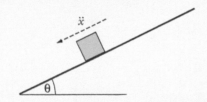

Figure 6.10 **Particle on an incline example**

Determine an expression for the acceleration of the particle $\ddot{x}$, and calculate the distance travelled in 4 s from a standing start when $\theta = 30°$. Assume that the acceleration due to gravity is 9.81 m s^{-2}.

The free body diagram of the particle showing both applied forces and accelerations is shown in Figure 6.11.

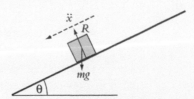

Figure 6.11 **Free body diagram of a particle on an inclined surface**

The applied forces are the weight mg of the particle and the contact force (see 'Coulomb (dry) friction between two surfaces') between the particle and the inclined surface. The contact force only has a normal reaction force component R because the surface is frictionless. The particle slides down the incline.

An expression involving the acceleration $\ddot{x}$ can be obtained by applying Newton's second law parallel to the surface (see 'Newton's laws')

$$\mathbf{F} = m\mathbf{a} \quad {}^{+}\!\!\!\swarrow \qquad mg \sin \theta = m\ddot{x}$$

Rearranging this equation, the acceleration is given by

$$\ddot{x} = g \sin \theta$$

as required.

It can be seen that the acceleration is independent of the mass of the particle, and if $\theta = 90°$, the particle will have acceleration g.

In this case the acceleration is constant, meaning that the results in 'Constant acceleration' can be used to analyse the distance travelled in a given time. The distance travelled in time t is given by

$$s = ut + \tfrac{1}{2}at^2$$

where u is the initial velocity and a is the acceleration. Using the obtained acceleration expression $a = \ddot{x} = g \sin\theta$ and noting that the particle moves from a standing start, such that $u = 0$, gives

$$s = \tfrac{1}{2}gt^2 \sin\theta$$

Substituting $g = 9.81 \text{ m s}^{-2}$, and $t = 4 \text{ s}$ into this expression, the distance travelled in 4 s from a standing start is 39.24 m, as required.

Simple crane

Figure 6.12 shows a crane drum of constant radius r rotating about a fixed axis through O. A cable with a particle of mass m attached at one end is wrapped around the drum. As the drum is rotated in an anticlockwise direction the particle is raised.

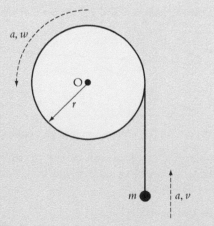

The crane drum is rotated so that the particle has constant acceleration a. Assuming that there is no slip between the drum and the cable, determine expressions for the tension in the cable and the angular velocity of the drum 2 s after being switched on.

Figure 6.12 **Simple crane drum example**

The tension in the cable is obtained by considering the applied forces and acceleration of the particle. The free body diagram of the particle showing both applied forces and acceleration is shown in Figure 6.13.

The applied forces are the weight mg of the particle and the tension T in the cable.

An expression involving the tension T can be obtained by applying Newton's second law vertically (see 'Newton's laws')

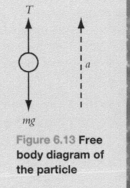

$$\mathbf{F} = m\mathbf{a} \qquad {}^{+}\!\!\uparrow \qquad T - mg = ma$$

Rearranging this equation, the tension in the cable is given by

$$T = m(g + a)$$

as required.

Figure 6.13 Free body diagram of the particle

Noting that the acceleration a is constant, the upwards velocity v of the particle at time t can be obtained using the results presented in 'Constant acceleration'; i.e.

$$v = u + at$$

where u is the initial velocity and a is the acceleration. Assuming that the initial velocity v (before being switched on) is zero, $u = 0$, the upwards velocity is given by

$$v = at$$

Substituting $t = 2\,\mathrm{s}$ into this expression, the velocity v after $2\,\mathrm{s}$ is $2a\,\mathrm{m\,s^{-1}}$. The angular velocity (see 'Angular (rotational) motion') ω of the drum is related to the vertical velocity of the particle by (see 'Relating rotation about a fixed axis to translational motion')

$$\omega = \frac{v}{r}$$

Using the expression for v and setting $t = 2\,\mathrm{s}$, the angular velocity is $\dfrac{2a}{r}\,\mathrm{rad\,s^{-1}}$ after $2\,\mathrm{s}$, as required.

6.3 Kinematics of a particle in a plane

Introduction

Many practical dynamics problems consider bodies moving in a plane and these are analysed using either Cartesian or polar coordinate systems. In this section expressions are derived for the velocity and acceleration of a particle moving in a plane using both of these coordinate systems. These kinematic expressions will be used extensively in later sections, particularly when applying Newton's laws of motion to particles (Section 6.4) and rigid bodies (Section 6.8), and when analysing linkage mechanisms (Section 6.6). Some examples are presented to demonstrate how the kinematic expressions are applied.

Cartesian coordinates

Consider the motion of a particle along a curved path (denoted by the dashed line) in the OXY plane as shown in Figure 6.14.

The OXY axes are our reference axes and are fixed so that they do not move. At time t, the Cartesian coordinates of the particle P relative to the reference axes are X in the X-direction and Y in the Y-direction. At time $t + \delta t$, the particle has moved to P' and its new coordinates are $X + \delta X$ in the X-direction and $Y + \delta Y$ in the Y-direction.

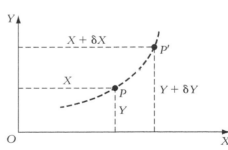

Figure 6.14 Motion of a particle along a curved path (Cartesian coordinates)

Velocity components

The instantaneous velocity of the particle in the X- and Y-directions can be determined by noting that during the time interval δt the average velocity of the particle is the displacement moved divided by the time interval. As the time interval approaches zero, the velocity approaches the instantaneous velocity of the particle (see 'Variable (non-constant) acceleration').

The instantaneous velocity in the X-direction is

$$V_X = \lim_{\delta t \to 0} \frac{\delta X}{\delta t} = \frac{dX}{dt} = \dot{X}$$

where the overdot represents differentiation with respect to time.

Similarly, the instantaneous velocity in the Y-direction is

$$V_Y = \lim_{\delta t \to 0} \frac{\delta Y}{\delta t} = \frac{dY}{dt} = \dot{Y}$$

Noting that velocity is a vector, the instantaneous velocity components in the X and Y-directions can be used to calculate the magnitude and direction of the velocity of the particle, as shown in Figure 6.15.

In this figure V is the magnitude of the velocity and its direction is indicated by angle α_V relative to OX. When stating the direction of a vector quantity it is important to state the angle relative to a reference direction.

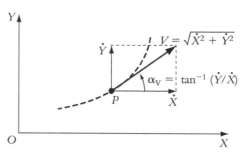

Figure 6.15 Velocity of a particle in Cartesian coordinates

Acceleration components

Using similar arguments to those used for the velocity components it can be shown that the instantaneous acceleration in the X and Y-directions are

$$a_X = \ddot{X}, \quad a_Y = \ddot{Y}$$

respectively. In these equations the double overdot represents double differentiation with respect to time. Figure 6.16 shows the acceleration components in Cartesian coordinates, together with the magnitude and direction of the acceleration.

In this figure a is the magnitude of the acceleration and its direction is indicated by angle α_a relative to OX.

Note that the resultant acceleration vector a does not necessarily act in a direction that is tangential to the curved path. This is because the acceleration depends on the rate of change of the velocity.

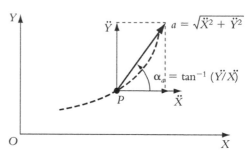

Figure 6.16 Acceleration of a particle in Cartesian coordinates

Polar coordinates

Consider the motion of the particle shown in Figure 6.14 along the same curved path, but this time use polar coordinates to identify the position of the particle, see Figure 6.17.

Here O is our fixed reference point. At time t, the polar coordinates of the particle P are r in the direction from O to P, and θ relative to the horizontal. At time $t + \delta t$, the particle has moved to P' and its new polar coordinates are $r + \delta r$ in the direction from O to P', and $\theta + \delta\theta$ relative to the horizontal.

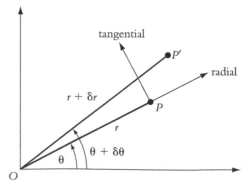

Figure 6.17 Motion of a particle along a curved path (polar coordinates)

Velocity components

The velocity of the particle at P in the radial and the tangential directions (see Figure 6.17) can be determined by noting that during the time interval δt the position of the particle relative to the radial and tangential directions changes. Assuming that the time interval is small and that δr and $\delta\theta$ are small, the changes in radial and tangential displacement are

radial $$(r + \delta r)\cos(\delta\theta) - r \approx \delta r$$

tangential $$(r + \delta r)\sin(\delta\theta) \approx r\delta\theta$$

These approximations are valid, since, $\cos(\delta\theta) \approx 1$, $\sin(\delta\theta) \approx \delta\theta$, and $\delta r\delta\theta \approx 0$.

Noting that the average velocity of the particle is the displacement moved divided by the time interval, and that as the time interval approaches zero the velocity approaches the instantaneous velocity, the instantaneous radial velocity V_R is given by

$$V_R = \lim_{\delta t \to 0} \frac{\delta r}{\delta t} = \frac{dr}{dt} = \dot{r}$$

and the instantaneous tangential velocity (or transverse velocity) V_T is given by

$$V_T = \lim_{\delta t \to 0} r\frac{\delta\theta}{\delta t} = r\frac{d\theta}{dt} = r\dot{\theta}$$

In these equations the radial velocity $\dot{r}$ is the velocity arising from a change in the distance of P from O, while the tangential velocity $r\dot{\theta}$ depends on the radius r and the angular velocity $\dot{\theta}$. Note: the angular velocity is often denoted by the symbol ω (see 'Angular (rotational) motion').

The radial and tangential components of velocity are shown in Figure 6.18.

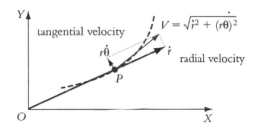

Figure 6.18 **Velocity of a particle in polar coordinates**

The magnitude and direction of the resultant velocity V is identical to that obtained in Figure 6.15.

Acceleration components

To determine the acceleration components it is necessary to redraw Figure 6.15 to show the positions and velocity components of the particle at times t and $t + \delta t$, see Figure 6.19.

Following a similar procedure to that for the velocity components, the changes in the velocities in the radial and tangential directions in the small time interval δt are

radial $$(\dot{r} + \delta\dot{r})\cos(\delta\theta) - (r\dot{\theta} + \delta(r\dot{\theta}))\sin(\delta\theta) - \dot{r} \approx \delta\dot{r} - r\dot{\theta}\delta\theta$$

tangential $$(r\dot{\theta} + \delta(r\dot{\theta}))\cos(\delta\theta) + (\dot{r} + \delta\dot{r})\sin(\delta\theta) - r\dot{\theta} \approx \delta(r\dot{\theta}) + \dot{r}\delta\theta$$

These approximations are valid, since

$$\cos(\delta\theta) \approx 1, \sin(\delta\theta) \approx \delta\theta, \delta(r\dot{\theta})\delta\theta \approx 0, \delta\dot{r}\delta\theta \approx 0.$$

Using these results, the instantaneous radial acceleration a_R is given by

$$a_R = \lim_{\delta t \to 0} \frac{(\delta\dot{r} - r\dot{\theta}\delta\theta)}{\delta t} = \ddot{r} - r\dot{\theta}^2$$

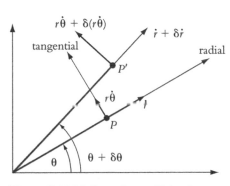

Figure 6.19 **Motion of a particle along a curved path (polar coordinates)**

In this equation (i) $\ddot{r}$ is due to the distance of P from O changing and is independent of angle θ, and (ii) $r\dot{\theta}^2$ is the centripetal acceleration which acts towards the centre of rotation and is independent of rate of change of radius r.

Similarly, it can be shown that the instantaneous tangential acceleration (or transverse acceleration) a_T is given by

$$a_T = \lim_{\delta t \to 0} \frac{\delta(r\dot{\theta}) + \dot{r}\,\delta\theta}{\delta t} = r\ddot{\theta} + 2\dot{r}\,\dot{\theta}$$

In this equation: (i) $r\ddot{\theta}$ is the product of the radius r and the angular acceleration $\ddot{\theta}$ and is independent of any change in radius r, and (ii) $2\dot{r}\,\dot{\theta}$ is the so-called Coriolis acceleration. Note that the angular acceleration is often denoted by the symbol α (see 'Constant acceleration'). The radial tangential components of acceleration are shown in Figure 6.20.

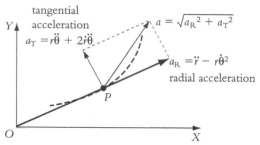

Figure 6.20 **Acceleration of a particle in polar coordinates**

The magnitude and direction of the resultant acceleration a is identical to that obtained in Figure 6.16.

Summary of velocity and acceleration components for motion in a plane

Figure 6.21 summarizes the velocity and acceleration results obtained in Cartesian coordinates and polar coordinates.

Note the *sign conventions* used in these figures. In Cartesian coordinates, all components of displacement, velocity and acceleration are assumed to be positive in the left to right or upwards directions. In polar coordinates, the angular rotation, velocity and acceleration are all assumed to be positive in the anticlockwise direction, and dictate the direction of the positive tangential components.

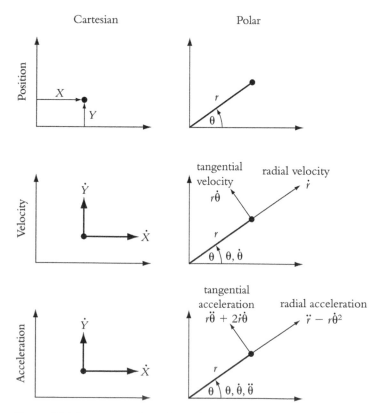

Figure 6.21 **Summary of position (displacement), velocity and acceleration components in Cartesian and polar coordinates**

Circular motion of a particle

A particle moves on a circular path of constant radius R about a fixed point O with constant angular velocity ω. Determine expressions for the radial and tangential components of velocity and acceleration of the particle.

The velocity and acceleration are obtained using the results summarized in Figure 6.21.

Noting that $r = R$ (constant), it is clear that $\dot{r} = \ddot{r} = 0$. The radial velocity '$\dot{r}$' of the particle P is zero, while the tangential velocity V_T is $R\omega$. Figure 6.22 indicates the velocity of the particle at two instances in time. In both cases, the particle is shown to move tangential to the circular motion with velocity $R\omega$.

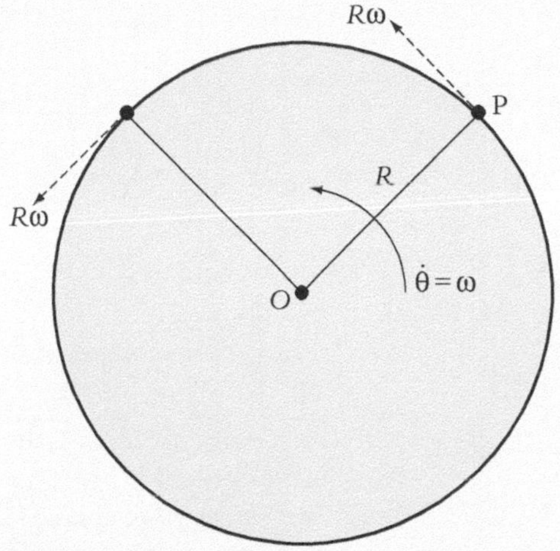

Similarly, it is found that $\ddot{r}$ and the Coriolis acceleration $2\dot{r}\dot{\theta}$ are both zero since $\dot{r} = \ddot{r} = 0$. Noting that the angular velocity $\dot{\theta}$ is constant, the motion is 'steady' and the tangential acceleration $r\ddot{\theta}$ is zero. The only non-zero term is the

Figure 6.22 **Velocity of a particle moving at constant speed in a circular motion**

centripetal acceleration $R\omega^2$. Figure 6.23 indicates the acceleration of the particle at two instances in time. In both cases, the particle is shown to have a centripetal acceleration $R\omega^2$ towards the centre of rotation.

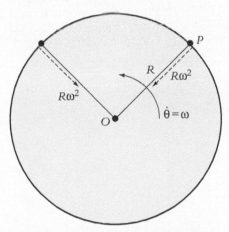

Figure 6.23 **Acceleration of a particle moving at constant speed in a circular motion**

The radial acceleration acts in a direction towards the centre of rotation O as shown. This is despite the angular motion being constant and comes about because the direction of the velocity vector changes as the particle moves. A heuristic explanation of this follows. The direction of the tangential velocity of P at any instant indicates that P does not want to follow a circular path, and that it is effectively trying to 'escape' from its circular motion. In order for P to remain on the prescribed circular path, P needs to move (accelerate) towards the centre of rotation O. This is taken into account by the centripetal acceleration '$r\dot{\theta}^2$' term which always acts towards the centre of rotation.

Velocity and acceleration in polar coordinates

Find the velocity and acceleration (both *magnitude* and *direction* (relative to the rod)) of the collar C which is sliding outwards along the rod which rotates about O as shown in Figure 6.24. The collar (at the instant shown) is located 0.5 m from the pivot and is sliding at constant velocity 1 m s^{-1} *relative to the rod* which is rotating at constant angular velocity 10 rad s^{-1}.

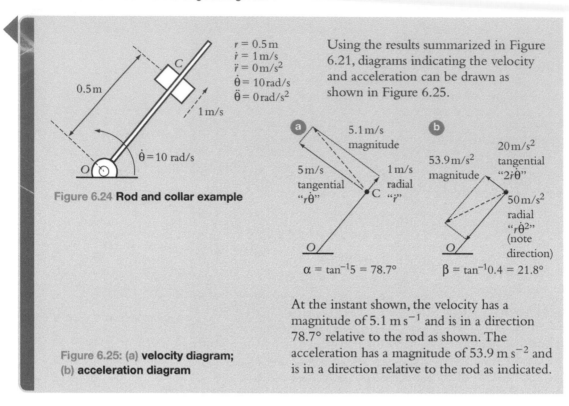

$r = 0.5\,\text{m}$
$\dot{r} = 1\,\text{m/s}$
$\ddot{r} = 0\,\text{m/s}^2$
$\dot{\theta} = 10\,\text{rad/s}$
$\ddot{\theta} = 0\,\text{rad/s}^2$

Using the results summarized in Figure 6.21, diagrams indicating the velocity and acceleration can be drawn as shown in Figure 6.25.

Figure 6.24 Rod and collar example

At the instant shown, the velocity has a magnitude of $5.1\,\text{m s}^{-1}$ and is in a direction $78.7°$ relative to the rod as shown. The acceleration has a magnitude of $53.9\,\text{m s}^{-2}$ and is in a direction relative to the rod as indicated.

Figure 6.25: (a) velocity diagram; (b) acceleration diagram

6.4 Kinetics of a particle in a plane

Introduction

In this section the equations of motion governing the dynamics of a particle moving in a plane are reconsidered. Newton's second law is used to determine the relationship between externally applied forces and the motion (acceleration) of the particle. Particular attention is given to understanding the differences between applying Newton's second law and using d'Alembert's Principle. Some examples are presented that are solved using Newton's second law and d'Alembert's Principle.

Motion of a particle in a plane

Consider a particle of mass m in a plane. This is the simplest possible rigid body and requires only two coordinates (say X, Y) to define its position, see Figure 6.26.

The application of forces $\mathbf{F}_i$ to the particle cause the mass to move with acceleration $\mathbf{a}$ in the direction shown.

Applying Newton's second law to the particle gives

$$\sum_i \mathbf{F}_i = m\mathbf{a} \qquad \text{(vector equation)} \qquad (6.3)$$

This is a vector equation and requires the magnitude and direction of the forces acting to be taken into account. For the purposes of performing calculations, it is convenient to write this equation in components

$$\xrightarrow{\;+\;} \sum_i F_{Xi} = m\ddot{X}$$

$$\uparrow^{+} \sum_i F_{Yi} = m\ddot{Y}$$

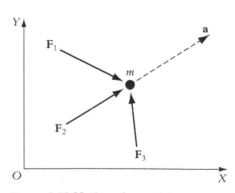

Figure 6.26 Motion of a particle

where F_{Xi} and F_{Yi} are the X and Y-components of the ith force respectively, and $\ddot{X}$ and $\ddot{Y}$ are the X and Y-components of the acceleration respectively (see 'Newton's laws').

These equations can be used to determine how a particle accelerates under the action of forces of known magnitude and direction. Conversely, if the magnitude and direction of the acceleration are known from kinematic considerations, then the magnitude and direction of the resultant force acting on the particle can be determined.

D'Alemberts Principle

Newton's second law relates the applied force to the resultant acceleration and is not a static force balance equation, unless there is no acceleration.

D'Alemberts Principle allows a dynamics problem to be analysed as a statics problem. Although most students will have met d'Alemberts Principle in the past, few have a clear understanding of its application. This issue is one of the main sources of error when solving dynamics problems. In this Chapter, the direct application of Newton's second law is preferred to d'Alembert's Principle because it clearly distinguishes applied forces from the resulting acceleration. Even though d'Alembert's Principle will tend not be used in later sections (except Section 6.9), it is important to understand its origins and how it relates to Newton's second law.

Newton's second law (equation (6.3)) can be rewritten as follows

$$\sum_i \mathbf{F}_i - m\mathbf{a} = 0 \qquad \text{(vector equation)} \qquad (6.4)$$

It is important to distinguish equation (6.4) from equation (6.3). Equation (6.3) arises from considering the applied forces and the resulting acceleration of a particle. In contrast, equation (6.4) is simply a rearranged version of equation (6.3).

In d'Alemberts Principle, the '$- m\mathbf{a}$' term is considered to be an imaginary inertia force, having magnitude ma acting in a direction opposite to that of the acceleration of the particle. By doing this all of the terms on the left-hand side can be considered to be applied forces and because these forces sum to zero, the particle can be considered to be in static equilibrium.

Circular motion of a particle

A particle of mass m is attached via an inextensible string of length R to a fixed point O and moves on a horizontal circular path with constant angular velocity ω. Determine an expression for the tension in the string.

'Circular motion of a particle' considered a similar situation and determined an expression for the acceleration (magnitude and direction) of the particle using kinematic considerations. There it was found that the particle experienced a centripetal acceleration acting towards the centre of rotation, as indicated in Figure 6.23.

The free body diagram of the particle showing both applied forces and accelerations is shown in Figure 6.27.

The only force acting on the particle is the tension force T. In this case it is clear that the tension in the string acts towards O the centre of rotation. In other cases the direction of the forces

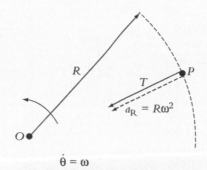

Figure 6.27 Free body diagram of a particle in circular motion-tension force (solid line); radial acceleration (dashed line)

may not be so clear. In these cases it is worth noting that if the actual force is in the opposite direction to that indicated on the free body diagram, the calculation will yield a negative value. In contrast, the direction of the acceleration here is specified from kinematic considerations, and must be positive towards the centre of rotation (see 'Circular motion of a particle').

Applying Newton's second law to the particle along the string gives

$$T = m(R\omega^2)$$

This equation defines the tension in the string, as required.

This problem can also be solved using d'Alemberts Principle, in which the free body diagram is redrawn as shown in Figure 6.28.

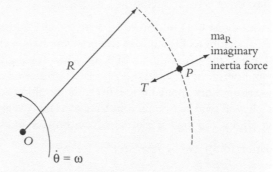

In this figure an imaginary inertia force of magnitude ma_R has been included, where $a_R = R\omega^2$. This force is in the opposite direction to the actual acceleration of the particle.

The tension force is obtained by considering the particle to be in static equilibrium, i.e.

Figure 6.28 Free body diagram of a particle in circular motion using d'Alemberts Principle

$$T - mR\omega^2 = 0$$

This equation defines the tension force $T = mR\omega^2$, as required.

Point mass in a slot

A particle A of mass m fits loosely in a smooth slot with vertical sidewalls cut into a disc mounted in the horizontal plane, as shown in Figure 6.29. The disc is rotating at constant angular speed Ω about its fixed centre O as shown. The particle is held in position by an inextensible cord having one end secured at B. Determine an expression for the tension in the cord.

As the disc rotates, the particle remains at A and as a consequence rotates at a constant radius R about O. To determine the tension in the inextensible cord, it is necessary to consider the forces that act on the particle to make it move. The free body diagram for the particle is shown in Figure 6.30.

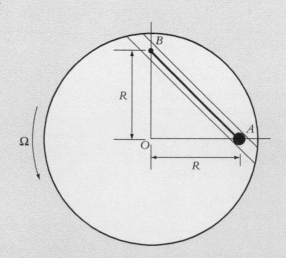

Figure 6.29 Schematic of particle in a slot

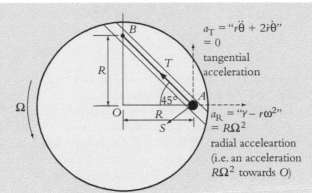

Figure 6.30 **Free body diagram of particle in a slot: forces (solid lines); accelerations (dashed lines)**

The components of acceleration are summarized in Figure 6.21, and must comply with the sign convention indicated there. The fact that the radius of motion is constant has been taken into account when determining the radial and tangential components of acceleration.

In this case it is clear that the tension force T acts along the cord towards B, and the reaction force S acts normal to the slot in the direction shown, reflecting the fact that the particle is being pushed inwards by the outer side of the slot. As in the previous example the directions of these forces are not important, as the sign of the calculated value will indicate their true directions. The weight of the point mass does not appear in Figure 6.30 because the disc is horizontal.

Applying Newton's second law to the particle in the directions indicated gives

$$T\cos45° + S\cos45° = m(R\Omega^2)$$

$$T\sin45° - S\sin45° = 0$$

The above equation implies that $T = S$. Using this fact in the previous equation gives

$$2T\cos45° = mR\Omega^2$$

Hence the tension in the cord is given by

$$T = \frac{mR\Omega^2}{\sqrt{2}}$$

as required.

It is interesting to note that the direction of rotation does not affect the answer because the direction of the centripetal acceleration is independent of the direction of rotation.

This problem can be re-solved using d'Alemberts Principle, in which the free body diagram is redrawn as shown in Figure 6.31.

In this figure an imaginary inertia force of magnitude $mR\Omega^2$ has been applied to the particle. This force is in the opposite direction to the actual acceleration of the particle.

The tension force is obtained by considering the particle to be in static equilibrium. i.e.

$$T - mR\omega^2\cos45° = 0$$

This equation defines the tension force

$$T = \frac{mR\omega^2}{\sqrt{2}},$$ as required.

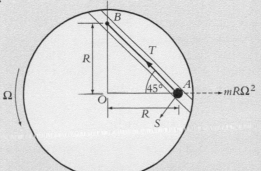

Figure 6.31 **Free body diagram of a point mass in a slot using d'Alemberts Principle**

6.5 Kinematics of rigid bodies in a plane

Introduction

In Section 6.3, the kinematics of a single particle moving in a plane was considered. In practice, engineering components have finite dimensions and it is necessary to take this into account when analysing their motions. Throughout this chapter these components are considered to be rigid bodies moving in a plane, and a consequence of this is that the position of any point within the body is fixed relative to another point in the same rigid body. This fact is used here to analyse the kinematics of a rigid body in a plane in terms of translation motions of a point in the body and rotations of the body. The concept of analysing the motions of a rigid body in terms of these translations and rotations will be used extensively in later sections to analyse linkage systems (Section 6.6) and apply Newton's laws of motion to rigid bodies (Section 6.8). The kinematics of connected rigid bodies are also analysed.

Degrees of freedom (DOF)

To define the position of a rigid body in a plane it is necessary to specify the location of a point in the body and the orientation of any line fixed in the body, such as AB, as shown in Figure 6.32.

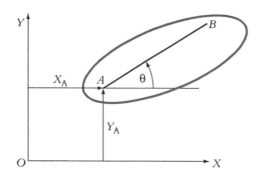

Figure 6.32 **Defining the position of a rigid body**

The three coordinates, X_A, Y_A, θ uniquely define the position of a rigid body. i.e. independent of how the body moves, it is only necessary to know X_A, Y_A and θ to be able to define the precise location and orientation of a body. This fact arises as a consequence of the body being rigid, which ensures that the relative position of any two points within a body is fixed.

As will be seen in 'Rigid body motion', the displacement of a rigid body can be expressed in terms of changes in (X_A, Y_A, θ). For this reason, a planar body is said to have three degrees of freedom (DOF) (two translations and one rotation). A rigid body in three-dimensional space has 6DOF (three translations and three rotations).

Degrees of freedom of connected bodies

It follows from the previous section that a rigid body that is free to move in a plane (see Figure 6.33(a) has 3DOF. In contrast a single body pivoted (pin jointed) to ground (see Figure 6.33(b)) has only 1DOF, because it is only necessary to specify θ to define the position of the body. In this case the constraint provided by the pivot has removed the two translational DOF.

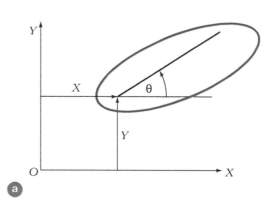

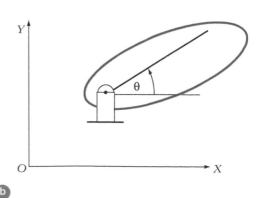

Figure 6.33(a) **free rigid body (3DOF); (b) pin jointed to ground (1DOF)**

Machine dynamics

A system consisting of two rigid bodies (in a plane) that are not connected in any way have 6DOF (i.e. 3DOF for each body). Consider the mechanical system shown in Figure 6.34 which consists of two rigid bodies connected by a pin joint.

This system has 4DOF (X_A, Y_A, θ_1, θ_2), because the pin joint has again removed 2DOF.

Other types of constraint are commonly found in machinery and mechanisms. A sliding joint, such as a piston in a cylinder removes 2DOF and leaves one translational DOF free. A 'pin in a slot' removes one translational DOF and leaves one translational DOF and one rotational DOF free.

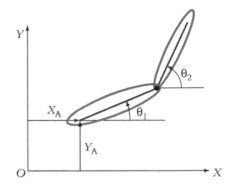

Figure 6.34 Two rigid bodies connected by a pin joint

Rigid body motion

Consider a rigid body moving in a plane as shown in Figure 6.35.

In the figure, points A and P are arbitrarily chosen and can be seen to move to points A_1 and P_1 respectively. The displacement of the rigid body can be seen to consist of two parts:

(i) Translation — All points move linearly a distance AA_1 (i.e. movement to the dashed line)

(ii) Rotation about A — The angle of rotation will be the same wherever A and P are taken.

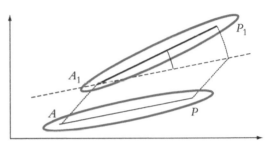

Figure 6.35 Motion of a rigid body in a plane

A consequence of this is that the instantaneous velocity of a point P in the rigid body is dependent on the translational velocity V of a point in the body (say A) and the angular velocity $\omega(=\dot{\theta})$ of the body about A. This is shown schematically in Figure 6.36.

In general, V and ω will vary with time, such that there are translational accelerations a and angular accelerations $\alpha(=\ddot{\theta})$. Consequently, the instantaneous acceleration of a point P in the rigid body is dependent on a and α. This is shown schematically in Figure 6.37.

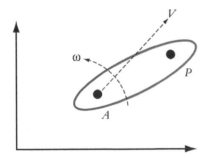

Figure 6.36 Linear and rotational velocity of a rigid body

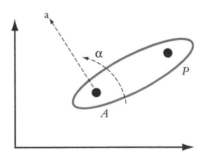

Figure 6.37 Linear and rotational acceleration of a rigid body

The velocity of any point P in the body relative to A (written V_{P_A}) due to rotation about A can be obtained by using the polar coordinate results considered in 'Polar coordinates', and summarized in Figure 6.21. However, since the body is rigid, the radius of motion is constant, such that $\dot{r} = 0$, and V_{P_A} will be in a direction tangential to the line AP, i.e. there will be no radial component of velocity along AP; see Figure 6.38.

Hence

$$V_{P\_A} = AP\omega \qquad \text{i.e. tangential to } AP$$

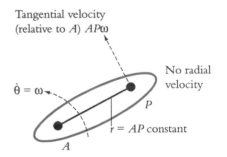

Figure 6.38 Relative velocity of a rigid link (compare with polar velocity result in Figure 6.38)

423

The velocity of P relative to the ground (referred to as the absolute velocity and written V_{P_GROUND}) is the vector sum of the velocity of P relative to A and the velocity of A relative to the ground, i.e.

$$V_{P\_GROUND} = V_{P\_A} + V_{A\_GROUND} \qquad \text{(vector equation)}$$

The acceleration of any point P in the body relative to A (denoted a_{P_A}) due to rotation about A can be obtained similarly by using the polar coordinate results considered in 'Polar coordinates'. Taking into account that the body is rigid such that $\dot{r} = \ddot{r} = 0$, it can be shown that a_{P_A} has components radial and tangential to the line AP; see Figure 6.39.

The $AP\omega^2$ term is the centripetal acceleration term, and ensures that the radial acceleration of point P is directed towards the centre of rotation A.

Hence,

$$a_{P\_A} = AP\alpha(\text{tangential to } AP) - AP\omega^2(\text{radially outwards})$$

Similarly, the acceleration of P relative to the ground (referred to as the absolute acceleration and denoted by a_{P_GROUND}) is the vector sum of the acceleration of P relative to A and the acceleration of A relative to the ground, i.e.

$$a_{P\_GROUND} = a_{P\_A} + a_{A\_GROUND} \qquad \text{(vector equation)}$$

In general, a_{P_A} (unlike V_{P_A}) will have radial and tangential components. It is important to take account of the vector form of the velocity and acceleration equations, by carefully considering their magnitude and direction when adding them together. This topic will be considered in more detail in Section 6.6 when velocity and acceleration diagrams of planar mechanisms are considered more formally.

Figure 6.39 Relative acceleration of a rigid link (compare with polar acceleration result in Figure 6.21)

Slider mechanism

Figure 6.40 shows part of a slider mechanism in which collar A moves along a fixed horizontal track with velocity V. The link AB has length 100 mm and rotates with angular velocity $\dot{\theta}$.

Figure 6.40 Slider mechanism

At the instant shown, link AB is at an angle of 30° to the track, $\dot{\theta} = 20$ rad s^{-1}, and $V = 1$ m s^{-1}. Determine the magnitude of the velocity of B relative to fixed point O at the instant shown, and show that the direction of this velocity is vertically upwards.

The velocity of A relative to O (V_{A_O}) and the velocity of B relative to A (V_{B_A}) are shown diagrammatically in Figures 6.41(a) and (b) respectively.

The magnitude and direction of velocity V_{B_A} was obtained using Figure 6.38, and depends on the tangential velocity only because rod AB is of fixed length. A consequence of this is that the

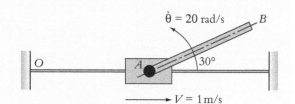

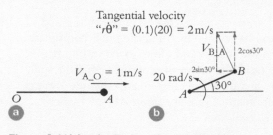

Figure 6.41(a) velocity V_{A_O} (b) velocity V_{B_A}

velocity is in a direction perpendicular to link AB, in accordance with the indicated sense of the angular velocity $\dot{\theta} = 20$ rad s^{-1}. Figure 6.41 is only valid at the instant shown.

The velocity of B relative to $O(V_{B\_O})$ is obtained using the vector equation:

$$V_{B\_O} = V_{B\_A} + V_{A\_O}$$

Using the component velocities indicated in Figure 6.41, the velocity vector V_{B_O} can be constructed as a velocity diagram, as shown in Figure 6.42.

This figure is constructed by adding together the velocity vectors V_{A_O} and V_{B_A} (see 'Vector quantities'), and then constructing the resultant vector V_{B_O}. In Figure 6.42, vector V_{A_O} is constructed first, and vector V_{B_A} second. The order in which the vectors are added is not important, since the resultant vector is independent of the order.

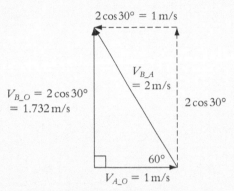

From Figure 6.42 it is found that the velocity of B relative to O is vertically upwards with magnitude 1.732 m s^{-1}, at the instant shown. As required.

Figure 6.42 Velocity diagram used to calculate V_{B_O}

Robotic arm

Figure 6.43 shows a robotic arm consisting of two rigid rods OA and AP. Rod OA pivots about fixed point O, while rod AP is connected to rod OA at the common pivot point at A. In practice, motors located at O and A are used to rotate the rods about these points.

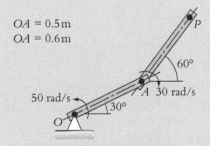

Figure 6.43 Robotic arm example

At the particular instant shown, OA is rotating in an anticlockwise direction with angular velocity 50 rad s^{-1}, while AP is rotating in a clockwise direction with angular velocity 30 rad s^{-1}, as shown. Determine the horizontal and vertical components of the velocity of P relative to O at the instant shown. The lengths of the rods are provided in Figure 6.43.

The velocity of A relative to $O(V_{A\_O})$ and the velocity of P relative to $A(V_{P\_A})$ are shown diagrammatically in Figures 6.44(a) and (b) respectively.

These components are obtained using Figure 6.21, and only depend on the tangential velocity because OA and AP have fixed length. Each component acts perpendicular to the links shown, in the sense indicated by the angular velocity.

Tangential velocity $\qquad$ Tangential velocity
"$r\dot{\theta}$" = (0.5)(50) = 25 m/s $\qquad$ "$r\dot{\theta}$" = (0.6)(30) = 18 m/s

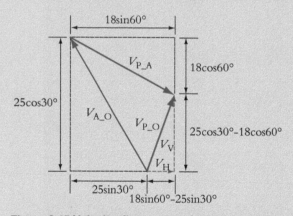

Figure 6.44(a) velocity V_{A_O} (b) velocity V_{P_A}

The velocity of P relative to $O(V_{P\_O})$ is obtained using the vector equation:

$$V_{P\_O} = V_{P\_A} + V_{A\_O}.$$

Using the component velocities indicated in Figure 6.44, the resultant velocity V_{P_O} can be constructed as a velocity diagram, as shown in Figure 6.45.

Figure 6.45 Velocity diagram used to calculate V_{P_O}

This figure is constructed by adding together the velocity vectors V_{A_O} and V_{P_A} (see 'Vector quantities'), and then constructing the resultant vector V_{P_O}. In Figure 6.45 velocity vector V_{A_O} was constructed first, and velocity vector V_{P_A} second. Finally vector V_{P_O} was constructed.

The magnitude and direction of the horizontal (V_H) and vertical (V_V) components of the resultant velocity V_{P_O}, at the instant shown, are indicated in Figure 6.45, where:

$$V_H = 18\sin60° - 25\sin30° = 3.09 \text{ m s}^{-1}$$

$$V_V = 25\cos30° - 18\cos60° = 12.65 \text{ m s}^{-1}$$

These results could have been obtained directly from Figure 6.44 by combining the components of the velocity vectors algebraically. Care must be taken when using this approach to take account of the direction of the vectors.

6.6 Kinematics of linkage mechanisms in a plane

Introduction

Linkage mechanisms are used widely to provide different types of motion in machinery (metal forming, printing, packaging, textile etc.). The design of mechanisms involves synthesis, kinematic analysis and force analysis. Synthesis of the mechanism is required to give the required motion, and is a complex process that requires (like all design) a mixture of experience and intuition. Kinematic analysis is needed to determine the linear and angular velocities and accelerations of each of the linkage members. This forms the focus of this section. Once a kinematic analysis has been performed a force (or kinetic) analysis is needed to determine the forces required to drive the mechanism, the forces acting on individual links (so that stress calculations can be performed), and the forces transmitted to the machine frame. This is tackled by applying Newton's laws of motion to rigid bodies, which will be discussed in Section 6.8.

Engineers who design mechanisms in a professional capacity will almost certainly use software packages to perform the required kinematic and force analyses. The aim of this section is to present the basic principles behind the kinematic analysis performed in these packages and to develop some engineering insight. A graphical approach will be used to obtain the required information, based on the concepts of space diagrams, velocity diagrams and acceleration diagrams.

Space diagram

The space diagram defines the geometry of the mechanism at a particular instant in time – i.e. the instant at which the analysis is performed – and is used to calculate the angles between different links and some reference datum. Figure 6.46 shows the space diagram for part of an example mechanism.

In what follows it will be necessary to use a formal notation to describe the vector position of a point on the mechanism relative to some other point. The vector position of B relative to A is written as r_{B_A}, while the vector position of A relative to B is written as r_{A_B}. These vectors are shown schematically in Figure 6.47.

The magnitude of the vectors corresponds to the straight line distance between the two points of interest, while the direction is defined by the orientation of the vector relative to some datum axis. This notation was introduced in 'Rigid body motion' to describe the velocity and acceleration of a single rigid body. For the purposes of analysing a planar mechanism, it is important to realize that each link of the mechanism is assumed to be a rigid body, and the analysis presented in 'Rigid body motion' can be applied to each link.

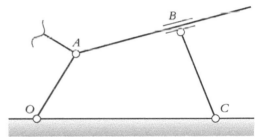

Figure 6.46 Space diagram for part of an example mechanism

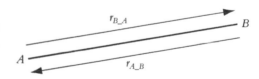

Figure 6.47 Vector notation used to describe the relative position of different points in the space diagram

Velocity diagram

Basics

The velocity diagram indicates the relative velocity between two points on a mechanism. The notation V_{B_A} (first introduced in 'Rigid body motion') is used to describe the velocity of B relative to A. It follows from Figure 6.21 that the velocity V_{B_A} has radial and tangential components, such that

$$V_{B\_A} = \dot{r}_{B\_A} \text{ (radial to } AB) + r_{B\_A}\dot{\theta}_{AB} \text{ (tangential to } AB)$$

For a rigid link, the magnitude of r_{B-A} (i.e. the length AB) is constant, meaning that $\dot{r}_{B-A} = 0$ (see 'Rigid body motion'). If link AB is increasing in length $\dot{r}_{B-A} = \dot{A}B$ is positive, while if link AB is reducing in length $\dot{r}_{B-A}$ is negative.

Single link example

Consider the linkage system shown in Figure 6.46. In particular, consider the situation when link AB has angular velocity $\dot{\theta}_{AB} = \Omega_{AB}$ and the slider at B is moving away from A along AB with positive velocity $\dot{r}_{B-A} = \dot{A}B$, as shown in Figure 6.48.

To determine the velocity V_{B-A}, it is necessary to consider the radial and tangential components of velocity, and then construct the velocity diagram for V_{B-A}.

If AB is increasing in length at rate $\dot{A}B$ and AB is rotating clockwise at rate Ω_{AB} ($= \dot{\theta}_{AB}$), then the radial component

$\dot{r}_{B-A}$ has magnitude $\dot{A}B$

     and direction   parallel to AB ($\| AB$)

and the tangential component

$\dot{r}_{B-A} \, \dot{\theta}_{AB}$ has magnitude $AB\Omega_{AB}$

     and direction   perpendicular to AB ($\perp AB$)

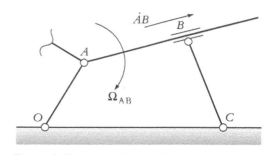

Figure 6.48 Space diagram for part of an example mechanism

The arrows above denote the positive sense of the components, and the directions 'parallel to AB' and 'perpendicular to AB' are obtained by analysing the geometry of the space diagram.

The radial and tangential components of velocity V_{B-A} are shown on the space diagram in Figure 6.49.

These components are used to draw the velocity diagram for V_{B-A}, as shown in Figure 6.50.

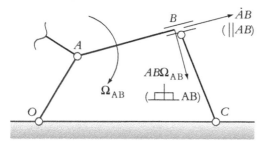

Figure 6.49 Radial and tangential components of the velocity V_{B-A}

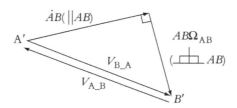

Figure 6.50 Constructing a velocity diagram for V_{B-A}

In Figure 6.50, point A' is the starting point for drawing the velocity diagram. The radial component is constructed first, followed by the tangential component. (The order in which the components are added is not important). Point B' is the finish point.

The vector from A' to B' represents the velocity vector V_{B-A}. The velocity vector V_{A-B} can be obtained from the velocity diagram, and is represented by the vector from B' to A' as shown.

Acceleration diagram

Basics

The acceleration diagram indicates the relative acceleration between two points on a mechanism. The notation a_{B-A} (first introduced in 'Rigid body motion') is used to describe the

Machine dynamics

acceleration of B relative to A. It follows from Figure 6.21 that the acceleration a_{B_A} has radial and tangential components, such that

$$a_{B\_A} = (\ddot{r}_{B\_A} - r_{B\_A}\dot{\theta}^2_{AB}) \text{ (radial to } AB) + (r_{B\_A}\ddot{\theta}_{AB} + 2\dot{r}_{B\_A}\dot{\theta}_{AB}) \text{ (tangential to } AB)$$

For a rigid link, the magnitude of r_{B_A} (i.e. the length AB) is constant, meaning that $\dot{r}_{B\_A} = \ddot{r}_{B\_A} = 0$ (see 'Rigid body motion'), but a_{B_A} still has radial and tangential components.

Single link example

Reconsider the linkage system shown in Figure 6.46. In particular consider the case when link AB has positive angular velocity $\dot{\theta}_{AB} = \Omega_{AB}$ in the clockwise direction and positive angular acceleration $\ddot{\theta}_{AB} = \dot{\Omega}_{AB}$ in the anticlockwise direction (i.e. AB rotates in the clockwise direction, but its rotational speed in the clockwise direction is decreasing). The slider at B is moving away from A along AB with positive velocity $\dot{r}_{B\_A} = \dot{A}B$, and B has positive acceleration $\ddot{r}_{B\_A} = \ddot{A}B$ (i.e. the speed at which B moves away from A is increasing). This situation is shown in Figure 6.51.

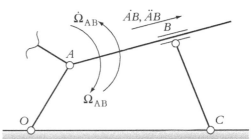

Figure 6.51 Space diagram for an example mechanism

To determine the acceleration a_{B_A}, it is necessary to consider the radial and tangential components of acceleration, and then construct the acceleration diagram for a_{B_A}.

Using the information provided, the radial component

$\ddot{r}_{B\_A}$ has magnitude $\ddot{A}B$

           and direction   parallel to AB ($\| AB$)

and

$-r_{B\_A}\dot{\theta}_{BA}^2$ has magnitude $AB\Omega^2_{AB}$

           and direction   parallel to AB towards A ($\| AB$)

The tangential terms

$r_{B\_A}\ddot{\theta}_{BA}$ has magnitude $AB\dot{\Omega}_{AB}$

           and direction   perpendicular to AB ($\perp AB$)

and

$2\dot{r}_{B\_A}\dot{\theta}_{BA}$ has magnitude $2\dot{A}B\Omega_{AB}$

           and direction   perpendicular to AB ($\perp AB$)

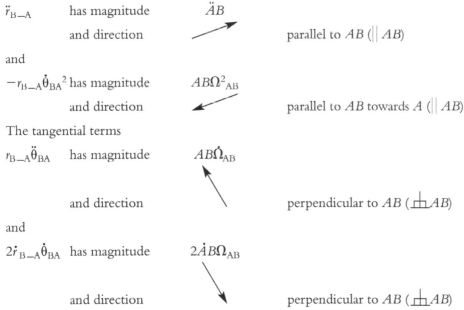

The magnitude and direction of the different components of acceleration a_{B_A} are shown in Figure 6.52.

Care should be taken identifying the direction of each component.

These radial and tangential components are used to draw the acceleration diagram for a_{B_A}, as shown in Figure 6.53.

In Figure 6.53, point A'' is the starting point for drawing the acceleration diagram for a_{B_A}. The acceleration diagram has been constructed in two different ways, by adding the four components in a different sequence. This is possible because the resultant vector is independent of the order in which they are added together.

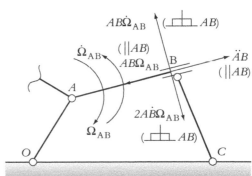

Figure 6.52 Radial and tangential components of the acceleration a_{B_A}

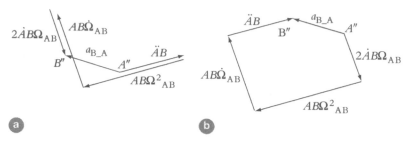

Figure 6.53 Acceleration diagram for a_{B_A}

General relationships

In the previous two sections, the velocity and acceleration diagrams for a single link were considered. These results need to be extended to consider the velocity and acceleration of different points on mechanisms consisting of connected links. This is achieved by noting that velocity and acceleration are vector quantities, and that it is always possible to write down vector expressions relating different points on a mechanism. For example, for any three points (O, A and B) on a mechanism, the vector relationships between their velocities and accelerations can be expressed as follows

$$V_{B\_O} = V_{B\_A} + V_{A\_O} \qquad \text{(vector equation)}$$

$$a_{A\_O} = a_{B\_A} + a_{A\_O} \qquad \text{(vector equation)}$$

Velocity and acceleration diagrams are based on relationships of this form. These relationships are identical to those introduced in 'Rigid body motion'.

Example 1 – slider-crank mechanism

Consider the slider–crank shown in Figure 6.54.

The geometry of the mechanism is specified at the instant shown in Figure 6.54. This means that lengths (AB, BC and AC) and orientations of the links are known, or can be calculated using trigonometry. In addition, for the example considered the angular velocity $\dot{\theta}_{AB}$ of crank AB is assumed to be constant.

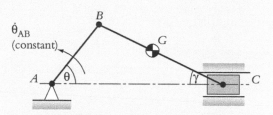

Figure 6.54 Slider-crank mechanism (space diagram) (not drawn to scale)

For the purposes of numerical calculation, the following numerical values are used: $AB = 80$ mm, $BC = 240$ mm, $BG = 120$ mm, $\theta = 45°$, $\dot{\theta}_{AB} = 100$ rad s^{-1}.

Determine the absolute velocity and acceleration of the slider C, and the absolute acceleration of G.

Figure 6.54 defines the geometry of the mechanism, and corresponds to the space diagram when it is drawn to scale. For the numerical values given, it can be shown either by construction or by using trigonometry that angle $\gamma = 13.6°$.

The absolute velocity of C is the velocity of C relative to the ground, and is the velocity of C relative to the fixed point A (i.e. V_{C_A}). This velocity is obtained using the following vector relationship for the velocity

$$V_{C\_A} = V_{B\_A} + V_{C\_B}$$

where V_{B_A} is the velocity of B relative to A and V_{C_B} is the velocity of C relative to B. In what follows, a vector diagram will be constructed that is consistent with this relationship. The key to achieving this is to identify the information that is known about each of the vectors in this relationship, taking into account the constraints imposed on the system. This information is summarized below:

V_{C_A} has unknown magnitude

and is in the horizontal direction (in the left or right direction) ⟷

V_{B_A} has no radial component because AB has constant length

V_{B_A} has a tangential component with *known* magnitude
$AB\dot{\theta}_{AB} = (80)(100) = 8000 \text{ mm s}^{-1}$

and direction ⟍ perpendicular to AB (⊥ AB)

V_{C_B} has no radial component because BC has constant length

V_{C_B} has a tangential component with unknown magnitude $BC\dot{\theta}_{BC}$

and direction ⟋ perpendicular to BC (⊥ BC).

Using the above information, the velocity diagram can be drawn as shown in Figure 6.55.

The basic strategy used to construct the velocity diagram is to construct the velocity vectors whose magnitude and direction are known first, and then construct the remaining velocity vectors in conjunction with the known velocity vector relationship.

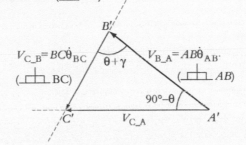

$V_{C\_B} = BC\dot{\theta}_{BC}$ (⊥ BC) $\theta + \gamma$ $V_{B\_A} = AB\dot{\theta}_{AB}$ (⊥ AB) $90° - \theta$ V_{C_A}

Figure 6.55 **Velocity diagram for slider-crank mechanism (not drawn to scale)**

The starting point for constructing Figure 6.55 is Point A'. Vector V_{B_A} is constructed first with a known magnitude (length) of 8000 mm s^{-1} and direction perpendicular to AB. This defines Point B'. The direction of vector V_{C_A} is either to the right or to the left because the piston can only move in these directions, and for this reason V_{C_A} is constructed as a horizontal line through A' of unknown length. Similarly, because the direction of V_{C_B} is known, it can be constructed as a line of unknown length through B' in a direction perpendicular to BC. Point C' lies at the interception of the two construction lines for V_{C_A} and V_{C_B}. Figure 6.55 forms a closed vector triangle such that:

$$V_{C\_A} = V_{B\_A} + V_{C\_B}$$

Provided that the velocity diagram has been drawn to scale (using a ruler, compass and protractor), the magnitude (length) of the velocity of the slider C (relative to A) V_{C_A} can be measured and its direction identified from Figure 6.55. For the numerical example considered it can be shown that the magnitude of V_{C_A} is 7029 mm s^{-1} and the direction of V_{C_A} acts from right to left. Referring back to Figure 6.54, this means that slider C moves towards A with velocity 7029 mm s^{-1} at the instant shown. This is the required absolute velocity of the slider C.

In addition, the magnitude (length) of vector V_{C_B} can be measured, and the result used to calculate $\dot{\theta}_{BC} = \dfrac{V_{C\_B}}{BC} = 24.25 \text{ rad s}^{-1}$. The direction of vector V_{C_B} indicates that the angular velocity of link CB is in the clockwise direction; i.e. in Figure 6.54 link CB rotates with an angular velocity of 24.25 rad s^{-1} in the clockwise direction at the instant shown. This information is used below when the accelerations are considered.

The absolute acceleration of C is the acceleration of C relative to the ground, and is the acceleration of C relative to the fixed point A (i.e. a_{C_A}). This acceleration is obtained using the following vector relationship for the acceleration

$$a_{C\_A} = a_{B\_A} + a_{C\_B}$$

where a_{B_A} is the acceleration of B relative to A and a_{C_B} is the acceleration of C relative to B.

The acceleration diagram is formed in a similar way to the velocity diagram. The information needed to construct the acceleration diagram is summarized below:

a_{C_A} has unknown magnitude

and is in the horizontal direction (in the left or right direction) ⟷

a_{B_A} has a radial component with *known* magnitude $AB\dot{\theta}^2_{AB} = (80)(100)^2$ mm s^{-2}

and direction ⟍ parallel to AB (acting towards A) ($\parallel AB$)

a_{B_A} has no tangential component because AB rotates at constant angular velocity

a_{C_B} has a radial component with *known* magnitude $BC\dot{\theta}^2_{BC} = (240)(24.25)^2$ mm s^{-2}

and direction ⟍ parallel to BC (acting from C towards B) ($\parallel BC$)

a_{C_B} has a tangential component of unknown magnitude BC

and direction ⟍ perpendicular to BC (BC).

Using this information, the acceleration diagram can be drawn as shown in Figure 6.56.

The starting point for constructing Figure 6.56 is Point A''. Vector a_{B_A} is constructed first with a known magnitude (length) direction parallel to AB. This defines Point B''. Vector a_{C_B} has two components (radial and tangential), but only the radial component has known magnitude and direction. Starting at B'', the radial component of a_{C_B} is constructed in a direction parallel to BC (acting from C towards B). Since only the direction of the tangential component of a_{C_B} is known, it is constructed as a line of unknown length perpendicular to BC as shown in Figure 6.56. Since vector a_{C_A} must be in the horizontal left or right direction, a_{C_A} is constructed as a horizontal line through A'' of unknown length. Point C'' lies at the interception of the two construction lines formed for a_{C_B} and a_{C_A}, and ensures that:

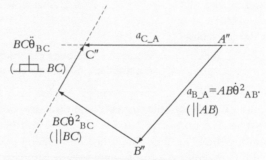

Figure 6.56 Acceleration diagram for slider-crank mechanism (not drawn to scale)

$$a_{C\_A} = a_{B\_A} = a_{C\_B}.$$

In Figure 6.56, the vector from B'' to C'' represents the acceleration vector a_{C_B}.

Provided that the acceleration diagram has been drawn to scale, the magnitude of the acceleration of slider C (relative to A) a_{C_A} can be measured and its direction identified from Figure 6.56. For the numerical example considered it can be shown that the magnitude of a_{C_A} is 574 s^{-2} and the direction of a_{C_A} is in the right to left direction. Referring back to Figure 6.53, this means that slider C accelerates towards A with an acceleration 574 m s^{-2} at the instant shown. This is the required absolute acceleration of the slider C.

In addition, the magnitude of the tangential component of a_{C_B} (i.e. $BC\ddot{\theta}_{BC}$) can be measured, and used to calculate the angular acceleration $\ddot{\theta}_{BC} = 2283$ rad s^{-2}. The direction of vector $BC\ddot{\theta}_{BC}$ indicates that angular acceleration is in the anticlockwise direction at the instant shown.

The acceleration of the centre of mass G of link BC can be obtained by repeating the above process for point G, i.e. by using

$$a_{G\_A} = a_{B\_A} + a_{G\_B}.$$

It is worthwhile noting that this equation is very similar to that considered for a_{C_A}, except that C has been replaced by G. Thus, the main difference with the earlier case is that a_{C_B} has been replaced by a_{G_B}. For this reason, only this extra term needs to be considered below.

a_{G_B} has a radial component with *known* magnitude $BG\dot{\theta}_{BC}^2 = (120)(24.25)^2$ mm s^{-2}

        and direction   parallel to BC (acting towards B) ($\|\ BC$).

a_{G_B} has a tangential component of *known* magnitude $BG\ddot{\theta}_{BC} = (120)(2283)$ mm s^{-2}

        and direction   perpendicular to BC ($\perp BC$).

If required $\ddot{\theta}_{BC}$, can be calculated from the earlier acceleration diagram.

The radial and tangential components of a_{G_B} have a similar form to a_{C_B}, except that length BC has been replaced by BG. Noting that the radial and tangential terms of a_{C_B} are both proportional to length BC, it can be concluded that vector a_{C_B} (i.e. the vector from B'' to C'' in Figure 6.56) is also proportional to BC. It follows that the radial and tangential terms of a_{G_B} are both proportional to BG, and that vector a_{G_B} (i.e. the vector from B'' to G'') is proportional to BG. A consequence of this is that Figure 6.56 can be modified easily to include G'', the acceleration of G. This is achieved by drawing a straight line between B'' and C'', and identifying the point G'' on this line such distance $B''G''$ (the distance from B'' to G'') is given by:

$$B''G'' = \frac{BG}{BC} B''C''.$$

Figure 6.57 shows a modified version of Figure 6.56 with G'' indicated.

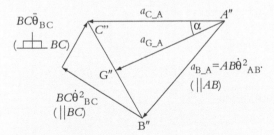

Figure 6.57 Acceleration diagram for slider-crank mechanism showing a_{G_A} (not drawn to scale)

Provided that Figure 6.57 has been drawn to scale, the magnitude of the acceleration of G (relative to A) a_{G_A} is indicated by the vector from A'' to G'' and can be measured. It can be shown that the acceleration magnitude is 636 m s^{-2} in a direction that is $\alpha = 26°$ beneath the horizontal. This is the required absolute acceleration of G.

Example 2 – Four-bar linkage mechanism

Consider the four-bar linkage mechanism $OABC$ shown in Figure 6.58.

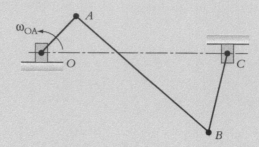

Figure 6.58 **Four-bar linkage mechanism (not drawn to scale)**

The geometry of the mechanism is specified at the instant shown in Figure 6.58. This means that lengths (OA, AB, BC and OC) and orientations of the links are known, or can be calculated using trigonometry. In addition, for the example considered, crank OA is driven anticlockwise at constant angular velocity ω_{OA}.

Determine the absolute velocity and acceleration of B relative to the fixed point O.

Figure 6.58 defines the geometry of the mechanism, and corresponds to the space diagram when it is drawn to scale.

The absolute velocity of B is the velocity of B relative to the ground, and is the velocity of B relative to a fixed point. In this problem there are two fixed points, O and C. Hence the required velocity is given by V_{B_O} or V_{B_C}. The required velocity is obtained using the following vector relationship

$$V_{B\_O} = V_{B\_A} + V_{A\_O} = V_{B\_C}$$

where V_{B_A} is the velocity of B relative to A and V_{A_O} is the velocity of A relative to O.

The velocity diagram is formed in a similar way to the velocity diagram in 'Example 1 – slider-crank mechanism'. The information needed to construct the velocity diagram is summarized below:

V_{A_O} has no radial component because OA has constant length

V_{A_O} has a tangential component of *known* magnitude $OA\omega_{OA}$
and direction perpendicular to OA ($\perp OA$)

V_{B_A} has no radial component because AB has constant length

V_{B_A} has a tangential component of unknown magnitude $AB\omega_{AB}$
and direction perpendicular to AB ($\perp AB$)

V_{B_C} has no radial component because BC has constant length

V_{B_C} has a tangential component of unknown magnitude $BC\omega_{BC}$
and direction perpendicular to BC ($\perp BC$).

In this example it is convenient to consider the velocity vector V_{B_C} when constructing the velocity diagram. This is because link BC has known length and moves (rotates) in a straightforward way.

Using this information, the velocity diagram can be drawn as shown in Figure 6.59.

The starting point for constructing Figure 6.59 is Point O', which lies at the same point as C' because O and C are both fixed points, and the velocity of C relative to O (and vice-versa) is zero. Vector V_{A_O} is constructed first with a known magnitude (length) of $OA\omega_{OA}$ and direction acting perpendicular to OA. This vector defines Point A'. The direction of vector V_{B_A} is perpendicular to AB and V_{B_A} is constructed as a line through A' of unknown length in this direction. Since the direction of vector V_{B_C} is known, it is constructed as a line of unknown length through C' in a direction perpendicular to BC. Point B' lies at the interception of the two construction lines formed for V_{B_A} and V_{B_C}. Figure 6.59 forms a closed vector triangle such that

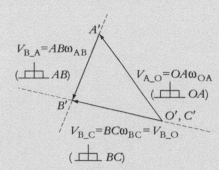

Figure 6.59 Velocity diagram for four-bar linkage mechanism

$$V_{B\_C} = V_{B\_O} = V_{A\_O} + V_{B\_A}$$

Provided that the velocity diagram has been drawn to scale, the magnitude and direction of the velocity of B (relative to O) V_{B_O} can be measured, as required. In addition, the magnitude of vector V_{B_O} can be measured and used to calculate the angular velocity ω_{BC} of link BC. From Figure 6.59 it can be deduced that B moves clockwise about C. In a similar way, the velocity of B (relative to A) V_{B_A} and the angular velocity ω_{AB} of link AB can be measured and calculated (B moves clockwise about A).

The absolute acceleration of B is the acceleration of B relative to either of the ground points O and C. Hence the required acceleration is given by a_{B_O} or a_{B_C}. The required acceleration is obtained using the following vector relationship

$$a_{B\_O} = a_{B\_A} + a_{A\_O} = a_{B\_C}$$

where a_{B_A} is the acceleration of B relative to A and a_{A_O} is the acceleration of A relative to O.

The acceleration diagram is formed in a similar way to the acceleration diagram in 'Example 1 – slider-crank mechanism'. The information needed to construct the acceleration diagram is summarized below:

a_{A_O} has a radial component with known magnitude $OA\omega_{OA}^2$

        in direction   parallel to OA ($\parallel OA$)

a_{A_O} has no tangential component because OA has constant length and angular velocity ω_{OA} is constant

a_{B_A} has a radial component with *known* magnitude $AB\omega_{AB}^2$

        in direction   parallel to AB ($\parallel AB$)

a_{B_A} has a tangential component with unknown magnitude $AB\dot{\omega}_{AB}$

        in direction   perpendicular to AB ($\perp AB$)

a_{B_C} has a radial component with *known* magnitude $BC\omega_{BC}^2$

        in direction   parallel to BC ($\parallel BC$)

a_{B_C} has a tangential component with unknown magnitude $BC\dot{\omega}_{BC}$

        in direction   perpendicular to BC ($\perp BC$)

Using this information, the acceleration diagram can be drawn as shown in Figure 6.60.

The starting point for constructing Figure 6.60 is Point O'', which lies at the same point as C'' because O and C are both fixed point, and the acceleration of C relative to O (and vice-versa) is zero. Vector a_{A_O} is constructed first with a known magnitude (length) of $OA\omega_{OA}^2$ and direction acting parallel to OA. This vector defines Point A''. Vector a_{B_A} has two components (radial and tangential), but only the radial component has known magnitude and direction. Starting at A'', the radial component of a_{B_A} is constructed with a magnitude of $AB\omega_{AB}^2$ and direction parallel to AB

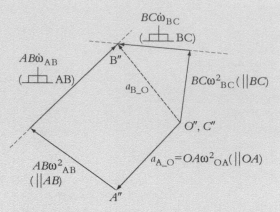

Figure 6.60 **Acceleration diagram for four-bar linkage mechanism**

(acting from B towards A). Since only the direction of the tangential component of a_{B_A} is known, it is constructed as a line of unknown length perpendicular to AB as shown in Figure 6.60. Vector a_{B_C} has two components (radial and tangential), but only the radial component has known magnitude and direction. Starting at C'', the radial component of a_{B_C} is constructed with a magnitude of $BC\omega_{BC}^2$ and direction parallel to BC (acting from B towards C). Since only the direction of the tangential component of a_{B_C} is known, it is constructed as a line of unknown length perpendicular to BC as shown in Figure 6.60. Point B'' lies at the interception of the two construction lines formed for the tangential components of a_{B_A} and $a_{B\_C}(= a_{B\_O})$, and ensures that:

$$a_{B\_O} = a_{B\_A} + a_{A\_O} = a_{B\_C}.$$

In Figure 6.60, the vector from A'' to B'' represents the acceleration vector a_{B_A}.

Provided that the acceleration diagram has been drawn to scale, the acceleration of B (relative to O) a_{B_O} can be measured (B accelerates in the direction indicated by the vector from O'' to B''). In addition, the magnitude of component $AB\dot{\omega}_{AB}$ can be measured and used to calculate the angular acceleration $\dot{\omega}_{AB}$. Similarly, the magnitude of vector $BC\dot{\omega}_{BC}$ can be measured and used to calculate the angular acceleration $\dot{\omega}_{BC}$. From Figure 6.60 it can be deduced that link AB has an angular acceleration $\dot{\omega}_{AB}$ that is positive in the anticlockwise direction, and link BC has an angular acceleration $\dot{\omega}_{BC}$ that is positive in the clockwise direction.

6.7 Mass properties of rigid bodies

Introduction

To analyse the dynamics of rigid bodies it is necessary to use the concepts of centre of mass and mass moment of inertia. Centre of mass plays an important role when analysing the translational motion of a rigid body, and it is always necessary to know the location of the centre of mass of a rigid body when performing dynamic calculations. Mass moment of inertia is needed to analyse the rotational motion of a body. In this section, both of these quantities are defined, and various results are then developed that aid their calculation. Numerous examples are presented for regular shaped engineering components and composite bodies.

Centre of mass

The centre of mass is a point at which the whole mass of the body may be considered to be concentrated, and corresponds to the average position of its mass. It is frequently used in the analysis of static bodies, when representing the weight of a body, and is used in Section 6.8 to analyse the dynamics of rigid bodies.

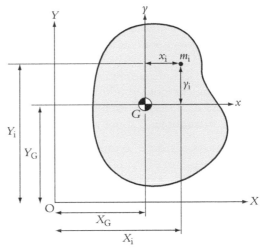

Figure 6.61 **Definition of the centre of mass**

Definition

Consider the planar rigid body shown in Figure 6.61, where m_i is a typical particle in the body.

The centre of mass G of the body has coordinates (X_G, Y_G) and is defined as the point about which the first moment of mass is zero; i.e. taking the first moment of mass for all particles about the Gx axis gives

$$\sum_i m_i y_i = 0$$

and taking the first moment of mass for all particles about the Gy-axis gives

$$\sum_i m_i x_i = 0$$

To determine the location of the centre of mass of the body relative to some arbitrarily chosen axes, mass moments are taken about these axes. Taking the first moment of mass about axes OX and OY, and noting that $X_i = X_G + x_i$ and $Y_i = Y_G + Y_i$ gives

$$\sum_i m_i Y_i = \sum_i m_i (Y_G + y_i)$$

$$\sum_i m_i X_i = \sum_i m_i (X_G + x_i)$$

Using the definition of the centre of mass, and noting that $M = \sum_i m_i$ is the total mass of the body gives

$$\sum_i m_i Y_i = M Y_G$$

$$\sum_i m_i X_i = M X_G$$

These equations indicate that the sum of the mass moments about a particular axis is equal to the mass moment (about the same axis) obtained by considering the mass to be concentrated at the centre of mass.

Rearranging these equations, the coordinates of the centre of mass are given by

$$X_G = \frac{\sum_i m_i X_i}{M}, Y_G = \frac{\sum_i m_i Y_i}{M}$$

These equations indicate that the coordinates are the average of the particle positions, weighted by their normalized masses.

To apply these equations to continuous bodies they are rewritten as

$$X_G = \frac{\int_B X \, dm}{M}, Y_G = \frac{\int_B Y \, dm}{M}$$

where the integrations are over the whole body B.

Composite bodies

The location of the centre of mass is calculated by integrating over the whole body B. Most engineering components are composite bodies, composed of an assembly of simple regular shaped bodies. For these components the integration over B can be simplified by treating it as a summation of integrations over the simple bodies. A consequence of this is that the centre of mass of the composite body can be obtained by calculating the centres of mass of each simple body, and then treating each simple body as a point mass concentrated at its centre of mass, and taking mass moments about a suitable axis.

The centres of mass of symmetrical uniform bodies (such as rectangles and spheres) lie at the geometric centre of the body, and can be identified easily. The centres of mass of some simple bodies are considered in the worked examples later.

Moment of inertia

Moments of inertia are needed to analyse the inertia properties of a body rotating about a particular axis. Physically the moment of inertia I of a body is a measure of how difficult it is to change the angular velocity of a body ('$L = I\alpha$', where L is the applied torque α and is the angular acceleration). It is analogous to the mass m of a body, which is a measure of how difficult it is to change the translational velocity ('$F = ma$', where F is the applied force and a is the acceleration). An important difference is that the moment of inertia depends not only on the mass of the body, but also on its mass distribution.

The moment of inertia is often called the mass moment of inertia by mechanical engineers to help distinguish it from the second moment of area, used in solid mechanics. The second moment of area is also denoted by the symbol I and is often referred to as the moment of inertia by structural engineers. The easiest way to differentiate between these quantities is through their units. The mass moment of inertia has units $kg\,m^2$, while the second moment of area has units m^4.

Definition

By definition, the moment of inertia of a point mass m about axis OZ is mr^2, where r is the perpendicular distance of the point mass from the axis considered; see Figure 6.62.

This definition can be used to derive expressions for the moment of inertia of a rigid body about any axis. As such, the moment of inertia about axis OZ of a point mass dm, which lies at a perpendicular distance r_Z from axis OZ, is $dm\,r_Z{}^2$, see Figure 6.63.

It follows that the moment of inertia of the (continuous) body about axis OZ is given by

$$I_{OZ} = \int_B r_Z^2 dm = \int_B (X^2 + Y^2) dm$$

where the integration is taken over the whole body B.

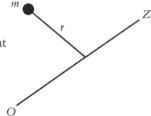

Figure 6.62 Moment of inertia of a point mass

The moments of inertia I_{OX} and I_{OY} of the body about axes OX and OY respectively can be written in a similar fashion by changing the subscripts X, Y and Z; i.e.

$$I_{OX} = \int_B (Y^2 + Z^2) dm$$

$$I_{OY} = \int_B (X^2 + Z^2) dm$$

In this chapter planar problems are analysed such that the motion parallel to a plane (say OXY) is considered. In this situation, I_{OZ} is written as I_O instead.

Composite bodies

The moment of inertia is calculated by integrating over the whole body B. Most engineering components are composite bodies, composed of an assembly of simple regular shaped bodies. For these components the

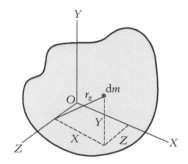

Figure 6.63 Moment of inertia of a particle in a three-dimensional rigid body

integration over B is simplified by treating it as a summation of integrations over the simple bodies. The moment of inertia for the composite body can then be obtained by combining together (adding or subtracting) the moments of inertia for the simple bodies.

Radius of gyration

For a body of mass M with moment of inertia I_O about a particular axis, the radius of gyration k of the body about that axis is defined such that

$$I_O = Mk^2$$

where k is the distance from the axis at which the body mass should be concentrated, as a point mass, to give the same moment of inertia as the body; see Figure 6.64.

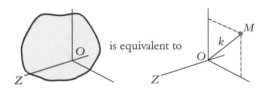

Figure 6.64 Physical interpretation of the radius of gyration k

Notice that the radius of gyration k is not the distance to the centre of mass.

Perpendicular axis (flat plate) theorem

Figure 6.65 shows a body that is flat and thin (i.e. a lamina) and lies in plane OXY.

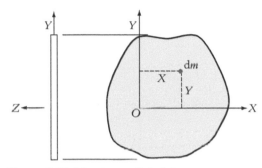

Figure 6.65 A lamina or flat plate

Given that the Z-dimension of the body is effectively zero, the moments of inertia about the OX and OY axes for mass dm are $(Y^2 + Z^2)dm = Y^2\,dm$ and $(X^2 + Z^2)dm = X^2\,dm$ respectively. It follows that the moments of inertia for the lamina about the OX and OY axes are given by

$$I_{OX} = \int_B (Y^2 + Z^2)dm = \int_B Y^2\,dm$$

$$I_{OY} = \int_B (X^2 + Z^2)dm = \int_B X^2\,dm$$

Noting that $I_{OZ} = \int_B (X^2 + Y^2)dm$, it is found that

$$I_{OZ} = I_{OY} + I_{OX}$$

> This is the Perpendicular Axis Theorem. It enables the moment of inertia of a body about an axis perpendicular to a flat body (I_{OZ}) passing through a particular point O to be calculated in terms of the moments of inertia about two mutually orthogonal axes lying in the plane of the flat body (I_{OX} and I_{OY}) passing through the same point O.

Parallel axis Theorem

Figure 6.66 shows a rigid body of mass M having centre of mass G, and two sets of axes $OXYZ$ and $O_1X_1Y_1Z_1$ (OZ and O_1Z_1 are perpendicular to the figure). The two sets of axes are fixed in the body and are parallel to each other. The coordinates of O_1 relative to O are $X = a$, $Y = b$.

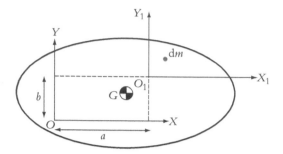

Figure 6.66 **Using different coordinate axes**

The coordinates of a point in the $O_1 X_1 Y_1 Z_1$ axes can be expressed in terms of coordinates in the $OXYZ$ axes by noting that the axes are shifted relative to each other by a constant amount, i.e.

$$X_1 = (X - a), \; Y_1 = (Y - b)$$

Using these shifted axes, it follows that the moment of inertia of the body about axis $O_1 Z_1$ can be expressed as

$$I_{Z1} = \int_B (X_1^2 + Y_1^2)\,\mathrm{d}m = \int_B ((X - a)^2 + (Y - b)^2)\,\mathrm{d}m$$

Expanding this equation gives

$$I_{Z1} = \int_B (X^2 + Y^2)\,\mathrm{d}m + (a^2 + b^2)\int_B \mathrm{d}m - 2a\int_B X\,\mathrm{d}m - 2b\int_B Y\,\mathrm{d}m$$

Using the definitions of the centre of mass G (see 'Definition' of centre of mass) and the moment of inertia I_Z about axis OZ (see 'Definition' of moment of inertia) gives

$$I_{Z1} = I_Z + M(a^2 + b^2) - 2a(MX_G) - 2b(MY_G)$$

This equation expresses the moment of inertia of the body about axis $O_1 Z_1$ in terms of the moment of inertia of the body about axis OZ and some additional terms related to the mass of the body and the location of the centre of mass. This equation can be simplified as follows. Firstly, if axes $OXYZ$ are chosen such that G lies on O, (in other words, if the original axis OZ passes through the centre of mass of the body), then $X_G = 0$, $Y_G = 0$. Also, if c is the perpendicular distance between the axes OZ and $O_1 Z_1$ (where $c^2 = a^2 + b^2$) then

$$I_{Z1} = I_G + Mc^2$$

> This is the Parallel Axis Theorem. It allows the moment of inertia of a body about any axis to be calculated from knowledge of the moment of inertia I_G about the parallel axis through the centre of mass, the mass of the body (M), and the perpendicular distance between the axes (c). It is very useful for calculating the moment of inertia of a body that is composed of an assembly of simple components, as will be demonstrated in the worked examples of later.

It is worth noting that because $c^2 \geqslant 0$, the Parallel Axis Theorem guarantees that the moment of inertia of a body about any axis that does not pass through the centre of mass G will be greater than or equal to the moment of inertia about the parallel axis passing through G.

Thin uniform rod

Consider a thin uniform rod of mass M and length L, as shown in Figure 6.67.

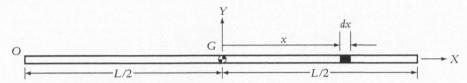

Figure 6.67 Thin uniform rod example

The centre of mass G lies at the midpoint of the rod because the rod is symmetrical and uniform.

Determine: (i) the moment of inertia about an axis through the centre of mass G perpendicular to the rod; and (ii) the moment of inertia about an axis through end O perpendicular to the rod.

(i) Taking into account that the rod is thin, the moment of inertia about an axis through G is given by

$$I_G = \int x^2 \, dm$$

where dm is the mass of the shaded element. Noting that the shaded element has length

dx, $dm = \dfrac{M}{L} dx$, and the moment of inertia is given by

$$I_G = \int_{-L/2}^{L/2} x^2 \left(\frac{M}{L}\right) dx = \frac{ML^2}{12}$$

as required.

(ii) The moment of inertia about an axis through O perpendicular to the rod can be found by using the Parallel Axis Theorem, such that

$$I_O = I_G + M\left(\frac{L}{2}\right)^2 = \frac{ML^2}{3}$$

as required.

Uniform thin ring (or thin cylinder)

Consider a thin uniform ring of radius R, thickness t, and axial length L, as shown in Figure 6.68.

The centre of mass lies at O because the ring is symmetrical and uniform.

Determine the moment of inertia about the polar axis passing through O of the uniform thin ring.

Taking into account that the ring is thin, the moment of inertia about O is given by

$$I_O = \int R^2 \, dm$$

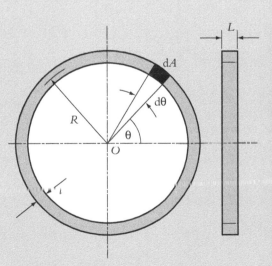

Figure 6.68 Uniform thin ring example

where dm is the mass of the shaded element. Noting that the shaded element has cross-section area $dA = tR\,d\theta$ then

$$dm = \rho L\,dA = \rho LtR\,d\theta$$

where ρ is the material density. Using this equation, the moment of inertia is given by

$$I_O = \int_0^{2\pi} \rho LtR^3\,d\theta = (\rho 2\pi RLt)R^2 = MR^2$$

where M is the mass of the ring, such that $M = \rho 2\pi RLt$, as required.

This result is not surprising because all of the mass of the ring is located at a distance R from the axis considered.

Uniform solid cylinder (or disc)

Consider a uniform solid cylinder of radius R and axial length L, as shown in Figure 6.69.

The centre of mass lies at G because the cylinder is symmetrical and uniform.

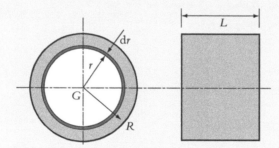

Figure 6.69 **Uniform solid cylinder example**

Determine the moment of inertia about the polar axis passing through G of the uniform solid cylinder.

By considering the solid cylinder to consist of many thin cylindrical elements, the moment of inertia about the axis through G is given by

$$I_G = \int r^2\,dm$$

where dm is the mass of the shaded cylindrical element. Noting that the shaded element has cross-section area $dA = 2\pi r\,dr$, then

$$dm = \rho L\,dA = \rho L\,2\pi r\,dr$$

where ρ is the material density. Using this equation, the moment of inertia about the polar axis passing through the centre of mass G is given by

$$I_G = \int_0^R \rho L 2\pi r^3\,dr = \frac{\rho \pi R^4 L}{2} = \tfrac{1}{2}MR^2$$

where M is the mass of the cylinder, such that $M = \rho \pi R^2 L$, as required.

Hollow thick uniform cylinder (or disc)

Consider a hollow thick uniform cylinder with outer radius R_2, inner radius R_1 and axial length L, as shown in Figure 6.70.

The centre of mass lies at O because the ring is symmetrical and uniform.

Determine the moment of inertia about the polar axis of the hollow cylinder.

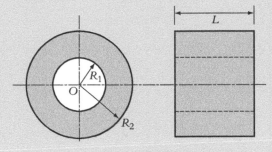

Figure 6.70 Hollow thick cylinder example

This can be tackled *either* by integrating over the body or by treating the body as a composite body in which a solid cylinder of radius R_1 is 'subtracted' from a solid cylinder of radius R_2. The latter approach is used here.

The moment of inertia of a solid cylinder was determined in 'Uniform solid cylinder (or disc)', and using this result the moments of inertia of the two solid cylinders about the same axis through O are

$$I_1 = \tfrac{1}{2}M_1R_1^2$$

$$I_2 = \tfrac{1}{2}M_2R_2^2$$

where I_1 and I_2 are the moments of inertia of solid cylinders of radius R_1 and R_2, and M_1 and M_2 are the masses of solid cylinders of radius R_1 and R_2.

The moment of inertia of the hollow cylinder is given by

$$I_O = I_2 - I_1 = \tfrac{1}{2}M_2R_2^2 - \tfrac{1}{2}M_1R_1^2$$

i.e.

$$I_O = \tfrac{1}{2}(\rho\pi R_2^2L)R_2^2 - \tfrac{1}{2}(\rho\pi R_1^2L)R_1^2$$

This expression can be rearranged to give

$$I_G = \frac{M(R_1^2 + R_2^2)}{2}$$

where $M = M_2 - M_1$ is the mass of the hollow cylinder, as required.

Thin rectangular plate.

Consider a thin rectangular plate with sides of length a and b, as shown in Figure 6.71.

The centre of mass G lies at the geometrical centre of the plate.

Determine the moments of inertia about the GX, GY and GZ axes of the thin plate.

The moment of inertia of the plate about the GX axis is obtained by first considering the moment of inertia of the shaded elemental strip about the GX axis, and then generalizing this result for the whole plate.

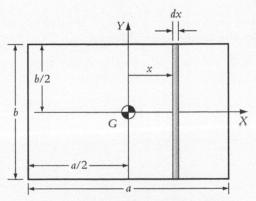

Figure 6.71 Thin rectangular plate example

The moment of inertia of the strip about the GX axis is obtained using the thin rod result described in 'Thin uniform rod', i.e.

$$I_{GX(STRIP)} = \tfrac{1}{12}(\mu b dx)b^2 = \tfrac{1}{12}\mu b^3 dx$$

where $\mu = \dfrac{M}{ab}$ is the mass per unit area, and $\mu b dx$ is the mass of the elemental strip.

The moment of inertia of the plate about the GX axis is then obtained by integrating the elemental strip result wrt x, i.e.

$$I_{GX(PLATE)} = \tfrac{1}{12}\mu b^3 \int_{-a/2}^{a/2} dx = \tfrac{1}{12}\mu b^3 a$$

Noting that μba is the mass of the plate M, the moment of inertia of the plate about the GX axis can be expressed as

$$I_{GX(PLATE)} = \tfrac{1}{12}Mb^2$$

as required.

Similarly, the moment of inertia of the plate about the GY axis can be expressed as

$$I_{GY(PLATE)} = \tfrac{1}{12}Ma^2$$

as required.

The moment of inertia of the plate about the GZ axis can be found using the Perpendicular Axis Theorem, and is given by

$$I_{GZ(PLATE)} = I_{GX(PLATE)} = I_{GY(PLATE)} = \tfrac{1}{12}M(a^2 + b^2)$$

as required.

Uniform rod/circular disc composite body

Figure 6.72 shows a composite rigid body consisting of a uniform rod of length L and mass M_R and a circular disc of radius R and mass M_D.

In this figure G_1 and G_2 refer to the centres of mass of the rod and disc respectively.

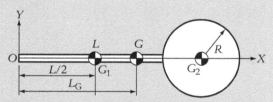

Figure 6.72 **Composite body example**

Determine the position of the centre of mass G of the composite rigid body and the moment of inertia of the composite rigid body about the OZ axis.

The required centre of mass is obtained by taking mass moments about OY for the two 'point' masses M_R and M_D. This gives

$$\frac{L}{2}M_R + (L + R)M_D = L_G(M_R + M_D)$$

Using this result the centre of mass is located a distance L_G from O, where

$$L_G = \frac{\dfrac{L}{2}M_R + (L + R)M_D}{(M_R + M_D)}$$

as required.

The first step to calculate the required moment of inertia is to calculate the moments of inertia of the rod and the disc about the OZ axis. The second step is to add these moments of inertia together.

Rod

The moment of inertia of the rod about an axis through G_1 is given by

$$I_{G1} = \tfrac{1}{12}M_R L^2$$

Using the Parallel Axis Theorem, the moment of inertia of the rod about the (parallel) OZ axis is given by

$$I_{OR} = \tfrac{1}{12}M_R L^2 + M_R\left(\frac{L}{2}\right)^2$$

Disc

The moment of inertia of the disc about an axis through G_2 is given by

$$I_{G2} = \tfrac{1}{2}M_D R^2$$

Using the Parallel Axis Theorem, the moment of inertia of the disc about the (parallel) OZ axis is given by

$$I_{OD} = \tfrac{1}{2}M_D R^2 + M_D(L + R)^2$$

Composite body

The moment of inertia of the combined rod and disc about the OZ axis is given by

$$I_O = I_{OR} + I_{OD} = \frac{M_R L^2}{3} + M_D\left(\frac{3R^2}{2} + 2RL + L^2\right)$$

as required.

6.8. Kinetics of a rigid body in a plane

Introduction

Newton's laws of motion are only applicable to a particle (see 'Kinetics of a particle in a plane'). However, by considering a rigid body to be composed of a number of particles arranged in a fixed pattern, Newton's laws of motion can be used to determine relationships between externally applied forces and translational motion of the body, and a relationship between externally applied moments and rotational motion of the body. This section derives these relationships for the case of a rigid body in general planar motion, and then considers the simplified cases of rigid bodies moving in pure translation and pure rotation. Numerous examples are presented using the obtained relationships to analyse the forces and motions of mechanical systems.

Analysing rigid body motion in a plane

When analysing the equations of motion of a machine element exhibiting planar motion it is convenient to consider: (i) pure translation; (ii) pure rotation; and (iii) general plane motion separately. Situations (i) and (ii) are special cases of the general case (iii). These different types of motion are reviewed next.

Pure translation

Figure 6.73 shows a rigid body moving in pure translation. A body moving in pure translation does not rotate, and all points within the body have the same (translational) velocity. The path followed by each point may be a straight line (rectilinear translation) or a curve (curvilinear translation), but any line on the body always remains parallel to its original direction.

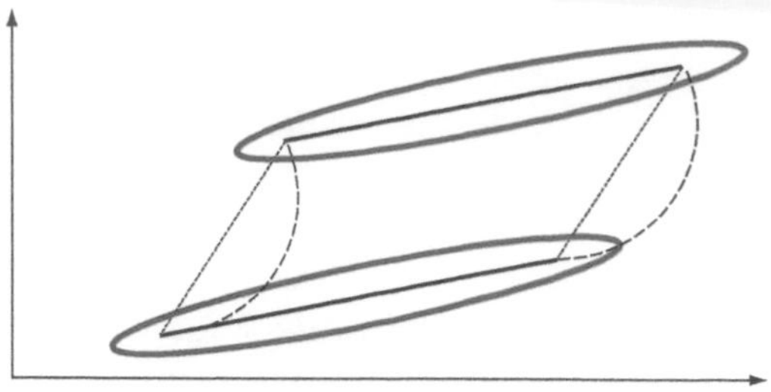

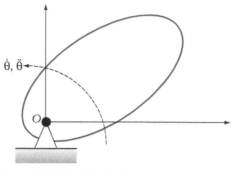

Figure 6.73 Rigid body in pure translation

Pure rotation

Figure 6.74 shows a rigid body moving in pure rotation. The body rotates about an axis through the fixed point O. This case is widely applicable to machinery (shafts, gears, cams, crank, etc.).

$\dot{\theta}, \ddot{\theta}$

O

Figure 6.74 Rigid body in pure rotation

General plane motion

Figure 6.75 shows a four-bar-linkage in which the coupler moves in general plane motion, while the crank and rocker move in pure rotation (see 'Pure rotation'). The motion of the coupler consists of a combination of translation and rotation.

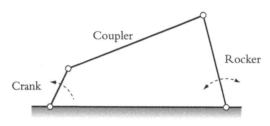

Coupler

Rocker

Crank

Figure 6.75 Four-bar linkage

Motion of a rigid body in a plane

In what follows the equations of motion for general plane motion will be considered first. The equations of motion for pure translation and pure rotation are then derived as special cases.

General plane motion – simultaneous translation and rotation

The number of coordinates required to specify the position of a rigid body in a plane is three (i.e. it has 3DOF). For this reason three equations of motion – two translational, one rotational – are needed.

Translational equations of motion

Consider the rigid body shown in Figure 6.76, where G is the centre of mass of the body, m_i is a typical particle within the body, and the total mass of the body is M.

In this figure, point P is an arbitrary point, fixed within the body but not fixed in space.

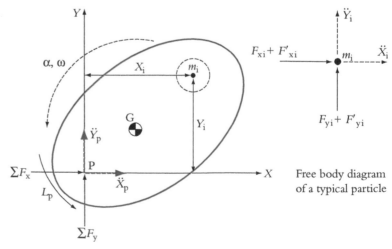

Figure 6.76 Rigid body in general planar motion (Cartesian coordinates)

Machine dynamics

It is assumed that the body is acted upon by a system of external forces, which causes the body to have both translational and rotational (angular) acceleration. All of the external forces can be represented in terms of the two force components ΣF_X and ΣF_Y acting at point P, and a torque L_P acting about P (see Figure 6.76). In accordance with 'Rigid body motion', the acceleration of the body is represented by the translational acceleration of P (in components $\ddot{X}_P$, $\ddot{Y}_P$) and the angular acceleration α. In addition the body has angular velocity ω.

Consider the free body diagram for the typical particle m_i shown in Figure 6.76. Each component of force acting on the particle consists of two parts: (i) F_{Xi} and F_{Yi} are the external forces acting on the particle and arise due to effects that are external to the body. e.g. gravity force, magnetic forces, and external contact and friction forces; (ii) F'_{Xi} and F'_{Yi} are the internal forces acting on the particle, and arise due to contact between adjacent particles within the body (which cause stresses within the material of the body).

Applying Newton's second law to the particle in the X and Y-directions gives

$$F_{Xi} + F'_{Xi} = m_i \ddot{X}_i$$
$$F_{Yi} + F'_{Yi} = m_i \ddot{Y}_i$$

Summing these equations over every particle in the body gives

$$\sum_i F_{Xi} + \sum_i F'_{Xi} = \sum_i m_i \ddot{X}_i$$
$$\sum_i F_{Yi} + \sum_i F'_{Yi} = \sum_i m_i \ddot{Y}_i$$

It follows from Newton's third law that the internal forces are in equilibrium and that, for the whole body, the sum of the internal forces is zero. For this reason, $\sum_i F'_{Xi} = 0$ and $\sum_i F'_{Yi} = 0$.

The definition of the centre of mass (see 'Centre of mass') ensures that $\sum_i m_i X_i = M X_G$ and $\sum_i m_i Y_i = M Y_G$. Differentiating these equations twice with respect to time it follows that $\sum_i m_i \ddot{X}_i = M \ddot{X}_G$ and $\sum_i m_i \ddot{Y}_i = M \ddot{Y}_G$, where M is the mass of the body and X_G and Y_G are the coordinates of G relative to P.

Using the above results gives

$$\Sigma F_X = M \ddot{X}_G \tag{6.5}$$
$$\Sigma F_Y = M \ddot{Y}_G \tag{6.6}$$

i.e. the sum of the external forces in the OX and OY directions is equal to the mass multiplied by the corresponding components of the acceleration of the centre of mass G in the OX and OY directions respectively. Note that the acceleration of the centre of mass G must be used in these equations.

Equations (6.5) and (6.6) govern the translational motion of a rigid body that is in general planar motion. These equations also apply to the special cases when the motion consists of pure translation or pure rotation (see 'Pure translation' and 'Pure rotation').

Rotational equation of motion

Consider the rigid body shown in Figure 6.77. This figure is identical to Figure 6.76, except that polar coordinates have been used to define the geometry and the free body diagram of the typical particle.

It follows from Figure 6.77 that the absolute acceleration of the typical particle m_i can be expressed as the sum of the acceleration of point P and the acceleration of m_i relative to P (see 'Rigid body motion'). These two components of acceleration are considered separately as

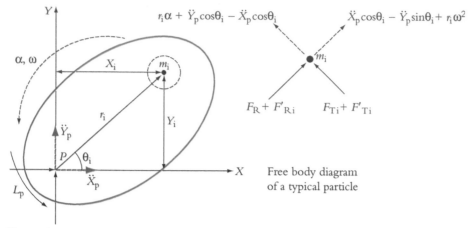

Figure 6.77 Rigid body in general planar motion (polar coordinates)

follows. The Cartesian components of the acceleration of P (denoted by $\ddot{X}_P$, $\ddot{Y}_P$) can be resolved into components in the radial and tangential directions, and are shown on the free body diagram of the particle. The acceleration of m_i relative to P has radial and tangential components $r_i\omega^2$ and $r_i\alpha$ (see Figure 6.21) as shown on the free body diagram.

Applying Newton's second law to the particle in the tangential direction gives

$$F_{Ti} + F'_{Ti} = m_i(r_i\alpha + \ddot{Y}_P \cos\theta_i - \ddot{X}_P \sin\theta_i)$$

Noting that $\sin\theta_i = \dfrac{Y_i}{r_i}$ and $\cos\theta_i = \dfrac{X_i}{r_i}$, the above equation can be written as

$$F_{Ti} + F'_{Ti} = m_i\left(r_i\alpha + \ddot{Y}_P\frac{X_i}{r_i} - \ddot{X}_P\frac{Y_i}{r_i}\right)$$

Multiplying both sides of this equation by r_i (which is equivalent to taking moments about point P), and then summing the equations for every particle in the body gives

$$\sum_i (F_{Ti} + F'_{Ti})r_i = \sum_i m_i(r_i^2\alpha + \ddot{Y}_P X_i - \ddot{X}_P Y_i)$$

As discussed earlier, Newton's third law ensures that the internal forces sum to zero. For this reason, $\sum_i F'_{Ti}r_i = 0$. The sum of the moments arising from the externally applied forces is the externally applied torque L_P, i.e. $L_P = \sum_i F_{Ti}r_i$. Using these equations, the above equation can be rewritten as

$$L_P = \sum_i m_i r_i^2\alpha + \ddot{Y}_P\sum_i m_i X_i - \ddot{X}_P\sum_i m_i Y_i$$

In this equation, $\sum_i m_i r_i^2$ is the moment of inertia of the body about the axis through P (see 'Moment of inertia') and from the definition of the centre of mass (see 'Centre of mass') $\sum_i m_i X_i = MX_G$ and $\sum_i m_i Y_i = MY_G$, where M is the mass of the body and X_G and Y_G are the coordinates of G relative to P. Using these results and denoting the moment of inertia by the symbol I_P, the above equation can be written as

$$L_P = I_P\alpha + \ddot{Y}_P MX_G - \ddot{X}_P MY_G \tag{6.7}$$

This equation governs the relationship between the externally applied moments about P, which was initially chosen *arbitrarily*, and the resulting motion of the rigid body.

When the rigid body is static, the terms on the right-hand side of equation (6.7) disappear, and the result reduces to the moment condition presented in 'Newton's laws'. For this situation the externally applied moment about any point P is zero.

Equation (6.7) can be simplified by choosing point P to lie at the centre of mass of the body G. In this case the coordinates $X_G = 0$ and $Y_G = 0$ and equation (6.7) becomes

$$L_G = I_G \alpha \qquad (6.8)$$

i.e. the resultant of the externally applied moments about G is equal to the moment of inertia of the body about G multiplied by the angular acceleration of the body.

Translational and rotational equations of motion

Equations (6.5), (6.6) and (6.8) govern the dynamics of a rigid body undergoing general motion in a plane. It is important to notice the role of the centre of mass G in these equations. Equations (6.5) and (6.6) govern the translation of the body and are similar to those used in 'Motion of a particle in a plane' to analyse the dynamics of a particle. The only difference is that the rigid body is treated as a point mass having the same mass as the actual body located at the centre of mass. Equation (6.8) governs the rotational motion of the body about the centre of mass. This equation has an identical form to equations (6.5) and (6.6), where the externally applied torque is analogous to the applied force, the moment of inertia is analogous to the mass, and the angular acceleration is analogous to the translational acceleration. Examples that use equations (6.5), (6.6) and (6.8) to analyse bodies in general planar motion are presented in 'Rolling drum (general plane motion)' and 'Linkage mechanism' later.

Based on equations (6.5), (6.6) and (6.8) it is a relatively simple matter to consider the two special cases of pure translation and pure rotation, which occur frequently in practice.

Pure translation

The translational motion of the body is governed by equations (6.5) and (6.6). A body in pure translation does not have any rotational motion, so its angular acceleration is zero. In this case, equation (6.8) simplifies to

$$L_G = 0 \qquad (6.9)$$

i.e. the sum of the externally applied moments about the centre of mass is zero. This relationship applies only to G. Equation (6.7) must be used if moments are taken about any other point.

In summary, the motion of a body in pure translation is governed by equations (6.5), (6.6) and (6.9). Examples that use these equations to analyse bodies in pure translation are presented in 'Door on rollers (pure translation)', 'Rear-wheel drive car (pure translation)', 'Simple crane (pure rotation)' and 'Linkage mechanism' later.

Pure rotation (rotation about a fixed axis)

This case occurs frequently in practice when considering machinery that contains components attached to rotating shafts (e.g. motor drives, turbines, compressors, etc.).

Figure 6.78 shows a rigid body in pure rotation about a fixed axis through O.

This figure is similar to Figures 6.76 and 6.77, except the arbitrary point P has been replaced by the fixed point O. The fact that O is fixed means that it has no acceleration, and for this reason no acceleration components are included for O in Figure 6.78. The externally applied forces ΣF_X and ΣF_Y include the bearing reaction forces and other external forces. The applied torque L_O is the sum of the externally applied moments exerted about O by all external forces.

The translational motion of the centre of mass G is governed by equations (6.5) and (6.6). Using the acceleration expressions derived in 'Polar coordinates', the acceleration of G (denoted by $\ddot{X}_G$, $\ddot{Y}_G$) can be expressed in terms of ω, α and the distance of G from O.

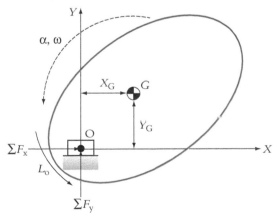

Figure 6.78 **Rigid body in pure rotation**

The equation that governs the rotational motion of the body is obtained from equation (6.7), which is repeated below for convenience

$$L_P = I_P\alpha + \ddot{Y}_P M X_G - \ddot{X}_P M Y_G \qquad \textbf{(6.7)} \text{ (repeated)}$$

In accordance with Figure 6.78, point P is replaced by O and it is noted that and are both zero, because O is a fixed point. Using these results the above equation simplifies to

$$L_O = I_O\alpha \qquad \textbf{(6.10)}$$

In summary, the motion of a body in pure rotation about a fixed axis through O is governed by equations (6.5), (6.6) and (6.10). Examples that use these equations are presented in 'Rotating bar (pure rotation)', 'Simple crane (pure rotation)', and 'Linkage mechanism' later.

Summary – dynamics of rigid bodies in plane motion

The important results derived in 'Motion of a rigid body in a plane' are summarized below.

Pure translation (equations (6.5), (6.6) and (6.9))

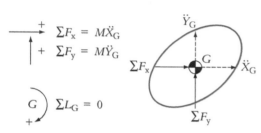

Figure 6.79

where ΣF_X and ΣF_Y are the algebraic sum of the components of the applied external forces in the X and Y-directions respectively, M is the mass of the body, $\ddot{X}_G$ and $\ddot{Y}_G$ are the components of the acceleration of the centre of mass G of the body in the X and Y-directions respectively, and ΣL_G is the sum of the externally applied moments about the centre of mass G.

Pure rotation (about fixed axis O) (equations (6.5), (6.6) and (6.10))

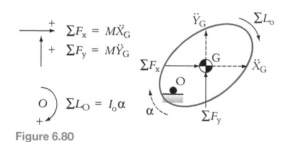

Figure 6.80

where ΣF_X and ΣF_Y are the algebraic sum of the components of the applied external forces in the X and Y-directions respectively, M is the mass of the body, $\ddot{X}_G$ and $\ddot{Y}_G$ are the components of the acceleration of the centre of mass G of the body in the X and Y-directions respectively, ΣL_O is the sum of the externally applied moments about the centre of rotation O, I_O is the moment of inertia of the body about the fixed axis of rotation through O, and α is the angular acceleration of the body.

General plane motion (equations (6.5), (6.6) and (6.8))

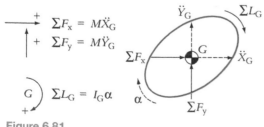

Figure 6.81

where ΣF_X and ΣF_Y are the algebraic sum of the components of the applied external forces in the X and Y-directions respectively, M is the mass of the body, $\ddot{X}_G$ and $\ddot{Y}_G$ are the components of the acceleration of the centre of mass G of the body in the X- and Y-directions respectively, ΣL_G is the sum of the externally applied moments about the centre of mass G, I_G is the moment of inertia of the body about the centre of mass G, and α is the angular acceleration of the body.

Door on rollers (Pure translation)

Figure 6.82 shows a door supported on rollers at A and B that is resting on a horizontal track.

The rollers are frictionless and a constant force P is applied to the door as shown. The combined mass of the door and rollers is M. Determine expressions for the acceleration of the door and the reaction forces at A and B.

The free body diagram is shown in Figure 6.83.

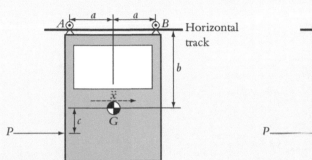

Figure 6.82 Door on rollers

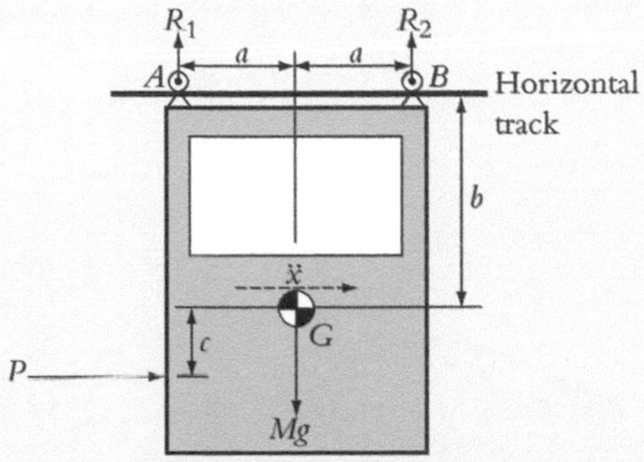

Figure 6.83 Free body diagram of door on rollers

R_1 and R_2 are normal reaction forces due to contact with the horizontal track. Mg is the weight of the door including the rollers. No friction forces are included because the rollers are frictionless.

Since the door moves in pure translation, equations (6.5), (6.6) and (6.9) must be used. Applying these equations gives

$$\Sigma F_X = M\ddot{X}_G \qquad \xrightarrow{+} \qquad P = M\ddot{x}$$

$$\Sigma F_Y = M\ddot{Y}_G \qquad \downarrow{+} \qquad Mg - R_1 - R_2 = 0$$

$$\Sigma L_G = 0 \qquad \overset{+}{\circlearrowright} G \qquad R_1 a - R_2 a - Pc = 0$$

The first of these equations ensures that acceleration $\ddot{x}$ is given by

$$\ddot{x} = \frac{P}{M}$$

as required. The acceleration is constant, provided that P and M are constant. A consequence of this is that the position and velocity of the door at any instant in time can be obtained using the expressions described in 'Constant acceleration'.

Solving the last two equations for R_1 and R_2 gives

$$R_1 = \frac{Mga + Pc}{2a}$$

$$R_2 = \frac{Mga - Pc}{2a}$$

as required. Assuming that $Pc > 0$, the reaction force R_1 is always greater than R_2. Also $R_1 = R_2$ only if $P = 0$ and/or $c = 0$.

Rear-wheel drive car (Pure translation)

Figure 6.84 shows a car of mass M in which the rear wheels only are used to produce motion in the left to right direction.

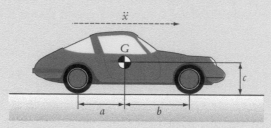

Assume that the front wheels are frictionless and the inertia of the wheels is negligible. Determine an expression for the maximum acceleration of the car, and the

Figure 6.84 Rear-wheel drive car

acceleration required to cause the front wheels to lift off the ground (i.e. a wheelie).

The free body diagram is shown in Figure 6.85.

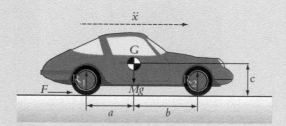

Figure 6.85 Free body diagram of rear wheel drive car

In this case the driven rear wheels produce the friction force F which causes the car to accelerate (see 'Coulomb (dry) friction between two surfaces' to understand the direction of the friction force). Mg is the weight of the vehicle. No friction force is shown at the front wheel because it is frictionless. (If the front wheels were not frictionless the friction force would oppose the motion (see 'Coulomb (dry) friction between two surfaces')). R_1 and R_2 are normal reaction forces due to contact with the ground.

Since the car moves in pure translation, equations (6.5), (6.6) and (6.9) must be used. Applying these equations gives

$$\Sigma F_X = M\ddot{X}_G \qquad \xrightarrow{\;+\;} \qquad F = M\ddot{x}$$

$$\Sigma F_Y = M\ddot{Y}_G \qquad \Big\downarrow{\scriptstyle +} \qquad Mg - R_1 - R_2 = 0$$

$$\Sigma L_G = 0 \qquad {\scriptstyle +}\,\circlearrowleft G \qquad R_1 a - R_2 b - Fc = 0$$

The first of these equations indicates that the acceleration $\ddot{x} = \dfrac{F}{M}$ and confirms that the friction force generated at the rear-wheel accelerates the car. Consequently the maximum acceleration occurs when the maximum friction force is available. From 'Coulomb (dry) friction between two surfaces', $F \leqslant \mu R_1$ so the maximum friction force occurs when $F = \mu R_1$. Setting $F = \mu R_1$ in the moment equilibrium equation, and using the vertical and moment equilibrium equations, it can be shown that

$$R_1 = \frac{Mgb}{(a + b - \mu c)}$$

Hence the maximum friction force is $\dfrac{\mu Mgb}{(a + b - \mu c)}$ and the maximum acceleration is given by

$$\ddot{x}_{\max} = \frac{\mu gb}{(a + b - \mu c)}$$

as required.

To determine the acceleration to cause a wheelie, it is important to notice that a wheelie occurs when $R_2 = 0$, and that the friction required to achieve this does not necessarily occur at limiting friction μR_1.

The amount of friction required to achieve a wheelie is obtained by solving the vertical and moment equilibrium equations for F when $R_2 = 0$. This gives:

$$F = \frac{Mga}{c}.$$

Using the equation of motion in the horizontal direction, the friction force produces an acceleration $\dfrac{ga}{c}$. Hence the acceleration required to cause a wheelie is given by:

$$\ddot{x}_{\text{wheelie}} = \frac{ga}{c}.$$

For accelerations greater than or equal to this value, a wheelie will occur. This is the required result.

It is worth noting that a wheelie will not occur if $\ddot{x}_{\text{wheelie}}$ is greater than the maximum possible acceleration $\ddot{x}_{\max}$, i.e. no wheelie will occur if

$$\frac{ga}{c} > \frac{\mu gb}{(a + b - \mu c)}.$$

Rotating bar (pure rotation)

Figure 6.86 shows a uniform bar rotating under the action of gravity in a vertical plane about a fixed (horizontal) axis through the pivot at O. The bar has mass M and length L.

Determine expressions for the angular acceleration and the reaction force at O at any angle to the vertical.

The free body diagram is shown in Figure 6.87.

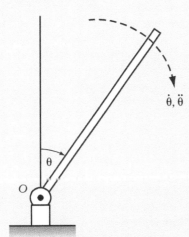

Figure 6.86 Rotating bar example

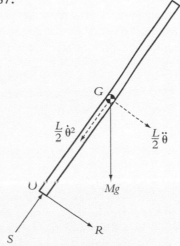

Figure 6.87 Free body diagram for rotating bar example

R and S are components of the reaction force acting at O. Although they have been chosen to act in directions parallel and perpendicular to the bar, it is equally valid to choose components that act horizontally and vertically. Alternatively, the resultant reaction force could have been represented as a vector with unknown magnitude and direction. Mg is the weight of the bar.

Since the bar is moving in pure rotation, equations (6.5), (6.6) and (6.10) *must* be used. In equations (6.5) and (6.6) it is necessary to consider the acceleration of the centre of mass G, where G is located midway along the length of the bar. The radial and tangential components of this acceleration are obtained using the results presented in Figure 6.21, and are indicated in Figure 6.87.

Applying equations (6.5), (6.6) and (6.10) gives

$$\Sigma F_X = M\ddot{X}_G \qquad \qquad Mg\cos\theta - S = M\frac{L}{2}\dot{\theta}^2$$

$$\Sigma F_Y = M\ddot{Y}_G \qquad \qquad R + Mg\sin\theta = M\frac{L}{2}\ddot{\theta}$$

$$\Sigma L_O = I_O\alpha \qquad \qquad Mg\frac{L}{2}\sin\theta = I_O\ddot{\theta}$$

In the last equation I_O is the moment of inertia about the pivot axis through O of the bar. The moment of inertia for a uniform bar about its end was considered in 'Thin uniform rod' and is given by

$$I_O = \frac{ML^2}{3}$$

Using this expression and solving the previous three equations for the angular acceleration $\ddot{\theta}$ and the reaction forces R and S gives

$$\ddot{\theta} = \frac{3g}{2L}\sin\theta$$

$$R = -\frac{Mg}{4}\sin\theta$$

$$S = Mg\cos\theta - \frac{ML}{2}\dot{\theta}^2$$

as required.

In this case it is worth noting that the angular acceleration $\ddot{\theta}$ is variable (non-constant) because it depends on θ. For this reason it is necessary to use the equations presented in 'Variable (non-constant) acceleration', not 'Constant acceleration'.

Simple crane (pure rotation)

Figure 6.88 shows a crane drum used to raise a load mass.

The drive torque L_D is applied to the drum in the counter clockwise direction, and the rotation of the drum is opposed by a frictional resistance torque L_F. The drum has radius r and moment of inertia I_O about the central axis through O, and is attached via a cable to a mass m.

Assuming that there is no slip between the drum and the cable, determine an expression for the drive torque L_D required to raise the load mass m with acceleration a.

The free body diagrams for the crane drum and load mass are shown in Figure 6.89.

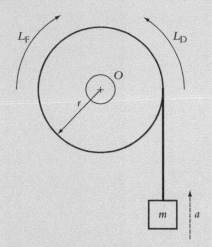

Figure 6.88 Simple crane drum example

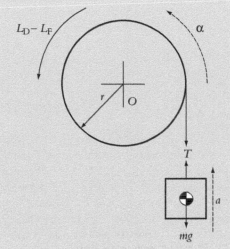

Figure 6.89 Free body diagram for crane drum and load mass

T is the tension in the cable and mg is the weight of the load mass.

In this example, the crane drum is in pure rotation and the load mass is in pure translation. For the purposes of analysis it is only necessary to consider the rotational motion of the drum and the vertical motion of the load mass.

Applying equation (6.10) to analyse the pure rotation of the crane drum about O gives

$$\Sigma L_O = I_O \alpha \qquad \qquad L_D - L_F - Tr = I_O \alpha$$

Applying equation (6.6) to analyse the vertical motion of the load mass gives

$$\Sigma F_Y = M\ddot{Y}_G \qquad \qquad T - mg = ma$$

Provided that the cable does not slip on the drum, the translational acceleration a of the mass is dependent on the angular acceleration α of the drum. The relationship between these accelerations is $a = r\alpha$ (see 'Relating rotation about a fixed axis to translational motion') or $\alpha = \dfrac{a}{r}$. Using this relationship, and eliminating the tension T between the previous two equations, it can be shown that the drive torque is given by

$$L_D = L_F + mgr + (I_O + mr^2)\frac{a}{r}$$

as required.

Rolling drum (general plane motion)

Figure 6.90 shows a cylinder of radius r and mass M rolling down an inclined plane.

Assuming that there is no slip between the roller and the plane, determine: (i) a general expression for the translational acceleration $\ddot{x}$ of the cylinder and (ii) the maximum translational acceleration.

The free body diagram is shown in Figure 6.91.

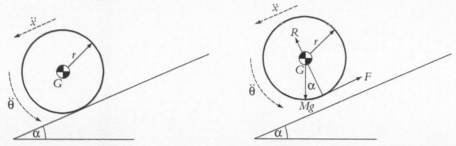

Figure 6.90 Rolling drum   **Figure 6.91 Free body diagram of rolling drum**

Mg is the weight of the cylinder, and the component of weight ($Mg \sin\alpha$) acting parallel to the inclined plane causes the roller to move down the slope. The friction force on the roller acts parallel to the inclined plane and causes the roller to rotate with angular acceleration $\ddot{\theta}$ in an anticlockwise direction. As a consequence the friction force F on the rolling drum acts up the slope as shown in Figure 6.91 ('Coulomb (dry) friction between two surfaces' discusses the direction of the friction force for the analogous situation when a wheel rolls on a horizontal surface under the action of a horizontal applied force). The reaction force R acts normal to the slope in a direction passing through the centre of mass G of the roller.

Since the roller is moving in general plane motion (translation down the slope and rotation about G), equations (6.5), (6.6) and (6.8) *must* be used. Applying these equations gives

$$\Sigma F_X = M\ddot{X}_G \qquad Mg \sin\alpha - F = M\ddot{x}$$

$$\Sigma F_Y = M\ddot{Y}_G \qquad Mg \cos\alpha - R = 0$$

$$\Sigma L_G = I_G\alpha \qquad Fr = I_G\ddot{\theta}$$

In the last equation I_G is the moment of inertia about the polar axis passing through the centre of mass G of the roller.

If there is no slip between the roller and the plane, then the translational acceleration of the roller $\ddot{x}$ is directly related to the angular acceleration $\ddot{\theta}$ of the roller. The relationship between these accelerations is $\ddot{x} = r\ddot{\theta}$ (see 'Relating rotation about a fixed axis to translational motion') or $\ddot{\theta} = \dfrac{\ddot{x}}{r}$. Using this relationship in the moment equation, it can be shown that the friction force is

$$F = \frac{I_G \ddot{x}}{r^2}$$

Substituting for F into the governing equation for motion parallel to the slope, it can be shown that the acceleration is given by

$$\ddot{x} = \frac{Mg \sin\alpha}{\left(M + \dfrac{I_G}{r^2}\right)}$$

The moment of inertia I_G was considered in 'Uniform solid cylinder (or disc)' and is given by

$$I_G = \tfrac{1}{2}Mr^2$$

Substituting for I_G, the acceleration is given by

$$\ddot{x} = \frac{2g\sin\alpha}{3}$$

as required.

The expressions determined for the translational acceleration are dependent on the frictional force F. The expression for the friction force obtained earlier $\left(\text{i.e. } F = \dfrac{I_G \ddot{x}}{r^2} \right)$ is based on the assumption that the roller does not slip, and is only valid when $F \leqslant \mu R$ (see 'Coulomb (dry) friction between two surfaces'), where μ is the coefficient of friction. Using the force equation perpendicular to the slope, the reaction force R is given by

$$R = Mg\cos\alpha$$

Hence the maximum allowable friction force is $\mu Mg\cos\alpha$. Using this friction force in the force equation parallel to the slope, it can be shown that the maximum acceleration is given by

$$\ddot{x}_{\max} = g(\sin\alpha - \mu\cos\alpha)$$

as required.

Linkage mechanism

Consider the slider-crank mechanism shown in Figure 6.92.

The crank AB is pinned at A and B, the connecting rod BC is pinned at B and C, and the piston at C moves inside a cylinder. At the instant shown, the crank is orientated at an angle θ from top-dead-centre.

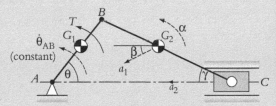

Figure 6.92 **Slider-crank mechanism**

The crank AB rotates about the fixed point A, has mass M_1 and moment of inertia I_A about A. At the instant shown the crank rotates at constant angular velocity $\dot{\theta}_{AB}$.

The connecting rod has mass M_2 and moment of inertia I_2 about G_2. At the instant shown, the centre of mass G_2 has acceleration a_1 and angular acceleration α in the anticlockwise direction, as shown.

The piston has mass m and moves in a frictionless cylinder such that the centre line of the piston through C is horizontal and coincident with A. The position of pivot C coincides with the centre of mass of the piston G_3. At the instant shown, the piston has acceleration a_2 in the horizontal direction as shown.

For the instant when the mechanism is in the position shown, determine expressions for the torque T applied to crank AB, the reaction forces acting at pin-joints A, B and C, and the vertical reaction force between the piston and cylinder wall.

The free body diagrams for the crank, connecting rod and piston are shown in Figure 6.93.

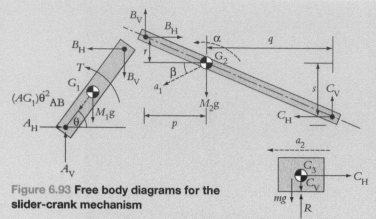

Figure 6.93 Free body diagrams for the slider-crank mechanism

The distances p, q, r, s can be calculated easily from the geometry of the mechanism. $M_1 g$, $M_2 g$ and mg are the weights of the crank, connecting rod and piston respectively.

A kinematic analysis of the system is used to determine the centripetal acceleration of the crank $[(AG_1)\dot\theta_{AB}^2]$, the translational acceleration (magnitude a_1 and direction β) and angular acceleration α of the connecting rod, and the acceleration a_2 of the piston. The latter calculations were performed in 'Example 1 – Slider-crank mechanism' for an identical slider-crank mechanism. It is assumed that this kinematic analysis has been performed and that all of the accelerations are known.

A_H and A_V are the required horizontal and vertical components of the reaction force at A respectively; B_H and B_V are the required horizontal and vertical components of the reaction force at B respectively; C_H and C_V are the required horizontal and vertical components of the reaction force at C respectively; R is the required vertical reaction force acting on the piston and is assumed to act through G_3; and T is the required applied torque. There are eight unknowns, all other quantities are known.

In this example, the crank is in pure rotation, the connecting rod is in general plane motion, and the piston is in pure translation.

Applying equations (6.5), (6.6) and (6.10) to the crank gives

$$\Sigma F_X = M\ddot{X}_G \qquad (B_V + M_1 g - A_V)\sin\theta + (B_H - A_H)\cos\theta = M_1(AG_1)\dot\theta_{AB}^2$$

$$\Sigma F_Y = M\ddot{Y}_G \qquad (B_V + M_1 g - A_V)\cos\theta - (B_H - A_H)\sin\theta = 0$$

$$\Sigma L_A = I_A \alpha \qquad +\,A\,) \quad T + AB\,B_H \sin\theta - (AB\,B_V + AG_1\,M_1 g)\cos\theta = 0$$

Using equations (6.5), (6.6) and (6.8) to analyse the connecting rod gives

$$\Sigma F_X = M\ddot{X}_G \qquad C_H - B_H = M_2 a_1 \cos\beta$$

$$\Sigma F_Y = M\ddot{Y}_G \qquad M_2 g - C_V - B_V = ma_1 \sin\beta$$

$$\Sigma L_G = I_G \alpha \qquad +\,A\,) \quad qC_V - pB_V - sC_H - rB_H = I_2 \alpha$$

Using equations (6.5) and (6.6) to analyse the piston gives

$$\Sigma F_X = M\ddot{X}_G \qquad -C_H = ma_2$$

$$\Sigma F_X = M\ddot{X}_G \qquad R - mg - C_V = 0$$

The rotational equation for the piston provides no additional information because all forces acting on the piston pass through the centre of mass G_3.

The eight equations above can be used to calculate the 8 unknowns ($A_H, A_V, B_H, B_V, C_H, C_V, R$ and T), as required.

6.9. Balancing of rotating masses

Introduction

For rotating machinery the location of the centre of mass relative to the axis of rotation has a significant influence on the reaction forces acting at the supporting bearings. It is vitally important that these reaction forces are kept to a minimum in order to maximize the life and prolong the reliability of the machinery. These forces can be minimized by designing the machinery so that the centre of mass coincides with the axis of rotation. However, in practice the centre of mass rarely coincides with the centre of rotation due manufacturing tolerances (variations), causing potentially large reaction forces to occur, particularly for high-speed rotating machinery. The process of modifying the rotating machinery post-manufacture to eliminate these forces is known as balancing.

This section introduces the ideas behind the balancing of rotating systems by considering the out-of-balance forces produced by imperfect discs mounted on a rotating rigid shaft. To aid understanding, the simplest case of a single imperfect disc is discussed first before generalizing the analysis to the multiple disc case. It is shown that the concepts of static and dynamic balance are fundamental to understanding balancing problems, and graphical techniques are used for their solution. Examples illustrating the application of the graphical technique are provided.

Single imperfect disc

Figure 6.94 shows a single thin disc of mass M mounted on a rigid shaft rotating at constant angular velocity $\omega(\text{rad s}^{-1})$. The shaft is supported in bearings at A and B. Due to manufacturing tolerances the centre of mass G of the disc will lie at some (small) eccentric distance e from the axis of rotation as shown.

The centre of mass G moves in a circular motion of radius e. Given that the shaft rotates at constant angular velocity ω, the centre of mass will have a centripetal acceleration $e\omega^2$ directed towards the axis of rotation (see 'Circular motion of a particle'). Using d'Alembert's Principle (see 'd'Alemberts Principle') the free body diagram of the disc can be drawn as shown in Figure 6.95.

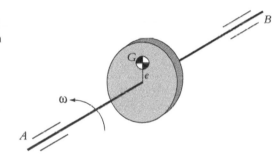

Figure 6.94 Single thin disc rotating on a rigid shaft

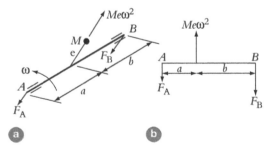

Figure 6.95 Free body diagram of thin disc using d'Alembert's Principle: (a) three dimensional; (b) plane *AMB*

In this figure the bearing reaction forces F_A and F_B and the out-of-balance force $Me\omega^2$ are coplanar – this plane is referred to as the plane AMB. The out-of-balance force is constant in magnitude, but its direction rotates with angular velocity ω. Similarly the bearing reaction forces rotate with the shaft, and are dependent on the orientation of the shaft at the particular instant in time considered. In Figure 6.95, the weight (Mg) of the disc has been neglected. This assumption is often used in practical balancing problems, and is justified on the basis that the out-of-balance force is usually much larger than the weight.

In accordance with d'Alembert's Principle, the rotor bearing system is in static equilibrium, and the system can be analysed using static analysis techniques.

Force equilibrium in plane AMB gives

$$Me\omega^2 - F_A - F_B = 0$$

Moment equilibrium in plane AMB about A gives

$$F_B(a + b) - Me\omega^2 a = 0$$

Solving the above two equations it can be shown that

$$F_A = \frac{Me\omega^2 b}{(a + b)}; \qquad F_B = \frac{Me\omega^2 a}{(a + b)}$$

The existence of non-zero bearing reaction forces F_A and F_B causes the structure supporting the bearings to vibrate and shake. To reduce this effect the out-of-balance force must be reduced by modifying the structure in some way, and this is achieved in practice by adding or removing mass in some way. A mass that is added or removed is referred to as a balancing mass. In what follows, some insights into the balancing problem are gained by investigating the use of single and multiple balancing masses that lie either in the 'same plane' as the original out-of-balance, or in a 'different plane'.

Balancing using a single 'same plane' balancing mass

Figure 6.96 considers the situation when a balancing mass M_1 is placed at radius r_1 diametrically opposite to M in the same plane as the thin disc.

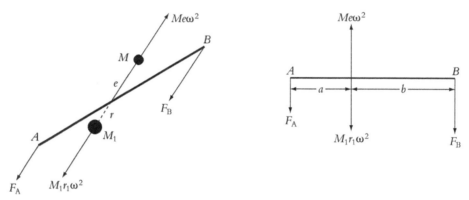

Figure 6.96 **Free body diagram for balancing with a single 'same plane' balancing mass using d'Alembert's Principle: (a) three dimensional; (b) plane AMB**

In this case, force equilibrium in plane AMB gives

$$Me\omega^2 - M_1 r_1 \omega^2 - F_A - F_B = 0$$

Moment equilibrium in plane AMB about A gives

$$F_B(a + b) - Me\omega^2 a + M_1 r_1 \omega^2 a = 0$$

If M_1 and r_1 are chosen such that

$$M_1 r_1 \omega^2 = Me\omega^2, \qquad \text{i.e. } M_1 r_1 = Me$$

then the original out-of-balance force will be cancelled out (i.e. there is zero net out-of-balance force), and it can be shown that $F_A = F_B = 0$. i.e. the system is balanced.

Balancing using multiple 'same plane' balancing masses

As an alternative to using a single 'same plane' balancing mass, a series of 'same plane' balancing masses can be used to achieve zero net out-of-balance force on the shaft. This is possible provided that the individual magnitudes and directions of the out-of-balance forces ($Mr\omega^2$) are chosen so that the vector sum of out-of-balance forces is zero in the plane. i.e. the out-of-balance force polygon is closed. Figure 6.97 shows the free body diagram for a system of two coplanar balancing masses (M_1 and M_2), together with the out-of-balance force polygon in three dimensions.

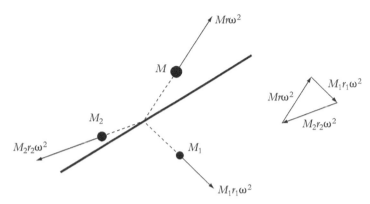

Figure 6.97 Free body diagram for balancing with two 'same plane' balancing masses using d'Alembert's Principle

Balancing using a single 'different plane' balancing mass

For practical reasons, it may not be feasible to place the balancing mass in precisely the same plane as the original out-of-balance mass. For this reason it may be necessary to place the balancing mass in a different plane, as shown Figure 6.98.

In this figure, the balancing mass M_1 is placed diametrically opposite to M at radius r_1, but in a different plane to the original out-of-balance mass.

In this case, force equilibrium in plane AMB gives

$$Me\omega^2 - M_1r_1\omega^2 - F_A - F_B = 0$$

Moment equilibrium in plane AMB about A gives

$$F_B(a + b) - Me\omega^2a + M_1r_1\omega^2(a - c) = 0$$

For zero net out-of-balance force, M_1 and r_1 are chosen such that

$$M_1r_1\omega^2 = Me\omega^2, \quad \text{i.e. } M_1r_1 = Me$$

and it can be shown that

$$F_B = \frac{Me\omega^2c}{(a + b)}; \quad F_A = \frac{-Me\omega^2c}{(a + b)}$$

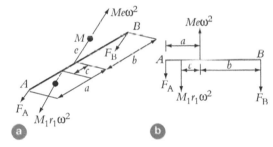

Figure 6.98 Free body diagram for balancing with a single 'different plane' balancing mass using d'Alembert's Principle: (a) three-dimensional; (b) plane *AMB*

These bearing reaction forces are equal in magnitude, but opposite in direction. Thus, although there is no net out-of-balance force, the net moment arising from the out-of-balance forces is not zero, giving rise to equal and opposite bearing reaction forces. This indicates that to eliminate the unwanted bearing reaction forces it is necessary to eliminate the net out-of-balance force *and* the net out-of-balance moment simultaneously. These conditions are fundamental to understanding the balancing problem and form the basis of the balancing process. For the case of a single balancing mass, zero net out-of-balance force and zero net out-of-balance moment can only be achieved by placing the balancing mass in the same plane as the original out-of-balance mass, i.e. $c = 0$.

Balancing using two 'different plane' balancing masses

If it is not possible to place the balancing mass in the same plane as the original eccentric mass, then it is necessary to use a minimum of two balancing masses to balance the system. In practice it is usual to use two balancing masses M_2 and M_3. This situation is considered here.

Consider balancing masses M_2 and M_3 to be placed diametrically opposite to M, but at different positions along the shaft, as shown in Figure 6.99.

For zero net out-of-balance force it is required that

$$Me\omega^2 = M_2r_2\omega^2 + M_3r_3\omega^2$$

In practice it is convenient to represent this equation using a vector diagram of the out-of-balance forces, as shown in Figure 6.100.

When forming the closed force polygon it is not necessary to include the angular velocity ω term because it is common to all components. For this reason, the force polygon is often referred to as the 'Mr' polygon. For there to be zero net out-of-balance force, the out-of-balance force polygon must be 'closed'.

For zero net out-of-balance moment, the vector sum of the moments arising from the individual out-of-balance forces, taken about any point along the shaft, must be zero. Using the moment equilibrium equation in plane AMB about A, it can be shown that there is zero net out-of-balance moment when

$$Me\omega^2 a = M_2r_2\omega^2 a_2 + M_3r_3\omega^2 a_3$$

In this equation it is important to take account of the sense of each moment. In practice it is convenient to represent the above out-of-balance moment equation using a vector diagram of the out-of-balance moments, as shown in Figure 6.101.

In the moment polygon each moment vector indicates the magnitude and direction (sense) of the moment. Also when forming the moment polygon it is not necessary to include the angular velocity ω term because it is common to all components. For this reason, the moment polygon is often referred to as the 'Mry' polygon. For there to be zero net out-of-balance moment, the moment polygon must be 'closed'.

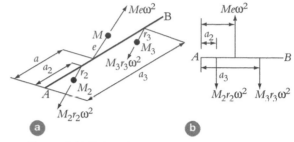

Figure 6.99 **Balancing using two 'different plane' masses**

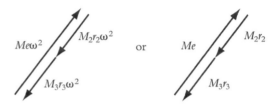

Figure 6.100 **Out-of-balance force polygon when using two 'different plane' masses**

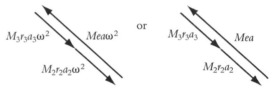

Figure 6.101 **Closed out-of-balance moment polygon when using two 'different plane' masses**

Static and dynamic balance

In summary, it has been shown that it is necessary to satisfy two conditions to balance a rotating system. The two conditions are: (i) there is zero net out-of-balance force, and (ii) there is zero net out-of-balance moment.

A rotor system for which condition (i) is satisfied is said to be statically balanced. This occurs if the non-rotating shaft is in force equilibrium in any angular configuration and is achieved if the polygon of out-of-balance forces (or 'Mr' polygon) is closed.

A system for which both conditions (i) and (ii) are satisfied is said to be dynamically balanced. This occurs if there is no rotating reaction force at either bearing and is achieved if the polygon of out-of-balance forces (or 'Mr' polygon) is closed and the polygon of out-of-balance moments (or 'Mry' polygon) is closed.

Multiple imperfect discs

In many practical cases the rotating shaft may carry several discs (or gears, pulleys, etc.), all of which are unbalanced. In these cases the total unbalance may consist of several unbalanced masses lying at different angles in different planes. This situation is considered here.

Consider a rigid shaft with three eccentric masses located along its length, as shown in Figure 6.102.

In this figure the centre of mass of each disc is identified relative to the reference axes $OXYZ$, which rotate with the shaft, by the coordinates r_i, θ_i and y_i, where r_i is the eccentricity of the ith disc, θ_i is the angular position of the centre of mass of the ith disc measured relative to the Z axis, and y_i is the position from O along the length of the shaft.

Noting that the out-of-balance force acting on the shaft due to the ith mass is $F_i = M_i r_i \omega^2$, the free body diagram (using d'Alembert's Principle) for the shaft showing all of the out-of-balance forces can be drawn as shown in Figure 6.103.

For the rotor system to be in static balance (see 'Static and dynamic balance'), the vector sum of the out-of-balance forces must equal zero, and the force 'Mr' polygon must close. If the force polygon closes then the bearing reaction forces will either be zero, or in equal and opposite pairs forming a couple. For the free body diagram shown in Figure 6.103, the force ('Mr') polygon can be drawn as shown in Figure 6.104.

The starting point for constructing the force polygon is O, with vectors $M_1 r_1$, $M_2 r_2$ and $M_3 r_3$ constructed in turn. The order in which these vectors are constructed is not important, and they can be added together in any sequence. If the system of three masses is not statically balanced, then the resultant out-of-balance force (vector) is F_4. To balance the system statically it is necessary to add a balancing mass such that its 'Mr' value is $M_4 r_4$ acting at an angular position θ_4 relative to the Z-axis.

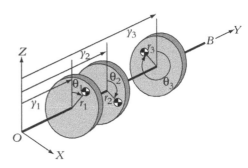

Figure 6.102 **Multiple thin discs mounted on a rotating rigid shaft**

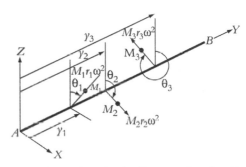

Figure 6.103 **Free body diagram showing the out-of-balance forces for a shaft with multiple discs using d'Alemberts Principle**

For the rotor system to be in dynamic balance (see 'Static and dynamic balance'), the vector sum of the out-of-balance moments must equal zero, and the moment 'Mry' polygon must close. To form the moment polygon care must be taken to understand the direction (sense) of the moment. Figure 6.105 considers the moment produced by mass M_1 only.

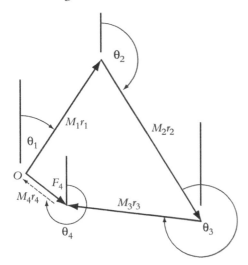

Figure 6.104 **Force ('Mr') polygon for the system shown in Figure 6.102**

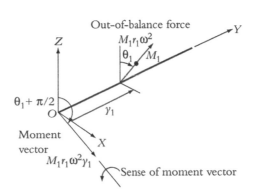

Figure 6.105 **Moment out-of-balance arising from mass M_1**

In this figure it can be seen that the out-of-balance force produces a moment vector of magnitude $M_1 r_1 \omega^2 y_1$. The direction of the moment vector is in accordance with the right-hand screw rule which takes account of the rotational sense (direction) of the moment, as indicated in the figure. The orientation of the moment vector is at an angle $\theta_1 + \dfrac{\pi}{2}$ relative to the Z axis.

Noting this, the moment ('Mry') polygon for the complete system shown in Figure 6.102 can be drawn as shown in Figure 6.106.

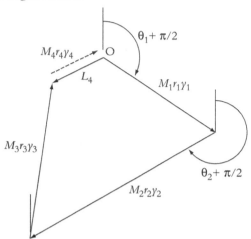

Figure 6.106 Moment ('Mry') polygon for the system shown in Figure 6.103

The starting point for constructing the moment polygon is O, with vectors $M_1 r_1 y_1$, $M_2 r_2 y_2$ and $M_3 r_3 y_3$, constructed in turn. As for the force polygon, the order in which these vectors are constructed is not important, and they can be added together in any sequence. If the system of three masses is not dynamically balanced, then the resultant out-of-balance moment is L_4. Since all of the vectors in the moment polygon are 90° (clockwise) to the corresponding out-of-balance forces, it is convenient to use Dalby's convention and rotate the moment polygon by 90° (anticlockwise) to align the corresponding force and moment vectors. The result of applying Dalby's convention is shown in Figure 6.107.

Note that when the reference point O is *not* at one end of the shaft, some of the y_i values may be negative. In these cases, the direction of the moment vector must be reversed relative to the corresponding force vector. Conversely, if the direction of an unknown $M_i r_i y_i$ value turns out to be negative, then y_i must be negative.

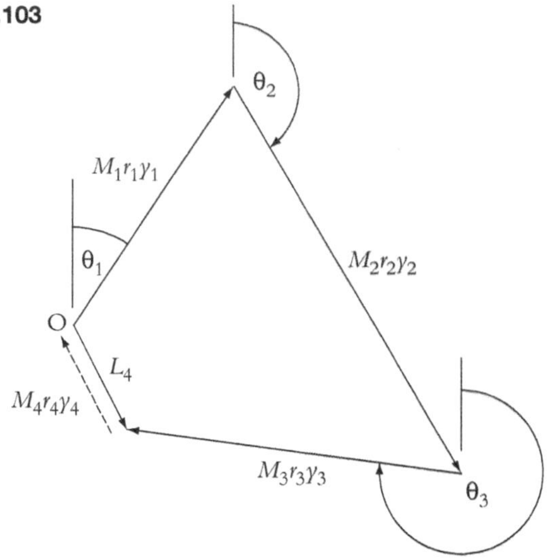

Solving balancing problems

In general, a rotating system will not be balanced. To ensure that the system is dynamically balanced it is necessary to add (or remove) two balancing masses (see 'Balancing using two "different plane" balancing masses'). For each of these balancing masses it is necessary to specify three quantities: (i) the Mr value; (ii) the angular orientation θ relative to a convenient datum; and (iii) the position along the shaft y from a convenient fixed location. These quantities are required to specify the magnitude and direction of the balancing forces and moments. Since there are three quantities per balancing mass and two balancing masses, it is necessary to consider six quantities in total.

Figure 6.107 Moment ('Mry') polygon for the system shown in Figure 6.103 using Dalby's convention

In most practical balancing problems, two of the six quantities are usually specified. For example, the positions of the two balancing masses along the shaft are known if the balancing planes are specified. The remaining four quantities need to be calculated, and this can be achieved graphically using the force ('Mr') polygon and the moment ('Mry') polygon. Worked examples showing typical calculations are provided next.

Example 1

Four eccentric masses (A,B,C,D) with 'mr' values of 10, 15, 12, 20 units respectively are to be arranged on a rigid shaft. Masses A and B are set at 90° to each other and are located a distance L apart. Determine the axial position and orientation of masses C and D to give static and dynamic balance.

Figure 6.108 shows a space diagram summarizing the information available, in which the reference datum has been chosen to coincide with mass A. All axial positions and orientations of the masses are measured relative to this datum.

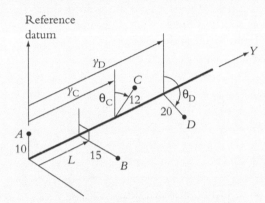

Figure 6.108 Space diagram for Example 1

The order of masses C and D may be different to that indicated in Figure 6.108. The information can be summarized in tabular form as shown in Table 6.1.

Plane	Mr	θ (degrees)	Y	Mry
A	10	0	0	0
B	15	90°	L	$15L$
C	12	θ_C	y_C	$12y_C$
D	20	θ_D	y_D	$20y_D$

Table 6.1 Summary of information provided for Example 1

In this table the quantities y_C, y_D, θ_C, θ_D are the unknowns and specify the axial position and orientation of masses C and D.

The force ('Mr') polygon can be constructed graphically as shown in Figure 6.109.

The starting point for constructing the force polygon is O. A vector of magnitude 10 is drawn vertically upwards from O to represent the out-of-balance force from mass A. It is in a vertical direction because it has been assumed that mass A is coincident with the reference datum (see Figure 6.108). A vector of magnitude 15 is then drawn at an angle of 90° (clockwise) from the vertical to represent the out-of-balance force from mass B, which is assumed to be at an angle of 90° from A. It is not

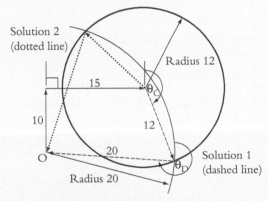

Figure 6.109 Force ('Mr') polygon for Example 1 (must be drawn to scale)

possible to draw the out-of-balance forces arising from masses C and D directly, because angles θ_C and θ_D are not specified. However, construction of the force polygon can be completed by noting that the force polygon must have four sides, because it is in static balance and there are four masses, and the lengths of the remaining two sides of the polygon are known. The last two vectors are completed by drawing two arcs corresponding to the magnitudes of the out-of-balance forces for masses C and D. In Figure 6.109, an arc of radius 12, centred on the end of the force vector arising from mass B, is drawn, together with an arc of radius 20, centred on starting point O. (Given that the order of the vectors in the force polygon does not matter, an alternative solution can be obtained by interchanging the arc centres.) Two solutions are indicated in Figure 6.109 corresponding to the two intersection points of the arcs. The required angles θ_C and θ_D for Solution 1 are indicated in Figure 6.109, and can be measured from the force polygon, provided that it has been drawn to scale.

The moment ('Mry') polygon can be constructed graphically as shown in Figure 6.110.

The starting point for constructing the moment polygon is O. Given that mass A does not provide a moment about O, it does not contribute to the moment polygon. A vector of magnitude 15L is drawn from O at an angle of 90° (clockwise) from the vertical to represent the out-of-balance moment arising from mass B. It is not possible to draw the out-of-balance moments arising from masses C and D directly, because lengths γ_C and γ_D are not specified.

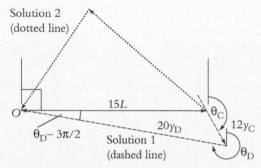

Figure 6.110 Moment ('Mry') polygon using Dalby's convention for Example 1 (must be drawn to scale)

However, construction of the moment polygon can be completed by noting that the moment polygon must have four sides, because it is in dynamic balance and there are four masses, and the orientations of the remaining two sides (θ_C and θ_D) are known from the force polygon. The last two vectors are completed by constructing two straight lines at orientations corresponding to the out-of-balance moments for masses C and D. In Figure 6.109, a line of unknown length at angle θ_C is drawn, together with a line at angle $\theta_D - \dfrac{3\pi}{2}$ through O. (Two possible solutions are possible, but only Solution 1 is considered here.) The intersection point of these lines defines the remaining two sides of the moment polygon, and the two remaining moment vectors have length $12\gamma_C$ and $20\gamma_D$, as shown. These lengths can be measured from Figure 6.110, provided that the figure has been drawn to scale, and used to calculate the required lengths γ_C and γ_D.

Example 2

Three discs, with known masses (M_1, M_2, M_3), radii (r_1, r_2, r_3), orientation angles (θ_1, θ_2, θ_3) and axial positions (γ_1, γ_2, γ_3), are to be balanced by two masses in known planes L and R, as shown in Figure 6.111.

Determine the magnitude ('Mr' values) and angular orientations of the masses in the L and R planes to give static and dynamic balance.

Figure 6.111 shows a space diagram summarizing the information available, in which the reference datum has been chosen to coincide with the mass in plane L. All axial positions and orientations of the masses are measured relative to this datum.

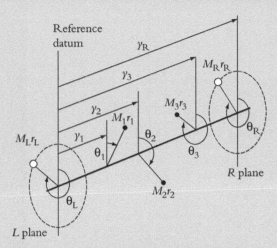

Figure 6.111 **Space diagram for Example 2**

The information provided can also be summarized in tabular form as shown in Table 6.2.

Plane	Mr	θ	y	Mry
1	$M_1 r_1$	θ_1	y_1	$M_1 r_1 y_1$
2	$M_2 r_2$	θ_2	y_2	$M_2 r_2 y_2$
3	$M_3 r_3$	θ_3	y_3	$M_3 r_3 y_3$
R	$M_R r_R$	θ_R	y_R	$M_R r_R y_R$
L	$M_L r_L$	θ_L	0	0

Table 6.2: Summary of information provided for Example 2

In this table the quantities $M_L r_L$, $M_R r_R$, θ_L, θ_R, are not known and correspond to the required magnitudes and orientations.

The moment ('Mry') polygon can be constructed graphically as shown in Figure 6.112.

The starting point for constructing the moment polygon is O. The moment vectors arising from masses M_1, M_2, M_3 are constructed in turn – the order in which these vectors are added together is not important. The moment vector $M_R r_R y_R$ is obtained by closing the moment polygon – this is possible because mass M_L makes zero contribution to the moment polygon and the system is in dynamic equilibrium. Using the moment polygon it is possible to measure the length of vector $M_R r_R y_R$ and the required θ_R angles, provided that it has been drawn to scale. Knowledge of y_R enables the required $M_R r_R$ value to be calculated.

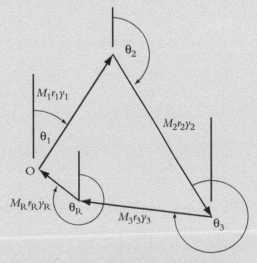

Figure 6.112 **Moment ('Mry') polygon using Dalby's convention for Example 2 (must be drawn to scale**

467

The force ('Mr') polygon can be constructed graphically as shown in Figure 6.113.

The starting point for constructing the force polygon is O. The force vectors arising from masses M_1, M_2, M_3, M_R are constructed in turn – the order in which these vectors are added together is not important. The force vector $M_L r_L$ is obtained by closing the force polygon – this is possible because the system is in static equilibrium. Using the force polygon it is possible to measure the length of $M_L r_L$ and the required θ_L value, provided that it has been drawn to scale. Knowledge of γ_R enables the required $M_R r_R$ value to be calculated.

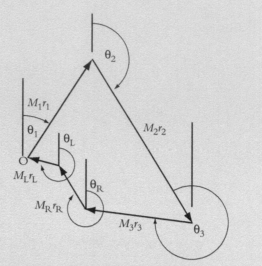

Figure 6.113 Force ('Mr') polygon for Example 2 – must be drawn to scale

6.10 Geared systems

Introduction

Gears perform important roles in many mechanical engineering systems. This section provides some introductory material related to the analysis of systems containing gears, including gear ratios and gear efficiency. Some of these relationships can be applied to other coupled systems, such as belt and chain drives, but attention will be restricted to gears here.

A few examples that use these results are also presented.

Equivalent pitch circles

For the purposes of analysis, a pair of meshing gear wheels can be represented as two equivalent pitch circles rolling against each other without slip at the point of contact, as shown in Figure 6.114. Gear 1 on the left has radius r_1, and N_1 teeth. Gear 2 on the right has radius r_2 and N_2 teeth.

Gear ratio

Gear ratio n is the ratio of the angular velocities of a pair of meshing gears. i.e.

$$n = \frac{\omega_1}{\omega_2}$$

where ω_1 is the angular velocity of the input gear, and ω_2 is the angular velocity of the output gear.

Assuming that the pitch-circles do not slip at the point of contact (i.e. roll without slip), the tangential displacements, velocities and accelerations of the pitch circles at the point of contact are the same on both circles. It follows that

displacement	$r_1\theta_1 = r_2\theta_2$	
velocity	$r_1\omega_1 = r_2\omega_2$	
acceleration	$r_1\alpha_1 = r_2\alpha_2$	

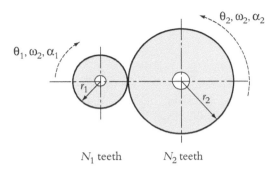

Figure 6.114 Schematic of a pair of meshed gear wheels

and

$$n = \frac{\omega_1}{\omega_2} = \frac{\alpha_1}{\alpha_2} = \frac{r_2}{r_1} = \frac{N_2}{N_1}$$

The final result arises from the fact that gear wheels are designed so that the number of teeth N is proportional to the wheel radius of the gear.

Simple gear train

Consider the simple gear train shown in Figure 6.115 consisting of three gears (A, B and C).

The tangential velocity of each pitch circle at the point of contact between gears A and B is given by

$$v = r_A\omega_A = r_B\omega_B$$

Rearranging this equation gives

$$\frac{\omega_A}{\omega_B} = \frac{r_B}{r_A} = \frac{N_B}{N_A}$$

Again, the final result arises from the fact that gear wheels are designed so that the number of teeth N is proportional to the wheel radius of the gear.

Similarly, for the point of contact between gears B and C it can be shown that

$$\frac{\omega_B}{\omega_C} = \frac{r_C}{r_B} = \frac{N_C}{N_B}$$

Combining the last two equations gives

$$\frac{\omega_A}{\omega_C} = \frac{r_C}{r_A} = \frac{N_C}{N_A}$$

This result indicates that the gear ratio is independent of the idler gear B.

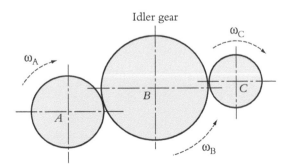

Figure 6.115 Simple gear train

Compound gear train

Consider the compound gear train shown in Figure 6.116.

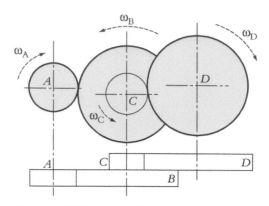

A compound gear train contains two or more gears mounted/fixed to the same shaft so that they rotate at the same speed. Compound gear trains offer a number of advantages over simple gear trains, including higher gear ratios and/or a more compact size, and the possibility to have the input and output shafts arranged coaxially.

For the case shown in Figure 6.116, it can be shown easily that

$$\frac{\omega_A}{\omega_B} = \frac{r_B}{r_A} = \frac{N_B}{N_A}$$

and

Figure 6.116 **Compound gear train**

$$\frac{\omega_D}{\omega_C} = \frac{r_C}{r_D} = \frac{N_C}{N_D}$$

Noting that $\omega_B = \omega_C$, the last two equations can be combined to give

$$\frac{\omega_A}{\omega_D} = \frac{r_B r_D}{r_A r_C} = \frac{N_B N_D}{N_A N_C}$$

The ability to choose different numbers of teeth on the compound gear gives a flexible choice of overall ratio.

Efficiency of gears

Power is lost in gear transmissions due to friction between the surfaces of the meshing gears, and in the bearings that support the shafts on which the teeth are mounted. The efficiency η of the gearbox is defined as

$$\eta = \frac{\text{power output}}{\text{power input}}$$

Power is the instantaneous rate at which work is done (see 'Theorem of work' later).

For a geared system, the power input is the instantaneous rate at which work is done by the applied torque L_1 acting on the input gear, and is given by

$$\text{power input} = L_1 \omega_1$$

where ω_1 is the angular speed of the input gear. The power output is the instantaneous rate at which work is done by the torque L_2 produced by the output gear, and is given by

$$\text{power output} = L_2 \omega_2$$

where ω_2 is the angular speed of the output gear.

Using these results gives

$$\eta = \frac{L_2 \omega_2}{L_1 \omega_1}$$

Using the definition of gear ratio n it can be shown that

$$\eta = \frac{1}{n} \frac{L_2}{L_1}$$

It should be noted that, because the speed ratio n of the gears is fixed, a power loss gives rise to a loss in torque, not a reduction in speed ratio.

Example 1

For the simple gear train shown in Figure 6.117, Gears 1 and 2 have moments of inertia I_1 and I_2 about axes through their centres O_1 and O_2 respectively.

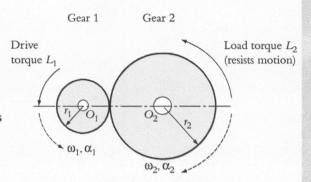

Determine the equivalent moments of inertia of the gear train:
(i) referred to the axis through Gear 1; and (ii) referred to the axis through Gear 2. In addition, determine expressions for angular acceleration α_1 and angular acceleration α_2.

Figure 6.117 Dynamics of a simple gear train

The free body diagrams for the two gear wheels are shown in Figure 6.118.

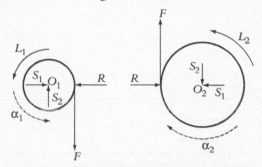

Figure 6.118 Free body diagrams for Example 1

F is the tangential contact force at the meshing point and is responsible for transmits the torque between the gears. R is the radial contact force at the meshing point, arising due to the shape of the gear teeth. These contact forces cause reaction forces S_1 and S_2 to occur at the bearings of each gear.

For the purposes of analysis it is only necessary to consider the rotational motion of the gears, which are in pure rotation.

The equation governing the rotation of the left-hand gear about O_1 is given by (see equation (6.10))

$$\Sigma L_O = I_O \alpha \quad +\circlearrowright O_1 \quad L_1 - Fr_1 = I_1 \alpha_1$$

The equation governing the rotation of the right-hand gear about O_2 is given by (see equation (6.10))

$$\Sigma L_O = I_O \alpha \quad +\circlearrowleft O_2 \quad Fr_2 - L_2 = I_2 \alpha_2$$

Eliminating F between the last two equations gives

$$L_1 r_2 - L_2 r_1 = I_1 \alpha_1 r_2 + I_2 \alpha_2 r_1$$

Using the gear ratio expressions

$$n = \frac{\alpha_1}{\alpha_2} = \frac{r_2}{r_1}$$

in the previous equation, it can be shown that:

$$L_1 - \frac{L_2}{n} = \left(I_1 + \frac{I_2}{n^2} \right) \alpha_1$$

This equation has the form '$L_e = I_e\alpha_1$' and indicates the influence that Gear 2 has on the system, referred to the axis through Gear 1. Referred to this axis, the equivalent moment of inertia $I_e = \left[I_1 + \left(\dfrac{I_2}{n^2} \right) \right]$, as required.

An equation referring to the axis through Gear 2 can be obtained in a similar way and is given by

$$- L_2 + nL_1 = (I_2 + n^2 I_1)\alpha_2$$

This equation has the form '$L_e = I_e\alpha_2$' and demonstrates the influence that Gear 1 has on the system, referred to the axis through Gear 2. Referred to this axis, the equivalent moment of inertia ($I_e = I_2 + n^2 I_1$), as required.

The above two equations can be rearranged to obtain expressions for acceleration and angular acceleration α_2, such that

$$\alpha_1 = \frac{L_1 - \dfrac{L_2}{n}}{\left(I_1 + \dfrac{I_2}{n^2} \right)}$$

$$\alpha_2 = \frac{nL_1 - L_2}{(I_1 n^2 + I_2)}$$

as required.

Example 2

A crane drum that is used to raise a load mass M is shown in Figure 6.119.

The motor develops a torque L and is coupled via gears to a crane drum of radius R. Rotation of the crane drum is opposed by constant drag torque L_D. A cable with the load mass attached to its end is wrapped around the crane. As the crane drum rotates the load mass is raised with acceleration a.

Assuming that the gears have an efficiency η and that the inertia of the gears is incorporated in the moment of inertia of the motor I_1 and the moment of inertia of the crane drum I_2, determine an expression for the acceleration a of the load in terms of the applied motor torque L.

The free body diagrams for the motor, gears, crane drum and load mass are shown in Figure 6.120.

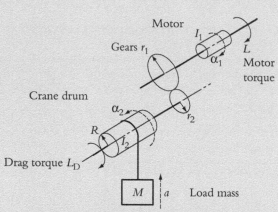

Figure 6.119 **Schematic of a motorized crane drum**

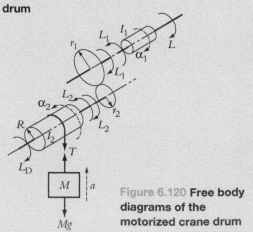

Figure 6.120 **Free body diagrams of the motorized crane drum**

In the free body diagram the bearing reaction forces are not shown.

For the purposes of analysis it is only necessary to consider the rotational motion of the motor and crane drum, and the translational motion of the load mass.

The equation governing the rotation of the motor about the motor axis is given by (see equation (6.10))

$$\Sigma L_O = I_O \alpha \quad +) \quad L - L_1 = I_1 \alpha_1$$

The equation governing the rotation of the crane drum about the crane drum axis is given by (see equation (6.10))

$$\Sigma L_O = I_O \alpha \quad +) \quad L_2 - L_D - TR = I_2 \alpha_2$$

The gear efficiency

$$\eta = \frac{1}{n} \frac{L_2}{L_1}$$

defines the relationship between the input torque L_1 and output torque L_2 to the gear box.

The gear ratio

$$n = \frac{\alpha_1}{\alpha_2}$$

defines the relationship between the angular accelerations of the motor and the crane drum (α_1 and α_2 respectively).

The equation governing the vertical motion of the load mass is given by (see equation (6.5))

$$\Sigma F_Y = M\ddot{Y}_G \quad +\uparrow \quad T - Mg = Ma$$

Using the above equations and noting that

$$a = R\alpha_2$$

it can be shown that

$$a = \frac{(\eta n L - L_D - MgR)R}{(I_2 + \eta n^2 I_1 + MR^2)}$$

as required.

6.11 Work and energy

Introduction

Energy concepts play an important role in solving problems in dynamics. Solutions can often be found more easily by using energy methods than by using equations of motion based on Newton's laws of motion, and this is often the case for complex systems. The main reason for this is that calculating the kinetic energy requires expressions for velocities and these are often much simpler than the expressions for acceleration. A fundamental difference between energy and Newton's laws of motion is that energy is a scalar quantity (the unit of energy is a joule $[1 \text{ kg m}^2 \text{ s}^{-2} = 1 \text{ Nm} = 1 \text{ J}]$), while Newton's laws of motion are based on vector quantities, such as forces and accelerations.

This section presents fundamental definitions and useful theorems relating to work and energy applied to particles and rigid bodies. Some examples are presented to illustrate their application to practical engineering problems, including flywheel design.

Kinetic energy

Kinetic energy of a particle

For a particle of mass m moving with velocity v, the kinetic energy T is given by

$$T = \tfrac{1}{2}mv^2$$

Kinetic energy of a rigid body

When analysing the equations of motion of a machine element exhibiting planar motion (Section 6.8) it was convenient to consider: (i) pure translation; (ii) pure rotation; and (iii) general plane motion separately. It is similarly convenient to consider these cases when considering the kinetic energy of a rigid body.

Pure translation

Figure 6.121 shows a rigid body moving in pure translation with velocity V.

Every point within the body has the same velocity V (magnitude and direction) (see 'Pure translation'), and the kinetic energy of the body is given by

$$T = \tfrac{1}{2}MV^2$$

where M is the total mass of the body.

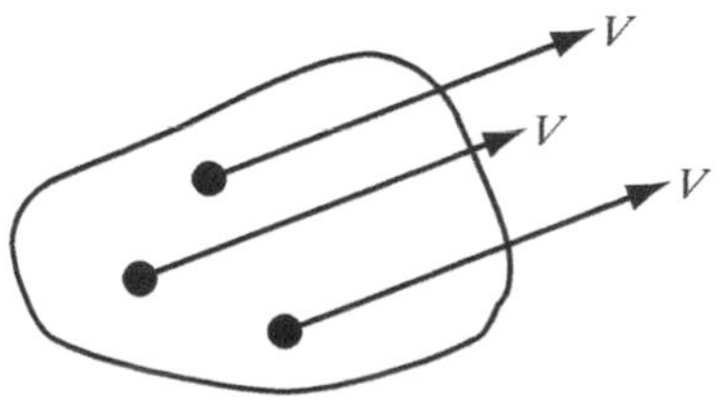

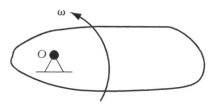

Figure 6.121 Rigid body moving in pure translation

Pure rotation

Figure 6.122 shows a rigid body rotating about a fixed axis at O with angular velocity ω.

Different points within the body have different velocities depends on their distance from O and the angular velocity, and the kinetic energy of the rigid body is given by

$$T = \tfrac{1}{2}I_o\omega^2$$

where I_0 is the moment of inertia of the body about an axis through O.

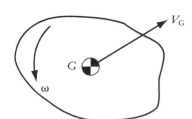

Figure 6.122 Rigid body rotating about a fixed axis

General plane motion

Figure 6.123 shows a rigid body moving in general plane motion, such that the centre of mass G of the body has velocity V_G and the body rotates with angular velocity ω.

For a rigid body in general plane motion the kinetic energy is given by

$$T = \tfrac{1}{2}MV_G^2 + \tfrac{1}{2}I_G\omega^2$$

where M is the total mass of the body and I_G is the moment of inertia of the body about an axis through G.

The first term on the right-hand side corresponds to the kinetic energy when all of the mass is concentrated at G. The second term corresponds to the kinetic energy due to rotation of the body about G. This relationship *only* applies to the centre of mass G and does not apply to any other point in the body.

Figure 6.123 Rigid body moving in general plane motion

Theorem of work

Figure 6.124 shows a particle of mass M moving in a horizontal plane under the action of a force F.

The work done WD by the force to move the particle through displacement dx is given by

$$WD = F\,dx$$

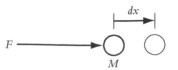

Figure 6.124 Particle of mass M under the action of a force F

Machine dynamics

The rate at which the force does work P is given by

$$P = F\frac{dx}{dt}$$

This is the power generated by the force and has units $\text{Nm s}^{-1} = \text{W}$ (watt).

Using Newton's second law $F = M\left(\dfrac{d^2x}{dt^2}\right)$, the power P can be rewritten as follows

$$P = F\frac{dx}{dt} = M\frac{d^2x}{dt^2}\frac{dx}{dt} = \frac{d}{dt}\left[\tfrac{1}{2}M\left(\frac{dx}{dt}\right)^2\right] = \frac{d}{dt}\left[\tfrac{1}{2}MV^2\right]$$

i.e.

$$P = \frac{d}{dt}\left[\tfrac{1}{2}MV^2\right]$$

This equation is only valid at a particular instant in time, and indicates that the instantaneous power generated by the input force is equal to the rate of change of kinetic energy. This result can be used to derive a particularly useful result for the more general case when the particle moves through a finite distance, as shown in Figure 6.125.

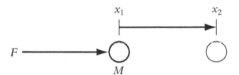

Figure 6.125 **Particle of mass *M* under the action of a force *F***

In this general case the force acts on the particle as it moves from its initial position x_1 (at time t_1) to final position x_2 (at time t_2), and the velocity V of the particle changes.

Integrating the expression for the instantaneous power generated by the force over the time taken for the particle to move from x_1 to x_2 gives

$$\int_{t_1}^{t_2} P\,dt = \int_{t_1}^{t_2}\frac{d}{dt}\left[\tfrac{1}{2}MV^2\right] = \tfrac{1}{2}M(V(t_2)^2) - V(t_1)^2) = T_2 - T_1$$

where $V(t_1)$ and $V(t_2)$ are the initial and final velocities of the particle respectively, and T_1 and T_2 are the initial and final kinetic energies of the particle respectively.

The left-hand side of the above equation can be expressed as

$$\int_{t_1}^{t_2} P\,dt = \int_{t_1}^{t_2} F\frac{dx}{dt}\,dt = \int_{x_1}^{x_2} F\,dx$$

Combining the last two equations gives

$$\int_{x_1}^{x_2} F\,dx = T_2 - T_1$$

This equation indicates that the work done by the force to move the particle from its initial position to its final position is equal to the change in kinetic energy. This is an extremely powerful result that provides an efficient way to solve some dynamics problems.

Although the above result was derived for a particle, it also applies to the work done by external forces acting on systems of particles and on rigid bodies in translation, and to the work done by external torques applied to rigid bodies that are rotating about a fixed point or are in general plane motion. The relevant results for rigid bodies that are rotating in some way are summarized below.

Pure rotation

Figure 6.126 shows a rigid body rotating about a fixed axis at O with angular velocity ω, that is acted on by torque L_O about O.

The torque L_O acts on the rigid body as it rotates from its initial orientation θ_1 (at time t_1) to its final orientation θ_2 (at time t_2), causing the angular velocity to change from its initial value ω to its final value ω_2.

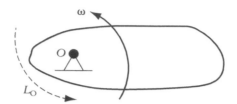

Figure 6.126 Rigid body rotating about a fixed axis

The work done is related to the change in kinetic energy as follows

$$\int_{\theta_1}^{\theta_2} L_O \mathrm{d}\theta = \tfrac{1}{2} I_O(\omega_2^2 - \omega_1^2)$$

where I_O is the moment of inertia of the body about an axis through O.

General plane motion

Figure 6.127 shows a rigid body moving in general plane motion, such that the centre of mass G of the body has velocity V_G and the body rotates with angular velocity ω. The rigid body is acted on by a resultant external applied force F and resultant external torque L_G about G. For simplicity it is assumed in what follows that force F does not change direction.

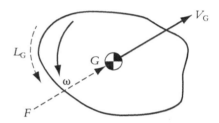

Figure 6.127 Rigid body moving in general plane motion

The force F acts on the rigid body through the centre of mass G as it moves (in a straight line) from initial position x_1 to its final position x_2, and the velocity of G changes from its initial value V_{G1} to its final value V_{G2}.

The torque L_G acts on the rigid body as it rotates from its initial orientation θ_1 (at time t_1) to its final orientation θ_2 (at time t_2), causing the angular velocity to change from its initial value ω_1 to its final value ω_2.

For this case separate expressions can be written relating the work done to the change in kinetic energy for translational and rotational motions of the body.

For translation of the body:

$$\int_{x_1}^{x_2} F \mathrm{d}x = \tfrac{1}{2} M(V_{G2}^2 - V_{G1}^2),$$

where M is the total mass of the body.

For rotation of the body:

$$\int_{\theta_1}^{\theta_2} L_O \mathrm{d}\theta = \tfrac{1}{2} I_G(\omega_2^2 - \omega_1^2),$$

where I_G is the moment of inertia of the body about an axis through G.

Potential energy

Potential forces are forces which change the potential of the system to do work by changing the amount of energy stored in the system in a reversible way. It is sometimes useful to consider the work done by these potential forces, and this work done is referred to as the potential energy U. The most important of these forces for mechanical systems are gravity and spring forces and the work done by these force are referred to as the gravitational potential energy and the strain energy in a spring respectively.

Machine dynamics

Gravitational potential energy

Consider a particle of mass M acted on by (constant) gravity force Mg, as shown in Figure 6.128.

The gravitational potential energy U is the work done by the applied force needed to move the particle vertically upwards (from a reference level) through a height h, and is given by

$$U = WD = \int F dx = \int_0^h Mg dx = Mgh$$

i.e.

$$U = Mgh$$

A particle of mass M that has moved upwards through a height h is said to have gained potential energy $U = Mgh$ (relative to the reference level used).

Note: if the particle is moved vertically *downwards* through a height h, then the gravitational potential energy is $U = -Mgh$. In this case the negative sign indicates that the particle has lost potential energy.

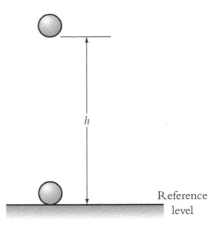

Figure 6.128 Gravitational Potential Energy of a particle

Strain energy in a spring

Consider the (linear) elastic spring of stiffness K shown in Figure 6.129(a), with a force F applied to its free end.

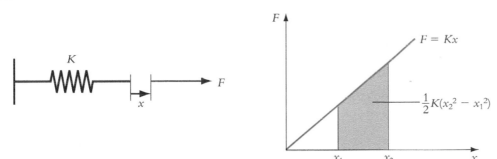

Figure 6.129 **(a) Linear spring; (b) Hooke's law**

According to Hooke's law (see Figures 6.129(a) and (b)) when the force F is applied to the free end of the spring it extends by a distance x from its natural length, such that $F = Kx$ and the spring extension x is proportional to the applied force F.

Consider the case when a force is applied to the spring which increases the spring extension from an initial extension x_1 to a final extension x_2. The strain energy in the spring U is depicted by the shaded region in Figure 6.129(b) and is given by

$$U = WD = \int F dx = \int_{x_1}^{x_2} Kx dx = \tfrac{1}{2}K(x_2^2 - x_1^2)$$

i.e.

$$U = \tfrac{1}{2}K(x_2^2 - x_1^2)$$

For the case when the initial extension is zero, such that the spring is initially at its natural length, and $x_2 = x$ is the final extension of the spring (from its natural length), the strain energy in the spring U is given by

$$U = \tfrac{1}{2}Kx^2$$

Conservation of energy

A system is said to be conservative if all of the forces acting are potential forces. A direct consequence of this is that the total energy of the system is constant. For most practical mechanical systems this implies that the sum of the kinetic energy T and potential energy U of the system is constant, i.e.

$$T + U = \text{CONSTANT}$$

A system is said to be non-conservative if energy is added or removed by external (non-potential) forces such as friction. In this case the work done by the external (non-potential) forces in a given interval of time is equal to the change in kinetic energy plus the change in potential energy. i.e.:

$$WD = (T_2 - T_1) + (U_2 - U_1)$$

where WD is the work done by the external (non-potential) forces, T_1 and T_2 are the initial and final kinetic energies, and U_1 and U_2 are the initial and final potential energies of the system. This result is known as the Theorem of Work.

Figure 6.130 shows part of a simple epicyclic gear system in which a planet wheel with pitch circle diameter (PCD) 30 mm, mounted on a planet carrier OA of length 65 mm, runs on a fixed internal ring gear with PCD 160 mm. The plane containing the gears lies in the vertical plane. The planet carrier OA has mass $M_C = 1$ kg and its centre of mass is located at G where $OG = 40$ mm. The moment of inertia of the planet carrier about the axis through O is $I_{CO} = 0.0015$ kg m^2.

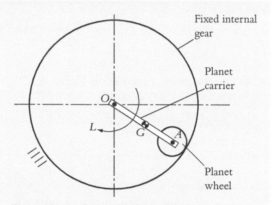

Figure 6.130 **Epicyclic Gear System**

The planet wheel has a mass $M_P = 0.3$ kg and the moment of inertia about its polar axis at A is $I_{PA} = 0.00003$ kg m^2.

Neglecting any losses due to fiction, calculate the angular velocity of the planet carrier OA when it reaches the vertical after being released from rest in the horizontal position and under the action of gravity and a constant drive torque, $L = 1$ Nm, as shown in Figure 6.130.

The gear system is a non-conservative system because the applied torque L increases the energy in the system and is not conservative. In this case the Theorem of Work is applicable and states that

$$WD = (T_2 - T_1) + (U_2 - U_1)$$

where WD is the work done by the external (non-potential) forces, T_1 and T_2 are the initial and final kinetic energies, and U_1 and U_2 are the initial and final potential energies of the system.

Work done

The work done by the applied torque as OA rotates from the (initial) horizontal to the (final) downwards vertical is given by

$$WD = \int_0^{\pi/2} L d\theta = \frac{L\pi}{2}$$

Note: the work done by the gravity force is taken into account in the potential energy terms because the gravity force is a potential force.

Kinetic energy

Since the system starts at rest in the (initial) horizontal position, the initial kinetic energy is zero, i.e.

$$T_1 = 0$$

After the system is released both the planet carrier and planet wheel begin to move and contribute to the kinetic energy of the system. The planet carrier is in pure rotation about O, while the planet wheel is in general plane motion, as it exhibits translational and rotational motion. The (final) kinetic energy when the planet carrier is vertical is given by

$$T_2 = \tfrac{1}{2}I_{CO}\omega_C^2 + \tfrac{1}{2}M_P V_A^2 + \tfrac{1}{2}I_{PA}\omega_P^2$$

where ω_C is the angular velocity of the planet carrier, ω_A is the translational velocity of A, and ω_P is the angular velocity of the planet wheel.

The translational velocity of A is related to the angular velocity of the planet carrier as follows

$$V_A = \omega_C(OA)$$

In addition, because the planet wheel rolls without slipping on the internal gear, the translational velocity of A can also be expressed in terms of the angular velocity of the planet wheel as follows

$$V_A = \omega_P r_P$$

Using the previous two equations, it can be shown that

$$\omega_P = \frac{\omega_C(OA)}{r_P}$$

Using these results the final kinetic energy can be expressed as follows

$$T_2 = \tfrac{1}{2}\omega_C^2\left(I_{CO} + M_P(OA)^2 + I_{PA}\frac{(OA)^2}{r_P^2}\right)$$

where the angular velocity of the planet carrier ω_C is the only unknown in this equation.

Potential energy

The change in height of the centres of mass of the planet carrier and planet wheel are the only sources of potential energy. In the horizontal starting position, the centres of mass of the planet carrier and planet wheel lie on a horizontal line through O. This horizontal line is chosen as the reference level about which the change in potential energy is calculated. Relative to this reference level the initial (gravitational) potential energy is zero, i.e.

$$U_1 = 0$$

When the planet carrier is vertical, the centre of mass of the planet carrier lies a distance OG beneath O, while the centre of mass of the planet wheel lies a distance OA beneath O. Relative to the chosen reference level, the final potential energy is given by

$$U_2 = -[M_C g(OG) + M_P g(OA)]$$

In this equation, the negative sign indicates that potential energy has been lost (relative to the chosen reference level).

Substituting the above results into the Theorem of Work

$$WD = (T_2 - T_1) + (U_2 - U_1),$$

gives

$$\frac{L\pi}{2} = \tfrac{1}{2}\omega_C^2\left[I_{CO} + M_P(OA)^2 + I_{PA}\frac{(OA)^2}{r_P^2} \right] - [M_Cg(OG) + M_Pg(OA)].$$

Substituting numerical values into this equation and rearranging, it can be shown that $\omega_C = 36\ \mathrm{rad\ s^{-1}}$, as required.

Application to flywheel design

Flywheels are used as energy storage devices to help reduce speed fluctuations in machinery when either the drive torque changes (e.g. as in a reciprocating engine), or the load torque fluctuates (e.g. as in a punch, press or reciprocating pump). As an example, consider the torque output from a single cylinder four-stroke engine. The variation in output torque $T(\theta)$ with crank angle (θ) over two revolutions of the engine, has the general form indicated in Figure 6.131.

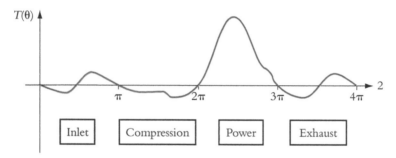

Figure 6.131 **Output torque against crank angle $T(\theta)$ for two revolutions of a four-stroke engine**

In practice the engine needs to be coupled in some way to a load which will have its own torque requirements. If the engine is coupled directly to the load (as shown in Figure 6.132(a) and the torque required by the load does not precisely match the torque output from the engine, the rotational speed (angular velocity) Ω of the combined system will fluctuate cyclically. These fluctuations can be significant if the combined moment of inertia of the engine and load (I_O) is relatively small. These fluctuations in rotational speed are normally undesirable and the flywheel found on most engines is designed to reduce the speed fluctuations to within acceptable limits. Figure 6.132(b) shows the coupled system with a flywheel incorporated.

In many situations the load requires a constant torque T_O so that it can be driven at constant speed. In this case the load torque must be equal to the mean value of engine drive torque $T(\theta)$.

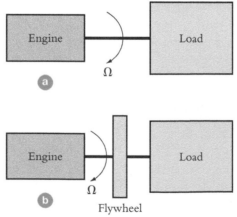

Figure 6.132 **Coupled engine-load system (a) without flywheel; (b) with flywheel**

Machine dynamics

In general, for a coupled engine-load system with flywheel, the equation of motion is given by (see equation (6.10))

$$T(\theta) - T_O = (I_f + I_O)\frac{d\Omega}{dt}$$

where $T(\theta) - T_O$ is the net applied torque, I_O is the combined moment of inertia of the engine and load, I_f is the moment of inertia of the flywheel, and $\frac{d\Omega}{dt}$ is the angular acceleration of the coupled system. Throughout what follows it will be assumed that $I_f \gg I_O$ and that the moment of inertia of the coupled system is I_f.

From the above equation of motion it can be deduced that if the net torque $[T(\theta) - T_O]$ is positive, then $\frac{d\Omega}{dt}$ will be positive and the rotational speed will increase. In contrast, if the net applied torque is negative, then $\frac{d\Omega}{dt}$ will be negative and the rotational speed will decrease.

These characteristics are shown diagrammatically in Figure 6.133.

In Figure 6.133 it can be seen that the minimum rotational speed Ω_1 occurs when $\theta = \theta_1$ corresponding to the interception of $T(\theta)$ and its mean value T_O. It can also be seen that the maximum rotational speed Ω_2 occurs when $\theta = \theta_2$ corresponding to the other interception of $T(\theta)$ and its mean value T_O.

The Theorem of Work applied to a system rotating about a fixed axis states that

$$WD = (T_2 - T_1)$$

where WD is the work done on the flywheel, and T_1 and T_2 are the initial and final kinetic energies. There are no sources of potential energy.

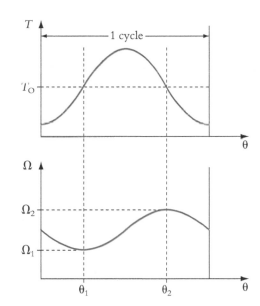

Figure 6.133 **Plots of drive torque *T* and rotational speed Ω against rotation angle θ of the coupled system**

The work done on the flywheel is the work done by the net applied torque, and is the area under the plot of $T(\theta) - T_O$ against θ. This corresponds to the energy transferred to (or from) the flywheel.

For the speed change from Ω_1 to Ω_2 the Theorem of Work gives

$$WD = \tfrac{1}{2}I_f(\Omega_2^2 - \Omega_1^2)$$

where WD is the net area under $[T(\theta) - T_O]$ between the crank angles θ corresponding to the maximum and minimum rotational speeds. This is the basic relationship used in flywheel problems.

The previous equation can be written as

$$WD = I_f \frac{(\Omega_1 + \Omega_2)}{2}(\Omega_2 - \Omega_1)$$

Often the main design requirement for a flywheel is to limit the rotational speed fluctuations to a small percentage of the mean rotational speed. If the speed fluctuation is small, the mean rotational speed can be approximated as follows

$$\Omega_0 \approx \frac{(\Omega_1 + \Omega_2)}{2}$$

Using this result in the previous equation gives

$$WD \approx I_f \Omega_0 (\Omega_2 - \Omega_1)$$

where can be expressed easily as a percentage of the nominal speed Ω_0.

Example 1

Figure 6.134 shows the torque generated by a motor that is used to drive a load requiring constant torque T_O at a mean rotational speed of 1500 rev min $^{-1}$.

Neglecting all losses, calculate the moment of inertia of the flywheel required to limit speed fluctuations to within 0.5% of the mean rotational speed.

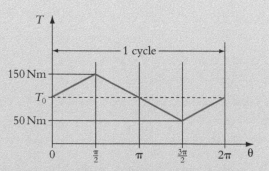

Figure 6.134 **Torque diagram for the motor in Example 1**

By neglecting the losses, the Theorem of Work derived in 'Application to flywheel design' can be applied.

By inspection of the torque diagram, it can be shown easily that the mean torque $T_O = 100$ Nm.

The shaded area under the torque diagram corresponds to the situation when the net applied torque is positive and depicts the part of the cycle when the rotational speed increases from its minimum value to its maximum value. i.e. the minimum rotational speed Ω_1 occurs when $\theta = 0$ and the maximum rotational speed Ω_2 occurs when $\theta = \pi$. The work done by the flywheel corresponds to the shaded area under the torque diagram and is given by

$$WD = \tfrac{1}{2}(\pi)(50) = 78.54 \, J$$

The mean rotational speed of the system is

$$\Omega_2 = 1500\left(\frac{2\pi}{60}\right) = 157.1 \text{ rad s}^{-1}.$$

For a speed fluctuation of ± 0.5 per cent of the mean rotational

$$\Omega_2 - \Omega_1 = 157.1 \text{ rad s}^{-1}$$

Using the relationship derived earlier, based on the Theorem of Work, i.e.

$$WD \approx I_f \Omega_0 (\Omega_2 - \Omega_1)$$

it can be shown easily that the moment of inertia of the flywheel is

$$I_f = \frac{78.54}{(157.1)(1.57)} = 0318 \text{ kgm}^2$$

as required.

Example 2

Figure 6.135 shows a schematic of a flywheel punch, used for punching holes in thin sheet metal.

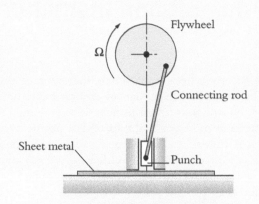

The flywheel is driven by an electric motor which delivers a constant drive torque. The load torque is effectively zero, except when the punch is cutting a hole in the sheet metal. This load torque can be represented as shown in Figure 6.136.

Figure 6.135 Schematic of a flywheel punch

Neglecting friction and the inertia of all parts except the flywheel, calculate the moment of inertia of the flywheel required to keep the running speed within 5% of the nominal running speed of 2000 rev min^{-1}.

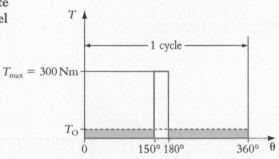

By neglecting the friction and inertia (except the flywheel), the Theorem of work derived in 'Application to flywheel design' can be applied. In 'Application to flywheel design' the drive torque varied with time, while the load torque was constant. In this

Figure 6.136 Load torque diagram for the hole punch in Example 2

example, the drive torque is constant with the load torque varying with time. Despite this difference the same approach can be used, such that the work done on the flywheel is equal to the change in kinetic energy.

The shaded area under the torque diagram corresponds to the part of the cycle when the drive torque is greater than the load torque and the rotational speed is increasing. The maximum rotational speed Ω_2 of the system occurs when $\theta = 150°$, while the minimum rotational speed Ω_1 occurs when $\theta = 180°$.

The required drive torque is the mean load torque T_O given by

$$T_O = \left(\frac{180 - 150}{360}\right)300 = 25 \, \text{Nm}$$

The *WD* on the flywheel when the net applied torque is positive and the flywheel accelerates from Ω_1 to Ω_2 corresponds to the shaded area in Figure 6.136 and is given by

$$WD = T_O(360 - (180 - 150))\frac{\pi}{180} = 25(330)\frac{\pi}{180} = 144 \, \text{J}$$

For speed fluctuations of ± 5 per cent from the mean rotational speed of 2000 rev min^{-1}

$$\Omega_2 - \Omega_1 = (0.1)2000\left(\frac{2\pi}{60}\right) = 20.9 \, \text{rad/s}$$

Recalling that

$$WD \approx I_f \Omega_0(\Omega_2 - \Omega_1)$$

it can be shown easily that the required moment of inertia of the flywheel is

$$I_f = 0.033 \, \text{kg m}^2$$

as required.

6.12 Impulse, impact and momentum

Introduction

Engineering systems involving impacts between different bodies can be difficult to solve using Newton's laws of motion. The reason for this is that the contact forces between the impacting bodies vary with time and are dependent on various quantities, including the geometry of the contact and material properties. For most practical impacts it is not necessary to have a detailed knowledge of the contact forces. Instead it is only necessary to specify the impulse of the force. In this section the impulse of the force is defined and used as the basis for analysing rigid body impact problems. Other concepts introduced include: coefficient of restitution, impulse torque and angular momentum. Some examples are presented to illustrate the application of these concepts, including a basic analysis of a clutch.

Impulse and momentum for linear motion

Figure 6.137 shows a particle of mass M moving with velocity V that is acted upon by an external force F.

Using Newton's second law, in its most general form, gives

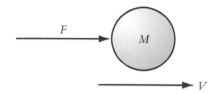

$$\longrightarrow^{+} \quad F = \frac{d(MV)}{dt}$$

Figure 6.137 **Particle of mass M moving with velocity V acted on by a force F**

Over a short time interval dt the above equation can be written in the following form

$$\longrightarrow^{+} \quad F\delta t = \delta(MV)$$

i.e. in the short time interval δt the quantity $F\delta t$ is equal to the change in momentum $\delta(MV)$. The quantity $F\delta t$ is the area under a plot of force against time, and is denoted by the symbol J. J is referred to as the impulse of the force (or simply 'impulse'), and has units of Ns. Figure 6.138(a) shows a plot of force against time in which the force F is constant over the interval dt. The impulse of this idealized force is depicted by the area J. Figure 6.138((b) shows a plot of a typical short duration impact force against time in which the force F varies during the interval δt. The actual shape depends on the material properties and the geometry of the contact. The impulse of this force is depicted similarly by the area J.

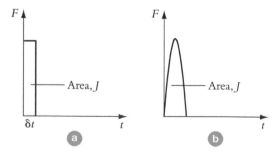

Figure 6.138 **Plots of force against time:
(a) constant force; (b) realistic impact force**

The impulse of the force can be written as

$$\longrightarrow^{+} \quad J = \delta(MV) = M(V' - V)$$

where V' and V are the final and initial velocities respectively, and the mass of the particle is constant. When applying this equation it is important to recall that as it was derived from Newton's second law and that the force and velocity were assumed to be positive in the left to right direction, the impulse must be positive in the left to right direction.

Machine dynamics

The above equation is expressed in words as follows

impulse of the force = change in linear momentum

This relationship holds no matter how long the force acts, but the above form is most useful for describing short duration forces, such as those that occur during impacts. For most rigid-body impact problems, the precise shape of the force is not important, and it is the area under the force time history that dictates what happens.

Although the previous result for the impulse was developed for a single particle, it is valid for a system of particles subjected to external and internal impulses. For external forces the impulse J is replaced by the resultant of the impulses. For internal forces within a system of particles, the impulses occur in equal and opposite pairs, yielding a zero net contribution to the resultant impulse. For this reason it is not necessary to consider internal impulses in calculations. A similar situation arises for zero net external impulsive forces (e.g. particle collisions). In this case collisions give rise to equal and opposite forces, and the impulses occur in equal and opposite pairs on the colliding bodies. In this case the resultant impulse is iven by:

$$\longrightarrow^+ \quad J = \sum_i J_i = 0.$$

In this situation it is clear from the definition of impulse that $\delta(mv) = 0$. This indicates that there is no change in the total momentum of the system. This is the Principle of Conservation of Momentum. The first worked example uses the definition of impulse to analyse a particle impact problem that can be analysed using the Principle of Conservation of Momentum.

Coefficient of restitution

When two bodies (e.g. particles) collide in such a way that the impulse is directed along the line joining their mass centres, it is observed to good approximation that their relative velocity along the line of mass centres after impact is in a fixed ratio to the relative velocity along that line before impact, and is of opposite sign. The ratio depends on the material of the bodies and the geometry of contact and is called the Coefficient of Restitution e. This empirical coefficient has a value between 0 (zero) and 1 (one) and is a measure of energy loss during the impact.

Figure 6.139 shows a system consisting of two impacting particles.

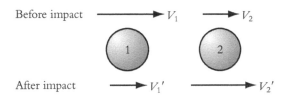

Figure 6.139 **Two-particle impact problem**

Before impact masses 1 and 2 have velocities V_1 and V_2 respectively, while after impact masses 1 and 2 have velocities V'_1 and V'_2 respectively

For this impact problem, the Coefficient of Restitution e is given by

$$e = -\frac{(V'_1 - V'_2)}{(V_1 - V_2)}$$

The case when $e = 1$ corresponds to the collision being perfectly elastic.

The case when $e = 0$ corresponds to the collision being perfectly inelastic, and implies that the relative velocity after impact is zero. This is achieved if the bodies 'stick together' and move with the same velocity after the impact.

Section 6.12 provides a worked example that uses the concept of Coefficient of Restitution to aid the analysis of particle velocities in a particle impact problem.

Impulse torques and angular momentum

Consider a particle of mass M under the action of an impulse J as shown in Figure 6.140

The definition of impulse implies that

$$\longrightarrow^{+} \quad J = \delta(MV)$$

Taking moments about O gives

$$+ \circlearrowright \quad rJ = r\delta(MV)$$

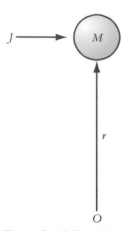

Figure 6.140 **Particle under the action of an impulse**

Note that the sense of the moments indicates that moment quantities are positive in the clockwise direction. The term on the left-hand side is the moment produced by the external impulse about O and is referred to as the impulse torque. The term on the right-hand side corresponds to the change in the moment of linear momentum about O.

Although the above result was derived for a particle it also applies to rigid bodies. Consider a rigid body rotating with angular velocity ω about a fixed axis O as shown in Figure 6.141.

In this figure m_i, represents a typical particle in the body. The particle is located at a distance r_i from O and is subjected to an impulse J_i acting perpendicular to the line joining O and m_i (an impulse acting parallel to this line would only affect the bearing reaction force at O).

From the previous result for a single particle

$$+ \circlearrowright \quad r_i J_i = r_i \delta(m_i V_i).$$

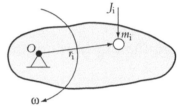

Figure 6.141 **Rigid body rotating about a fixed axis through O**

For a rigid body rotating with angular velocity ω about O, the velocity $V_i = r_i\omega$. Using this result and summing over all particles gives

$$\sum_i r_i J_i = \sum_i r_i \delta(m_i r_i \omega)$$

Rewriting gives

$$\sum_i r_i J_i = \sum_i m_i r_i^2 \delta(\omega)$$

Recalling that $I_O = \sum_i m_i r_i^2$ is the moment of inertia of the body about O (see 'Moment of inertia'), then

$$+ \circlearrowright \quad \sum_i r_i J_i = I_O(\omega' - \omega)$$

The term on the left-hand side is the resultant impulse torque. Referring to the quantity $I_O\omega$ as the angular momentum of the body, the term on the right-hand side is the change in angular momentum (i.e. $\delta(I_O\omega)$), i.e.

$$\text{impulse torque} = \text{change in angular momentum}$$

This result is directly analogous to the definition of the impulse of the force described in 'Impulse and momentum for linear motion' (i.e. impulse of the force is equal to the change in linear momentum). In a similar manner, the above equation can be used to analyse the influence of external and internal impulse torques, and for cases where internal impulse torques are applied or there is zero external impulse torques applied to a system, the change in angular momentum is zero (i.e. $\delta(I_O\omega) = 0$). This is the Principle of Conservation of Angular Momentum.

Worked Example 2 uses the definition of impulse torque to analyse a particle-rod impact problem, while Example 3 provides an example that uses the Principle of Conservation of Angular Momentum to analyse a system in pure rotation.

Example 1

Figure 6.142 shows two masses m_1 and m_2 moving along the same line of action with velocities V_1 and V_2 respectively, where $V_1 > V_2$.

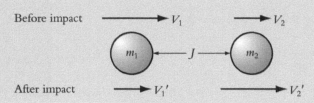

Figure 6.142 **Particle impact problem**

When the bodies impact, equal and opposite impulses act at the point of collision. After the collision has taken place masses m_1 and m_2 have velocities V_1' and V_2' respectively

Determine expressions for velocities V_1' and V_2'.

The definition of the impulse of the force can be applied to each mass to obtain expressions involving the velocities and impulse.

For mass m_1

$$\xrightarrow{\hspace{1cm}}{}^{+} \quad -J = m_1(V_1' - V_1)$$

For mass m_2

$$\xrightarrow{\hspace{1cm}}{}^{+} \quad J = m_2(V_2' - V_2)$$

When forming these equations it is important to take into account the direction of the impulse.

Eliminating the impulse force between these equations by adding gives

$$m_1 V_1' + m_2 V_2' = m_1 V_1 + m_2 V_2$$

This equation could have been written directly by using the Principle of Conservation of Momentum (i.e. applying $\delta(MV) = 0$). This is possible because no external impulse is applied to the system, and the impulses that are present act as equal and opposite pairs.

The above equation has two unknowns (V_1' and V_2') and can only be used to determine an expression for V_1' in terms of V_2' and an expression for V_2' in terms of V_1'. To determine explicit expressions for V_1' and V_2' another equation is needed, and this equation is provided here by the definition of the Coefficient of Restitution. Applying the definition of the Coefficient of Restitution gives

$$e = \frac{(V_1' - V_2')}{(V_1 - V_2)}$$

The above two equations can be combined and solved for V_1' and V_2' to give

$$V_1' = \frac{[V_1(1 + e)m_1 + V_2(m_2 - em_1)]}{m_1 + m_2}$$

$$V_2' = \frac{[V_1(1 + e)m_2 + V_1(m_1 - em_2)]}{m_1 + m_2}$$

as required.

In this example momentum is conserved during impact but, in general, energy is not conserved. The loss in kinetic energy as a result of the impact is given by

$$\text{KE LOSS} = \tfrac{1}{2}m_1(V_1^2 + V_2^2) - \tfrac{1}{2}m_2(V_1'^2 + V_2'^2)$$

Using the expressions for V_1' and V_2' given earlier it can be shown that

$$\text{KE LOSS} = \frac{m_1 m_2(1 - e^2)(V_1 - V_2)^2}{2(m_1 + m_2)}$$

This equation is useful because it indicates that the kinetic energy loss is zero when $e = 1$. This situation corresponds to a perfectly elastic collision.

The equation also indicates that maximum kinetic energy loss occurs when $e = 0$. This last case corresponds to a perfectly inelastic collision in which the two masses combine so that they both have the same velocity after impact. Using the above expressions it can be shown that in this case

$$V_1' = V_2' = \frac{(m_1 V_1 + m_2 V_2)}{(m_1 + m_2)}$$

Example 2

A thin uniform rod of mass M and length L is suspended vertically from a frictionless pivot at O and hangs initially at rest (i.e. angular velocity $\omega = 0$), as shown in Figure 6.143(a). A particle of mass m moving with horizontal velocity v impacts the lower end of the rod. After the impact the rod rotates with angular velocity ω' and the particle moves with velocity ω', as shown in Figure 6.143(c).

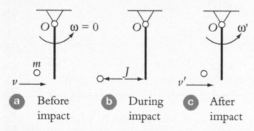

| | a | Before impact | b | During impact | c | After impact |

Figure 6.143 **Particle impacting a suspended uniform rod**

Assuming that there is no loss of energy during impact, determine an expression for the angular velocity of the rod ω' immediately after impact.

During impact, impulses J act on the particle and the rod as shown in Figure 6.143(b).

The definition of impulse can be applied to the particle to obtain an expression involving the velocity of the particle and the impulse. i.e.

$$\longrightarrow^+ \quad -J = m_1(v' - v)$$

The definition of impulse torque can be applied to the rod to obtain an expression involving the angular velocity of the rod and the impulse. i.e.

$$+\circlearrowleft O \quad JL = I_O(\omega' - 0)$$

Eliminating J between the last two equations gives

$$v' = v - \frac{(I_0\omega')}{(mL)}$$

If energy is conserved, then the kinetic energy before impact is equal to the kinetic energy (immediately) after impact, such that

$$\tfrac{1}{2}mv^2 = \tfrac{1}{2}mv'^2 + \tfrac{1}{2}I_O\omega'^2$$

Noting that the moment of inertia for a rod of mass M and length L rotating about its end is $I_O = \tfrac{1}{3}ML^2$ (see 'Thin uniform rod'), and using the last two equations it can be shown that the angular velocity immediately after impact is

$$\omega' = \frac{6mv}{L(3m + M)}$$

as required.

Example 3

Consider the simple motor-clutch-load system shown in Figure 6.144.

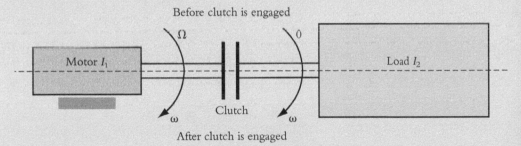

Figure 6.144 **Simple motor-clutch-load system during clutch engagement**

Before the clutch is engaged, the motor, which has moment of inertia I_1, rotates with angular speed Ω, while the load, which has moment of inertia I_2, is stationary. After the clutch is engaged, the combined motor and load system rotates with angular speed ω.

Determine the common shaft speed after the clutch is engaged and the loss in kinetic energy during clutch engagement.

To simplify the solution to this problem is necessary to note that no external impulse torque is applied to the motor-clutch-load system, and that the Principle of Conservation of Angular Momentum applies. (It is also interesting to note that because the motor shaft and load shaft have the same velocity after engagement, then the effective coefficient of restitution is zero.)

Applying the Principle of Conservation of Angular Momentum (i.e. $\delta(I_0\omega) = 0$), such that the angular momentum before the clutch is engaged is equal to the angular momentum after the clutch is engaged gives

$$I_1\Omega = (I_1 + I_2)\omega$$

Rearranging gives

$$\omega = \frac{I_1 \Omega}{(I_1 + I_2)}$$

as required.

The loss in kinetic energy is given by

$$\text{KE LOSS} = \frac{1}{2} I_1 \Omega^2 - \frac{1}{2}(I_1 + I_2)\left(\frac{I_1 \Omega}{I_1 + I_2}\right)^2$$

Rearranging this equation it can be shown that

$$\text{KE LOSS} = \frac{1}{2}\left(\frac{I_1 I_2}{I_1 + I_2}\right)\Omega^2$$

as required.

Questions

Solid mechanics

1. (a) A 100 kg block rests on an inclined plane. Draw the free body diagram for the block and determine the maximum angle, θ, to which the incline may be raised, before the block slips down the plane. The coefficient of static friction may be assumed to be 0.3.

   (b) The inclined plane is now set at an angle of 15°, and a force P is applied to the block parallel to the plane in the upwards direction. Draw the free body diagram when the applied force P is such that the block is just about to move up the plane. Determine the magnitude of the force P required to cause this upward movement.

   (c) If the inclined plane is now set at the angle calculated in part (a):

   (i) comment on the magnitude of P which will prevent the block slipping down the plane; and

   (ii) determine the magnitude of P which will just cause the block to move up the plane.

2. A pin-jointed structure, ABCD, comprises, in part, a horizontal member, BCD, 1 metre long, pin-jointed to a wall at B. The end D of this member carries a vertical load, W = 5 kN. BCD is supported by a further member AC, length 1.25 metres, pin-jointed at the wall at A, which is above B. AC is pin-jointed to BCD at C, midway between B and D. Determine the magnitudes and directions of the reaction forces at A, B and C.

3. (a) A stepped aluminium bar comprises two parts. The left-hand part is 100 mm long and 20 mm diameter and the right hand part is 200 mm long and 30 mm diameter. Both parts are joined rigidly end on. The bar is subjected to a uniaxial load of 25 kN at either end. The Young's modulus for aluminium, $E_{aluminium}$ = 70 GPa. Determine the total extension of the bar.

   (b) If the two ends of the bar are constrained against movement, and the load of 25 kN is applied (to the right) at the shoulder where the two parts of the bar meet, determine the movement of the shoulder i.e. the point at which the 25 kN load is applied.

4. A compound bar, 400 mm in length, comprises a left-hand part made from steel and a right-hand part made from aluminium alloy, both parts being firmly attached at the centre of the bar. The cross-section of the steel half is a solid square section and has dimensions 20 mm × 20 mm. The aluminium is a similar square cross-section with dimensions 30 mm × 30 mm. The bar is fixed at either end between two rigid supports.

   The bar is subjected to a temperature rise of 50 °C and, simultaneously, the right support is moved horizontally to its right by a displacement of 0.4 mm with respect to the left support.

   Determine the stresses in the steel and aluminium parts of the bar.

   [E_{steel} = 210 GPa; $E_{aluminium}$ = 70 GPa; α_{steel} = 11.10^{-6}/° C; $\alpha_{aluminium}$ = 23.10^{-6}/° C]

5. A plane element is subjected to the following stresses:

   σ_x = 40 MPa
   σ_y = − 20 MPa
   τ_{xy} = − 30 MPa

   (a) Sketch the Mohr's Circle for the element.

   (b) Determine the following:

   (i) The principal stresses;

   (ii) The maximum positive shear stress;

   (iii) The stresses on a plane at a 45° anti-clockwise angle to the x-plane.

6. A closed-end, thin-walled, cylindrical pressure vessel is pressurised to 2 MPa. Before pressurisation, the length of the vessel is 2 metres, the mean diameter, 1 metre, and the wall thickness, 10 mm.

   Determine:

   (i) the axial and hoop stresses

(ii) the change in length

(iii) the change in mean diameter

(iv) the maximum shear stress in the wall

[Assume E = 210 GPa and ν = 0.29]

7. A beam, 2 metres in length, is simply supported at both ends. The beam is subjected to a uniformly distributed load of 1 kNm^{-1} over its full span, a point load of 2 kN, 0.5 metres from the left hand support and a clockwise applied moment of 2 kNm, 0.5 metres from the right hand support.

    (a) Draw the shear force diagram for the beam.

    (b) Draw the bending moment diagram for the beam.

    (c) Determine the magnitude and position of the maximum bending moment.

8. An I-section beam has a top flange measuring 200 mm (wide) × 50 mm (deep) and a bottom flange measuring 100 mm (wide) × 50 mm (deep). The flanges are connected together by a vertical web measuring 50 mm (wide) × 200 mm deep. The whole cross-section is symmetrical about a vertical centre line.

    (a) Determine:
        (i) the vertical position of the neutral axis;
        (ii) the second moment of area about the neutral axis.

    (b) A cantilever beam, 4 m in length, carries a point load of 50 kN at the free end. Given that the beam has an I-section as above, determine the positions and magnitudes of the maximum tensile and compressive stresses.

9. An inverted T-section comprises a lower flange, 30 mm (wide) × 10 mm (deep) and a vertical web, 10 mm (wide) × 40 mm (deep), on top of the flange. The section is symmetric about a vertical centre line.

    (a) Determine:
        (i) The vertical position of the neutral axis;
        (ii) The second moment of area about the neutral axis.

    (b) A positive bending moment of 1000 Nm is applied to the section. Determine the maximum tensile and compressive stresses in the section.

10. A 1-metre-long hollow shaft is to transmit a torque of 400 Nm. The outer diameter of the shaft is 25 mm. The relative rotation of the two ends of the shaft, i.e. twist angle, is to be limited to 0.375 radians. The maximum shear stress in the shaft must not exceed the allowable shear stress for the material chosen (see data table below). The shaft is to be made from either titanium alloy or aluminium alloy.

    (a) Determine the maximum inner diameter **to the nearest mm** for both a titanium and an aluminium shaft, such that both the twist angle and shear stress criteria are satisfied.

    (b) For the calculated inner diameters in part (a), which material results in the lighter shaft?

Material	G (GPa)	$T_{allowable}$ (MPa)	Density (kg/m^3)
Titanium Alloy	36	450	4400
Aluminium Alloy	28	150	2800

Materials and processing

1. (a) Define what is meant by the fracture toughness of a material.

    (b) What is the critical crack size for a material with a fracture toughness of 54 MNm$^{-3/2}$ for an applied stress of 90 MNm^{-2}?

    (c) A material contains a 0.1 mm long crack which propagates under fatigue loading according to the relationship:

$$\frac{da}{dN} = A(\Delta K)^4$$

where A = 4×10^{-13}. If the mean stress is 90 MNm^{-2} and the material is subjected to an alternating stress, $\Delta\sigma$, of 180 MNm^{-2}. Given that $\Delta K = \Delta\sigma(\pi a)^{1/2}$ calculate the number of cycles to failure.

    (d) What determines the fatigue life of the component?

2. The phase diagram of the system A-B is shown in Figure Q2. Three single phase fields are shown, namely the solid phases α and β and the liquid phase, L. Using the phase diagram, answer the following questions concerning an alloy of composition 75wt%A-25wt%B (normally written as A-25wt%B).

    (a) What phases are present in the three unlabelled regions?

    (b) What is the temperature at which the first solid phase is formed on cooling from the liquid phase?

    (c) Identify the compositions of the two phases present when the alloy is held at 490°C. Identify also the proportions of those two phases.

(d) Identify the compositions and proportions of the two phases present *just* above 290°C.

(e) Identify the compositions and proportions of the two phases present *just* below 290 °C.

(f) Comparing you answers for parts (d) and (e) identify the wt% of a in the eutectic mixture.

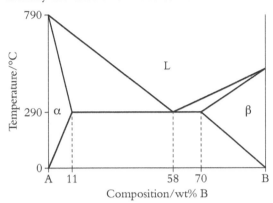

Figure Q2 Phase diagram of the A-B system

3. This question concerns two types of paper clip. Find two paper clips (one of each of the two types illustrated below). We will refer to the paper clip types as Type A and Type B (as marked in Figure Q3).

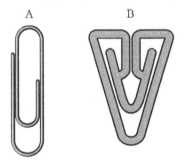

**Figure Q3 Two types of paper clip
Type A (left) and Type B (right)**

The operation of a paper clip

paper clips are small objects that serve to hold several sheets of paper together as a single unit. In effect, a paper clip consists of two surfaces that are pressed against one another by the elasticity of the material from which the paper clip is made. As you distort the paper clip away from its equilibrium shape, by spreading the two surfaces apart, it experiences restoring forces. These forces tend to return the paper clip to its equilibrium shape and push the two surfaces together. Because the paper clip behaves like a spring, the restoring forces are proportional to the distance separating the two surfaces. When several sheets of paper are placed between the two surfaces, the restoring forces on the metal surfaces cause them to exert inward, compressing forces on the paper sheets. The force between each sheet and its neighbours

means that high forces have to be applied to each sheet to cause it to move relative to its neighbours to overcome the frictional forces. As such, the sheets cannot slide easily relative to each other, and the sheets of paper are therefore held together as a stack.

(a) Identify the basic material types from which Type A and Type B paper clips have been made.

(b) The starting material for type A paper clips is continuous wire coiled onto a drum. Describe a manufacturing process by which Type A paper clips may be made. Ensure that all of the main stages in the process are included.

(c) Describe and explain differences in the mechanical properties (and structure) of a Type A paper clip in a straight section and a curved section.

(d) Type B paper clips are made by injection moulding. Make a sketch of the paper clip and mark on it the position of the gate, where the ejection pin contacts the moulding and positions where weld lines are likely to form. Justify your answers. (You will need to examine your paper clip).

(e) Keeping the thickness of all the sections of a Type B paper clip small results in a high rate of production being achievable. Explain why this is so.

(f) It was observed that when 40 sheets of paper were clamped by the paper clips and held for many weeks on the window ledge of a hot office, the paper was loose in the Type B clip but remained tightly clamped in the Type A clip. When the paper was removed and the clips were examined, no noticeable change was observed for the Type A clip but the Type B clip was permanently distorted. How can you account for this behaviour and why was there such a difference in behaviour between the two types of paper clip?

(g) Now you can test the clips to destruction. What did you need to do to get the clip to fail and why? Describe the failure mechanisms of both paper clip types.

(h) Both types of paper clip are available in the market, indicating that both have certain advantages over the other. Make a list of features which you would highlight in a marketing document attempting to increase the market share of the Type B clip.

4. Con-rods are being designed for a rally-car. The con-rods convert the linear motion of the pistons into a rotary motion. However, as they rapidly accelerate and decelerate with each stroke, there is a need to minimise their mass to improve performance.

A novel design is being considered. The length of the rod, *l,* is fixed. The main part of the rod consists of a solid forged cylinder of radius *r.* The first constraint is that the con-rods must not buckle.

The critical load for Euler Buckling, F_{crit}, for this geometry is given by:

$$F_{crit} = \frac{4\pi^2 EI}{l^2}$$

where E is the Young's Modulus of the material and I is the second moment of area of the component, which for a cylinder is $\pi d^4/64$.

To cope with the design constraints, the diameter of the piston, d, can be varied freely for different materials (i.e. d is the free variable).

(a) Write down an equation for the mass of the con-rod in terms of the dimensions given in the question and other relevant material properties. Formulate an equation that describes the critical load in terms of the rod dimensions given in the question and other relevant material properties.

(b) Write down an equation for the mass of the con-rod with the free variable eliminated.

(c) Three candidate materials for the rod are being considered; the materials and some relevant mechanical properties are listed in Table Q4. The design is for a maximum axial compressive load on the rod of 20 kN with a con-rod length of 250 mm. Calculate the mass of the rod for each material. Thus identify the optimum material given this constraint.

Material	ρ / kg m^{-3}	E / GPa	σ_e / MPa
Steel	7800	210	590
Aluminium alloy	2700	70	180
Titanium alloy	4400	115	530

Table Q4 **Mechanical properties of candidate materials**

In addition to resisting bucking, the columns must not fail by fatigue (i.e. the maximum stress must be lower than the endurance limit, s$_e$).

(d) Sketch a typical S–N curve and explain clearly what is meant by the endurance (or fatigue) limit. Why it is important to design so that the stress on the component is below this limit?

(e) Formulate an equation that describes the endurance limit in terms of the rod dimensions given in the question and other relevant material properties. Write down an equation for the mass of the con-rod with the free variable eliminated.

(f) Considering the candidate materials in Table Q4 and the same loading and dimensions as in (c), calculate the mass of the rod for each material to satisfy this constraint and identify the optimum material.

(g) Considering both constraints, explain which material is optimum. What is the mass of the rod made from this material?

5. (a) Creep deformation is 'time-dependent'. Explain what this means and sketch how creep strain varies with time.

(b) Failures due to creep can be a result of either creep strains causing seizure of components or when the creep strain exceeds the creep ductility. Explain fully these two failure methods, identifying what causes failure and how the creep life might be predicted in each case.

(c) The creep strain rate during diffusional flow (at 873K) can be described by the equation below where C is a constant, α is the stress (in MPa) and d is the grain size of the material (in m). If the maximum permissible strain before seizure is 0.1%, the applied stress is 30MPa and C is 2×10^{-18} m^{2}s^{-1}(MPa)$^{-1}$, calculate the time taken for a material with a grain size of 1×10^{-4}m to achieve this strain.

$$\dot{\varepsilon} = \frac{C\sigma}{d^2}$$

(d) Figure Q5 shows a creep rupture diagram. The middle curve is the rupture behaviour at 873 K. Comment on how such a plot of applied stress versus time to failure might have been constructed. Identify the creep life for the conditions detailed in (c) and comment on the likely mode of failure when compared with (c).

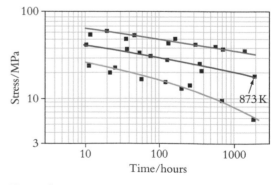

Figure Q5

(e) How might the creep life of a material undergoing creep by diffusional flow be improved through alloy design or processing?

Questions

6. Figure Q6 is a schematic diagram of the iron–carbon phase diagram (**drawn to scale**). Pure iron is represented by 0 wt% carbon. You may assume that all phase boundaries are straight lines.

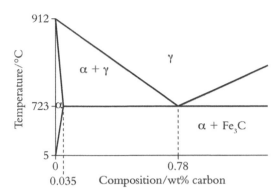

Figure Q6 Schematic diagram of the iron–carbon phase diagram (drawn to scale)

A 0.45 wt% carbon steel is taken and heated to 900 °C (i.e. into the austenite (γ) region) and then **slowly** cooled.

(a) At what temperature will the first ferrite (α) phase begin to appear?

(b) Calculate the compositions and proportions of the two phases present at 761°C.

(c) Calculate the compositions and proportions of the two phases present at $(723+\delta)$°C (where δ is a *very small* positive number). What is the special name of the composition of the austenite (γ) phase at this point?

(d) Describe and explain the microstructural changes on cooling of the steel from $(723+\delta)$°C to $(723-\delta)$°C. As part of your answer, describe the microstructural constituents of the steel at $(723-\delta)$°C.

The same steel was heated again to 900°C (i.e. into the austenite (γ) region) and then **rapidly** cooled (for example, by plunging a sample into a bucket of cold water).

(e) Identify the microstructure of the steel following this heat treatment, and the mechanical properties that would be expected of such a material.

(f) A further heat treatment of such a structure at 600 °C can be employed to alter the mechanical properties and make this into a useful engineering material. Describe the processes occurring, and the changes in properties that would be expected.

7. (a) Describe the process of 'work hardening' (sometimes referred to as 'work strengthening') during the deformation processing of metallic components. On a dislocation level, explain the mechanisms by which work hardening occurs.

(b) Two processes are being considered for the production of a component from an aluminium alloy, namely machining (turning and milling in this case) and cold closed-die forging. Describe the two processes being considered. Describe also the differences in component mechanical properties which would be expected to result from the two processing routes.

When considering the costs of competing processes, the cost per component, C, can be formulated as follows:

$$C = \left[\frac{C_m}{1-f}\right] + \left[\frac{C_t}{n}\right] + \frac{\dot{C}_{oh}}{\dot{n}}$$

where C_m is the cost of the material per component, f is the scrap fraction (the proportion of the starting material which does not make its way into the final component), C_t is the dedicated tool cost (cost of tooling bought specifically for the batch of components being made), n is the batch size (number of components required), $\dot{C}_{oh}$ is the overhead rate (cost per unit time of ruining the process) and $\dot{n}$ is the production rate (number of components produced per unit time).

(c) Approximate process and cost data for the two processes being considered are listed in Table Q7. Using the assumptions outlined in part (c) of this question, identify the minimum batch size required to make cold forging a cheaper process than machining per component manufactured.

(d) Without calculation, identify whether the batch size required to make cold forging cheaper than machining (in terms of cost per component produced) would increase, decrease or stay the same compared to your answer in part (d) in the following cases:

    (i) the material cost doubles in both cases (all other things stay the same);

    (ii) the overhead rate doubles in both cases (all other things stay the same);

    (iii) the production rate doubles in both cases (all other things stay the same).

	Machine	Cold forge
Material cost	1	1
Material scrap fraction in processing	0.2	0.05
Overhead rate, (hr^{-1})	150	150
Tool cost, C_t	30	3000
Production rate, n (hr^{-1})	10	60

Table Q7 Costs and process data associated with component production by two candidate processes. (All costs normalised to material cost)

Fluids

In all questions assume:

The density of water is $1000\,\text{kg/m}^3$
Acceleration due to gravity is $9.81\,\text{m/s}^2$
Atmospheric pressure is 1.01 bar
Specific gas constant for air is 287.1 J/kgK
The dynamic viscosity of water is 0.001 Pas

1. A rectangular tank is separated into two compartments by a vertical wall as shown in Figure Q1. One side of the tank is filled with water to a depth of 1.25m. The other side of the tank is filled with oil to a depth of 1.65m. The tank is 2 m wide.

    What is the magnitude of the horizontal force that must be applied to the wall and at what height above the base of the tank must it act to ensure that the wall is in equilibrium?

    The density of water is $1000\,\text{kg/m}^3$
    The density of the oil is $750\,\text{kg/m}^3$

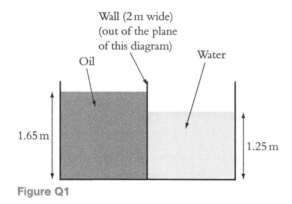

Figure Q1

2. A horizontal venturimeter has a throat diameter of 50 mm and is placed in a pipe of 110 mm diameter. The discharge coefficient for the venturimeter is 0.985. Air at a temperature of 35 °C and a pressure of 1.35 bar flows through the pipe. An inclined tube manometer is connected between a pressure tapping 200 mm upstream of the throat and a pressure tapping at the throat. The manometer is filled with a fluid of density $850\,\text{kg/m}^3$ and is inclined at an angle of 15° to the horizontal. The reservoir on the manometer has a cross-sectional area of $350\,\text{mm}^2$ and the inclined tube has a cross-sectional area of $12\,\text{mm}^2$.

    (a) When air flows through the pipe, the water level in the manometer moves by 46 mm along the inclined tube. Calculate the mean air velocity at the throat of the venturimeter. Assume that air density does not vary through the venturimeter.

    (b) Calculate the air mass flow rate through the venturimeter.

    (c) If the venturimeter were mounted vertically explain how the reading on the manometer would change, assuming that the air flow rate does not change.

    The specific gas constant for air is 287.1 J/kgK.

3. A hot air balloon has a volume of $2500\,\text{m}^3$ when fully inflated. The maximum temperature that the air can be heated to is 100°C. The mass of the balloon, basket, fuel and burner is 300 kg, and three people each of 75 kg in mass travel in the balloon.

    (a) To what temperature must the air be heated to lift the balloon off the ground at sea level, where the ambient temperature is 17°C and the air pressure is 1.01 bar?

    (b) The density of the atmosphere ρ decreases with height according to the expression:

    $$\rho/\rho_0 = 1 - 8 \times 10^{-5}h$$

    Where: ρ_0 is the density of the atmosphere at sea level.
    h is the altitude of the balloon in metres above sea level.

    Atmospheric air pressure also decreases with height according to the expression:

    $$p/p_0 = 1 - 9 \times 10^{-5}h$$

    Where: p_0 is the atmospheric air pressure at sea level.
    h is the altitude of the balloon in metres above sea level.

    What is the maximum height that the balloon can reach?

    State any assumptions that you make.

    For air the specific gas constant is 287.1 J/kgK

4. A fountain is to be installed in the middle of a lake as shown in Figure Q4. It is fed from a reservoir with a surface 75 m above the level of the lake. The pipe feeding the fountain is 1.5 km long and 100 mm in diameter. The nozzle of the fountain has a diameter of 25 mm and points vertically upwards. The roughness of the supply pipe is 0.25 mm. The entry loss factor at the inlet to the supply pipe from the reservoir is 0.5. Assume no significant losses in the pipe or nozzle other than pipe friction and the inlet loss.

    (a) Calculate the mass flow rate of water from the nozzle of the fountain.

    (b) To what height will the water be projected from the fountain nozzle above the surface of the lake?

Questions

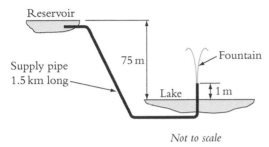

Figure Q4

5. Water of density 1000 kg/m^3 flows through the pipe bend illustrated in Figure Q5 at a rate of $0.21 \text{ m}^3/\text{s}$. The diameter of the pipe at the inlet is 200 mm and the outlet, which is 0.5 m higher than the inlet, has a diameter of 100 mm. Frictional losses in the bend can be accounted for by a loss factor of 0.3 applied to the *inlet* velocity head. The static pressure at the bend inlet is 847 kPa gauge and the volume of water in the bend is 0.9 m^3.

(a) Calculate the mean inlet and outlet velocities for the bend.

(b) Calculate the pressure at the bend outlet.

(c) Calculate the horizontal and vertical components of the force acting on the bend and resolve to find the magnitude and direction of the resultant force.

In your calculations neglect the mass of the pipework in the bend.

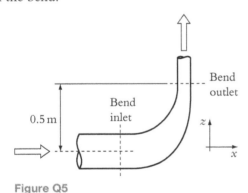

Figure Q5

Thermodynamics

1. What is the absolute pressure, in SI units, of a fluid at a gauge pressure of 1.5 bar if atmospheric pressure is 1.01 bar?

2. Convert $-25°C$ to a temperature in degrees Kelvin.

3. Calculate the following:

(a) the kinetic energy of a body which has a mass of 5 kg and a velocity of 10m/s;

(b) the change in potential energy of a mass of 5 kg when it is raised a height of 3 m;

(c) the strain energy stored in a spring compressed by 18 mm from its free length if the spring constant is 1.50 MN/m;

(d) the increase in internal energy of a gas in a closed system during a process in which -100 J of heat transfer and 400 J of work transfer take place.

4. The piston in Figure 4.12 on page 221 has no mass and is free to move. The point force F acting on the piston is applied gently so that the piston moves slowly to the left. As the gas in the cylinder is compressed, the product of pressure and system volume, pV, remains constant. The piston area is 0.02 m^2 and when x is x_1 the system volume is 0.006 m^3. Atmospheric pressure is 1 bar. Calculate the distance $(x_2 - x_1)$ moved by the piston as F increases from 0 to 70N.

5. Calculate the work input to the closed system undergoing the cycle shown in Figure 4.13 on page 222 if the pressure during process (2) to (3) is raised from 1000 kPa to 1400 kPa.

6. (a) Calculate the efficiency of a reversible heat engine operating between a hot reservoir at 900 K and a cold reservoir at 500 K.

(b) The temperature of one of the heat reservoirs can be changed by 100 degrees kelvin up or down. What is the highest efficiency that can be achieved by making this temperature change?

7. A perfect gas at a pressure of 58 bar and a temperature of 450 K has a density of 50 kg/m^3. The ratio of specific heats γ is 1.48.

(a) Calculate the values of molar mass $\tilde{m}$, specific heat at constant pressure c_p and specific heat at constant volume c_v.

(b) Calculate the change in specific entropy of the gas if the pressure is raised to 100 bar and the temperature is lowered to 400 K.

8. Calculate the dryness fraction x of saturated steam if it has a specific enthalpy of 800 kJ/kg and at the same pressure and temperature, the specific enthalpy of saturated liquid is 505 kJ/kg and the latent heat of vaporization is 2202 kJ/kg.

9. A perfect gas with a γ value of 1.4 undergoes an expansion process from a pressure of 600 kPa. The ratio of specific volumes v_2/v_1 is 3.0. Calculate the pressure p_2 at the end of the process if this is:

(a) polytropic with an index n of 1.6;

(b) isothermal;

(c) isentropic.

10. A perfect gas with a γ value of 1.5 can be compressed in either a non-flow or a steady flow process between the same initial and final states. Neglecting any change in kinetic and potential energy, what is the ratio of the specific work done is the two cases if the process is:

   (a) polytropic with an index n of 1.4;

   (b) isothermal;

   (c) isentropic?

11. Calculate the efficiency of the following ideal cycles when undergone by a perfect gas with a γ value of 1.4:

   (a) a Stirling cycle operating between a hot reservoir at 600 K and a cold reservoir at 300 K;

   (b) a Brayton cycle with a pressure ratio of 8;

   (c) an Otto cycle with a compression ratio of 8;

   (d) a Diesel cycle with a compression ratio of 12 and a cut-off ratio of 2.

12. A perfectly insulated, rigid tank with a volume of $0.2\,\mathrm{m}^3$ contains a perfect gas which has a molar mass of 18 kg/mol and a ratio of specific heats of 1.45. Initially the pressure and temperature in the tank are 9 bar and 320 K respectively. A fan inside the tank is spun at 3600 rev/min for 20 seconds. The torque required to turn the fan is 30 Nm. Calculate the following:

   (a) The R, c_p and c_v values of the gas and the mass of gas in the tank.

   (b) The work input (given by torque $\times$ angular rotation in radians) to the gas from the fan.

   (c) The final temperature of the gas. (Explain briefly why the temperature continues to rise for a short time after the fan has stopped rotating.)

   (d) The increase in entropy of the gas.

13. A closed system containing argon undergoes a reversible isothermal process from an initial state (1) where $p_1 = 50$ bar, $V_1 = 0.03\,\mathrm{m}^3$ and $T_1 = 450$ K to state (2). The work done during the process is -100 kJ. The system is then heated reversibly at constant volume to final state (3). The total heat transferred during the two processes is 170 kJ. Treat argon as a perfect gas with $c_p = 520$ J/kgK and molar mass $= 40$ kg/kmol. Calculate the following:

   (a) The mass of argon in the system.

   (b) The heat transferred during the constant volume process (2)–(3).

(c) The final temperature, T_3.

(d) Sketch the processes on a pressure–volume diagram.

14. A closed system containing steam undergoes a reversible constant pressure process during which 400 kJ/kg of heat transfer takes place. Initially the steam has a dryness fraction, x, of 1.0 and a temperature of 357°C. Using the tables, and using linear interpolation where necessary, determine:

   (a) The specific enthalpy, temperature and specific internal energy of the steam at the end of the process.

   (b) The specific work transfer.

   (c) State whether the process is an expansion or compression process and give two reasons to support your choice.

15. Air flows through an open system at the steady mass flow rate of 5 kg/s. At inlet the air velocity is negligible, the pressure is 1 bar and the temperature is 15°C. The air flow is compressed isentropically and leaves the system at a pressure of 5 bar through a pipe with an internal diameter of 80 mm. Calculate:

   (a) the temperature of the air as it leaves the system.

   (b) the velocity in the exit pipe.

   (c) the power input to the air.

   (d) Explain why the power input is not given by

$$\dot{W} = \dot{m} \int_1^2 v \, dp$$

   Assume for air, $R = 287$ J/kgK and $\gamma = 1.4$.

16. (a) Determine the rate of heat rejection from a reversible heat engine operating between a hot reservoir at 900 K and a cold reservoir at 400 K if the engine produces a power output of 40 kW.

   (b) A pipe with an inside diameter of 80 mm, a wall thickness of 30 mm and a thermal conductivity of 0.09 W/mK is lagged with a 20 mm thick layer of lagging with a thermal conductivity of 0.02 W/mK. The ambient temperature is 24°C and the inner surface of the pipe wall is maintained at 180°C. The heat transfer coefficient at the outer surface of the lagging is 18 W/m²K. Calculate:

   (i) the rate of heat transfer per unit length of pipe;

   (ii) the temperature at the inner surface of the lagging.

Electronics and electronic systems

1. Calculate the resistance of a copper wire 50 m long and 2.5 mm diameter at 40 °C.

2. For the circuit shown in Figure Q2 calculate flowing through the 5W the current resistor.

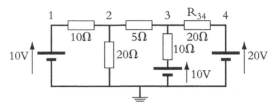

Figure Q2

3. A strain guage bridge comprises two 150 Ω resistors, one active gauge and one unstrained gauge for temperature compensation. The two gauges have unstrained resistances of 150 Ω and a strain gauge factor of 2.5. The bridge supply voltage is 5 V. Calculate the strain when the voltmeter reading is 2 mV.

4. A toroid is made from steel with the following B vs H characteristic.

B (T)	1.85	1.89	1.91	1.93	1.95	1.97	1.99
H (A/m)	500	1000	2000	3000	4000	5000	6000

The toroid has a mean diameter of 8 cm and a cross-sectional area of 1 cm². It is wrapped with a coil of 1500 turns through which a current of 0.7 A flows.

Calculate the magnetic field strength, the flux density, the flux and the inductance.

5. The toroid shown in Figure Q5 is made from steel with the following B versus H characteristic.

B (T)	1.90	1.94	1.96	1.98	2.00	2.02	2.04
H (A/m)	500	1000	2000	3000	4000	5000	6000

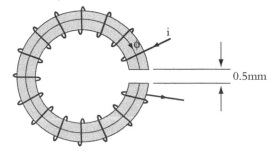

Figure Q5

The toroid has a mean diameter of 5 cm and a cross-sectional area of 2 cm². The radial cut is 0.5 mm wide. The toroid is wrapped with a coil of 4000 turns

through which a current flows producing a magnetic flux of 0.4 mWb in the steel core. Calculate the flux density, the magnetic field strength in the steel, the mmf drop in the steel, the magnetic field strength in the air gap, the mmf drop across the air gap and the current flowing in the coil.

6. Two C-shaped steel cores are made of silicon steel with the B vs H characteristic shown in Figure 5.12 on page 300. The two cores are clamped together with 1mm thick non-magnetic spacers in each limb as shown in Figure Q6. The magnetic flux path length in each core is 50 cm and the cross sectional area is 10 cm². A coil with 4000 turns is wound around the cores. Calculate the current necessary to produce a flux density of 1.9 T and the coil inductance.

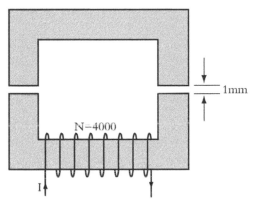

Figure Q6

7. A 50 V dc supply is connected in series with a 100 mH inductor, a 50 Ω resistor and an open switch. Calculate the current 0.5 ms, 1 ms, 2 ms and 5 ms after the switch is closed.

8. 2 nF and 8 nF capacitors are connected in parallel and the combination connected in series with a 5 nF capacitor. Calculate the equivalent circuit capacitance.

9. A 10 V battery is connected in series with a 1 μF capacitor, a 100 kΩ resistor and an open switch. The capacitor is initially uncharged. Calculate the voltage across the capacitor 50 ms, 100 ms, 200 ms and 500 ms after the switch is closed.

10. A 50 V 20 kHz single-phase ac supply feeds a 120 Ω resistor and a 1 mH inductance connected in series. Calculate the inductive reactance, the circuit impedance, the current magnitude and phase angle and the voltage across the inductance.

11. A 10 V 1000 Hz supply feeds a 3000 Ω resistor in series with a 39.8 nF capacitor. Calculate the capacitive reactance, the circuit impedance, the current magnitude and phase angle and the voltage across the capacitor.

12. A 240 V 10 kHz single-phase supply feeds a 50 Ω resistor, a 0.25 mH inductor and a 0.4μF capacitor connected in series. Calculate the inductive reactance, the capacitive reactance, the circuit impedance, the current magnitude and phase angle, the voltage across the inductance and the voltage across the capacitor.

13. A 1 V single phase supply feeds a 4 mH inductor, 25 Ω resistor and 15.83 nF capacitor connected in series. Calculate the resonant frequency. The supply frequency is set at the resonant frequency. Calculate the current magnitude and phase angle, and the voltages across the resistor, inductor and capacitor.

14. A 5 V 15 kHz supply feeds an ac bridge like that shown in Figure 5.39 on page 338. R_1 is 1 MΩ and R_2 is 5 MΩ. C_T is a capacitance transducer of unknown value. C_1 is set at 100 pF to balance the bridge. Calculate the capacitance of the transducer. The capacitance transducer is similar to that shown in Figure 5.40(a) on page 338 with a plate area of 20 cm² . The dielectric between the plates is air. Calculate the plate separation.

15. A balanced three-phase 415 V 50 Hz star-connected generator supplies a balanced star-connected load, of which each phase comprises a 40 Ω resistor in series with a 133 mH inductor. Calculate the generator phase voltage, the magnitude of the load phase current, the total power dissipated in the load.

16. A balanced three-phase 210 V (line) 60 Hz star-connected generator supplies a balanced delta-connected load, each phase of which comprises a 50 Ω resistor in series with a 66 μF capacitor. Calculate the load impedance per phase, the magnitude of the current in each phase of the load, the magnitude of the line current and the total power dissipated in the load.

17. A 240 V 50 Hz supply feeds the thyristor bridge shown in Figure 5.56 on page 355. The thyristors are triggered 30° after every voltage zero. Calculate the average dc output voltage.

18. The average dc output from the smoothed uncontrolled half wave rectifier shown in Figure 5.57 on page 355 is 12 V, the ripple voltage is 0.8 V and the load resistance is 1000 Ω. The supply frequency is 60 Hz. Calculate the supply rms voltage, the load current and capacitance.

19. A bipolar transistor with the collector current (I_C) versus collector –emitter voltage (V_{CE}) characteristic shown in Figure Q19 is used in the circuit shown in Figure 5.64 on page 364. The supply voltage is 12 V.

When the transistor conducts the base-emitter voltage (V_{BE}) is 0.7 V. The circuit is to be used as a simple amplifier. Find suitable values for resistors R_C and R_B What will be the maximum amplitude of the input signal current to produce an undistorted output voltage?

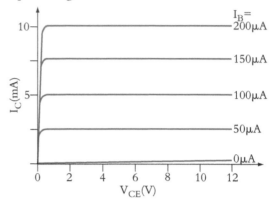

Figure Q19

20. In a transistor amplifier like that shown in Figure 5.64 on page 364 R_C is 1800 Ω and R_B is 140 kΩ. The transistor input resistance, h_{ie} is 1200 Ω and the transistor current gain, h_{fe} is 40. The peak base current is 80 μA. Calculate the peak base current and peak collector-emitter voltage.

21. An operational amplifier (op amp) with infinite input resistance, negligible output resistance and an open loop gain A like that shown in Figure 5.67(b) on page 366 is used in the circuit shown in Figure 5.68 on page 367. R_1 is 200 kΩ and R_f is 500 kΩ. Calculate the closed loop gain, G at frequencies 100 Hz, 1 kHz, 10 kHz, 100 kHz and 1 MHz.

22. An operational amplifier (op-amp) with very large open loop voltage gain negligible output resistance and infinite input resistance is incorporated in the circuit shown in Figure 5.71 on page 370. Resistance, R_1 is 100 kΩ, capacitor, C is 0.1 μF and the input voltage, V_i is constant at 2 V. Calculate the output voltage after 40 ms.

23. Three sensors A, B, and C are fitted to a milling machine to prevent it from being operated unsafely.
- Sensor A checks the safety guard is in position. It produces a high output only when the guard is fitted correctly.
- Sensor B monitors the flow of cooling fluid. The output is normally high but goes low when the flow stops.
- Sensor C detects the height of the cutter. It normally produces a high output in the safe condition but goes low when the cutter is in contact with the mill base.

The output from each sensor is fed to a controller that only produces a high output when all the sensors indicate that the machine is safe to operate. Find the Boolean expression for the controller output.

24. A builder's lorry is fitted with a small crane for loading and unloading materials. The vehicles are fitted with four sensors to ensure that these processes are carried out safely.

- Sensor A checks the lorry is on level ground. It gives a low output (= 0) when the ground is level, otherwise the output is high (= 1).

- Sensor B monitors that the flashing light on the crane is illuminated. It only produces a high output when the crane light is on.

- Sensor C monitors if the lorry hazard warning lights are on. The output from C is only high when the lights are flashing.

- Sensor D checks that the lorry stabilizer legs are extended and in contact with the ground. D produces a high output only when the legs are pulled out and touching the ground.

The output from each sensor is fed to a controller that only produces a high output when all the sensors indicate that the lorry is on level ground, that the stabilizer legs are fully extended and touching the ground and either the crane light is flashing or the lorry hazard warning lights are on or both. Find the Boolean expression for the controller output.

25. Figure Q25 shows a synchronous counter. The Q outputs of all the bistables in the network are initially reset to 0. Calculate the bistable outputs after the first 10 clock pulses and hence find the counter modulo.

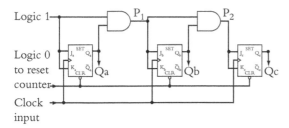

Figure Q25

26. A 6-bit ladder D/A converter has a full range output of 12 V. Calculate the output voltage when the binary input is $101\ 101_2$.

27. A transformer has a primary winding with 2400 turns and a secondary winding with 600 turns. A 40 Ω resistor in series with 100 mH inductor is connected across the secondary winding. The primary winding is connected to a 120 V 60 Hz supply. Calculate the primary current.

28. A 5 kVA 3.3 kV : 250 V 50 Hz single phase transformer dissipates 100 W when it is operated with its secondary open circuited. The resistance of the primary winding (high voltage) is 40 Ω and the resistance of the secondary winding is 0.2 Ω. The secondary supplies full load current at 0.8 power factor lagging. Calculate the secondary current, the primary current, the total I^2R losses and the efficiency.

29. A three-phase, 415 V (line) 50 Hz, two–pole induction motor has an effective stator:rotor turns ratio of 2:1. The rotor resistance is 0.1 Ω per phase and the rotor reactancce at standstill is 0.4 Ω per phase. The motor runs at 2900 rev min^{-1}. Calculate the torque generated.

30. A three-phase 210 V(line) six-pole 60 Hz induction motor has an effective stator: rotor turns ratio of 3:1, rotor resistance 0.1 Ω/phase and rotor standstill reactance 0.5 Ω/phase. Calculate the starting torque. The motor drives a constant load torque of 8 Nm. Calculate the rotor speed and the mechanical output power.

Machine dynamics

1. Figure Q1 shows a sphere of mass M and diameter d which can slide in a frictionless slot in a horizontal disc which is rotating about a vertical axis through its centre O. The sphere is restrained by a massless spring of free length L_0 and stiffness K, one end of which is fixed to the disc at O, as shown. Show that the steady state extension x of the spring when the disc is rotating at constant speed W is:

$$x = \frac{M\left(L_o + \dfrac{d}{2}\right)\Omega^2}{(K - M\Omega^2)}$$

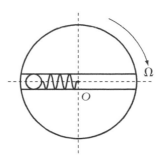

Figure Q1

2. Figure Q2 shows a slider-crank mechanism *ABC* representing a single cylinder engine. At the instant shown, the crank *AB* is rotating at a constant angular speed of 100 rad/s in the anticlockwise direction.

   By drawing velocity and acceleration diagrams, calculate the instantaneous magnitude and direction of:

   (a) the velocity of the slider *C* and the angular velocity of the connecting rod *BC*.

   (b) the angular acceleration of *BC* and the acceleration of the slider *C*

   (c) the acceleration of the centre of mass *G* of the connecting rod *BC*.

   All relevant dimensions are shown in the figure.

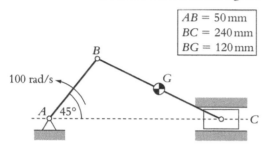

| $AB = 50$ mm |
| $BC = 240$ mm |
| $BG = 120$ mm |

Figure Q2 (Not to scale)

3. Figure Q3 shows a schematic of a pendulum system used to perform impact tests.

   The pendulum takes the form of a heavy uniform rectangular plate *ABCD* (width *w* and height *h*) which has mass *M*. The plate is hinged at its upper left and lower left corners to two parallel rods O_1A and O_2B which are each of length *L*. The rods are attached via frictionless hinges to a rigid vertical support at their ends and have negligible mass. The plate can swing in the vertical plane so that edges *AB* and *DC* remain vertical. The pendulum is released from rest when $\theta = 60°$. At the instant when the plate is released:

   (a) Determine the magnitude and direction of the acceleration of the centre of mass of the plate;

   (b) Determine expressions for the forces in the supporting rods.

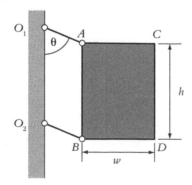

Figure Q3

4. Figure Q4 shows a non-uniform rigid beam freely pivoted about a horizontal, frictionless hinge *O* with a point mass *m* attached at *A*, which is at a distance *L* from *O*. The beam has mass *M* and radius of gyration *K* about *O*.

   (a) Determine the moment of inertia of the composite structure about an axis through *O*.

   (b) The composite structure has a centre of mass *G* located a distance *l* from *O*. Draw a free body diagram for the composite structure, showing all forces and accelerations.

   (c) If the composite structure is released from rest with *OG* horizontal, find the angular acceleration immediately after release.

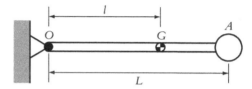

Figure Q4

5. Figure Q5 shows a schematic representation of part of a printing machine which transfers paper from a roll to a geared drum. The drive torque L_m is applied to the smaller gear by an electric motor. The larger gear forms part of the drum onto which the paper is wound from the roll. The effective radii (r_1, r_2, r_3, r_4) of the gears, drum and roll are shown in the Figure. The effective moments of inertia of the small gear, large gear/drum, and paper roll are I_m, I_D and I_R respectively, and L_F is a drag torque which acts about the roll axis. α_m and α_R are the angular accelerations of the motor and roll respectively.

   (a) Draw clear free body diagrams of the small gear, large gear/drum and the roll of paper indicating all of the forces that act, and write down the equations for the motion of each component about its axis of rotation.

   (b) Calculate the tension in the paper between the drum and the roll when the angular acceleration of the motor is $\alpha_m = 20$ rad/s². Assume that $r_1 = 0.2$ m, $r_2 = 2$ m, $r_3 = r_4 = 1$ m, $I_R = 50$ kgm², and $L_F = 50$ Nm.

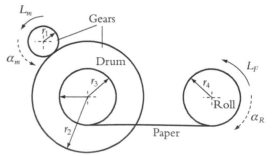

Figure Q5

Questions

6. Figure Q6 shows the shaft of a multi-stage compressor. It is supported on bearings at A and E and carries uniformly spaced discs of mass 15 kg at B, 10 kg at C and 10 kg at D as shown. The centres of mass of the impeller discs at B, C and D are offset from the axis of the shaft by 1 mm, 2 mm and 2.5 mm respectively, and the angular positions of the centres of mass are arranged so that the system is in static balance.

   (a) Determine the angular orientations of the centres of mass of the discs at C and D relative to disc B, and calculate the bearing reaction forces at A and E when the shaft speed is 2500 rev/min.

   Balancing masses are to be placed in planes A and E at a radius of 50mm to achieve dynamic balance. Determine the magnitude of the balancing masses that must be placed at A and E, and their angular orientations relative to the centre of mass of disc B.

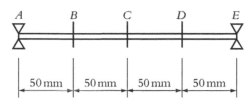

   Figure Q6

7. Figure Q7 shows a schematic of a pile driver in which a hammer H of mass 200 kg is dropped vertically through a distance of 0.5 m onto the top of a pile P of mass 800 kg.

   (a) Assuming that the hammer and pile remain in contact, determine the velocity of the pile immediately after impact takes place.

   (b) Calculate the kinetic energy lost during the impact.

   (c) Assuming that the constant resistance force against the pile is 20 kN, calculate the distance that the pile moves.

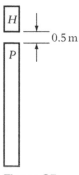

   Figure Q7

Index

A

absolute pressure 138-139, 215
absolute velocity 424
ac circuits *See* alternating current circuits
ac induction motors 392-403
   See also torque
   construction 393
   equivalent circuit 396-397
   mechanical output power 397-398
   operating principles 393-394
acceleration
   constant 410
   variable (non-constant) 409-410
acceleration diagram 428-430
activity gradient 111
adhesive bonding 119
adiabatic processes 254
aerofoil 170
alloys, solidification 96-99
alternating current bridges 338
alternating current circuits 318-340
alternating currents, rectifying 351-352
alternating voltage supply
   capacitance connected to 325-326
   inductance connected to 322-323
   resistance and capacitance in series with 330-333
   resistance connected to 320-321
   resistance, inductance and capacitance in series with 333-336
   resistance and inductance in series with 326-330
alumina 65
aluminium alloys 64, 118
ambient pressure 215
amplifiers 360-371
   operational 365-370
analogue-to-digital (A/D) converters 384-385
AND gate 372
angular momentum 486
angular motion See rotational motion
anisotropy 106

annealing of metals 107-108
anode 350
armature 308
asynchronous binary counter 379-380
atomic structure 60-61
austenite 116
axial stress 47

B

balancing of rotating masses 459-468
   multiple 'same plane' balancing masses 460-461
   single 'different plane' balancing mass 461
   single 'same plane' balancing mass 460-461
   solving balancing problems 464-468
   two 'different plane' balancing masses 461
balloons 165-166
battery 294-295
beam bending 26-41
   theory 32-34
beam-bending equation 34
beam problems, solving 38
bending moment 27-32
   and shear force 30-31
bending stress 33-34
bends, flow around 200, 207-208
Bernoulli equation 174-175
   applicability 184-185
   applications 176-185
   extended (EBE) 194-195
binary notation 378
binary system 372
bipolar transistor 360-361
bistables 378-382
black body 261
blow moulding 105
Bohr model 60-61
Boolean algebra 372
borosilicate glasses 65
boundary 214
boundary layer 189
boundary work 221

Brayton cycle 276-277
brazing 120
bridge measurements 293-294
built-in support 27
bulk modulus 48-49
buoyancy 163-166
buoyancy force 163-164
burglar alarm 374-375

C

cantilever beams 26
   material selection 89-91
capacitance 311-318
   measurement 338
capacitive reactance 325
capacitors
   connected in parallel 313
   connected in series 313-314
   energy stored in 317-318
   in series with resistance 316-317
   series/parallel combination 315
car, rear-wheel drive 452-453
carbon equivalent value (CEV) 122
carbon fibre 65
Carnot cycle 271-272, 274-275
Carnot efficiency 228-229, 271-275
Carnot principles 228
Cartesian form 327
castability 102
casting 95-103
   defects 103
   design rules 103
   processes 100-102
cathode 350
cations 61, 64
cementite 116
centre of mass 437-438
centroid (centre of area) 153-154
ceramics 64-65
   bonding and packing in 62-63
   powder processing 113
charge 312
circular motion, of particle 417, 419-420

Clausius inequality 229-230
claywares 113
closed loop gain 366
closed systems 214, 220-232
    open systems versus 233
    work done during reversible process 255-256
coefficient of restitution 485
coefficient of static friction 5
coefficient of thermal expansion (CTE) 20, 85-86
cohesion 136
coil
    with inductance and resistance 303-305
    wound around core 310-311
cold deformation 110
cold-worked structures, heat treatment 115
colloidal suspension 67
combinational logic 371-373
    in circuit construction 374-375
compact disc (CD), material selection 91-92
comparator 385
compatibility condition 19
complex impedance See impedance
composite materials 66
composite shaft, torque in 55
compositions, of phases 97
compound assembly, heating 21-22
compound gear train 470
compressibility 141-142
compression process 214
compression ratio 278
concentrated moment, on simply supported beam 31
conceptual diagrams 214
conduction 258-259
conservation of angular momentum, principle of 486
conservation of energy 223, 478
    law of 174, 192
conservation of mass 170, 234
conservation of momentum, principle of 485
conservative system 478
constant-pressure process 253, 256
constant-volume process 253, 256
continuity equation 170-171
contraction 14
control volumes 192, 204
controlled full-wave rectification 354-355
controlled half-wave rectifier 353

convection 259-260
convective heat transfer coefficient 260
copper alloys 64
copper losses 391
core loss 391
cost of materials 87-88
Coulomb (dry) friction, between two surfaces 408
counter-ramp A/D converter 384-385
couple 2
coupling of shafts 55
covalent bonding 62-63
crane drum 412-413, 454-455, 472-473
creep 130-133
critical point 247
crystals 67
current 285
    lagging 323, 328
    leading 325, 331
cut-off ratio 280
cutting 114
cycles 219, 271-281
cylinders
    hollow thick uniform 443
    thin 46-47, 50
    uniform solid 442

D

Dalby's convention 464
d'Alembert's Principle 419
Darcy, Henri 195
Darcy equation 195
Darcy-Weisbach equation 196
Darcy-Weisbach friction factor 196-197
dc circuits See direct current circuits
dc motor, two-pole 308-309
de Havilland Comet 128
deflection, geometry of 33
deformation processing 107-111
degree of superheat 246
degrees of freedom (DOF) 422
    of connected bodies 422-423
delta connection 345-347
density 83-84, 137, 216
    See also specific gravity
depletion layer 349
design analysis 1-14
design principles 6-7
die casting 101-102
dielectrics 312
Diesel cycle 280
differential manometer 147-149

diffusion 111-112
diffusion coefficient 112
digital electronics 371-386
digital-to-analogue (D/A) converters 382-384
direct current circuits 284-296
disc brakes 263-264
discharge coefficient 181
dislocation density 109
dislocations 76-77
displacement measurement 338-339
door on rollers 451
double cage rotors 401
double-glazing 266-268
drag forces 187-188
drawing 109
driving torque 395
dryness fraction 246, 247
ductility 77-80
ducts
    flow in 171, 237
    incompressible flow in 195
    non-circular 197-200
dynamic balance 462
dynamic forces 204
dynamic viscosity 186

E

E-glass 65
eddies 189
effective radiant heat transfer coefficient 263
efficiency 228-229, 391-392
    of gears 470
elastic modulus 15-16, 69-72
elasticity 14-15
elbow, 90° 200
electrical torque 395
electromagnetic induction 301-302
electromagnetic systems 296-311
electrons 64
elongation 14
emissivity 261
encastré 27
energy 286
    See also conservation of energy
    stored in inductor 306
engine cycles 278
enthalpy 237
entropy 230-232
epicyclic gear system 478-480
epoxies 66
equilibrium, conditions of 3
equivalent impedance 389
equivalent pitch circles 468

Index

Ericsson cycle 273
ethylene 67
Euler equation 172-174
eutectic 99
eutectoid reaction 116
expansion joints 85
expansion processes 225
extended Bernoulli equation (EBE)
        194-195
extensive properties 216
extrusion 104, 109

F

failure of materials 17, 122-133
Fanning friction factor 196-197
Faraday, Michael 301-302
Faraday's Law 302, 318
fast fracture 125-127
fatigue 128-130
fatigue limit 128
ferrite 116
fibre forming 106
Fleming's Left-hand Rule 308, 319
Fleming's Right-hand Rule 318-319
flip-flops 378-382
floating bodies 163-166
fluid flows, representing 168-170
fluids 136
        See also viscous fluids
        at rest 143-167
        continuum 136-137
        in motion 167-185
                See also linear momentum
        phase changes 245-247
flux 298
flux density (B) 298
        magnetic field strength (H) versus
                299-300
flux linkage 302
flywheel design 480-481
flywheel punch 483
force 1
        on current-carrying conductor
                306-307
        direction 2
        frictional 5, 11
        magnitude 2
        resultant 3-4
force polygon 461-464
forced convection 260
forging 109
forward biased connection 350
four-stroke engine 480
Fourier's law 259
fracture 17

fast 125-127
fracture toughness 125-126
free body diagrams (FBDs) 5-6, 10,
        192
free surface flow 184
frequency, series resonant 335
friction 5, 11
friction factor 195
friction force 408
friction head loss 195
friction losses 194-195
frictionless flow 176
full-wave rectifiers 354
fusion zone 121

G

gas distribution system 235
gas turbine engines 276-277
gases
        See also perfect gases
        liquids versus 136
gate 352
gate valve 200
gauge pressure 138-139, 215
gear ratio 468
gear trains 469-472
geared systems 468-473
glasses 64-65, 67
grain boundary strengthening 77
grain growth 108
grain shape 109
gravitational forces 205
grey body 261
grey body view factor 262

H

hardenability 117, 121
hardening mechanisms 76
hardness 81-82
head 175
head form of Bernoulli equation 175
heat-affected zone (HAZ) 120-122
heat engines 227-229
        irreversible 229-230
        reversed 232
        reversible 229-230
heat reservoirs 227
heat supplied 228
heat transfer 214, 220, 222
        modes 258-264
        steady-state 264-266
        through walls and surfaces in series
                266-270
heat treatment of metals 115-119
hollow thick uniform cylinder 443

Hooke's law 15, 69
        Generalized 46
        two-dimensional 45
hoop stress 47
hot deformation 110
hydraulic diameter 197
hydrocarbon polymer 66
hydrostatic stress 48
hydrostatics 143-167

I

ideal fluids 167-168
ideal gas *See* perfect gas
impact 484-490
impedance 328, 332
        equivalent/referred 389-390
imperfect discs
        multiple 463-464
        single 459-462
impulse 484
impulse torques 486
inclined-tube manometer 149-151
inductance 302-303
inductive reactance 323
injection moulding 104
integral over an area 152-153
integrating amplifier 369-370
intensive properties 216
internal combustion engines
                276-277
        reciprocating 277-280
internal energy 223-226
interpolation 250-251
inverting amplifier 366-367
investment casting 100-101
inviscid fluids 167-168
ionic bonding 62
irreversible processes 217, 254
isenthalpic process 253-254
isentropic process 253-254, 256
isobaric process 253, 256
isochoric process 253, 256
isothermal process 253-254, 255,
                256

J

j operator 323-324
jet
        reaction of 209-210
        striking surface 210-211
jet engine 204
JK bistable 378-382
joining 119-122
Joule, James 220, 223
junction diode 349-351

K

Karnaugh Map 375
Kelvin, Lord 213
kevlar 66
kilomole 140
kinematic viscosity 187
kinematics of particle in plane
        413-418
    acceleration 414, 415-418
    Cartesian coordinates 413-414,
        416
    polar coordinates 414-416,
        417-418
    velocity 414, 415, 416-418
kinetic energy
    in closed system 226
    of particle 474
    of rigid body 474
kinetic friction 408
kinetics of particle in plane 418-421
Kirchhoff's Current Law 288-289
    modified 310
Kirchhoff's Voltage Law 289
    modified 309

L

laminar flow 188-192
laws of thermodynamics
    first 220-227
    second 227-232
least significant bit (LSB) 380
Liberty ships 125
limiting friction 408
line current 342, 345-347
line voltage 342, 345
linear elastic 15
linear interpolation 250-251
linear momentum 204-211, 484-485
linear momentum equation 204-205
    applications 205-211
linkage mechanisms in plane
    four-bar 434-436
    kinematics 427-436
    kinetics 457-458
liquid columns 146
liquids 67
    gases versus 136
logic gates 373
lubrication 187-188

M

machining 114-115
magnetic circuits 308-311
magnetic field, around conductor 296
magnetic field strength 299

flux density (B) versus 299-300
magnetomotive force (mmf) 298
    balance 388
manometers 146-152
martensite 117-118
mass 137
    conservation of 170, 234
    of system 216
mass flow continuity equation 234
mass flowrate equation 234
mass flowrates 176, 234
mass moment of inertia 438-445
mass transfer 214
materials
    classification 60
    failure 17, 122-133
    processing 95-122
    properties 69-87
    selection in engineering design
        89-94
    structure 61-68
mechanical fastening 120
membrane strains 47
membrane stresses 46
meniscus 151-152
mesh analysis 290-293
mesh connection 345-347
metal deformation 76-77
    See also deformation processing
metallic bonding 61-62
metals 65
    annealing 107-108
    heat treatment 115-119
    powder processing 113
microstructure 67-68
    development in two-phase region
        98
    of steels 116-117
minor losses 199
Mohr's circle for plane stress 43-45, 49
molar mass 140
moment 2
moment of inertia 438-445
moment polygon 462-464
momentum, linear See linear
        momentum
monomers 67
Moody, Lewis 196
Moody chart 196
most significant bit (MSB) 380
motion, of matter 233
motor-clutch-load system 489-490
moulding 103-107
    processes 104-106
'Mr' polygon See force polygon

'Mry' polygon See moment polygon
multi-pole motors 401-403

N

NAND gate 373
natural convection 260
necking 17
neutral 341
neutral axis (NA) 32
neutral surface 32
Newtonian fluids 186
Newton's law of cooling 260
Newton's law of viscosity 186
Newton's laws of motion
    first 406
    second 204, 406-407, 418-419
    third 1, 407
nickel superalloys 64
no-slip condition 189
non-conservative system 478
non-inverting amplifier 368-369
non-Newtonian fluids 186-187
NOR gate 373
NOT gate 373
nozzle 200
    force at 209-210
nozzle flow meter 180
nylon 66

O

Ohm's law 285
one-dimensional flow 168
open loop gain 365
open systems 214, 233-239
    closed systems versus 233
    work done during reversible
        process 255-258
OR gate 372
orifice plate meter 181-182
Otto cycle 278-279

P

paddle wheel 224
Parallel Axis Theorem 35, 439-440
parallel circuits 288
Paris' law 129
particle impact 487-488
particle on inclined surface,
        acceleration 411-412
particle in plane See kinematics of
        particle in plane; kinetics of
        particle in plane
particle-rod impact 488-489
pascal (Pa) 71
pascal second 187

Index

Pascal's law 143
pathlines 169
pearlite 116
per unit slip 395
perfect gas equation 139-141, 240, 241
perfect gases 139, 240-245
    change in entropy 244-245
    properties 240
    relationships 245
permanent mould casting 101-102
permeability 298-299
permittivity 312
Perpendicular Axis Theorem 35, 439
perspex 66
phase angle 328-329
phase changes 245-247
phase current 342, 346
phase diagrams 96
phase voltage 342, 345
phasor diagrams 341, 344, 345
phasors 321, 323-324
phenolics 66
piezometer 146, 183
piezometric pressure 175
pin-jointed structures 7-9, 12-13
pipe entry loss 199
pipe exit loss 199
pipes
    flow around bends in 200, 207-208
    flow in 171-172, 176-177, 191
    incompressible flow in 195
    lagged 269-270
    non-circular 197-200
    of uniform diameter 195-196
pistons 224-225
pitot-static probes 182-184
plane strain 45
plane stress 42-45
    general state 42
point load 27
    on simply supported beam 28-29
point moment 27
Poisson's ratio 16
polar form 328
polar second moment of area 35, 52, 53-54
polyethylene 63, 66, 68
polymerization 67
polymers 65–66
    bonding and packing in 63
polytropic index 252
polytropic process 252, 256
potential difference 285
potential energy 476-477
    in closed system 226

gravitational 477
powder processing 111-114
power 222, 285
power dissipation 336-337
    in delta connected load 347
    in star-connected load 344-345
power factor (p.f.) 329
power station 273-275
    reactor vessel 126-127
power transmission 54
    in shaft 56-57
precipitation strengthening 77, 118-119
press moulding 104
    measurement using manometers 146-151
pressure 137-138, 143-151, 205, 215-216
    absolute 138-139, 215
    ambient 215
    gauge 138-139, 215
    piezometric 175
    stagnation 182
    and temperature 253
    variation with elevation 144-145
pressure form of Bernoulli equation 175
pressure ratios 277
pressure vessels 124-125, 126-127
primary ferrite 117
primary winding 386
prime movers 227
process diagrams 217-219, 254, 272-274, 277-278
processes 217
    irreversible 217, 254
    reversible 217, 254-258
    types 251-258
properties, of systems 215
    extensive 216
    intensive 216
pumps, performance 200-203

Q

quench 117

R

R-2R ladder digital-to-analogue converter 383-384
radial stress 46
radiation 261-264
radius of curvature 34

radius of gyration 439
Rankine cycle with superheat 273-275
rate of shear *See* viscosity gradient
reaction force 408
real fluids *See* viscous fluids
recovery 108
recrystallization 108
rectifiers 350
    semiconductor 348-360
refrigerators 232
regeneration 272
relative permittivity 312
relative roughness 196
reluctance 298
reservoir walls/gates 156-157
residual stress 110
resistance 285
resistive heat expansion 21
resistivity 286
reverse biased connection 350
reversible processes 217, 254-258
Reynolds, Osbourne 188
Reynolds number (Re) 190
rigid bodies in plane
    general plane motion 446-450, 456-457, 474, 476
    kinematics 422-426
    kinetics 445-459
    mass properties 436-445
    motion 423-424
    pure rotation 446, 449-450, 452-455, 474, 475-476
    pure translation 446, 449, 450, 451-452, 474
ring, thin uniform 441-442
ripple voltage 356
robotic arm 425-426
rod, thin uniform 441
rolling 108
rolling drum 456-457
root mean square (rms) current 322
root mean square (rms) voltage 322
rotating bar 453-454
rotating masses, balancing *See* balancing of rotating masses
rotational motion 410-411
    relation to translational motion 411-413
rotational moulding 105
rotor 393
rotor emf 396
rotor frequency 395
rotor resistance to rotor standstill reactance ratio 400
rotor slip 394-395
rubbers 66

S

sand casting 100
saturated liquid line 246
saturated liquids 245
saturated mixture 246
saturated mixture region 246
saturated vapour 246
saturated vapour line 246
saturation temperature 245
second moments of area 34-37
    circular cross-section 36
    I-section, symmetrical 36-37
    polar 35, 52, 53-54
    rectangular cross-section 34-35
    T-section (nonsymmetrical) 37
    unsymmetric 40-41
secondary bonding 63
secondary winding 386
    impedance connected to 387-388
semiconductor rectifiers 348-360
semiconductors 348-349
sequential logic 378-379
series circuits 287
series resonance 335
shaft work 172, 221, 256-257
sharp bend 200
shear force 27-32
    and bending moment 30-31
shear modulus 17
shear strain 17, 23
shear stress 17, 23
    complementary 42
    in torsion 51, 52
shock absorbers 226
silicon 348-349
silicon nitride 65
simple gear train 469, 471-472
simply supported beams 26-32,
    38-39
single-phase alternating current
    generators 318-320
sintering 114
slider-crank mechanism 430-433,
    457-458
slider mechanism 424-425
slip 113, 395
slot, point mass in 420-421
sluice gates 161-163
smooth bend, 90° 200
smoothed supplies 355-357
soda-lime glasses 65
soldering 120
solenoid 296-297
solid solution strengthening 77
solidification 67-68

solutionizing 118
space diagram 427
specific energy 193
specific enthalpy 237
specific entropy 230
specific flow work 236
specific gas constant 140-141, 240
specific gravity (SG) 137
specific heat at constant pressure
    242-244
specific heat at constant volume
    242-244
specific internal energy 236
specific kinetic energy 236
specific latent heat of vaporization 246,
    248
specific potential energy 236
specific volume 216
    temperature and 252
spheres, thin 47-48
springs, material selection 93-94
squirrel cage rotor 393
stability 164
stagnation pressure 182
standard resistances 293
star connection 341-345
state
    equations of 240
    of system 216
state diagrams See process diagrams
state table 379, 381, 382
static balance 462
static equilibrium, of body in plane
    407-408
statically determinate systems 18-19
statically equivalent systems 6-7
statically indeterminate systems 19-20
stator 393
steady flow 168, 176
steady flow energy equation (SFEE)
    192-203, 235-239
    application to include pumps
        200-203
    derivation 192-194
steady flow problems 233-239
steady-state heat transfer 264-266
steam 245-249
steam power plant 273-275
steels 64
    heat treatment 115-118
    welding 121-122
Stefan-Boltzmann law for black body
    261
stepped shaft
    torque in 54

torsion in 55-56
stiff lightweight beam, material
    selection 89-91
Stirling cycle 272-273
strain
    See also membrane strain; plane
        strain; shear strain; volumetric
        strain
    direct 14, 22-23
    lateral 15
    longitudinal 15
    thermal 20-22, 23-26
    three-dimensional 46-49
    to failure 17
strain energy in spring 477
strain hardening 17
strain measurement 294-295
streaklines 169-170
streamlines 169
stress
    See also plane stress; shear stress;
        yield stress
    bending 33-34
    compressive 14
    direct 14, 22-23
    nominal 16
    residual 110
    tensile 14, 73-77
    thermal 20-22, 23-26
    three-dimensional 46-49
    ultimate tensile (UTS) 17, 73
stress overload 123-125
stress-strain
    curve 16-17
    solving problems 18-20
structural reaction forces 205
subcooled liquid 245, 247
submerged surfaces 152-163
    curved and inclined 159-163
    horizontal 154-155
    inclined 158-159
    vertical 155-157
sudden contraction 200
sudden enlargement 199
summing amplifier 368
supercritical steam 247
superheated steam 247
superheated vapour 246
superposition, principle of 20
surface heat transfer coefficient 260
surface tension 142-143, 151
    and manometers 151-152
surroundings 214
synchronous binary counter 380-382
synchronous speed 394

Index

T

temperature 216
  pressure and 253
  specific volume and 252
temperature coefficient of resistance 287
temperature-entropy diagram 271-272
temperature-specific volume diagram 246-248
tempering 117-118
tensile stress/strength 14, 73-77
thermal conductivity 259
thermal expansion 85-86
thermal resistance 265-268
  of cylinder 265
  of plane wall 265
  in series 266-268
  of surface 265
thermal strain 20-22, 23-26
thermal stress 20-22, 23-26
thermally insulated system 223
thermally isolated system 223
thermodynamic equilibrium, state of 217
thermodynamic systems 214
thermodynamics
  See also closed systems; cycles; heat transfer; laws of thermodynamics; open systems; perfect gases; processes; steam
  concepts 213-219
  notation 281
thermoplastics, property changes during deformation 106
thermosetting polymers, moulding 105-106
thin cylinders 46-47, 50
thin rectangular plate 443-444
thin spheres 47-48
thin uniform ring 441-442
thin uniform rod 441
three-dimensional strain 46-49
three-dimensional stress 46-49
three-force principle 7
three-phase circuits 340-348
three-phase connections 341-347
three-phase generator 340
thyristor 352-353
ticketing machines 376-378
torch light 284-285
toroid 297, 300-301, 303
  with radial air gap 309
torque 51

electrical/driving 395
  maximum 399-400
  in multi-pole motor 402-403
  starting 399
  in two-pole motor 398-400
torque-slip characteristics 400
torsion 51-57
torsion equation 53
total force 205
toughness 77-80
transducers 338-339
transformation ratio 386
transformers 386-392
  power loss in 391
transistor amplifier 361-364
  analysing 364-365
transistor-biasing currents 364
transition 189-191
translational motion, relation to rotational motion 411-413
transmissibility, principle of 6
trigonometry 10
true stress versus strain curve 17
truth table, for logic circuit 374
tungsten light bulb filaments 130-131
turbine blade, creep 132-133
turbulent flow 188-192
turns ratio 386
twisting 51-52
two-force principle 7
two-phase regions 96-99

U

U-tube manometer 147
ultimate tensile stress/strength (UTS) 17, 73
uncontrolled full-wave rectifier 354
uncontrolled half-wave rectifier 352
uniaxial loading 15
uniaxial tensile test 16
uniform flow 168
uniform rod/circular disc composite body 444-445
uniform solid cylinder 442
uniformly distributed load (UDL) 27
  I-section 40
  on simply supported beam 29-30
universal gas constant 240

V

vacuum flask 264
vanes 205-206, 211
vectors 406
velocity, absolute 424
velocity diagram 427-428

velocity profile 186
vena contracta 181
venturimeter 177-180
Vickers hardness test 81-82
viscosity 185-187
  and lubrication 187-188
viscosity gradient 185-186
viscous fluids 185-203
voltage 285
  See also line voltage; phase voltage; ripple voltage
voltage gain 365
voltage rating 390
voltage stabilizer 358-360
voltampere (VA) rating 390
volume, of system 216
volume flowrate 178
volumetric strain 48
vulcanization 67

W

water 245-251
weight 2, 137
weight of water 208
weldability 122
welding 120-122
Wheatstone bridge 293-294
wheels
  driven 408-409
  non-driven 409
Work, Theorem of 474-475, 478
work strengthening 76
work transfer 214, 220-222
working fluid 214
World Trade Center 130-131

X

XOR gate 373

Y

yield point 16
yield stress/strength 16, 73-77
yoke 308
Young's modulus 15-16, 69-72

Z

zener diode 357-358